Let's Review Regents:

Geometry 2020

Andre Castagna, Ph.D.
Mathematics Teacher
Albany High School
Albany, New York

Dedication

*To my loving wife,
Loretta, who helped
make this endeavor
possible with her
unwavering support;
my geometry
buddy Eva;
and my future
geometry buddies
Rose and Henry.*

Published by Kaplan, Inc., d/b/a Barron's Educational Series
750 Third Avenue
New York, NY 10017
www.barronseduc.com

ISBN: 978-1-5062-5402-9

Printed in Canada

10 9 8 7 6 5 4 3 2 1

Kaplan, Inc. d/b/a Barron's Educational Series print books are available at special
quantity discounts to use for sales promotions, employee premiums, or
educational purposes. For more information or to purchase books, please call the
Simon & Schuster special sales department at 866-506-1949.

TABLE OF CONTENTS

PREFACE   vii

CHAPTER 1   THE TOOLS OF GEOMETRY   1
             1.1   The Building Blocks of Geometry   1
             1.2   Basic Relationships Among Points,
                    Lines, and Planes   11
             1.3   Brief Review of Algebra Skills   19

CHAPTER 2   ANGLE AND SEGMENT RELATIONSHIPS   30
             2.1   Basic Angle Relationships   30
             2.2   Bisectors, Midpoint, and the Addition
                    Postulate   35
             2.3   Angles in Polygons   41
             2.4   Parallel Lines   47
             2.5   Angles and Sides in Triangles   56

CHAPTER 3   CONSTRUCTIONS   67
             3.1   Basic Constructions   67
             3.2   Constructions that Build on the Basic
                    Constructions   73
             3.3   Points of Concurrency, Inscribed
                    Figures, and Circumscribed Figures   78

CHAPTER 4   INTRODUCTION TO PROOFS   86
             4.1   Structure and Strategy of Writing Proofs   86
             4.2   Using Key Idea Midpoints, Bisectors,
                    and Perpendicular Lines   88
             4.3   Properties of Equality   94
             4.4   Using Vertical Angles, Linear Pairs,
                    and Complementary and Supplementary
                    Angles   102
             4.5   Using Parallel Lines   108
             4.6   Using Triangle Relationships   113

CHAPTER 5   TRANSFORMATIONS AND
             CONGRUENCE   121
             5.1   Rigid Motion and Similarity
                    Transformations   121
             5.2   Properties of Transformations   127

5.3	Transformations in the Coordinate Plane	136
5.4	Symmetry	149
5.5	Compositions of Rigid Motions	154

CHAPTER 6 TRIANGLE CONGRUENCE 159

6.1	The Triangle Congruence Criterion	160
6.2	Proving Triangles Congruent	169
6.3	CPCTC	181
6.4	Proving Congruence by Transformations	188

CHAPTER 7 GEOMETRY IN THE COORDINATE PLANE 193

7.1	Length, Distance, and Midpoint	193
7.2	Perimeter and Area Using Coordinates	202
7.3	Slope and Equations of Lines	208
7.4	Equations of Parallel and Perpendicular Lines	216
7.5	Equations of Lines and Transformations	223
7.6	The Circle	227

CHAPTER 8 SIMILAR FIGURES AND TRIGONOMETRY 235

8.1	Similar Figures	235
8.2	Proving Triangles Similar and Similarity Transformations	245
8.3	Similar Triangle Relationships	257
8.4	Right Triangle Trigonometry	271

CHAPTER 9 PARALLELOGRAMS AND TRAPEZOIDS 289

9.1	Parallelograms	289
9.2	Proofs with Parallelograms	295
9.3	Properties of Special Parallelograms	303
9.4	Trapezoids	311
9.5	Classifying Quadrilaterals and Proofs Involving Special Quadrilaterals	315
9.6	Parallelograms and Transformations	324

CHAPTER 10 COORDINATE GEOMETRY PROOFS 330

10.1	Tools and Strategies of Coordinate Geometry Proofs	330
10.2	Parallelogram Proofs	331
10.3	Triangle Proofs	338

CHAPTER 11   CIRCLES   **346**
    11.1   Definitions, Arcs, and Angles in Circles   346
    11.2   Congruent, Parallel, and Perpendicular
            Chords   360
    11.3   Tangents   370
    11.4   Angle-Arc Relationships with Chords,
            Tangents, and Secants   376
    11.5   Segment Relationships in Intersecting
            Chords, Tangents, and Secants   390
    11.6   Area, Circumference, and Arc Length   400

CHAPTER 12   SOLIDS AND MODELING   **412**
    12.1   Prisms and Cylinders   412
    12.2   Cones, Pyramids, and Spheres   418
    12.3   Cross Sections and Solids of Revolution   424
    12.4   Proving Volume and Area by
            Dissection, Limits, and Cavalieri's
            Principle   431
    12.5   Modeling and Design   443

ANSWERS AND SOLUTIONS TO PRACTICE EXERCISES   **455**

GLOSSARY OF GEOMETRY TERMS   **490**

**SUMMARY OF GEOMETRIC RELATIONSHIPS AND
FORMULAS**   **498**

ABOUT THE EXAM   **512**

APPENDIX I—THE GEOMETRY LEARNING STANDARDS   **518**

APPENDIX II—PRACTICE AND TEST-TAKING TIPS   **523**

June 2018 Regents Exam   **529**

June 2019 Regents Exam   **557**

INDEX   **583**

PREFACE

This book presents the concepts, applications, and skills necessary for students to master the Geometry curriculum. Topics are grouped and presented in an easy-to-understand manner similar to what might be encountered in the classroom. Both students preparing for the Regents exam and teachers planning daily classroom lessons will find this book a valuable resource.

Special Features of This Book

- *Aligned with the Current Learning Standards*
  This book has been rewritten to reflect the current curriculum—both in content and in degree of critical thinking expected. The Geometry curriculum has placed more emphasis on transformational geometry, especially as applied to congruence. This emphasis has been integrated throughout the book in each chapter. Example problems and practice exercises can be found throughout this book. These demonstrate higher-level thinking, applying multiple concepts in a single problem, making connections between concepts, and demonstrating understanding in words.
- *Easy-to-Read Format*
  Topics are arranged in a logical manner so that examples and practice problems build on material from previous chapters. This format complements the presentation of material a student may see in the classroom. The format makes this book an excellent resource both for improving understanding throughout the school year and for preparing for the Regents exam. Numerous example problems with step-by-step solutions are provided, along with many detailed figures and diagrams to illustrate and clarify the topic at hand. Each subsection begins with a "Key Ideas" summary of the major facts the student should take away from that section. "Math Facts" can be found throughout the book and provide further insight or interesting historical notes.
- *Review of Algebra Skills*
  The first chapter of this book contains a review of practice problems of the algebra skills required to work through some of the problems encountered in geometry. Although many geometry problems involve pure reasoning, logic, and application of geometric principles, students can expect to encounter numerical problems in which they must apply equation-solving skills learned in previous years.

- *An Introduction to Proofs*
  A step-by-step guide to writing accurate geometry proofs can be found in chapter 4. Students are provided with the opportunity to develop the skills needed to write proofs by starting with just a small handful of geometric concepts and tools. The scaffolding provided in this chapter will help students develop the skills and confidence needed to meet the demands of the learning standards successfully. Both the two-column and paragraph formats of proofs are presented.
- *A Wide Variety of Practice Problems and Two Actual Regents Exams*
  The practice problems at the end of each subsection feature a range of complexity. Basic application of skills, including applying a formula or recalling a definition, lead into multistep problem solving and critical analysis. Problems that ask students to put their understanding in writing can also be found in each chapter. These practice problems include a large number of multiple-choice questions, similar to what can be expected on the Regents exam. The answers to all "Check Your Understanding" problems are provided. Two actual Regents exams with answer keys are included. These give students valuable experience with the style, format, and length of the Geometry Regents exam.
- *A Detailed Description of the Exam Format and Study Tips*
  The format of the exam, point distribution among topics, and scoring conversion are thoroughly explained. Students can use this information to help focus their efforts and ensure that they thoroughly master topics with high point value. Study tips and advice for test day are provided to help students make the most of their study time.
- *The Learning Standards*
  A complete list of the Geometry learning standards can be found in Appendix I. All teachers should be thoroughly familiar with the content of these standards.

What's Not in This Book

This review book does not provide proofs for all the theorems found within it. In fact, it shows fewer proofs than the typical geometry textbook. The theorems and proofs that were included in this work are those that:
- illustrate the level of complexity expected of students
- demonstrate specific strategies and approaches that a student may be expected to apply
- are specifically required within the geometry learning standards
Students are strongly encouraged to read and understand the proofs that are included here carefully. They should be able to complete on their own any proof noted to be specifically required by the curriculum. Of course, rote

memorization of these proofs is strongly discouraged. Instead, students should familiarize themselves with the tools and strategies used in proofs and then be able to work through the proofs by applying their critical thinking and geometry skills.

Who Will Benefit from Using This Book

- Students who want to achieve their best-possible grade in the classroom and on the Regents exam will benefit. Students may use this book as a study guide for both their day-to-day lessons and for the Regents exam.
- Teachers who would like an additional resource when planning geometry lessons aligned to the current learning standards will benefit.
- Curriculum and district administrators who want to ensure their math department's curriculum is aligned to the current learning standards will benefit.

Chapter
One

THE TOOLS OF GEOMETRY

1.1 THE BUILDING BLOCKS OF GEOMETRY

KEY IDEAS

The building blocks of geometry are the point, line, and plane. The definitions of the other geometric figures can all be traced back to these three. We can think of the point, line, and plane as analogous to the elements in chemistry. All compounds are built up from the elements in the same way that the geometric figures are built up of points, lines, and planes. Along with definitions, we also look at the notation used for each. The ability to interpret vocabulary and notation is important for success in geometry.

Point, Lines, and Planes

A point is location in space. It is zero dimensional, having no length, width, or thickness. Points are represented by a dot and named with a capital letter, as shown by Point A in Figure 1.1a. Don't let the dot confuse you— points are infinitely small. Even the smallest dot you can draw is two-dimensional.

A line is a set of points extending without end in opposite directions. Lines can be curved or straight. In this book, we will use the term *line* to refer to straight lines. Lines are one dimensional. They have an infinite length but have no height or thickness. They are represented by a double arrow to indicate the infinite length. They are named with any two points on the line as shown in Figure 1.1b or with a lowercase letter as shown in Figure 1.1c. Three or more points may also be used if we want to indicate the line continues straight through multiple points as in Figure 1.1d.

A plane is a set of points that forms a flat surface. Planes are two-dimensional. They have infinite length and width but no height. A tabletop or wall can represent a portion of a plane. Remember, though, that the plane continues infinitely beyond the boundaries of the tabletop or wall in each direction. Planes are named with any three points that do not lie on the same line, as shown in Figure 1.1e, or with a capital letter, as shown in Figure 1.1f.

1

Figure 1.1 Points, lines, and planes

Example 1

Name the following line in 7 different ways.

Solution: $\overleftrightarrow{FG}, \overleftrightarrow{GH}, \overleftrightarrow{FH}, \overleftrightarrow{GF}, \overleftrightarrow{HG}, \overleftrightarrow{FGH}, \overleftrightarrow{HGF}$

Example 2

Name the plane in two different ways.

Solution: Plane *QRS*, plane *Z*

Example 3

How many points lie on $\overleftrightarrow{JKL}$?

Solution: An infinite number. Every line contains an infinite number of points. We just show a few of them when representing and naming a line.

Rays and Segments

A **ray** is a portion of a line that has one endpoint and continues infinitely in one direction. A ray is named by the endpoint followed by any other point on the ray. When naming a ray, an arrow is used. The endpoint of the arrow is over the endpoint of the ray. Figure 1.2 illustrates ray $\overrightarrow{AB}$ with endpoint A and ray $\overrightarrow{BA}$ with endpoint B.

Figure 1.2 Rays $\overrightarrow{AB}$ and $\overrightarrow{BA}$

When two rays share an endpoint and form a straight line, the rays are called **opposite rays**. We say the union of the two rays forms a straight line.

A **line segment** is a portion of a line with two endpoints. It is named using the two endpoints in either order with an overbar. Figure 1.3 illustrates segment $\overline{FG}$ or $\overline{GF}$. The length of a segment is the distance between the two endpoints. The length of $\overline{FG}$ can be referred to as FG or $|FG|$. In some situations, we may wish to specify a particular starting point and ending point for the segment by using a **directed segment**. For example, a person walking along directed segment FG would begin at point F and walk directly to point G.

Figure 1.3 Segment $\overline{FG}$ or $\overline{GF}$

Remember that an infinite number of points are on any line, ray, or segment even though they are not explicitly shown in a figure. Also remember that lines, rays, and segments can be considered to exist even though they are not explicitly shown in a figure.

Example 1

Name each segment and ray in the figure.

Solution: Segments $\overline{ST}$ and $\overline{TU}$, rays $\overrightarrow{UV}$ and $\overrightarrow{SR}$

Example 2

If $\overline{QP}$ has a length of 5, what is the length of $\overline{PQ}$?

Solution: $\overline{PQ}$ also has a length of 5 because $\overline{PQ}$ and $\overline{QP}$ are the same segment.

Angles

An **angle** is the union of two rays with a common endpoint. The common endpoint is called the vertex. Angles can be named using three points—a point on the first ray, the vertex, and a point on the second ray. The vertex is always listed in the middle. Alternatively, one can use only the vertex point or a reference number. Figure 1.4 shows the different ways to name an angle.

(a) ∠*RST*, ∠*TSR*, or ∠*S*   (b) ∠1

Figure 1.4 Naming angles

Angles are measured in degrees. One degree is defined as $\dfrac{1}{360}$ of the way around a circle. Halfway around the circle is 180°, and one-quarter around is 90°. The measure of an angle can be specified using the letter *m*. For example, m∠*RST* = 30°. Angles can be classified by their degree measure.

Acute angle—an angle whose measure is less than 90°.
Right angle—an angle whose measure is exactly 90°.
Obtuse angle—an angle whose measure is more than 90° and less than 180°.
Straight angle—an angle whose measure is exactly 180°.

Figure 1.5 shows examples of each type of angle. The square positioned at the vertex of the right angle is often used to specify a right angle.

(a) acute   (b) right   (c) obtuse   (d) straight

Figure 1.5 Classification of angles

MATH FACT

Our definition of the degree as $\frac{1}{360}$ of a rotation around a center point has been used since ancient times. No one knows for sure why $\frac{1}{360}$ was chosen. One theory is that it originated with ancient Babylonian mathematicians, who used a base-60 number system instead of the base-10 system we use today. They divided a circle into 6 congruent equilateral triangles with 60° central angles. Then the ancient Babylonians subdivided each central angle into 60 parts. Another theory is that the circle was divided into 360 parts because one year is approximately 360 days. Either way, 360 is a convenient number to partition the circle with because 360 is divisible by 1, 2, 3, 4, 5, 6, 8, 9, and 10.

Example 1

Name one angle and two rays.

    Solution: $\angle SRT$, $\overrightarrow{RS}$, $\overrightarrow{RT}$

Example 2

Name each angle in 3 ways,
and classify each angle.

    Solution:

    $\angle R$, $\angle SRT$, $\angle TRS$; acute angle

    $\angle E$, $\angle DEF$, $\angle FED$; obtuse angle

    $\angle I$, $\angle HIJ$, $\angle JIH$; right angle

Adjacent Angles

Angles that share a common ray and vertex but no interior points are **adjacent angles.** In Figure 1.6, ∠*ABC* and ∠*CBD* are adjacent angles. ∠*ABC* and ∠*ABD* are not to be considered adjacent because they share interior points in the region of ∠*CBD*.

To avoid confusion, always use three vertices or a reference number when naming adjacent angles. Using the vertex alone would be ambiguous.

Figure 1.6 Adjacent angles ∠*ABC* and ∠*CBD*

Example 1

Name 3 pairs of adjacent angles.

Solution: ∠*AOB* and ∠*BOC*, ∠*BOC* and ∠*COA*, ∠*COA* and ∠*AOB*

Polygons

A **polygon** is a closed figure with straight sides. They are named for the number of sides.

Be familiar with these common polygons.	
Triangle—3 sides	**Hexagon**—6 sides
Quadrilateral—4 sides	**Octagon**—8 sides
Pentagon—5 sides	**Decagon**—10 sides

The intersection of two sides in a polygon is called a vertex (plural is *vertices*). The vertices are used to name specific polygons by listing the vertices in order around the polygon. They can be called out either clockwise or counterclockwise but must always be stated in continuous order—no skipping allowed.

Figure 1.7 shows some polygons with their names. We can list the vertices of the triangle in any order since it would be impossible to skip a vertex. For the quadrilateral, the name *EFGD* is valid but *EFDG* is not. For triangles, we often precede the vertices with the triangle symbol, △, so triangle *ABC* would be referred to as △*ABC*.

6

Figure 1.7 Triangle *ABC* (or △*ABC*), quadrilateral *DEFG*, pentagon *HIJKL*

When all the sides of a polygon are congruent to one another (equilateral) and all the angles of the polygon are congruent to one another (equiangular), we refer to that polygon as regular. So a square is an example of a regular quadrilateral, while a rectangle may have two sides with lengths different from the other two.

Example

Sketch hexagon *RSTUVW*. Name each side and each angle.

   Solution:

   Sides $\overline{RS}, \overline{ST}, \overline{TU}, \overline{UV}, \overline{VW}, \overline{WR}$
   Angles ∠*R*, ∠*S*, ∠*T*, ∠*U*, ∠*V*, ∠*W*

Classifying Triangles

Triangles can be classified by their angle lengths and measures as shown in Figure 1.8 below.

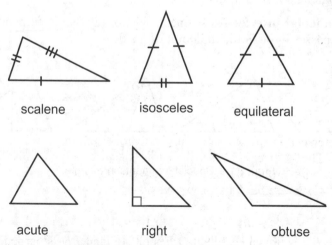

scalene   isosceles   equilateral

acute   right   obtuse

Figure 1.8 Triangle classifications

7

Classifying triangles by sides:
Scalene—no congruent sides, no congruent angles
Isosceles—at least two congruent sides, two congruent angles
Equilateral—three congruent sides, three congruent angles

Classifying triangles by angles:
Acute—all angles are acute
Right—one right angle
Obtuse—one obtuse angle

Example 1

$\triangle ABC$ has side lengths $AB = 2$, $BC = 1$, and $AC = 1.7$. $\angle A$ measures $30°$, $\angle B$ measures $60°$, and $\angle C$ measures $90°$. Classify the triangle.

Solution: $\triangle ABC$ is a right acute triangle.

Four special segments can be drawn in a triangle, and every triangle has three of each. These special segments are the altitude, median, angle bisector, and perpendicular bisector, shown in Figure 1.9.

Altitude—a segment from a vertex perpendicular to the opposite side

Median—a segment from a vertex to the midpoint of the opposite side

Angle bisector—a line, segment, or ray passing through the vertex of a triangle and bisecting that angle

Perpendicular bisector—a segment, line, or ray that is perpendicular to and passes through the midpoint of a side

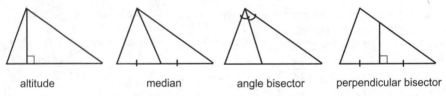

altitude   median   angle bisector   perpendicular bisector

Figure 1.9 Special segments in triangles

Example 2

In $\triangle ABC$, $\overline{BD}$ is drawn such that D lies on $\overline{AC}$ and $\overline{BD}$ is perpendicular to $\overline{AC}$. What special segment is $\overline{BD}$?

Solution: $\overline{BD}$ is an altitude. It has an endpoint at a vertex and is perpendicular to the opposite side of the triangle.

Example 3

In triangle $\triangle ABC$, $\overline{AD}$ is drawn such that $\angle BAD$ and $\angle CAD$ have the same measure. What special segment is $\overline{AD}$?

   Solution: $\overline{AD}$ is an angle bisector of $\triangle ABC$.

Check Your Understanding of Section 1.1

A. Multiple-Choice

1. In $\triangle FGH$, K is the midpoint of $\overline{GH}$. What type of segment is $\overline{FK}$?
   (1) median
   (2) altitude
   (3) angle bisector
   (4) perpendicular bisector

2. The side lengths of a triangle are 8, 10, and 12. The triangle can be classified as
   (1) equilateral
   (2) isosceles
   (3) scalene
   (4) right

3. The side lengths of a triangle are 12, 12, and 15. The triangle can be classified as
   (1) equilateral
   (2) isosceles
   (3) scalene
   (4) right

4. The angle measures of a triangle are 72°, 41°, and 67°. The triangle can be classified as
   (1) obtuse   (2) right   (3) isosceles   (4) acute

5. A pair of adjacent angles in the accompanying figure are
   (1) $\angle ABD$ and $\angle CBD$
   (2) $\angle ABC$ and $\angle CBA$
   (3) $\angle CBD$ and $\angle ABC$
   (4) $\angle ABD$ and $\angle ABC$

6. Which is *not* a valid way to name the angle?
   (1) $\angle STR$
   (2) $\angle RST$
   (3) $\angle TSR$
   (4) $\angle 1$

9

Chapter One THE TOOLS OF GEOMETRY

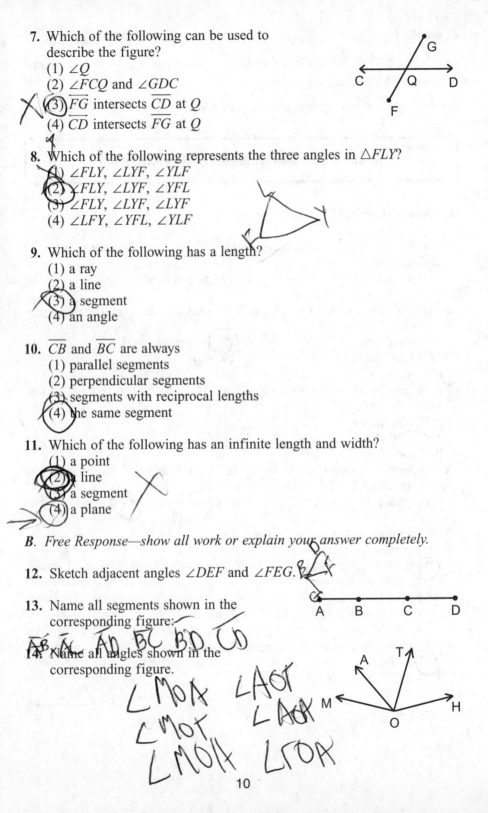

7. Which of the following can be used to describe the figure?
   (1) ∠Q
   (2) ∠FCQ and ∠GDC
   (3) $\overleftrightarrow{FG}$ intersects $\overleftrightarrow{CD}$ at Q
   (4) $\overleftrightarrow{CD}$ intersects $\overleftrightarrow{FG}$ at Q

8. Which of the following represents the three angles in △FLY?
   (1) ∠FLY, ∠LYF, ∠YLF
   (2) ∠FLY, ∠LYF, ∠YFL
   (3) ∠FLY, ∠LYF, ∠LYF
   (4) ∠LFY, ∠YFL, ∠YLF

9. Which of the following has a length?
   (1) a ray
   (2) a line
   (3) a segment
   (4) an angle

10. $\overline{CB}$ and $\overline{BC}$ are always
    (1) parallel segments
    (2) perpendicular segments
    (3) segments with reciprocal lengths
    (4) the same segment

11. Which of the following has an infinite length and width?
    (1) a point
    (2) a line
    (3) a segment
    (4) a plane

B. Free Response—show all work or explain your answer completely.

12. Sketch adjacent angles ∠DEF and ∠FEG.

13. Name all segments shown in the corresponding figure:

14. Name all angles shown in the corresponding figure.

1.2 BASIC RELATIONSHIPS AMONG POINTS, LINES, AND PLANES

KEY IDEAS

Geometric building blocks can be arranged in a number of ways relative to one another. These arrangements include parallel and perpendicular for lines and planes, and collinear for points. The relationships may be definitions, postulates, or theorems. A definition simply assigns a meaning to a word. A postulate is a statement that is accepted to be true but is not proven. A theorem is a true statement that can be proven.

Postulates and Theorems

A **definition** assigns a meaning to a word using previously defined words. For example, "A triangle is a polygon with three sides." Definitions provide only the minimum amount of information needed to define the word unambiguously. Properties that can be proven using the definition are not part of the definition. For example, in the definition of a triangle, we would not mention the fact that the angles in a triangle sum to 180°. That is a theorem that can be proven.

A **postulate** or an **axiom** is a statement that is accepted to be true but cannot be proven. When proving a theorem, we cannot rely entirely on previously proven theorems because we need to start somewhere. Postulates are that starting point. Some of the postulates may seem obvious, so obvious in fact that the best one could do is to restate the postulate in different words. For example, "Exactly one straight line may be drawn through two points" is a postulate. It is obviously true but cannot be proven using more fundamental postulates.

A **theorem** is a statement that can be proven true using a logical argument based on facts and statements that are accepted to be true. If points, lines, and planes are the building blocks of geometry, then theorems are the cement that binds them together. Theorems often express the relationships among the geometric figures and their measures that are the heart of geometry. An example of a theorem is "the diagonals of a square are perpendicular." When proving a theorem, we may call upon previously proven theorems, postulates, and definitions.

Congruent

The term *congruent* is similar to the term *equal*. However, *congruent* applies to geometric figures while *equal* applies to numbers. Figures that have the same size and shape are said to be congruent. The symbol for congruent is ≅.

As often happens in mathematics, there are different approaches to determining if two figures are congruent. Since congruent figures have the same size and shape, we can compare lengths and angle measures. Two segments are congruent if their lengths are equal. Two angles are congruent if their angle measures are equal. Polygons are congruent if all pairs of corresponding angles and sides have the same measure. Circles are congruent if their radii are congruent. Alternatively, congruence can be established through transformations. Two figures are congruent if a set of rigid motion transformations map one figure onto the other. The transformation point of view is one that is emphasized and discussed in detail in Section 5.

Keep in mind the difference in notation between congruent and equality. If two segments, $\overline{CD}$ and $\overline{EF}$, are congruent, we state that fact with $\overline{CD} \cong \overline{EF}$. Since the segments are congruent, we know their lengths are equal, which we state with $CD = EF$. Note the difference in symbol, ≅ versus =. In addition, we use the overbar when referring to the segment and just the endpoints when referring to its length.

Congruence of segments and angles can be specified in a sketch using tick marks for segments and arcs for angles. Sides with the same number of tick marks are congruent to one another, and angles with the same number of arcs are congruent to one another. Figure 1.10 shows a parallelogram with two pairs of congruent sides and two pairs of congruent angles. The pair of long sides each have one tick mark and are congruent, while the pair of short sides each have two tick marks and are congruent. The same is true for the two pairs of angles but using arcs. Figure 1.11 shows the congruent markings for a square. All four sides are congruent, so each side has one tick mark. The four angles are congruent, but they are also right angles, so the right angle marking can be used in place of the arcs.

Figure 1.10 Congruent markings in a parallelogram

Figure 1.11 Congruent and right angle markings in a square

12

Collinear and Coplanar

A set of points that all lie on the same line is described as **collinear**. Figure 1.12a illustrates collinear points L, M, N, O. Points that are not collinear are described as **noncollinear**. Points R, S, and T in Figure 1.12b are noncollinear. Any two given points will always be collinear since a straight line can always be drawn through two points. This is a consequence of our first postulate.

Postulate 1
There is one, and only one, line that contains two given points.

Extending to three dimensions, a set of points that all lie on the same plane is described as **coplanar**. Figure 1.13 illustrates coplanar points L, M, N, O. Points that do not lie on the same plane are **noncoplanar**. Any three given points will always be coplanar.

Postulate 2
There is one, and only one, plane that contains three given points.

(a) Collinear points L, M, N, O   (b) Noncollinear points R, S, T

Figure 1.12 Collinear and noncollinear points

Figure 1.13 Coplanar points L, M, N, O

In addition to points, lines may also be coplanar. Coplanar lines are lines that are completely contained within the same plane. Remember, both the plane and the lines continue forever in their respective dimensions.

Intersecting, Parallel, Perpendicular, and Skew

Coincide simply means to lie on top of one another. Two lines or planes that coincide are essentially the same. **Intersecting** means to cross one another.

13

Intersecting lines always cross at a single point, called the point of intersection. Figure 1.14 shows lines *r* and *s* intersecting at point M. The intersection of two planes is always a single line, called the line of intersection. Figure 1.15 shows planes *ABC* and *ABD* intersecting at $\overleftrightarrow{AB}$.

Figure 1.14 Intersecting lines   **Figure 1.15** Intersecting planes

Postulate 3
The intersection of two lines is a point.

Postulate 4
The intersection of two planes is a line.

Postulate 5
Intersecting lines are always coplanar.

Perpendicular

Perpendicular is a special case of intersecting, where the lines or planes intersect at right angles. The symbol for perpendicular is ⊥. In Figure 1.16, line *r* ⊥ line *s*. In Figure 1.17, plane *R* ⊥ plane *S*. The small square at the right angle in Figure 1.16 is a symbol for a right angle. Segments and rays are perpendicular if the lines that contain them are perpendicular. Note that our definition of perpendicular involves right angles, not a 90° measure. Perpendicular lines lead us to right angles, and the right angles lead us to the 90° measure.

Figure 1.16 Perpendicular lines

Figure 1.17 Perpendicular planes

Parallel

Parallel lines are lines that never intersect <u>and</u> are coplanar. You can recognize parallel lines by the way they run in the same directions like a pair of train tracks. We use the symbol ‖ for parallel. In Figure 1.18, line *r* ‖ line *s*. Segments and rays are parallel if the lines that contain them are parallel. The "and are coplanar" part of the definition is important because it distinguishes parallel from skew. Planes can be parallel as well, as shown in Figure 1.19. Parallel planes never intersect.

Figure 1.18 Line *r* ‖ line *s*

Figure 1.19 Plane *R* ‖ Plane *S*

Skew Lines

Skew lines are lines that are not coplanar. Like parallel lines, skew lines will never intersect. However, unlike parallel lines, skew lines run in different directions. In Figure 1.20, line $\overleftrightarrow{EA}$ and line $\overleftrightarrow{GF}$ are skew. We do not have a special symbol for skew.

Figure 1.20 Skew lines $\overleftrightarrow{EA}$ and $\overleftrightarrow{GF}$

15

Even though $\overline{EG}$ and $\overline{HF}$ are not connected with arrows in Figure 1.20, a line still exists that passes through each of the two pairs of points. Any two points can be used to specify a line. The same goes for planes. Plane *ACGE* slices diagonally through the prism even though we do not see the points connected in the manner seen in plane *EFG*. Any three points can be used to specify a plane.

If you look at any pair of lines, one and only one of the following must be true. They can coincide, intersect, be parallel, or be skew. Any pair of planes will coincide, be parallel, or intersect. We do not use the word *skew* to describe planes.

Examples

For examples 1–5, use the figure of the cube below.

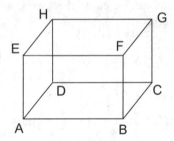

1. Identify 3 segments parallel to $\overline{EA}$.

2. Identify 4 segments perpendicular to $\overline{BC}$.

3. Identify 4 segments skew to $\overline{HD}$.

4. Identify 1 plane parallel to plane *EFG*.

5. Identify 4 planes perpendicular to plane *EFG*.

Solutions to examples 1–5:

1. $\overline{FB}$, $\overline{GC}$, and $\overline{HD}$ are parallel to $\overline{EA}$.

2. $\overline{AB}$, $\overline{CD}$, $\overline{BF}$, and $\overline{CG}$ are perpendicular to $\overline{BC}$.

3. $\overline{EF}$, $\overline{FG}$, $\overline{AB}$, and $\overline{BC}$ are skew to $\overline{HD}$.

4. Plane *ABC* is parallel to plane *EFG*.

5. Planes *EAB*, *FBC*, *GCD*, and *HDA* are perpendicular to plane *EFG*.

Check Your Understanding of Section 1.2

A. Multiple-Choice

1. Lines that are coplanar but do not intersect can be described as
   (1) perpendicular
   (2) parallel
   (3) skew
   (4) congruent

2. The intersection of two planes is
   (1) 1 point
   (2) 1 line
   (3) 2 points
   (4) 2 planes

3. Line *r* intersects parallel planes *U* and *V*. The intersection can be described as
   (1) 2 parallel lines
   (2) 1 line
   (3) 2 intersecting lines
   (4) 2 points

4. Points *A*, *B*, and *C* are not collinear. How many planes contain all three points?
   (1) one
   (2) two
   (3) three
   (4) an infinite number

5. In the figure of a rectangular prism, which of the following is true?
   (1) Points *E*, *H*, *D*, and *A* are coplanar and collinear.
   (2) $\overline{HD}$ is skew to $\overline{CG}$, and $\overline{CD} \perp \overline{CG}$.
   (3) $\overline{EA} \perp \overline{BC}$, and $\overline{AB} \parallel \overline{CD}$.
   (4) $\overline{EA} \parallel \overline{CG}$, and $\overline{EH}$ skew to $\overline{FB}$.

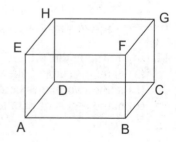

6. Which parts of the accompanying figure are congruent?

(1) $\overline{HI} \cong \overline{HG}$, $\angle I \cong \angle G$, and $\angle H \cong \angle F$

(2) $\overline{HI} \cong \overline{HG}$, $\overline{IF} \cong \overline{FG}$, and $\angle I \cong \angle G$

(3) $\overline{HI} \cong \overline{IF}$, $\overline{HG} \cong \overline{FG}$, and $\angle I \cong \angle G$

(4) $\overline{IF} \cong \overline{FG}$ and $\angle H \cong \angle F$

7. $\overleftrightarrow{MN}$ and $\overleftrightarrow{JK}$ intersect at point L. Which of the following is *not* true?

(1) Points J, K, and M are collinear.

(2) $\overleftrightarrow{MN}$ and $\overleftrightarrow{JK}$ are coplanar.

(3) Points J, K, and L are collinear.

(4) Points J, K, L, and M are coplanar.

8. Given points F, G, H, and I with no three of the points collinear, what is the maximum number of distinct lines that can be defined using points F, G, H, and I?

(1) 4   (2) 5   (3) 6   (4) 8

9. Lines r and s intersect at point A. Line t intersects lines r and s and points B and C, respectively. Which of the following is true?

(1) Lines r, s, and t must all be perpendicular.

(2) Line t must be skew to lines r and s.

(3) Points A, B, and C must be collinear.

(4) Lines r, s, and t must all be coplanar.

10. If $\angle J \cong \angle L$, which must be true?

(1) $m\angle J = m\angle L$

(2) $\angle J \perp \angle L$

(3) $\angle J \parallel \angle L$

(4) $m\angle J + m\angle L = 180°$

B. *Free Response—show all work or explain your answer completely.*

11. In the triangular prism,

(a) name a segment skew to $\overline{EF}$

(b) name two planes containing $\overleftrightarrow{AB}$

(c) name a pair of parallel planes

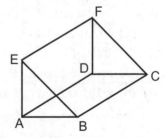

12. Points M, N, and P are contained in both planes S and T. Juan states that the three points must be collinear, but his friend Carla disagrees and says they do not have to be. Who is correct? Explain your reasoning.

13. A stool has three legs, but one of the legs is shorter than the other two. When the stool is placed on a flat floor, will all three legs touch the floor? Explain why or why not.

1.3 BRIEF REVIEW OF ALGEBRA SKILLS

KEY IDEAS

Certain algebra skills show up frequently in our study of geometry. They are tools used to complete the evaluation of geometric relationships. Procedures for operations with radicals, solving linear equations, multiplying polynomials, solving quadratic equations, and solving proportions as well as word problem strategies are briefly reviewed.

Operations with Radicals

The square root of a number is the number that when multiplied by itself results in the original number. It is represented with the square root, or radical, symbol $\sqrt{}$. The number under the radical symbol is the radicand. An example of a radical expression is $3\sqrt{5}$. The 3 is the coefficient, and the 5 is the radicand.

Adding and Subtracting Radicals: Add or subtract the coefficient if the radicands are the same, otherwise the radicals cannot be combined. For example, only the $\sqrt{5}$ terms can be combined in the following equation.

$$3\sqrt{5} + 4\sqrt{7} + 9\sqrt{5} = 12\sqrt{5} + 4\sqrt{7}$$

Multiplying and Dividing Radicals: Multiply or divide the coefficients and then multiply or divide the radicands.

$$8\sqrt{15} \cdot 2\sqrt{3} = 16\sqrt{45}$$
$$8\sqrt{15} \div 2\sqrt{3} = 4\sqrt{5}$$

Simplifying Radicals—A radical is said to be in simplest form when the following 3 conditions are met.

1) No perfect square factors appear in the radicand.

2) No fractions appear in the radicand.

3) No radicals appear in the denominator.

Remove perfect square factors from the radicand by factoring the radicand using the largest perfect square factor. Then take the square root of the perfect square factor. In the radical expression below, 12 is factored into $4 \cdot 3$. Then $\sqrt{4}$ is simplified to 2.

$$5\sqrt{12} = 5\sqrt{4 \cdot 3}$$
$$= 5 \cdot 2\sqrt{3}$$
$$= 10\sqrt{3}$$

Fractions in the radicand can be rewritten as the quotient of two radicals as shown below.

$$4\sqrt{\frac{3}{5}} = 4\frac{\sqrt{3}}{\sqrt{5}}$$

Remove radicals in the denominator by multiplying the numerator and denominator by the radical in the denominator. This is called "rationalizing the denominator."

$$\frac{\sqrt{3}}{\sqrt{5}} = \frac{\sqrt{3}}{\sqrt{5}} \cdot \frac{\sqrt{5}}{\sqrt{5}}$$
$$= \frac{\sqrt{15}}{5}$$

MATH FACT

Taking the square root and squaring are inverse operations. Taking the square root of a number and then squaring it results in the original number, as in $\sqrt{7} \cdot \sqrt{7} = 7$.

Example 1

Express $2\sqrt{6} \cdot 5\sqrt{3}$ in simplest radical form

Solution:

$$2\sqrt{6} \cdot 5\sqrt{3} = 10\sqrt{18}$$
$$= 10\sqrt{9 \cdot 2}$$
$$= 10 \cdot 3\sqrt{2}$$
$$= 30\sqrt{2}$$

Example 2

Express $\dfrac{\sqrt{12}}{\sqrt{5}}$ in simplest radical form

Solution:

$$\sqrt{\dfrac{12}{5}} = \dfrac{\sqrt{12}}{\sqrt{5}}$$

$$= \dfrac{\sqrt{12}}{\sqrt{5}} \cdot \dfrac{\sqrt{5}}{\sqrt{5}}$$

$$= \dfrac{\sqrt{60}}{5}$$

$$= \dfrac{\sqrt{4 \cdot 15}}{5}$$

$$= \dfrac{2\sqrt{15}}{5}$$

MATH FACT

Expressing radicals in simplest radical form make it easier to compare radical expressions. In geometry, we often want to determine if two measures are equal or satisfy a particular inequality. Once all the radicals have been completely simplified, comparing radicals is just a matter of comparing the coefficients. Radicals will frequently show up when working with solving quadratic equations, when using the Pythagorean theorem, or with the distance formula.

Solving Linear Equations

Linear equations are equations that involve the variable raised to the first power only. They can be solved using the following steps.

1) Apply the distributive property to terms with parentheses.

2) Eliminate fractions by multiplying both sides by the denominator of any fraction, or the greatest common denominator if there are several fractions.

3) Combine like terms on each side of the equal sign.

4) Isolate the variable by undoing additions/subtractions and then multiplications.

Example 1

Solve $8x + 6 = 2x + 4(x + 5)$

   Solution:

$$8x + 6 = 2x + 4(x + 5)$$
$$8x + 6 = 2x + 4x + 20 \quad \text{distribute}$$
$$8x + 6 = 6x + 20 \quad \text{combine like terms}$$
$$8x = 6x + 14 \quad \text{subtract 6 from both sides}$$
$$2x = 14 \quad \text{subtract } 6x \text{ from both sides}$$
$$x = 7 \quad \text{divide both sides by 2}$$

Example 2

Solve $3x + 3 = \dfrac{2}{3}x + 17$

   Solution:

$$3x + 3 = \frac{2}{3}x + 17$$
$$9x + 9 = 2x + 51 \quad \text{multiply both sides by 3 to eliminate the fraction}$$
$$9x = 2x + 42 \quad \text{subtract 9 from both sides}$$
$$7x = 42 \quad \text{subtract } 2x \text{ from both sides}$$
$$x = 6 \quad \text{divide both sides by 7}$$

Multiplying Polynomials

When multiplying monomials, multiply the coefficients and multiply the variables. When multiplying powers of the same variable, use the rule "keep the base and add the powers."

Example 1

Multiply $5x \cdot 12x^3$

   Solution:

$$5x \cdot 12x^3 = (5 \cdot 12)\,(x \cdot x^3) = 60x^4$$

When multiplying binomials, use the double distributive property by applying the vertical method, box method, FOIL (first-outer-inner-last), or any other technique you may have learned.

Example 2

Multiply $(4x + 2)(5x + 6)$

> *Solution:* Use FOIL.

$$\text{First} \quad 4x \cdot 5x = 20x^2$$
$$\text{Outer} \ 4x \cdot 6 = 24x$$
$$\text{Inner} \ \ 2 \cdot 5x = 10x$$
$$\text{Last} \quad 2 \cdot 6 = 12$$

$$20x^2 + 34x + 12$$

Example 3

Multiply $(3x - 7)(x + 2)$

> *Solution:* Use the vertical method.

$$3x - 7$$
$$\underline{x + 2}$$
$$6x - 14$$
$$\underline{3x^2 - 7x}$$
$$3x^2 - x - 14$$

Factoring and Solving Quadratic Equations

Quadratic equations have second-order, or x^2, terms as the highest power of x. Solving quadratic equations requires factoring. The procedure is as follows:

- Get all terms on one side.
- Factor.
- Apply the zero product rule.
- Solve for x.

MATH FACT

The zero product rule states that if a product of factors equals zero, then each factor may be individually set equal to zero and solved to find a solution to the equation.

Some Factoring Methods

Greatest Common Factor: If a common factor exists among all terms, divide all terms by that factor. Put the new terms inside parentheses, and move the divided factor outside the parentheses. If a factor is still not linear, use another method on that factor.

Example 1

Solve $3x^2 + 9x = 0$

   Solution:

$$3x^2 + 9x = 0$$
$$3x(x + 3) = 0 \qquad \qquad \text{greatest common factor is } 3x$$
$$3x = 0 \qquad x + 3 = 0 \qquad \text{zero product rule}$$
$$x = 0 \qquad \qquad x = -3$$

   Grouping with a = *1*: For the quadratic equation $x^2 + bx + c = 0$, find numbers r_1 and r_2 such that $r_1 + r_2 = b$ and $r_1 \cdot r_2 = c$. The equation factors to $(x - r_1)(x - r_2) = 0$. From here, apply the zero product theorem.

Example 2

Solve $x^2 - 5x + 6 = 0$

   Solution:

$$x^2 - 5x + 6 = 0 \qquad \qquad \text{the factors of 6 that sum to } -5$$
$$\text{are } (-2) \text{ and } (-3)$$

$$(x - 2)(x - 3) = 0$$
$$x - 2 = 0 \qquad x - 3 = 0 \qquad \text{apply the zero product theorem}$$
$$x = 2 \qquad \qquad x = 3$$

   Difference of Perfect Squares: Quadratics in the form $a^2x^2 - b^2$ factor into $(ax + b)(ax - b)$. Once factored, set each factor equal to zero and solve for x.

Example 3

Solve $4x^2 - 9 = 0$

   Solution:

$$4x^2 - 9 = 0$$
$$(2x + 3)(2x - 3) = 0$$
$$2x + 3 = 0 \qquad 2x - 3 = 0$$
$$x = -\frac{3}{2} \qquad \qquad x = \frac{3}{2}$$

   Completing the Square: Completing the square can be used on any trinomial with the form $x^2 + bx + c = 0$.

- Rewrite the equation as $x^2 + bx = -c$
- Add the quantity $\left(\frac{1}{2}b\right)^2$ to both sides of the equation.

- Factor the left side, which will be a perfect square.
- Take the square root of both sides, and solve for x.
- Remember, there will be a positive and negative root when taking the square root. So there will be two solutions.

Example 4

Solve $x^2 + 6x + 1 = 0$

   Solution:

$$x^2 + 6x = -1$$

$x^2 + 6x + 9 = -1 + 9$ $\left(\dfrac{1}{2}b\right)^2 = \left(\dfrac{1}{2}\cdot 6\right)^2 = 3^2 = 9$, add 9 to both sides

$x^2 + 6x + 9 = 8$ factor the perfect square

$(x + 3)^2 = 8$ take the square root

$x + 3 = \pm\sqrt{8}$ solve for x

$x = -3 \pm \sqrt{8}$

Quadratic Formula: The solution to any quadratic equation of the form $ax^2 + bx + c = 0$ can be found using the quadratic formula.

$$x = \frac{-b \pm \sqrt{b^2 - 4ac}}{2a}$$

Example 5

Solve $6x^2 - x - 2 = 0$

   Solution:

$$a = 6,\ b = -1,\ c = -2$$

$$x = \frac{-b \pm \sqrt{b^2 - 4ac}}{2a}$$

$$= \frac{1 \pm \sqrt{(-1)^2 - 4 \cdot 6(-2)}}{2 \cdot 6}$$

$$= \frac{1 \pm \sqrt{49}}{12}$$

$$= \frac{1 \pm 7}{12}$$

$$x = \frac{2}{3},\ x = -\frac{1}{2}$$

Some geometric relationships result in a quadratic equation that must be solved in order to find the measure of an angle or segment. The quadratic will give two solutions, and both must be checked for consistency with the problem. Lengths or angle measure in this course will always be positive. If either solution results in a negative length or angle, that solution is thrown out. If both solutions lead to an acceptable answer, the problem has two solutions. Two solutions often correspond to a situation where two different geometric configurations could lead to the relationship modeled in the equation.

Solving Proportions

A **proportion** is an equation involving two ratios. They can be solved using the fact that the cross products must be equal.

$$\text{If } \frac{a}{b} = \frac{c}{d} \text{ then } a \cdot d = b \cdot c$$

Example 1

Solve $\dfrac{2}{x+7} = \dfrac{3}{4x+3}$

Solution:

$$\frac{2}{x+7} = \frac{3}{4x+3}$$

$2(4x + 3) = 3(x + 7)$ set cross products equal

$8x + 6 = 3x + 21$ multiply

$5x = 15$ solve for x

$x = 3$

Example 2

Solve $\dfrac{2x+2}{x+19} = \dfrac{x}{x+5}$

Solution:

$$(x + 5)(2x + 2) = x(x + 19) \qquad \text{set cross products equal}$$
$$2x^2 + 2x + 10x + 10 = x^2 + 19x \qquad \text{multiply}$$
$$2x^2 + 12x + 10 = x^2 + 19x \qquad \text{simplify}$$
$$x^2 + 12x + 10 = 19x$$
$$x^2 - 7x + 10 = 0$$
$$(x - 5)(x - 2) = 0 \qquad \text{factor}$$
$$x - 5 = 0 \qquad x - 2 = 0 \qquad \text{apply zero product rule}$$
$$x = 5 \qquad x = 2$$

Word Problem Strategies

Word problems in geometry may involve phrases that describe a relationship between two figures or measures. Some common phrases and their algebraic translations are shown below.

Phrase	Algebra
x is two more than y x is two greater than y	$x = y + 2$
x is two less than y	$x = y - 2$
x is twice y x is double y	$x = 2y$
x is half y	$x = \dfrac{1}{2}y$
Three quantities are in a $1:2:3$ ratio	Represent the quantities as x, $2x$, and $3x$

The following are some good general strategies for solving word problems:

1) Make a sketch and label it.

2) Underline or highlight key words and definitions, such as bisector, midpoint, and so on.

3) Underline phrases to be translated into mathematical expressions.

4) Identify what the question is asking—the value of a variable, the measure of an angle or segment, an explanation or justification, and so on.

Example 1

Write an expression that represents "12 less than double a number."

Solution: Let the number equal x.

$$2x - 12$$

Example 2

Three integers are in a $4:7:9$ ratio. If their sum equals 60, what are the numbers?

Solution: Let the integers equal $4x$, $7x$, and $9x$.

$$4x + 7x + 9x = 60$$
$$20x = 60$$
$$x = 3$$

Check Your Understanding of Section 1.3

A. Multiple-Choice

1. $5\sqrt{18} - 3\sqrt{2}$ is equal to
   (1) $3\sqrt{16}$ (2) $12\sqrt{2}$ (3) $42\sqrt{2}$ (4) $10\sqrt{3}$

2. $6\sqrt{5} \cdot 11\sqrt{20} =$
   (1) 110   (2) 330   (3) 440   (4) 660

3. $\dfrac{\sqrt{7}}{\sqrt{8}}$ is equivalent to
   (1) $4\sqrt{14}$ (2) $\sqrt{14}$ (3) $\dfrac{\sqrt{14}}{2}$ (4) $\dfrac{\sqrt{14}}{4}$

4. The solution to $6x + 4 = 2(x + 6)$ is
   (1) $x = 0$ (2) $x = 1$ (3) $x = 2$ (4) $x = 3$

5. The solution to $\dfrac{1}{2}x + 5 = x - 1$ is
   (1) $x = 3$ (2) $x = 6$ (3) $x = 9$ (4) $x = 12$

6. The solution to the equation $x^2 - 12x + 20 = 0$ is
   (1) $x = -2$, $x = 10$ (3) $x = -4$, $x = -5$
   (2) $x = 2$, $x = 10$ (4) $x = -20$, $x = 12$

7. When factored, $x^2 - 36$ is equal to
   (1) $(x^2 + 6)(6 - 12)$ (3) $(x + 6)(x - 6)$
   (2) $(x - 6)^2$ (4) $(x + 6)^2$

8. The length of a segment is given by the solution to $x^2 + 5x - 50 = 0$.
   What are the possible lengths?
   (1) 5 only   (2) 5 or 10   (3) 5 or -10 (4) 10 only

9. $\sqrt{32} + \sqrt{2} =$
   (1) $\frac{1}{2}\sqrt{32}$ (2) $2\sqrt{2}$ (3) $5\sqrt{2}$ (4) 8

10. Which expression represents "6 less than twice the measure of $\angle 1$"?
    (1) $2 \cdot m\angle 1 - 6$ (3) $2 - 6 \cdot m\angle 1$
    (2) $6 - 2 \cdot m\angle 1$ (4) $6 \cdot m\angle 1 - 2$

B. Free Response—show all work or explain your answer completely. All answers involving radicals should be in simplest radical form.

11. Solve $3x^2 - 27 = 0$ by any method.

12. Solve $x^2 + 6x - 8 = 0$ by completing the square.

13. Is $3\sqrt{75}$ less than, greater than, or equal to $7\sqrt{27}$? Justify your answer.

14. If AB is 3 greater than four times CD and the sum of the lengths is 33, find each length.

15. The measures of $\angle A$ and $\angle B$ sum to 180°, and $m\angle A$ is 9° greater than one-half $m\angle B$. Find the measure of each angle.

16. The sides of a triangle are in a $3:5:6$ ratio. If the perimeter has a length of 56, what is the length of the shortest side?

17. The measures of two angles sum to 90°, and they are in a ratio of $2:3$. Find the measure of each angle.

18. Solve for x: $\dfrac{x+5}{2} = \dfrac{2x+4}{3}$

19. Solve for x: $\dfrac{x+2}{8} = \dfrac{9}{x+8}$

<table>
<tr><td>Chapter
Two</td><td># ANGLE AND SEGMENT RELATIONSHIPS</td></tr>
</table>

In this section, we explore some of the basic relationships involving the measures of angles and segments. Algebraic modeling and equation solving will be applied to find the measure of a missing angle or the measure of a segment. These relationships represent the building blocks that will be used to explore more complex problems, theorems, and proofs.

2.1   BASIC ANGLE RELATIONSHIPS

KEY IDEAS

Basic theorems and definitions used to solve problems involving angles include:

- The sum of the measures of adjacent angles around a point equals 360°.
- Supplementary angles have measures that sum to 180°.
- Complementary angles have measures that sum to 90°.
- Vertical angles are the congruent opposite formed by intersecting lines and are congruent.
- Angle bisectors divide angles into two congruent angles.
- The measure of a whole equals the sum of the measures of its parts.

Sum of the Angles About a Point

The measures of the adjacent angles about a point sum to 360°. In Figure 2.1, $m\angle 1 + m\angle 2 + m\angle 3 + m\angle 4 = 360°$.

$$m\angle 1 + m\angle 2 + m\angle 3 + m\angle 4 = 360°$$

Figure 2.1 Sum of the angles about a point = 360°

Example

Find the measure of each angle in the accompanying figure.

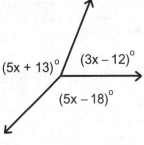

Solution:

$$(5x + 13) + (3x - 12) + (5x - 18) = 360$$
$$13x - 17 = 360$$
$$13x = 377$$
$$x = 29$$
$$5(29) + 13 = 158°$$
$$3(29) - 12 = 75°$$
$$5(29) - 18 = 127°$$

Supplementary Angles, Complementary Angles, and Linear Pairs

Supplementary angles are angles whose measures sum to 180°. The angles may or may not be adjacent. Two adjacent angles that form a straight line are called a **linear pair**. They are supplementary. Figure 2.2 shows a linear pair. Multiple adjacent angles around a line also sum to 180°, as shown in Figure 2.3.

Complementary angles are angles whose measures sum to 90°. As with supplementary angles, complementary angles may or may not be adjacent. Complementary angles are illustrated in Figure 2.4.

$$m\angle 1 + m\angle 2 = 180°$$

Figure 2.2 Linear pair

$$m\angle 1 + m\angle 2 + m\angle 3 + m\angle 4 = 180°$$

Figure 2.3 Adjacent angles around a line

$$m\angle 1 + m\angle 2 = 90°$$

Figure 2.4 Complementary angles

Example 1

$\overrightarrow{DB}$ intersects $\overleftrightarrow{ABC}$ at B. If m$\angle ABD =$
$(4x + 8)°$ and m$\angle CBD = (2x + 4)°$, find
the measure of each angle.

$\quad$ *Solution:* The angles form a linear pair
and are supplementary.

$$(4x + 8) + (2x + 4) = 180$$
$$6x + 12 = 180$$
$$6x = 168$$
$$x = 28$$
$$m\angle ABD = 4(28) + 8 = 120°$$
$$m\angle CBD = 2(28) + 4 = 60°$$

Example 2

In the accompanying figure, m$\angle 1 = (x + 16)°$ and
m$\angle 2 = (3x + 6)°$. Find the measure of each angle.

$\quad$ *Solution:* The angles are complementary.

$$(x + 16) + (3x + 6) = 90$$
$$4x + 22 = 90$$
$$4x = 68$$
$$x = 17$$
$$m\angle 1 = 17 + 16 = 33°$$
$$m\angle 2 = 3(17) + 6 = 57°$$

Vertical Angles

When two lines intersect, four angles are formed. Each pair of opposite
angles are congruent and are called **vertical angles**. In Figure 2.5, $\angle 1$ and
$\angle 3$ are vertical angles and are therefore congruent. Also, $\angle 2$ and $\angle 4$ are
vertical angles and are therefore congruent. Vertical angles show up frequently in geometry. So always be on the lookout for them. Once you know
the measure of any one of the four vertical angles formed by two intersecting
lines, you can easily calculate the measures of the other three using the
supplementary angle relationship for a linear pair.

Figure 2.5 Vertical angles

Example

In Figure 2.5, m∠1 = 150°. Find the measure of each of the other angles.

> *Solution:*   Vertical Angles   Linear Pair
> m∠3 = m∠1   m∠4 + m∠1 = 180°
> m∠3 = 150°   m∠4 + 150° = 180°
> Vertical Angles   m∠4 = 30°
> m∠2 = m∠4
> m∠2 = 30°

Check Your Understanding of Section 2.1

A. *Multiple-Choice*

1. Two complementary angles measure $(12x - 18)°$ and $(5x + 23)°$. What is the measure of the smaller angle?
   (1) 5°   (2) 42°   (3) 48°   (4) 90°

2. Two supplementary angles measure $(7x + 11)°$ and $(14x + 1)°$. What is the measure of the smaller angle?
   (1) 40.5°   (2) 67°   (3) 108°   (4) 123°

3. $\overrightarrow{AC}$ and $\overrightarrow{BD}$ intersect at O. If m∠DOC
   = $(8x - 30)°$ and m∠BOA =
   $(6x + 12)°$, what is m∠DOA?
   (1) 138°
   (2) 51°
   (3) 42°
   (4) 21°

4. What is the value of x in the
   accompanying figure?
   (1) 97°
   (2) 98°
   (3) 117°
   (4) 136°

5. Find the measure of ∠FIM.
(1) 58°
(2) 98°
(3) 102°
(4) 112°

$10x-12+6x+16$

$16\boxed{x}+4=180$

$(10x-12)°$ $(6x+16)°$

6. Two lines intersect such that a pair of vertical angles have measures of $(5x-57)°$ and $(3x+21)°$. Find the value of x.
(1) 39
(2) 27
(3) 12
(4) 6.5

$5x-57=3x+21$

$2x-57=21$

$(5x-57)°$

$(3x+21)°$

7. $\overline{BFA} \perp \overline{CF}$ and m∠CFE = 42°. Find m∠DFA.
(1) 132°
(2) 138°
(3) 142°
(4) 144°

8. Lines $\overleftrightarrow{AB}$ and $\overleftrightarrow{CD}$ intersect at P. Which of the following is *not necessarily* true?
(1) ∠APC and ∠CPB are supplementary.
(2) ∠BPC ≅ ∠APD
(3) ∠APC ≅ ∠BPD
(4) ∠APC ≅ ∠BPC

B. *Free Response—show all work or explain your answer completely.*

9. In the accompanying figure, $\overleftrightarrow{IOJ} \perp \overleftrightarrow{OK}$, $\overleftrightarrow{IOJ}$ intersects $\overleftrightarrow{GOH}$ at point O, and m∠JOH = 31°. Find the measures of ∠KOG and ∠IOH.

10. Two supplementary angles are congruent. What is the measure of each?

11. The measure of ∠*A* is 15° greater than twice the measure of ∠*B*. If ∠*A* and ∠*B* are complementary, what are the measures of ∠*A* and ∠*B*?

12. $\overline{QRS}$ intersects $\overline{TRU}$ at *R*. If the measure of ∠*QRU* increases by 15°, what is the change in the measure of ∠*SRT* and ∠*SRU*?

2.2 BISECTORS, MIDPOINT, AND THE ADDITION POSTULATE

KEY IDEAS

Whenever a segment or an angle is divided into two or more smaller parts, the sum of the smaller parts equals the original part. This is called the segment or angle addition postulate. When the part is divided exactly in half, we say it is bisected and the two parts are congruent to each other. An angle bisector is a line, segment, or ray that divides an angle into two congruent parts. A segment bisector intersects a segment at its midpoint, which divides it into two congruent parts.

Segment and Angle Addition

Any segment can be divided by locating a point between the two endpoints, creating two new segments. The **segment addition postulate** states a segment's length equals the sum of the two new segments formed by any point between the endpoints of the original segment. As shown in Figure 2.6, point *C* is located on $\overrightarrow{AB}$ between *A* and *B*. Therefore, $AC + CB = AB$.

A   C   B

$AC + CB = AB$

Figure 2.6 Segment addition

Example 1

Point *S* is located between points *R* and *T* on $\overline{RST}$. If $RS = (2x + 1)$, $ST = (3x + 4)$, and $RT = 35$, find the lengths of RS and ST.

35

Solution:

sketch

$$RS + ST = RT$$ segment addition postulate

$$(2x + 1) + (3x + 4) = 35$$ substitution

$$5x + 5 = 35$$ simplify and solve for x

$$5x = 30$$

$$x = 6$$

$$RS = 2(6) + 1 = 13$$ substitution

$$ST = 3(6) + 4 = 22$$

$$13 + 22 = 35$$ check that the sum of RS and ST equals RT

You should always sketch the figure and write the segment addition postulate in terms of segment names. This will give you the opportunity to check the postulate against the figure. If something does not make sense, you can make a correction before going further. Once you have substituted in algebraic expressions, it is much more difficult to check whether the equation makes sense from a geometry point of view.

The **angle addition postulate** is similar to the segment addition postulate. The measure of an angle is equal to the sum of the two adjacent angles formed by an interior ray with its endpoint at the vertex of the original angle.

Example 2

$m\angle ABC = (6x - 7)°$, $m\angle ABD = (3x + 3)$, and
$m\angle CBD = (2x + 12)$. Find $m\angle ABC$.

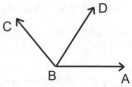

Solution:

$$m\angle ABC = m\angle ABD + m\angle CBD$$ angle addition postulate

$$(6x - 7) = (3x + 3) + (2x + 12)$$

$$6x - 7 = 5x + 15$$

$$x = 22$$

$$m\angle ABC = (6x - 7)°$$ substitute

$$= 6(22) - 7$$

$$= 125°$$

$$m\angle ABD = (3x + 3)$$

$$= 3(22) + 3$$

$$= 69°$$

$$m\angle CBD = (2x + 12)$$

$$= 2(22) + 12$$

$$= 56°$$

$$56 + 69 = 125$$ check

36

Besides addition, we also have postulates for **segment subtraction** and **angle subtraction**. These are used in a manner similar to segment and angle addition.

Example 3

Write an expression for the length of *EB* in terms of lengths *AB* and *AE*.

Solution:

$$EB = AB - AE$$

Angle Bisectors

A line, segment, or ray that divides an angle into two congruent angles is called an **angle bisector**. Figure 2.7 illustrates an angle bisector. Ray $\overrightarrow{CD}$ bisects $\angle ACB$. The two congruent angles formed are $\angle ACD$ and $\angle BCD$.

Figure 2.7 Angle bisector $\overrightarrow{CD}$

Example

$\overrightarrow{BC}$ bisects $\angle ABD$. If $m\angle ABD = (8x - 12)°$ and $m\angle ABC = (3x + 4)°$, find the value of x.

Solution:

$m\angle ABC + m\angle CBD = m\angle ABD$	angle addition
$m\angle ABC = m\angle CBD$	bisector forms 2 ≅ segments
$m\angle ABC + m\angle ABC = m\angle ABD$	substitution

$$(3x + 4) + (3x + 4) = (8x - 12)$$
$$6x + 8 = 8x - 12$$
$$-2x + 8 = -12$$
$$-2x = -20$$
$$x = 10$$

37

Midpoints and Segment Bisectors

In segments, the point that divides a segment into two congruent segments is called a **midpoint**. Any line, segment, or ray that intersects a segment at its midpoint is called a **segment bisector**, or simply **bisector**. Figure 2.8 illustrates a segment bisector. Line m bisects $\overline{RT}$ at its midpoint S, and $\overline{RS} \cong \overline{ST}$. Note that lines can bisect segments. However, lines cannot be bisected themselves because lines do not have a finite length.

Figure 2.8 Segment bisector m

=== **MATH FACT** ===

In geometry, the definition of a segment bisector, or simply bisector, is "a segment, line, or ray that passes through the midpoint of a segment." The definition says nothing about ending up with two congruent segments. It is the definition of a midpoint that tells us we end up with two congruent segments.

A midpoint on a segment will always give two relationships, the pair of congruent parts and the segment addition postulate, as seen in Figure 2.9. One or both relationships may be needed to solve a problem.

M   N   O

$MN = NO$ and $MN + NO = MO$

Figure 2.9 Segment relationships with a midpoint

Example 1

J is the midpoint of $\overline{GH}$. $GJ = (7x + 1)$ and $JH = 3x + 9$. Find GH.

Solution:

$$GJ = JH \qquad \text{midpoint forms two} \cong \text{segments}$$
$$7x + 1 = 3x + 9$$
$$4x = 8$$
$$x = 2$$
$$GJ = 7(2) + 1 = 15$$
$$JH = 3(2) + 9 = 15$$
$$GH = GJ + JH = 30$$

Example 2

$\overleftrightarrow{UV}$ bisects $\overline{PQ}$ at point W. If $PW = 2x + 7$ and $PQ = 3x + 24$, find the length of PW.

   Solution:

$PW + WQ = PQ$	segment addition
$PW = WQ$	midpoint creates two congruent segments
$PW + PW = PQ$	substitution
$(2x + 7) + (2x + 7) = 3x + 24$	
$4x + 14 = 3x + 24$	simplify and solve for x
$x + 14 = 24$	
$x = 10$	
$PW = 2x + 7$	substitute for x
$PW = 2(10) + 7$	
$= 27$	

A special case of a bisector is the **perpendicular bisector**. This is a segment, line, or ray that passes through a segment's midpoint at a right angle. Figure 2.10 shows perpendicular bisector $\overleftrightarrow{CD}$ passing through midpoint E at a right angle.

Figure 2.10 Perpendicular bisector $\overleftrightarrow{CD}$

39

Check Your Understanding of Section 2.2

A. *Multiple-Choice*

1. Point R is the midpoint of $\overline{UV}$, and point S is the midpoint of $\overrightarrow{RV}$. Which of the following is true?
   (1) $UR = RS$ (3) $UR + RS = SV$
   (2) $SV = \dfrac{1}{4}UV$ (4) $UR = 2(UV)$

2. In segment $\overline{CDF}$, $CD = 2x + 7$, $DF = 3x + 6$, and $CF = 9x - 11$. What is the length of CF?
   (1) 6   (2) 9   (3) 24   (4) 43

3. Point X is the midpoint of segment $\overline{BM}$. If $BX = 3x - 4$ and $XM = x + 12$, what is the length of BM?
   (1) 5   (2) 8   (3) 40   (4) 80

4. $m\angle ABD = (2x + 8)°$, $m\angle DBC = (3x + 12)°$, and $m\angle ABC = (8x - 1)°$. What is the measure of $\angle ABC$?
   (1) 55°
   (2) 57°
   (3) 61°
   (4) 63°

5. $\angle FGH$ is bisected by $\overrightarrow{GK}$. If $m\angle FGK = (4x + 5)°$ and $m\angle FGH = (14x - 20)°$, what is the value of x?
   (1) 1   (2) 2.5   (3) 3.5   (4) 5

6. In segment $\overline{RU}$, S is between R and U, and T is between S and U. Which relationship must be true?
   (1) $RS = ST$ (3) $RU - TU = RT$
   (2) $RS + ST = TU$ (4) $ST - TU = RS$

7. $\angle SAR$ is bisected by $\overrightarrow{AT}$. Which relationship must be true?
   (1) $m\angle SAT = m\angle SAR$ (3) $m\angle TAR = \dfrac{1}{2}m\angle SAT$
   (2) $m\angle TAR = m\angle RAS$ (4) $m\angle SAR = 2m\angle TAR$

8. Perpendicular bisector $\overleftrightarrow{XW}$ intersects $\overline{UV}$ at A. Which of the following is *not necessarily* true?

   (1) $\overline{UA} \cong \overline{AV}$

   (2) $\overline{XA} \cong \overline{AW}$

   (3) $\angle UAX \cong \angle XAV$

   (4) $\angle WAV \cong \angle UAX$

9. Given $\overleftrightarrow{YXW}$, m$\angle VXY$ = 144° and m$\angle ZXW$ = 138°. What is the measure of $\angle ZXV$?

   (1) 36°

   (2) 42°

   (3) 90°

   (4) 102°

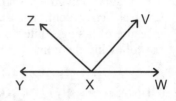

B. *Free Response—show all work or explain your answer completely.*

10. In $\triangle JKL$, $\overline{KM}$ is a median. If $JM = (8x - 21)$ and $ML = (2x + 33)$, find JL.

11. $\overleftrightarrow{APB}$ intersects $\overleftrightarrow{CPD}$ at P. If $AP = (7x - 14)$, $BP = (3x + 6)$, and m$\angle BPD = (8y - 6)°$, for what values of x and y is $\overleftrightarrow{CPD}$ the perpendicular bisector of $\overline{APB}$?

12. In the accompanying figure, m$\angle TRA$ = 62°, m$\angle ARP = (5x + 1)°$, and m$\angle PRT = (10x + 3)°$. Determine if $\overrightarrow{RA}$ is an angle bisector. Explain your reasoning.

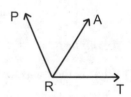

2.3 ANGLES IN POLYGONS

KEY IDEAS

There are relationships between the sides of a polygon with n sides and the measures of its interior and exterior angles. The sum of the exterior angles of a polygon always equals 360°. The sum of the interior angles of a polygon equals $180°(n - 2)$, where n equals the number of sides. A central angle of a regular polygon equals $360°/n$, where n equals the number of sides.

Interior Angles of Polygons

An interior angle of a polygon is the interior angle formed by the intersection of two consecutive sides. Figure 2.11 shows the 4 interior angles of a quadrilateral.

A regular polygon is a polygon in which all the interior angles and all the sides are congruent. The sum of the interior angles of any polygon, and measure of a single interior angle of a regular polygon can be found using the following theorem.

Interior Angle Theorem for Polygons
- The sum of the measures of the interior angles of a polygon with n sides is given by $180°(n - 2)$.
- The measure of one interior angle of a regular polygon with n sides is given by $\dfrac{180°(n-2)}{n}$.

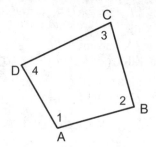

Figure 2.11 Interior angles in a quadrilateral

Example 1

What is the measure of each interior angle of a regular hexagon?

Solution: For a hexagon, $n = 6$. So each interior angle in a regular hexagon measures $\dfrac{180°(n-2)}{n} = 120°$

Example 2

The sum of the measures of the interior angles of a polygon equals $540°$. How many sides does the polygon have?

Solution:

$$\text{Sum} = 180°(n - 2)$$
$$540° = 180°(n - 2)$$
$$3 = n - 2$$
$$n = 5$$

Example 3

In regular hexagon *ABCDEF*, $\overline{BD} \perp \overline{AB}$.
Find m∠*CBD*.

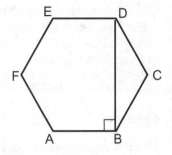

Solution:

$$\text{m}\angle ABC = \frac{180°(n-2)}{n} = 120°$$

m∠*ABD* = 90°

m∠*ABC* − m∠*ABD* = m∠*CBD*   angle subtraction

120° − 90° = m∠*CBD*

m∠*CBD* = 30°

MATH TIP

When faced with a complex problem, break it up into smaller pieces that you recognize. Then begin working your way toward your goal.

Exterior Angles in Polygons

An exterior angle of a polygon is formed by extending one side of the polygon. We consider only one extended side at each vertex, so a polygon with *n* sides has *n* exterior angles. Figure 2.12 illustrates the 5 exterior angles of a pentagon. Which side is extended does not matter because extending the second side at any vertex would result in a pair of vertical angles with equal measures.

Figure 2.12 Exterior angles in a pentagon

For regular polygons, the measure of one exterior angle equals $\dfrac{360°}{n}$, where *n* equals the number of sides.

Exterior Angle Theorem for Polygons

- The sum of the measures of the exterior angles of a polygon with n sides equals 360°.
- The measure of one exterior angle of a regular polygon with n sides equals $\dfrac{360°}{n}$.

Example

What is the measure of one exterior angle of a regular decagon?

Solution: Each exterior angle measures $\dfrac{360°}{n}$ and $n = 10$.

$$\frac{360°}{10} = 36°$$

Central Angles

A **central angle** is an angle with its vertex at the center of the polygon and rays through consecutive vertices, as shown in Figure 2.13. Central angles of a polygon are found in the same way as exterior angles.

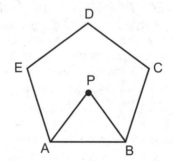

Figure 2.13 Central angle $\angle APB$

Central Angles Formula

- The sum of the central angles in a polygon is 360°.
- The measure of each central angle in a regular polygon equals $\dfrac{360°}{n}$, where n is the number of sides.

| MATH FACT |

If a regular polygon has an even number of sides, the center is the point of concurrency of segments with opposite vertices as their endpoints. If the polygon has an odd number of sides, the center is the point of concurrency of segments with a vertex at one endpoint and the midpoint of the opposite side as the other endpoint.

Example

Find the measures of angles x and y in a regular pentagon with center P, as shown in the accompanying figure.

Solution: In a pentagon, $n = 5$. So each central angle measures $\dfrac{360°}{5}$ and $y = 72°$.

Each interior angle of the pentagon measures $\dfrac{180°(5-2)}{5} = 108°$. m∠$x$ is $\dfrac{1}{2}$ the measure of an interior angle, or $54°$.

Check Your Understanding of Section 2.3

A. *Multiple-Choice*

1. What is the sum of the interior angles of a pentagon?
   (1) 108°   (2) 360°   (3) 540°   (4) 720°

2. How many sides does a polygon have whose interior angles sum to 1,800°?
   (1) 6   (2) 8   (3) 10   (4) 12

3. What is the measure of one central angle of an equilateral triangle?
   (1) 60°   (2) 90°   (3) 120°   (4) 150°

4. What is the measure of one central angle of a regular octagon?
   (1) 45°   (2) 60°   (3) 72°   (4) 144°

5. What is the measure of one exterior angle of a regular hexagon?
   (1) 60°   (2) 120°   (3) 135°   (4) 144°

6. The difference between one interior and one exterior angle of a regular polygon is 90°. How many sides does the polygon have?

(1) 4   (2) 5   (3) 6   (4) 8

7. In a regular polygon, adjacent interior and exterior angles are congruent. What is the name of the polygon?

(1) triangle   (2) rectangle   (3) square   (4) hexagon

8. A regular octagon and a regular hexagon have equal side lengths and share a side. What is the measure of the angle indicated by x in the figure?

(1) 10°
(2) 15°
(3) 20°
(4) 33°

B. *Free Response—show all work or explain your answer completely.*

9. Point O is the center of the accompanying figure. Find the measure of angle x.

10. Find the value of x in the accompanying figure.

11. In the accompanying hexagon, two pairs of sides are extended. Using the angle measures provided, find the value of x.

46

12. A regular pentagon and a regular octagon have equal side lengths and share a side as shown. What is the value of x?

2.4 PARALLEL LINES

When two or more parallel lines are intersected by a transversal, any pair of angles formed are either congruent or supplementary. We can determine if two lines are parallel by looking at the measures of these pairs. Two lines are parallel if and only if the following are true:

- Alternate interior angles are congruent.
- Same side interior angles are supplementary.
- Corresponding angles are congruent.

Named Angle Pairs

Whenever two lines are intersected by another line, called a *transversal*, the eight angles shown in Figure 2.14 will be formed. Some of the angle pairs have names to make referencing them easier. Notice that the transversal divides the figure into two regions, on alternate sides of the transversal. In Figure 2.14, angles 1, 4, 5, and 8 are on one side. Angles 2, 3, 6, and 7 are on the other side. Next, the two lines intersected by the transversal form an interior region and an exterior region. Angles 3, 4, 5, and 6 are interior angles, while angles 1, 2, 7, and 8 are exterior angles.

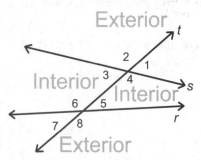

Figure 2.14 The 8 angles formed when two lines are intersected by a transversal

47

Using these definitions, the named pairs are:

- *Alternate interior angles* are ∠3 and ∠5 and also ∠4 and ∠6.
- *Same side interior angles* are ∠4 and ∠5 and also ∠3 and ∠6.
- *Corresponding angles* are ∠4 and ∠8, are ∠1 and ∠5, are ∠2 and ∠6, and also ∠3 and ∠7.
- *Alternate exterior angles* are ∠1 and ∠7 and also ∠2 and ∠8.

Recognizing Angle Pairs

Alternate interior angles make a "Z" pattern that can help in recognizing them. The alternate interior angles are in the corners of the "Z," as shown in Figure 2.15. Same side interior angles make an "F" pattern. Those angles are in the corners of the "F" as shown in Figure 2.16.

Figure 2.15 Alternate interior angles in the corners of the "Z"

Figure 2.16 Same side interior angles in the corners of the "F"

Theorems Involving Parallel Lines and the Named Angle Pairs

When parallel lines are intersected by a transversal, a set of relationships is formed among the angles.

Corresponding Angle Postulate
Two lines intersected by a transversal are parallel if and only if the corresponding angles are congruent.

Alternate Interior Angle Theorem
Two lines intersected by a transversal are parallel if and only if the alternate interior angles are congruent.

Same Side Interior Angle Theorem
Two lines intersected by a transversal are parallel if and only if the same side interior angles are supplementary.

Alternate Exterior Angle Theorem
Two lines intersected by a transversal are parallel if and only if the alternate exterior interior angles are congruent.

All of the angle relationships involving parallel lines result in congruent pairs of angles or supplementary pairs of angles. Combining these three angle theorems with the linear pair and vertical angle postulates allows you to calculate the measure of all eight of the angles if the measure of just one angle is known.

Example 1

Line m ∥ line n and m∠1 = 118°. Find the measures of the other angles.

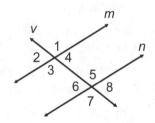

Solution:

$$\begin{aligned}
\text{m}\angle 1 + \text{m}\angle 2 &= 180° &&\text{linear pair is supplementary}\\
118° + \text{m}\angle 2 &= 180°\\
\text{m}\angle 2 &= 62°\\
\text{m}\angle 3 = \text{m}\angle 1 &= 118° &&\text{vertical angles are congruent}\\
\text{m}\angle 4 = \text{m}\angle 2 &= 62° &&\text{vertical angles are congruent}\\
\text{m}\angle 6 = \text{m}\angle 2 &= 62° &&\text{corresponding angles are congruent}\\
\text{m}\angle 7 = \text{m}\angle 3 &= 118° &&\text{corresponding angles are congruent}\\
\text{m}\angle 5 = \text{m}\angle 7 &= 118° &&\text{vertical angles are congruent}\\
\text{m}\angle 8 = \text{m}\angle 4 &= 62° &&\text{corresponding angles are congruent}
\end{aligned}$$

Example 2

Line $m \parallel$ line n. Find the value of x.

Solution: The indicated angles are supplementary, same side interior angles.

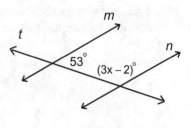

$53 + 3x - 2 = 180$
$\qquad\quad 3x = 129$
$\qquad\quad\ x = 43$

Example 3

Line $m \parallel$ line n. What is the value of x?

Solution:

$4x - 12 = x + 27$ alternate interior angles are congruent
$3x - 12 = 27$
$\qquad 3x = 39$
$\qquad\ x = 13$

Example 4

$\overline{AB} \parallel \overline{DCE}$, m$\angle A = 32°$, and m$\angle B = 38°$. Find m$\angle ACD$ and m$\angle BCE$.

Solution: Use the two pairs of alternate interior angles.

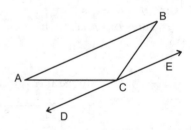

m$\angle DCA =$ m$\angle A = 32°$
m$\angle BCE =$ m$\angle B = 38°$

Using an Auxiliary Line

Sometimes we need to construct an additional line, called an auxiliary line, to help solve a parallel line problem.

Example 1

$\overleftrightarrow{AB} \parallel \overleftrightarrow{CD}$, m∠CDE = 38°, and m∠ABE = 32°. Find m∠BED.

Solution: We need to construct an auxiliary line, $\overleftrightarrow{FEG}$, through point E and parallel to $\overleftrightarrow{AB}$ and $\overleftrightarrow{CD}$.

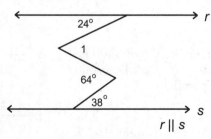

m∠FED = m∠EDC = 38°   congruent alternate interior angles
m∠BEF = m∠ABE = 32°   congruent alternate interior angles
m∠BED = m∠BEF + m∠FED   angle addition
      = 38° + 32°
      = 70°

Example 2

Lines *r* and *s* are parallel. Find m∠1.

r ∥ s

51

Solution: There are two "jogs" in the transversal. So we need two auxiliary lines, both parallel to lines *r* and *s*.

$r \parallel s$

m∠5 = 38°	alternate interior angles
m∠4 = 64° − 38° = 26°	angle subtraction
m∠3 = m∠4 = 26°	alternate interior angles
m∠2 = 24°	alternate interior angles
m∠1 = m∠2 + m∠3 = 50°	angle addition

Determining if Two Lines Are Parallel

The relationships between corresponding, alternate interior, and same side interior angles are true when read in either direction. For example, the corresponding angle postulate tells us two things:

> If two lines are parallel, then the corresponding angles are congruent.
> If the corresponding angles are congruent, then the lines are parallel.

The second statement can be used to determine whether two lines are parallel by comparing the measures of any pairs of corresponding angles. The same can be done for alternate interior and same side interior angles.

Example 1

Lines *m* and *n* are intersected by transversal *v*, m∠4 = 52°, and m∠5 = 128°. Determine if lines *m* and *n* are parallel.

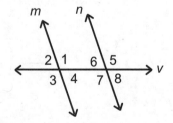

Solution:

m∠7 = m∠5 = 128°   vertical angles are congruent

Same side interior angles ∠4 and ∠7 are supplementary, so lines *m* and *n* are parallel.

Example 2

In quadrilateral *ABCD*, m∠*A* = 111°, m∠*B* = 69°, m∠*C* = 107°, and m∠*D* = 73°. How many pairs of parallel sides does *ABCD* have?

 Solution: $\overline{AD} \parallel \overline{BC}$ because same side interior angles are supplementary.

m∠*A* + m∠*B* = 69° + 111° = 180°

 $\overline{AB}$ and $\overline{CD}$ are not parallel because the same side interior angles are not supplementary.

m∠*A* + m∠*D* = 111° + 73° = 184° ≠ 180°

 So quadrilateral *ABCD* has 1 pair of parallel sides.

Example 3

For what value of *x* is $\overline{CD} \parallel \overline{AB}$?

 Solution: ∠*AEF* and ∠*CFH* are corresponding angles. They must be congruent if $\overline{CD} \parallel \overline{AB}$.

$$5x + 18 = 2x + 72$$
$$3x + 18 = 72$$
$$3x = 54$$
$$x = 18$$

Check Your Understanding of Section 2.4

A. *Multiple-Choice*

Use the following figure for problems 1–5.

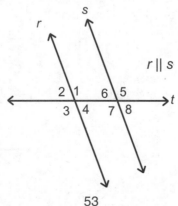

1. Line r ∥ line s, m∠2 = $(7x + 18)$°, and m∠8 = $13x$. What is m∠2?
   (1) 3°   (2) 13°   (3) 39°   (4) 141°

2. Line r ∥ line s, m∠4 = $(3x + 34)$°, and m∠6 = $(2x + 46)$°. What is m∠6?
   (1) 12°   (2) 20°   (3) 70°   (4) 86°

3. Line r ∥ line s and m∠3 = 118°. What is m∠6?
   (1) 32°   (2) 62°   (3) 118°   (4) 360°

4. Line r ∥ line s, m∠2 = $(4x - 27)$° and m∠6 = $(x + 12)$°. What are the measures of ∠2 and of ∠6?
   (1) m∠2 = 39° and m∠6 = 141°
   (2) m∠2 = 51° and m∠6 = 129°
   (3) m∠2 = 25° and m∠6 = 155°
   (4) m∠2 = 25° and m∠6 = 25°

5. Line r ∥ line s, m∠4 = $(4x + 35)$°, and m∠7 = $(6x + 25)$°. What is the measure of ∠4?
   (1) 5°   (2) 12°   (3) 55°   (4) 83°

6. Jay St. and Clayton St. run parallel to each other. Henry St. is straight and intersects both Jay St. and Clayton St. If the obtuse angle formed by Jay St. and Henry St. measures $(3x + 51)$° and the measure of the acute angle formed by Clayton St. and Henry St. is $(4x + 3)$°, what is the measure of the acute angle formed by Jay St. and Henry St.?
   (1) 18°   (2) 48°   (3) 52°   (4) 75°

7. $\overleftrightarrow{EF}$ intersects $\overleftrightarrow{AEB}$ and $\overleftrightarrow{CFD}$. If m∠AEF = $(2x + 5)$° and m∠CFE = $(2x + 35)$°, for what value of x are $\overleftrightarrow{AEB}$ and $\overleftrightarrow{CFD}$ parallel?
   (1) 4.5   (2) 6.5   (3) 35   (4) 69

8. Line m ∥ line n. Find the value of x.
   (1) 19°
   (2) 41°
   (3) 60°
   (4) 68°

9. Find the measure of ∠1.
   (1) 63°
   (2) 117°
   (3) 95°
   (4) 32°

10. Which of the following would be a justification for proving two lines are parallel?
    (1) A pair of corresponding angles are supplementary.
    (2) A pair of alternate interior angles are supplementary.
    (3) A pair of same side interior angles are supplementary.
    (4) A pair of same side interior angles are congruent.

B. *Free-Response—show all work or explain your answer completely.*

11. In the accompanying figure, $\overline{AB} \parallel \overline{CDE}$, $m\angle CDA = (4x + 10)°$, $m\angle ADB = (4x)°$, $m\angle DBA = (3x - 6)°$, and $m\angle DEB = (5x - 6)°$. Is $\overline{AD} \parallel \overline{BE}$? Justify your reasoning.

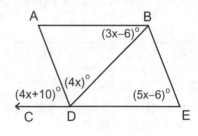

12. Lines s and t are parallel. Find $m\angle 1$.

13. Lines s and t are parallel. Find the measure of $\angle 1$.

14. Lines s and t are parallel. Find the measure of angle 1.

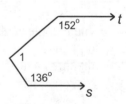

15. A pair of parallel lines intersected by a transversal forms same side interior angles that are in a $5:1$ ratio. What are the measures of two same side interior angles?

16. Are rays *a* and *b* parallel? Justify your reasoning.

2.5 ANGLES AND SIDES IN TRIANGLES

KEY IDEAS

A set of theorems relate angle measures in triangles.

- The sum of the measures of the interior angles of a triangle equals 180°.
- The measure of any exterior angle of a triangle equals the sum of the measures of the nonadjacent interior angles.
- If two sides of a triangle are congruent, then the angles opposite them are congruent.
- All interior angles of an equilateral triangle measure 60°.

Triangle Angle Sum Theorem

The angle sum theorem for triangles is simply the polygon interior angle theorem evaluated for a polygon with three sides. The sum of the interior angles of a triangle equals 180°.

MATH FACT

The triangle angle sum theorem can be used to help classify the type of triangle. The theorem is used to calculate the measures of each of the three angles in a triangle.

Example 1

The measures of the three angles in a triangle are $(3x + 12)°$, $(4x - 6)°$, and $(2x + 12)°$. Find the measure of each angle, and completely classify the type of triangle.

Solution:

Set the sum of the angle measures equal to 180°.

$$3x + 12 + 4x - 6 + 2x + 12 = 180$$
$$9x + 18 = 180$$
$$9x = 162$$
$$x = 18$$

The three angle measures are:

$$3(18) + 12 = 66°$$
$$4(18) - 6 = 66°$$
$$2(18) + 12 = 48°$$

The triangle is isosceles because two angles are congruent, and acute because all angles measure less than 90°.

Example 2

In $\triangle LOV$, m$\angle LOE = (5x - 68)°$, m$\angle VOE = (2x + 7)°$, m$\angle V = 28°$, and $\overrightarrow{OE}$ is an angle bisector. Find m$\angle L$.

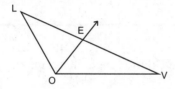

Solution:

m$\angle LOE = $ m$\angle VOE$ angle bisector forms two congruent angles
$5x - 68 = 2x + 7$
$3x = 75$
$x = 25$

By substituting x into the expressions for each angle, we find:

m$\angle LOE = 5(25) - 68 = 57°$
m$\angle VOE = 2(25) + 7 = 57°$
m$\angle LOV = $ m$\angle LOE + $ m$\angle VOE$ angle addition
m$\angle LOV = 57° + 57° = 114°$
$114° + 28° + $ m$\angle L = 180°$ angle sum theorem in $\triangle LOV$
m$\angle L = 38°$

Triangle Exterior Angle Theorem

The triangle exterior angle theorem relates the measure of an exterior angle of a triangle to the two nonadjacent interior angles. In Figure 2.17, m$\angle 1 = $ m$\angle 2 + $ m$\angle 3$.

This theorem is particularly useful when exterior angles are expressed in terms of a variable.

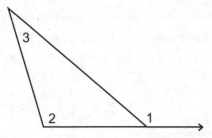

Figure 2.17 Exterior angle 1 and nonadjacent angles 2 and 3

Example 1

In △*ABC*, side $\overline{AB}$ is extended through point *A* to point *D*. If m∠*CAD* = 140° and m∠*ACB* = 95°, find m∠*ABC*. Is △*ABC* isosceles?

Solution: From the exterior angle theorem,

m∠*C* + m∠*B* = m∠*DAC*	exterior angle theorem
95° + m∠*B* = 140°	
m∠*B* = 45°	
∠*DAC* + ∠*BAC* = 180°	linear pair
140° + m∠*BAC* = 180°	
m∠*BAC* = 40°	

△*ABC* is not isosceles since no two angles in the triangle are congruent.

Example 2

In △*RST*, m∠*URT* = $(10x + 39)°$, m∠*RST* = $(7x + 4)°$, and m∠*STR* = $(6x + 8)°$. Find m∠*URT*.

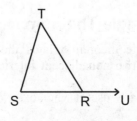

Solution: Apply the triangle exterior angle theorem,

$$m\angle URT = m\angle RST + m\angle STR$$
$$10x + 39 = 7x + 4 + 6x + 8$$
$$10x + 39 = 13x + 12$$
$$-3x + 39 = 12$$
$$-3x = -27$$
$$x = 9$$
$$m\angle URT = (10x + 39)°$$
$$m\angle URT = (10(9) + 39)°$$
$$= 129°$$

Isosceles Triangles

An isosceles triangle has two congruent sides, and the angles across from those sides are always congruent. The congruent angles are called the base angles, and the congruent sides are called the legs. The side that differs in length is called the base, and the angle that differs in measure is called the vertex angle. These parts are shown in Figure 2.18.

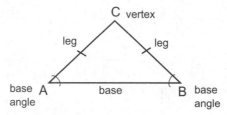

Figure 2.18 Isosceles triangle

Isosceles Triangle Theorem
If two sides in a triangle are congruent, then the angles opposite those sides are congruent.

Converse of the Isosceles Triangle Theorem
If two angles in a triangle are congruent, then the sides opposite those angles are congruent.

If any one angle of an isosceles triangle is known, then the other angles can be determined.

Example 1

In Figure 2.18, m∠C = 106°. Find m∠A and m∠B.

    Solution: Base angles ∠A and ∠B are congruent.

Use the triangle angle sum theorem,

$$m\angle A + m\angle B + m\angle C = 180°$$
$$2x + 106° = 180°$$
$$2x = 74°$$
$$x = 37°$$
$$m\angle A = m\angle B = 37°$$

Example 2

In Figure 2.18, m∠A = 42°. Find m∠C.

    Solution: Base angles ∠A and ∠B are congruent.

$$m\angle A + m\angle B + m\angle C = 180°$$
$$42° + 42° + m\angle C = 180°$$
$$84° + m\angle C = 180°$$
$$m\angle C = 96°$$

Example 3

In the accompanying figure, $\overleftrightarrow{DCE} \parallel \overline{AB}$ and $\overline{AC} \cong \overline{BC}$. If m∠1 = 68°, what is m∠2?

    Solution:

$m\angle A = m\angle 1 = 68°$
$m\angle B = m\angle A = 68°$
$m\angle A + m\angle B + m\angle 2 = 180°$ triangle angle sum theorem
$ 68° + 68° + m\angle 2 = 180°$
$ 136° + m\angle 2 = 180°$
$ m\angle 2 = 44°$

Example 4

In the accompanying figure, $\overline{AC} \cong \overline{BC}$, $m\angle ACD = (9x - 4)°$, and $m\angle ABC = (5x - 10)°$. Find $m\angle A$.

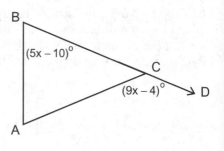

Solution: $\triangle ABC$ is isosceles with $\angle B \cong \angle A$. Use the exterior angle theorem,

$$m\angle DCA = m\angle CBA + m\angle BAC$$
$$9x - 4 = 5x - 10 + 5x - 10$$
$$9x - 4 = 10x - 20$$
$$x = 16$$
$$m\angle A = 5(16) - 10$$
$$= 70°$$

Equilateral Triangles

An equilateral triangle is one in which all three sides are congruent and all three angles measure 60°.

Example

$\triangle ABC$ is equilateral, and D lies on $\overline{BC}$. $m\angle ADC = 102°$. Find $m\angle BAD$.

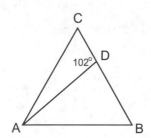

Solution:

$m\angle BAC = m\angle B = m\angle C = 60°$ angles in an equilateral triangle measure 60°

$m\angle DAC + 60° + 102° = 180°$ angle sum theorem in $\triangle ADC$
$m\angle DAC = 18°$

$m\angle BAD + m\angle DAC = 60°$ angle addition
$\angle BAD + 18° = 60°$
$m\angle BAD = 42°$

Pythagorean Theorem

In a right triangle, the sum of the squares of the legs is equal to the square of the hypotenuse, as shown in Figure 2.19. The hypotenuse is always the longest leg of the triangle, which is opposite the 90° angle. Remember, the Pythagorean theorem applies only to right triangles.

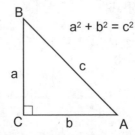

Figure 2.19 Pythagorean theorem

> In a right triangle with legs represented by lengths a and b and with hypotenuse c, $a^2 + b^2 = c^2$.

Example 1

In $\triangle RST$, $\angle S$ measures 90°, $RS = 6$, and $ST = 3$. Find RT.

Solution: Since $\angle S$ is the right angle, RT must be the hypotenuse.

$a^2 + b^2 = c^2$
$6^2 + 3^2 = RT^2$
$RT = \sqrt{45} = 3\sqrt{5}$

Example 2

In the accompanying figure, $m\angle E = 45°$, $EO = GO$, and $EG = 8$. Find EO and GO.

Solution:

$m\angle E = m\angle G = 45°$ isosceles triangle theorem
$m\angle O = 180° - 45° - 45° = 90°$ angle sum theorem
$a^2 + b^2 = c^2$ $\triangle GEO$ is a right triangle, so
                       Pythagorean theorem applies
$x^2 + x^2 = 8^2$ let $EO = GO = x$
$2x^2 = 64$
$x^2 = 32$
$x = 4\sqrt{2}$

Example 3

The length of a rectangle is 3 times its width. If the length of a diagonal is 10, what is the area of the rectangle?

Solution: The right triangle formed by the sides of the rectangle and a diagonal has legs equal to $3x$ and x and a hypotenuse equal to 10.

$$a^2 + b^2 = c^2$$
$$(3x)^2 + x^2 = 10^2$$
$$9x^2 + x^2 = 100$$
$$10x^2 = 100$$
$$x^2 = 10$$
$$x = \sqrt{10}$$
$$\text{Length} = 3\sqrt{10} \text{ and width} = \sqrt{10}$$
$$\text{Area} = \text{length} \times \text{width}$$
$$= 3\sqrt{10} \cdot \sqrt{10}$$
$$= 3 \cdot 10$$
$$= 30$$

Example 4

In $\triangle ABC$, $m\angle A = 90°$, $AC = 6$, $AB = 2x$, and $BC = x + 6$. Find the area of the triangle.

Solution:

$a^2 + b^2 = c^2$
$6^2 + (2x)^2 = (x + 6)^2$
$36 + 4x^2 = x^2 + 12x + 36$
$3x^2 - 12x = 0$
$3x(x - 4) = 0$
$x = 0, x = 4$ Side length cannot equal 0, so $x = 4$ and $AB = 8$.
$\text{Area} = \dfrac{1}{2}AC \cdot AB = 24$

63

Check Your Understanding of Section 2.5

A. Multiple-Choice

1. In the accompanying figure, $m\angle I = 32°$, $m\angle H = 76°$, and $m\angle G = 78°$. Find $m\angle E$.
   (1) 24°
   (2) 30°
   (3) 72°
   (4) 78°

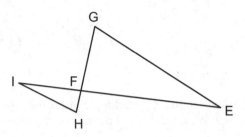

2. In $\triangle MNO$, side MN is extended to point P. $m\angle PNO = (6x + 41)°$, $m\angle NMO = (2x + 20)°$, and $m\angle MON = (x + 39)°$. What is the measure of $\angle MNO$?
   (1) 32°   (2) 45°   (3) 77°   (4) 103°

3. In $\triangle ABD$ with side $\overline{BCD}$, $\overline{AB} \cong \overline{AD}$, $m\angle BAC = 38°$, and $m\angle D = 61°$. Find $m\angle DAC$.
   (1) 20°
   (2) 23°
   (3) 29°
   (4) 71°

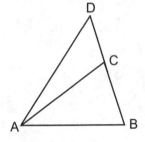

4. In $\triangle ABC$, $AC = BC$. If $m\angle A$ is four times $m\angle C$, find $m\angle B$.
   (1) 20°   (2) 60°   (3) 80°   (4) 144°

5. The perimeter of an equilateral triangle is $(6x - 12)$ inches. If each side measures 24 inches, what is the value of x?
   (1) 6   (2) 14   (3) 32   (4) 72

6. The measures of the angles of a triangle are in a $2:5:8$ ratio. The measure of the smallest angle is
   (1) 6°   (2) 12°   (3) 24°   (4) 96°

7. The measures of the angles of a triangle are $(3x - 4)°$, $(2x + 15)°$, and $(10x + 4)°$. The triangle can be classified as
   (1) isosceles acute   (3) scalene acute
   (2) isosceles obtuse   (4) scalene obtuse

8. In △*FIS*, side *IS* is extended to point *H*. Find the measure of ∠*F*.
   (1) 11°
   (2) 18°
   (3) 19°
   (4) 30°

9. In △*FGH*, m∠*F* = 35°, m∠*G* = (x + 21)°, and m∠*H* = (3x − 20)°. Find the measure of ∠*H*.
   (1) 18°   (2) 36°   (3) 57°   (4) 88°

10. Two legs of a right triangle measure 4 and 8. What is the length of the hypotenuse?
    (1) $8\sqrt{10}$ (2) $20\sqrt{2}$ (3) $4\sqrt{5}$ (4) $5\sqrt{4}$

B. *Free-Response—show all work or explain your answer completely.*

11. $\overleftrightarrow{DCE} \parallel \overleftrightarrow{AB}$, m∠*DCA* = 78°, and *AC = BC* to *AC = BC*. Find m∠*ACB*.

12. Is a triangle whose side lengths measure 5, 6, and 8 a right triangle? Justify your answer.

13. In the accompanying figure, m∠*GJI* = 132°, $\overline{IH} \cong \overline{IJ}$, and $\overline{HF} \cong \overline{HG}$. Find m∠*F*.

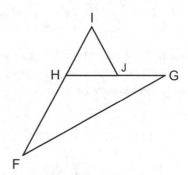

14. In rectangle *JUMP* with diagonal *JM*, *JU* = 8 and *JM* = 12. Find the length of *UM*. Give your answer in simplest radical form.

65

15. To get home from school, Stanasha has to walk 5 blocks north and 12 blocks east. Her school provides free bus passes for any student who lives more than $\frac{3}{4}$ mile from the school, as measured by the straight-line distance. If 20 blocks equals 1 mile, does Stanasha qualify for a free bus pass?

16. The hypotenuse of a right triangle measures $5\sqrt{3}$. If the length of one leg is 5, what is the length of the second leg? Give your answer in simplest radical form.

17. The area of an isosceles right triangle is 32. What is the length of a leg of the triangle?

18. The length of the diagonal of a square measures 10 cm². What is the perimeter of the square? Express your answer in simplest radical form.

19. Corey is building a new rectangular deck with a length of 24 feet and a width of 10 feet. He marks the location of the four corner support posts and uses a tape measure to confirm the lengths and widths are correct. Explain how he can use the tape measure to check if the angles are 90°.

20. Point P is the center of regular pentagon $ABCDE$. Find m$\angle PAC$.

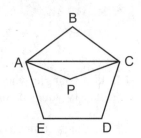

21. In the accompanying figure, $\overline{OP} \parallel \overline{MN}$, $\overline{NO} \perp \overline{OP}$, and $\triangle MNP$ is equilateral. Find m$\angle PNO$.

22. $\overline{JB} \cong \overline{BO}$, $\overline{JA} \cong \overline{AO}$, m$\angle JAO = 126°$, and m$\angle BOA = 18°$. Find m$\angle JBO$.

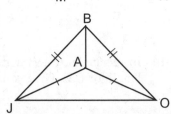

CONSTRUCTIONS

3.1 BASIC CONSTRUCTIONS

KEY IDEAS

A construction uses a straightedge and a compass to create a precise geometric figure. The compass is used to construct arcs whose points are all equidistant from the center point. Intersecting arcs locate a point equidistant from the two centers. Two intersecting circles create points equidistant from the two centers. The fundamental loci described in Section 3.1 are used to create a variety of constructions from these two building blocks. Some of the basic constructions are the equilateral triangle, perpendicular line, and parallel line.

Definition of Construction

A **construction** is a geometric drawing in which only a *straightedge* and *compass* may be used. The straightedge is used only for connecting two points in a straight line. If your straightedge is a ruler, you cannot use the length markings on it to measure length. When constructing an angle of a particular measure, such as a 45° angle, you cannot use a protractor to measure the angle. The only measuring device allowed is the compass, and we will use it to measure both length and angle opening.

The Compass

The compass is simply a tool for constructing circles. The point of the compass locates the center point P, and the distance the compass is opened defines the radius. As long as the opening of the compass does not change, every point it traces out is a fixed distance from the center, resulting in a circle.

You should purchase a good-quality compass. A compass that is too loose or not rigid enough will be very frustrating to use.

> **Tips for using a compass**
>
> - Adjust the height of the pencil so that the compass stands nearly vertical when the point and pencil are touching the paper.
> - Experiment with different techniques for holding the compass and making arcs. One technique worth trying is to hold the compass from only the point end and rotate the paper instead of the compass.
> - Some compasses have a screw at the pivot that can be tightened. Tighten this screw if you find the compass opening and changing as you make an arc.
> - All arcs shown in the following constructions are called the "marks of construction." You must include them when doing a construction on an exam; they are required. These marks demonstrate that you followed the proper procedure with the compass.

Copy a Segment

Given: $\overline{AB}$ and point C, construct $\overline{CD}$ congruent to $\overline{AB}$.
Procedure:

1) Place compass point on A and pencil on B; make a small arc.

2) With the same compass opening, place point on C and make an arc.

3) Use the straightedge to connect point C to any point D on the arc, $\overline{CD} \cong \overline{AB}$.

Copy an Angle

Given: Angle $\angle ABC$ and $\overrightarrow{ED}$, construct $\angle DEF$ congruent to $\angle ABC$.
Procedure:

1) Place compass point on B, and make an arc intersecting the angle at R and at S.

2) With the same compass opening, place point on E and make an arc intersecting $\overline{ED}$ at T.

3) Place compass point on R and pencil on S, and make a small arc.

4) With the same compass opening, place point on *T* and make an arc intersecting the previous one at *F*.

5) Use the straightedge to form $\overline{EF}$.

6) ∠*DEF* is congruent to ∠*ABC*.

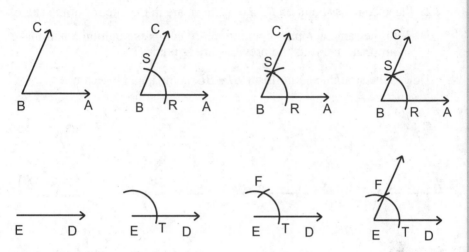

Equilateral Triangle

Given: Segment $\overline{AB}$, construct an equilateral triangle with side length *AB*.
Procedure:

1) Place compass point on *A* and pencil on *B*. Make a quarter circle.

2) Place compass point on *B* and pencil on *A*. Make a quarter circle that intersects the first at *C*.

3) Use a straightedge to form *AC* and *BC*. △*ABC* is an equilateral triangle.

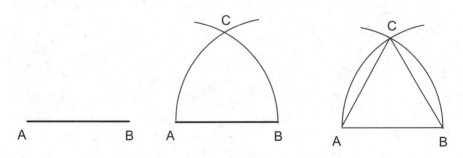

Angle Bisector

Given: Angle $\angle ABC$, construct angle bisector $\overrightarrow{BD}$.
Procedure:

1) Place compass point on B, and make an arc intersecting $\overrightarrow{BA}$ and $\overrightarrow{BC}$ at R and S, respectively.

2) Place compass point on R, and make an arc in the interior of the angle.

3) With the same compass opening, place compass point on S and make an arc that intersects the previous arc at point D.

4) Use the straightedge to form $\overrightarrow{BD}$. $\overrightarrow{BD}$ is the angle bisector.

Perpendicular Bisector

Given: Segment $\overline{AB}$, construct the perpendicular bisector of $\overline{AB}$.
Procedure:

1) With the compass open more than half the length of $\overline{AB}$, place the point at A and make a semicircle running above and below $\overline{AB}$.

2) With the same compass opening, place the point at B and make a semicircle running above and below $\overline{AB}$. Make sure the semicircle intersects the first semicircle at R and at S.

3) Use a straightedge to connect R and S. $\overline{RS}$ is the perpendicular bisector of $\overline{AB}$.

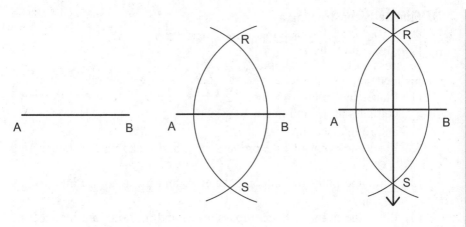

Check Your Understanding of Section 3.1

A. Multiple-Choice

1. Which of the following is *not necessarily* true in the construction shown in the accompanying figure?
   (1) m∠RAD = m∠CAD
   (2) m∠CAD + m∠RAD = m∠CAR
   (3) AC ≅ AD
   (4) AS ≅ AT

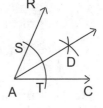

2. Which of the following must be true in the accompanying construction?
   (1) m∠C is greater than m∠A.
   (2) There is not enough information in the figure to determine the measure of ∠ABC.
   (3) Point C is equidistant from points A and B.
   (4) AB is greater than BC and AC.

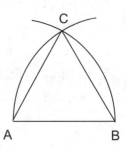

3. Frankie is constructing a copy of ∠*SAD*. The accompanying figure shows his progress so far. What should his next step be?

(1) Place the point of his compass at *M*, and make an arc intersecting $\overline{AD}$ at *D*.
(2) Place the point of his compass at *S*, and make an arc intersecting $\overline{AD}$ at *D*.
(3) Place the point of his compass at *A*, and make an arc intersecting $\overline{AD}$ at *N*.
(4) Place the point of his compass at *M*, and make an arc intersecting $\overline{AD}$ at *N*.

B. *Free Response—show all work or explain your answer completely.*

4. Construct $\overline{MN}$ congruent to $\overline{JK}$.

5. Construct a segment whose length is equal to 3*AR*.

6. Construct a segment whose length is equal to *MA* + *TH*.

7. Construct ∠*FGH* congruent to ∠*CDE*.

8. Construct an angle whose measure is twice that of ∠LMN.

9. Bisect ∠LMN above.

10. Construct a right angle.

11. Construct a 45° angle.

12. Construct an equilateral triangle with $\overline{BK}$ as one of its sides.

13. Construct a 30° angle.

14. Construct the midpoint of $\overline{TV}$.

3.2 CONSTRUCTIONS THAT BUILD ON THE BASIC CONSTRUCTIONS

KEY IDEAS

The basic constructions of the previous section can be combined to produce a number of other constructions.

Perpendicular to Line from a Point not on the Line

Given line $\overleftrightarrow{AB}$ and point P not on $\overleftrightarrow{AB}$, construct a line perpendicular to $\overleftrightarrow{AB}$ passing through P.

Procedure:

1. Place compass point at P, and make an arc intersecting $\overleftrightarrow{AB}$ at R and S. (Extend $\overleftrightarrow{AB}$ if necessary.)

2. Place compass point at R, and make an arc on the opposite side of the line as P.

73

3. With the same compass setting, place compass point at *S*, and make an arc intersecting the previous arc at *Q*.

4. Use the straightedge to connect *P* and *Q*. $\overleftrightarrow{PQ}$ is perpendicular to $\overleftrightarrow{AB}$ and passes through *P*.

Perpendicular to Line from a Point On the Line

Given line $\overleftrightarrow{AB}$ and point *P* on $\overleftrightarrow{AB}$ between *A* and *B*, construct a line perpendicular to $\overleftrightarrow{AB}$ and passing through *P*.
Procedure:

1. Place compass point at *P*, and make an arc intersecting $\overleftrightarrow{AB}$ at *R* and at *S*.

2. Place compass point at *R*, and make an arc below $\overleftrightarrow{AB}$.

3. With the same compass setting, place point at *S* and make an arc intersecting the previous arc at *Q*.

4. Use a straightedge to connect *P* and *Q*. $\overleftrightarrow{PQ}$ is perpendicular to $\overleftrightarrow{AB}$ and passes through *P*.

Parallel to Line from Point Off the Line

Given: Segment $\overleftrightarrow{AB}$ and point *P* not on $\overleftrightarrow{AB}$, construct a line parallel to $\overleftrightarrow{AB}$ and passing through *P*.
Procedure:

1. Use the straightedge to construct a line passing through *P* and intersecting $\overleftrightarrow{AB}$ at *R*.

2. With compass point at *R*, make an arc intersecting $\overrightarrow{PR}$ and $\overleftrightarrow{AB}$ at *S* and *T*, respectively.

74

3. With the compass at the same setting and compass point at P, make an arc intersecting $\overrightarrow{PR}$ at U.

4. With compass point at T, make an arc intersecting $\overrightarrow{PR}$ at S.

5. With the compass at the same setting and compass point at U, make an arc intersecting at V.

6. Use a straightedge to connect P and V. $\overrightarrow{PV}$ is parallel to $\overleftrightarrow{AB}$ and passes through point P.

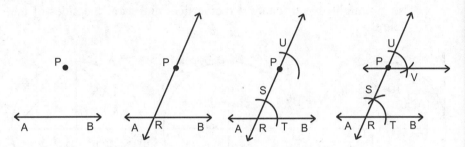

Inscribe a Regular Hexagon or an Equilateral Triangle in a Circle

Procedure:

1. Construct circle P of radius PA.

2. Using the same radius as the circle, place compass point on A and make an arc intersecting the circle at B.

3. With the same compass setting, place compass point at B and make an arc intersecting the circle at C. Continue making arcs intersecting at D, E, and F.

4. Using the straightedge, connect each point on the circle to form regular hexagon $ABCDEF$. Connecting every other point will form equilateral triangle ACE.

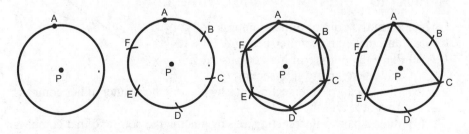

Inscribe a Square in a Circle

Procedure:

1. Construct a diameter through the center point T of a circle.

2. Construct the perpendicular bisector of the first diameter, which will also be a diameter.

3. The intersections of the diameters with the circle are the vertices of the square.

4. Use a straightedge to connect the vertices and form the sides of the square.

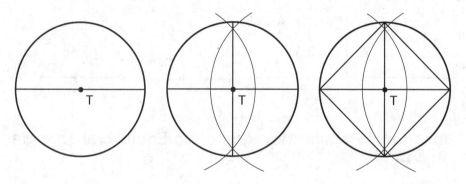

Check Your Understanding of Section 3.2

A. Multiple-Choice

1. Jamie is constructing a line perpendicular to $\overline{XY}$. Her progress so far is shown in the accompanying figure. What should be her next step?

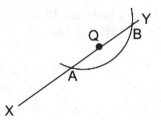

   (1) Place the pointer of her compass at X, and make an arc intersecting Q.
   (2) Place the pointer of her compass at Y, and make an arc intersecting Q.
   (3) Make a pair of intersecting arcs by placing the pointer of her compass first at A and then at B.
   (4) Make a pair of intersecting arcs by placing the pointer of her compass first at X and then at Y.

2. In the construction shown in the accompanying figure, which of the following must be true?
   (1) $HM \cong ME$
   (2) $WM \cong MV$
   (3) $WL \cong WP$
   (4) $WV \cong HE$

3. In the construction of a hexagon shown in the accompanying figure, side $\overline{SN}$ was constructed first. How was length SN determined?

   (1) SN equals the length of the radius of circle A.
   (2) SN equals the length of the diameter of circle A.
   (3) SN equals twice the length of the diameter of circle A.
   (4) SN equals one-half the length of the radius of circle A.

4. Claire wants to construct a 30° angle. Which of the following methods could she use?
   (1) Construct a right triangle, and then bisect one of its angles.
   (2) Construct an equilateral triangle, and then bisect one of its angles.
   (3) Construct a hexagon, and then bisect one of its angles.
   (4) Construct a square, and then bisect one of its angles.

B. Free Response—show all work or explain your answer completely.

5. Construct a line perpendicular to $\overleftrightarrow{QT}$ and passing through V.

6. Construct a line perpendicular to $\overleftrightarrow{AB}$ and passing through point V on line $\overleftrightarrow{AB}$.

7. Construct a line parallel to $\overleftrightarrow{GO}$ through point J not on line $\overleftrightarrow{GO}$.

8. Construct a regular hexagon.

9. Construct an equilateral triangle inscribed in a circle.

10. Construct a square inscribed in a circle.

11. Construct a triangle whose angles measure 30°, 60°, and 90°.

12. Construct a rectangle whose length and width are in a ratio of 2:1.

3.3 POINTS OF CONCURRENCY, INSCRIBED FIGURES, AND CIRCUMSCRIBED FIGURES

KEY IDEAS

Angle bisectors, medians, altitudes, and perpendicular bisectors in triangles have the special property of being concurrent, which means they intersect at a single point.

Vocabulary

Some new vocabulary is needed to discuss points of currency in triangles:

Concurrent lines—3 or more lines that intersect at a single point.

Tangent—a line coplanar with a figure that intersects it at exactly one point, or a segment or ray that lies on a tangent line.

*(We also use tangent as an adjective, as in $\overleftrightarrow{AB}$ is tangent to circle P.)

Figure 3.1 shows $\overline{AQB}$ tangent to circle P.

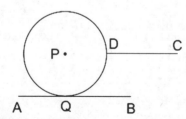

Figure 3.1 $\overline{AB}$ is tangent to circle P at Q. $\overline{CD}$ is not a tangent.

Point of tangency—the point at which a tangent intersects a circle.

Inscribed circle (or incircle)—a circle that is tangent to each of the sides of a polygon.

Circumscribed circle (or circumcircle)—a circle that intersects each of the vertices of a polygon.

Incenter

The three angle bisectors of any triangle are concurrent at a point called the **incenter**. The incenter is the center of the circle tangent to each of the three sides of the triangle. This circle is called an inscribed circle. To construct the incenter of any triangle, simply construct the angle bisectors of at least two of the angles. Constructing the third would be a check since the first two will always intersect. If the third angle bisector intersects the first at the same point, then you know the construction is correct. Figure 3.2 shows the construction of the incenter of △*ABC*.

Once the incenter is located, the point of tangency of the inscribed circle is located by constructing a line from the incenter perpendicular to one of the sides of the triangle. The theorem at work here is that a radius to a point on the circle is perpendicular to the tangent at that point. A circle can then be constructed using the incenter as the center and radius from the incenter to the point of tangency. Figure 3.3 shows the construction of the inscribed circle using the incenter.

Figure 3.2 Construction of the incenter of a triangle

Figure 3.3 Construction of the inscribed circle using the incenter

---

Theorem—Radius to a Point of Tangency
A radius to a point on a circle is perpendicular to the tangent at that point.

Since the incenter lies on the three angle bisectors, it is equidistant from the three sides of the triangle. The distance to each of the sides of the triangle is equal to the radius of the circle. The inscribed circle is also the largest circle that can fit inside a triangle, and the incenter is always inside the circle.

MATH FACT

Every triangle can have an inscribed circle, but not every polygon with four or more sides can have one. The incenter of a triangle is equidistant from the three sides of the triangle.

Circumcenter

The three perpendicular bisectors of the three sides of any triangle are concurrent at a point called the **circumcenter**. The circumcenter is also the center of the circle that passes through each of the three vertices of the triangle. This circle is called a circumscribed circle. To construct the circumcenter, construct at least two of the perpendicular bisectors and find the point of concurrency. The circumscribed circle is centered at the circumcenter. The distance to any vertex of the triangle is the radius, as shown in Figure 3.4.

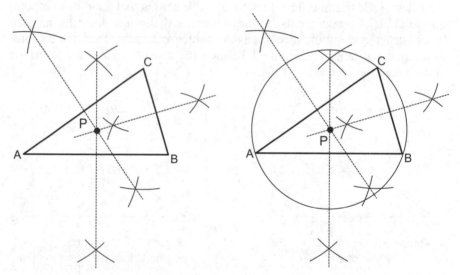

Figure 3.4 Construction of the circumcenter and circumscribed circle

The theorem at work here is that the points equidistant from two given points lie on the perpendicular bisector. The circumcenter is equidistant from each of the three vertices of triangle, so it must be the center of the circle containing those points.

Centroid

The three medians of a triangle are concurrent at a point called the **centroid**. Each median is constructed by first constructing the midpoint of each side of the triangle, using the perpendicular bisector construction. Then segments are drawn from each vertex to the midpoint of the opposite side. The centroid is the intersection of the three medians. Figure 3.5 shows the construction of centroid P and medians $\overline{AR}$, $\overline{BS}$, and $\overline{CT}$ in $\triangle ABC$. The centroid of any triangle will always be located inside the triangle.

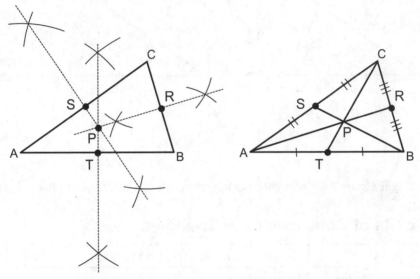

Figure 3.5 Construction of the median of a triangle

MATH FACT

The centroid of a triangle is also the triangle's center of gravity. The center of gravity is the point where the figure or object is perfectly balanced.

Orthocenter

The three altitudes of any triangle are concurrent at a point called the **orthocenter**. Figure 3.6 shows the construction orthocenter P of $\triangle ABC$. The orthocenter will be inside of an acute triangle, on a vertex of a right triangle, and outside an obtuse triangle.

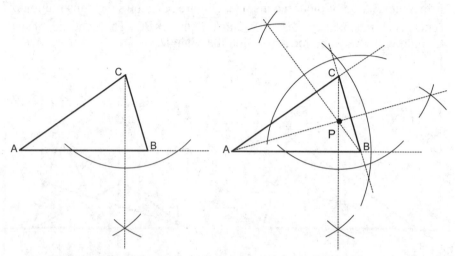

Figure 3.6 Construction of the orthocenter, P, of triangle $\triangle ABC$

Points of Concurrency in Triangles

Segment	Where Concurrent	Feature
Perpendicular bisector	Circumcenter	Center of the circumscribed circle; equidistant from each of the three vertices
Angle bisector	Incenter	Center of the inscribed circle; equidistant from the three sides of the triangle

Segment	Where Concurrent	Feature
Median	Centroid	Divides each median in a 2:1 ratio; is the center of gravity
Altitude	Orthocenter	Located inside acute triangles, on a vertex of right triangles, and outside obtuse triangles

MATH TIP

Mnemonic devices can be helpful for memorizing information. An example for points of concurrency is "**a**ll **o**f **m**y **c**hildren **a**re **b**ringing **i**n **p**eanut **b**utter **c**ookies."

AO—altitudes/orthocenter
MC—medians/centroid
ABI—angle bisectors/incenter
PBC—perpendicular bisectors/circumcenter

Check Your Understanding of Section 3.3

Multiple-Choice

1. A phone company wants to locate a cell phone tower equidistant from three cities. An appropriate strategy to find the correct location would be to construct a triangle on a map with the three cities as vertices and then locate the cell phone tower at
   (1) the intersection of the three perpendicular bisectors of the sides of the triangles.
   (2) the intersection of the three angle bisectors of the triangle.
   (3) the intersection of the three medians of the triangle.
   (4) the intersection of the three altitudes of the triangle.

2. Gina sketches a triangle and finds the midpoint of each side. She then constructs segments from each vertex to the midpoint of the opposite side. The point where the medians intersect is the
   (1) incenter
   (2) circumcenter
   (3) orthocenter
   (4) centroid

3. In $\triangle RST$, $\overline{RH} \perp \overline{ST}$, $\overline{SG} \perp \overline{RT}$, $RI = 6$, and $IJ = 8$. Find RJ.
   (1) 6
   (2) 8
   (3) 10
   (4) 12

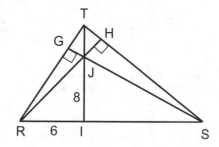

4. The center of circle Q is the point of concurrency of the three angle bisectors of scalene $\triangle ABC$. Which of the following circles can be constructed using Q as the center?
   (1) a circle passing through A and B but not C
   (2) a circle passing through A and C but not B
   (3) a circle passing through A, B, and C
   (4) a circle tangent to $\overline{AB}$, $\overline{BC}$, and $\overline{CA}$

5. In $\triangle DEF$, $\overline{DR}$, $\overline{ES}$, and $\overline{FT}$ are medians. If $DT = 4$, $ER = 5$, and $FS = 6$, what is the perimeter of $\triangle DEF$?
   (1) 10   (2) 15   (3) 30   (4) 60

6. In $\triangle ABC$, $\overline{AD}$ and $\overline{CE}$ are concurrent at the incenter P. If $m\angle B = 70°$ and $m\angle EAD = 30°$, what is $m\angle ECD$?
   (1) 25°   (2) 30°   (3) 70°   (4) 110°

7. In $\triangle JKL$, median $\overline{KP}$ is drawn from vertex K. Which of the following is *always* true?
   (1) $\overline{KP} \perp \overline{JL}$
   (2) $JP \cong PL$
   (3) $PK \cong PL$
   (4) $\angle JKP \cong \angle PKL$

8. In △ABC, $\overline{DF} \perp \overline{AFB}$, $\overline{DE} \perp \overline{AEC}$, $\overline{AF} \cong \overline{FB}$, and $\overline{AE} \cong \overline{EC}$. If m∠DBC = 18°, what is m∠BDC?
   (1) 36°
   (2) 102°
   (3) 144°
   (4) 162°

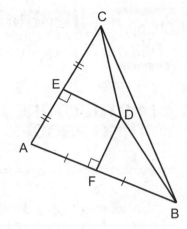

9. Circle Q passes through each of the vertices of △FGH. If m∠GHQ = 25°, what is m∠HGQ?
   (1) 25°   (2) 50°   (3) 130°   (4) 155°

10. In △BAT, circle H is tangent to $\overline{BA}$, $\overline{AT}$, and $\overline{BT}$. If m∠TAH = 24° and m∠TBH = 32°, what is m∠BTA?
    (1) 48°
    (2) 64°
    (3) 68°
    (4) 114°

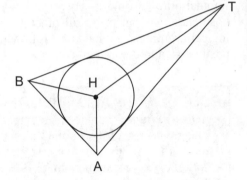

Chapter Four
INTRODUCTION TO PROOFS

4.1 STRUCTURE AND STRATEGY OF WRITING PROOFS

KEY IDEAS

A proof uses a set of definitions, theorems, and postulates to explain why a statement must be true. There must be a logical connection between each statement or sentence in the proof and the next. Making use of figures is an excellent way of planning your strategy when writing a proof.

What Is a Proof?

A proof is a logical argument that explains why a statement is true. It is the foundation of geometry. By starting with only a small number of accepted postulates, we can prove all other theorems in geometry. When making a logical argument, we have the following tools at our disposal: definitions, postulates, and theorems.

POSTULATES AND THEOREMS
- A **postulate** is an obvious statement of fact, such as "only one line can be drawn through two points on a plane." It is such a basic truth that we don't have a satisfactory way of proving it.
- A **theorem** is a statement that has been logically proven. Once we accept the proof of a theorem, we can use that theorem in proofs of other theorems. For example, we may use the angle sum theorem, which states the sum of the angles in a triangle equals 180°. Since it is an accepted and proven theorem, we do not need to prove it again when using it in another proof.

Two-Column Versus Paragraph Format

The two proof formats we will use in this book are the two-column proof and the paragraph proof. The two columns of a **two-column proof** are labeled "Statements" and "Reasons." In the statements column, we start with given information that was provided and assumed to be true. We then continue making statements that are true based on logical reasoning and the previous statements. Each statement is justified with a reason. The reasons can be:

- Given information
- Definitions
- Theorems
- Postulates

A **paragraph proof** includes the same information and degree of logical reasoning as a two-column proof. It is just presented in a more conversational tone. If you had to explain why some theorem is true to a friend, what you would say is what might be written in a paragraph proof. As a paragraph proof, we might write:

"Since $\overrightarrow{GH}$ bisects $\overline{AB}$ at C, we know C is a midpoint because bisectors pass through the midpoint of a segment. We also know $\overline{AC} \cong \overline{CB}$ because midpoints divide segments into two congruent segments."

Here are some general strategies for writing proofs:

1. Make and label a sketch. If a sketch is not provided, draw your own. Then label the sketch with any given information and any information you can infer from the sketch.

2. Make a plan. The most concise and elegant proofs start with an outline.

3. Think about the flow. Sometimes the order of statements matters. For example, you may not be able to conclude that two segments are congruent until you have stated a certain point is a midpoint.

4. Don't mix statements and reasons. Anything specific to the problem, such as named points, angles, segments, triangles, and so on, belong with the statements. Reasons should be information given in the problem, definitions, postulates, or theorems.

5. Do not use circular reasoning—you cannot use what you are trying to prove as part of the proof. Also, do not repeat statements as reasons.

6. Number your statements and reasons.

7. It's not over until it's over. There is no predetermined number of statements and reasons needed for a proof. Your last statement should be what you want to prove.

8. Ask yourself, "How do I know that is true?" If you don't know why a statement is true, it might not be.

4.2 USING KEY IDEA MIDPOINTS, BISECTORS, AND PERPENDICULAR LINES

The following table lists some of the elementary geometric relationships and how they might be used in a proof.

If You Know ...	You Can State ...	With the Reason ...
$\overrightarrow{CD}$ bisects $\overline{AB}$ at E	E is the midpoint of $\overline{AB}$	A bisector intersects a segment at its midpoint
E is the midpoint of $\overline{AB}$	$\overline{AE} \cong \overline{EB}$ or $AE = EB$ or $AE = \dfrac{1}{2} AB$	A midpoint divides a segment into two congruent segments or A midpoint divides a segment into two segments with equal measures or A midpoint divides a segment in half
$\overrightarrow{CD}$ is the perpendicular bisector of $\overline{AB}$	E is the midpoint of $\overline{AB}$ and $\overline{AB} \perp \overleftrightarrow{CD}$	A perpendicular bisector intersects a segment at its midpoint and is perpendicular to the segment

If You Know ...	You Can State ...	With the Reason ...
$\overrightarrow{BD}$ bisects $\angle ABC$ 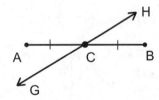	$\angle ABD \cong \angle CBD$ or $m\angle ABD = m\angle CBD$ or $m\angle ABD = \frac{1}{2}\,m\angle ABC$	An angle bisector divides an angle into two congruent angles or An angle bisector divides an angle into two angles with equal measures or An angle bisector divides an angle in half

Using Midpoints and Bisectors

Midpoints and bisectors are usually used to show that two segments or angles are congruent. Keep in mind the fact that the bisector points to the midpoint and the midpoint divides the segment into congruent segments. So if you are given a bisector, two separate statements are required to conclude that two segments are congruent.

Example 1

Given: $\overrightarrow{GH}$ bisects $\overline{AB}$ at C
Prove: $\overline{AC} \cong \overline{BC}$

Solution:

Step 1. It is usually a good idea to make a sketch if one is not provided. Then mark the sketch with congruent or other markings.

Step 2. Plan your strategy. C is a midpoint, which divides $\overline{AB}$ into congruent segments $\overline{AC}$ and $\overline{BC}$.

Step 3. Write your statements and reasons, numbering each statement and reason. Remember that anything specific to the problem goes in the statement column. Only definitions, postulates, and theorems can be reasons.

Statements	Reasons
1. $\overline{GH}$ bisects $\overline{AB}$ at C	1. Given
2. C is the midpoint of $\overline{AB}$	2. A bisector intersects a segment at its midpoint
3. $\overline{AC} \cong \overline{BC}$	3. A midpoint divides a segment into two congruent segments

---

Math Fact

If a statement relies on some other fact, then the other fact must be presented first as a statement. For example in the previous example, this is the chain of conditionals in the reasons column:

- If a segment is a bisector, then it intersects a midpoint.
- If a point is a midpoint, then it divides a segment into two congruent segments.

We start by establishing the hypothesis of the first conditional and end with the conclusion of the second conditional. We cannot switch the order of statements.

When writing proofs involving perpendicular lines, keep in mind that the definition of perpendicular lines is "lines intersecting at right angles" and the definition of a right angle is "an angle measuring 90°." Therefore, the flow of logic in a proof would be:

$$\text{perpendicular lines} \rightarrow \text{right angles} \rightarrow 90° \text{ (or congruent) angles}$$

Example 2

Given: Line $m \perp$ ray p
Prove: m$\angle 1 =$ m$\angle 2$

Statements	Reasons
1. Line $m \perp$ ray p	1. Given
2. $\angle 1$ and $\angle 2$ are right angles	2. Perpendicular lines form right angles
3. $\angle 1 \cong \angle 2$	3. All right angles are congruent

We sometimes need to classify a triangle as a right triangle. Use the definition that a right triangle is one that has a right angle.

Example 3

Given: $\triangle ABC$, $\overline{AB} \perp \overline{BC}$
Prove: $\triangle ABC$ is a right triangle

    Solution: As a paragraph proof: Since $\overline{AB} \perp \overline{BC}$, $\angle ABC$ is a right angle because perpendicular lines form right angles. $\triangle ABC$ must be a right triangle, because a triangle with a right angle is a right triangle.

    The special segments in triangles—altitude, median, perpendicular bisector, and angle bisector—can all show up in proofs. Angle bisectors and perpendicular bisectors are self-explanatory. For altitudes and medians, simply use the definition of the special segment as the reason to justify the relationship it creates:

- Altitude—a segment from a vertex perpendicular to the opposite side (or its extension) of a triangle.
- Median—a segment from a vertex of a triangle to the midpoint of the opposite side.

Example 4

Given: $\triangle RTQ$ with median $\overline{QS}$
Prove: $\overline{RS} \cong \overline{ST}$

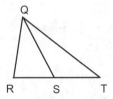

    Solution:

Statements	Reasons
1. $\triangle RTQ$ with median $\overline{QS}$	1. Given
2. S is the midpoint of $\overline{RT}$	2. A median intersects a side of a triangle at its midpoint
3. $\overline{RS} \cong \overline{ST}$	3. A midpoint divides a segment into two congruent segments

Example 5

Given: $\overrightarrow{OQ}$ bisects $\angle POR$
Prove: $\angle POQ \cong \angle QOR$

Solution: As a paragraph proof: $\overrightarrow{OQ}$ is given as a bisector of $\angle POR$. Since a bisector divides an angle into two congruent angles, $\angle POQ$ must be congruent to $\angle QOR$.

Check Your Understanding of Section 4.2

A. Multiple-Choice

1. Which of the following statements can be concluded from the fact that $\overline{RS}$ bisects $\overline{LM}$ at point O?

   (1) O is the midpoint of $\overline{LM}$
   (2) O is the midpoint of $\overline{RS}$
   (3) $\angle ROL$ and $\angle MOS$ are right angles
   (4) $LO = RO$ and $MO = SO$

2. Given the following statements:

   I $\angle ADB$ and $\angle CDB$ are right angles   III $\overline{BD} \perp \overline{AC}$
   II $\overline{BD}$ is an altitude of $\triangle ABC$ IV $\angle ADB \cong \angle CDB$

   Which is an appropriate order in which they might appear in a proof?
   (1) I, III, IV, II
   (2) III, II, I, IV
   (3) II, III, I, IV
   (4) III, II, I,VI

3. Kenny is writing a proof and knows that $\overline{PQ}$ is bisected by $\overline{XY}$ at C. Before Kenny can state two segments are congruent, what should he state?
   (1) C is the midpoint of $\overline{PQ}$
   (2) C is the midpoint of $\overline{XY}$
   (3) $\overline{PQ} \parallel \overline{XY}$
   (4) $\overline{PQ} \perp \overline{XY}$ at C

4. Which of the following reasons could justify why adjacent angles ∠MON and ∠NOP are congruent?
   (1) supplementary angles have measures that sum to 180°
   (2) complementary angles have measures that sum to 90°
   (3) $\overrightarrow{ON}$ bisects $\overline{MP}$
   (4) $\overrightarrow{ON}$ bisects ∠MOP

5. Jack is writing a proof. The second statement is $\overline{GKH} \perp \overline{JK}$, and the third statement is K is the midpoint of $\overline{GKH}$. Which could have been the first statement?
   (1) $\overline{JK}$ bisects $\overline{GKH}$ at K
   (2) $\overline{JK}$ is the perpendicular bisector of $\overline{GKH}$
   (3) $\overline{JK} \perp \overline{GKH}$
   (4) $\overline{GK} \cong \overline{KH}$

B. *Free-Response—show all work or explain your answer completely.*

6. Given: △XYZ, point D on $\overline{XY}$, and $\overline{XD} \cong \overline{DY}$
   Prove: $\overline{ZD}$ is a median

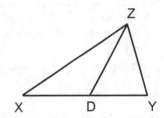

7. Given: △XYZ, point D on $\overline{XY}$, m∠YDZ = 90°
   Prove: $\overline{ZD}$ is an altitude

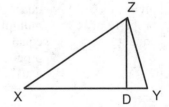

8. Given: $\triangle ABC$, X is the midpoint of $\overline{AB}$, Z is the midpoint of $\overline{AC}$, $AB = 10$, $AC = 12$, and $ZX = 4$.
Prove: The perimeter of $\triangle AXZ = 15$

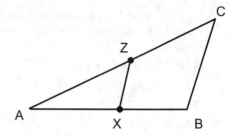

4.3 PROPERTIES OF EQUALITY

Besides the postulates we have already seen, the following postulates are useful tools when working out a proof. They show up frequently and you should be familiar with them.

Property	What It Means	Example
Reflexive property of equality / congruence	Any quantity is equal to itself; any figure is congruent to itself.	$\overline{CD} \cong \overline{CD}$
Addition/ subtraction property of equality	When equal quantities are added/ subtracted to equal quantities, the sums/ differences are equal.	If $AB = DE$ and $BC = ED$, then $AB + BC = DE + EF$
Multiplication/ division property of equality	When equal quantities are multiplied/divided by equal quantities, the products/quotients are equal.	If $AB = DE$ and $BC = EF$, then $AB \cdot BC = DE \cdot EF$

Property	What It Means	Example
Partition postulate of equality	The whole equals the sum of its parts.	A B C $\overline{AB} + \overline{BC} \cong \overline{AC}$ or $AB + BC = AC$
Substitution property of equality/ congruence	If quantity $b = a$, then b can be substituted for a in any equality. The same holds for congruent figures.	1 2 3 If $\angle 1$ and $\angle 2$ are complementary and if $\angle 2 \cong \angle 3$, then $\angle 1$ and $\angle 3$ are complementary
Transitive property of equality/ congruence	Given quantities a, b, and c, if $a = b$ and $b = c$, then $a = c$. The same holds for congruent figures.	A •———• B C •———• D E •———• F If $\overline{AB} \cong \overline{CD}$ and $\overline{CD} \cong \overline{EF}$, then $\overline{AB} \cong \overline{EF}$

Substitution and Transitive Properties of Equality

Substitution Property of Equality—Any quantity can be substituted for an equal quantity.
Transitive Property of Equality—Given quantities a, b, and c, if $a = b$ and $b = c$, then $a = c$.

The substitution and transitive properties of equality are useful when we have an indirect relationship between three different figures or quantities. It states that if two quantities are both equal to a third quantity, then the first two quantities are equal to each other. In a proof, the equality would be the statement and "substitution property" or "transitive property" would be the reason.

Example

Given: G is the midpoint of
$\overline{FH}$, $\overline{GH} \cong \overline{HI}$

Prove: $\overline{FG} \cong \overline{HI}$

Solution: Label the figure with congruent markings. The midpoint tells us that $\overline{FG} \cong \overline{GH}$. Combining that with the given information $\overline{GH} \cong \overline{HI}$ allows us to use the transitive property to link $\overline{FG}$ to $\overline{HI}$.

Statements	Reasons
1. G is the midpoint of $\overline{FH}$	1. Given
2. $\overline{FG} \cong \overline{GH}$	2. A midpoint divides a segment into two congruent segments
3. $\overline{GH} \cong \overline{HI}$	3. Given
4. $\overline{FG} \cong \overline{HI}$	4. Transitive property of congruence

Multiplication and Division Postulates

The Multiplicative Property of Equality—If equal quantities are multiplied by equal quantities, then the products are equal.
The Division Property of Equality—If equal quantities are divided by equal quantities, then the quotients are equal, provided the divisors are not equal to zero.

In terms of quantities a, b, c, and d where $a = b$ and $c = d$ (with c ≠ 0 and d ≠ 0), we have:

$$a \cdot c = b \cdot d \quad \text{(multiplicative property of equality)}$$
$$a/c = b/d \quad \text{(division property of equality)}$$

There is a special case for the division property of equality when dividing by 2: halves of equals quantities are equal.

Example

Given: $\angle RST$ is bisected by $\overrightarrow{SX}$,
m$\angle RST = (4x + 12)°$

Prove: m$\angle TSX = (2x + 6)°$

Solution:

Statements	Reasons
1. $\angle RST$ is bisected by $\overrightarrow{SX}$	1. Given
2. $m\angle TSX = \frac{1}{2}m\angle RST$	2. An angle bisector divides an angle in half
3. $m\angle RST = (4x + 12)°$	3. Given
4. $\frac{1}{2}m\angle RST = \frac{1}{2}(4x+12)°$	4. Division property of equality
5. $\frac{1}{2}m\angle RST = (2x+6)°$	5. Simplify
6. $m\angle TSX = (2x + 6)°$	6. Transitive property of equality

In the previous example, the reason for line 6 could also have been substitution instead of the transitive property of equality. The reason for line 5 is an algebraic simplification. The algebraic operation was applying the distributive property to the factor of $\frac{1}{2}$. We will take the shortcut of not justifying every step of an algebraic simplification unless the steps are not obvious.

Addition, Subtraction, and Partition Postulates

Addition/Subtraction Property of Equality
If equal quantities are added (or subtracted) from equal quantities, then the sums (or differences) are equal.
 In terms of quantities a, b, c, and d, if $a = b$ and $c = d$, then:
$$a + c = b + d$$
$$a - c = b - d$$

The addition postulate states that if equal quantities are added to equal quantities, then the sums are equal. Using angle measures, for example, we might have:

If $m\angle A = m\angle C$
       $m\angle B = m\angle D$
then $m\angle A + m\angle B = m\angle C + m\angle D$

This looks a lot like an algebraic system of equations. When we use these postulates in a proof, the proof will be easier to read if the two lines to be added follow one another. The quantities to be added can be segment measures, angle measures, or the figures themselves.

97

When using the addition postulate with figures instead of measures, we often follow it up with the partition postulate of equality/congruence. The partition postulate states that a whole figure can be broken up into a sum of consecutive parts or that a sequence of consecutive parts is equivalent to a single whole figure. The segment would have to be collinear with a common endpoint, and the angles would have to be adjacent.

Partition Postulate

If points A, B, and C are collinear with B between A and C:
- $AB + BC = AC$
- $\overline{AB} + \overline{BC} \cong \overline{AC}$

If $\angle ABD$ and $\angle CBD$ are adjacent angles:
- $\text{m}\angle ABD + \text{m}\angle CBD = \text{m}\angle ABC$
- $\angle ABD + \angle CBD \cong \angle ABC$

The addition symbol in the congruence statements implies the joining of two figures into a single figure, not an algebraic addition. Figure 4.1 shows the application of the partition postulate to ABC and $\angle ABC$.

$$\overline{AB} + \overline{BC} \cong \overline{AC} \qquad\qquad \angle ABD + \angle DBC = \angle ABC$$

$$AB + BC = AC \qquad\qquad \text{m}\angle ABD + \text{m}\angle DBC = \text{m}\angle ABC$$

Figure 4.1 Partition postulate applied to segments and angles

Example 1

Given: $\overline{AB} \cong \overline{DE}$

Prove: $DE + BC = AC$

Solution:

Statements	Reasons
1. $\overline{AB} \cong \overline{DE}$	1. Given
2. $AB = DE$	2. Congruent segments have equal measures
3. $AB + BC = AC$	3. Partition postulate
4. $DE + BC = AC$	4. Substitution Property

Example 2

Given: $\overline{RT}$ bisects $\angle SRA$, $\angle S \cong \angle A$

Prove: $m\angle S + m\angle SRT = m\angle A + m\angle ART$

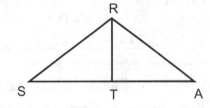

Solution: Strategy—Use the definition of angle bisector to show that $m\angle SRT = m\angle ART$

Statements	Reasons
1. $\overline{RT}$ bisects $\angle SRA$	1. Given
2. $\angle S \cong \angle A$	2. Given
3. $m\angle SRT = m\angle ART$	3. An angle bisector divides an angle into angles of equal measures
4. $m\angle S = m\angle A$	4. Congruent angles have equal measures
5. $m\angle S + m\angle SRT = m\angle A + m\angle ART$	5. Addition property of equality

Reflexive Property of Equality

> ### Reflexive Property of Equality/Congruence
> • Any quantity equal to itself.
> • Any figure congruent to itself.

This seems like an obvious statement. However, it is one that needs to be formally stated and justified in proofs.

Example

Given: $\overline{ABCD}$ and $\overline{AB} \cong \overline{CD}$
Prove: $\overline{AC} \cong \overline{BD}$

Strategy—We see from the figure that $\overline{AC}$ is comprised of $\overline{AB} + \overline{BC}$ and that $\overline{BD}$ is comprised of $\overline{BC} + \overline{CD}$. We want to establish two separate equalities and sum them to get $\overline{AB} + \overline{BC} = \overline{BC} + \overline{CD}$. The first equality is given, but the second equality requires the use of the reflexive property.

Statements	Reasons
1. $\overline{ABCD}$	1. Given
2. $\overline{AB} \cong \overline{CD}$	2. Given
3. $\overline{BC} \cong \overline{BC}$	3. Reflexive property of equality
4. $\overline{AB} + \overline{BC} \cong \overline{BC} + \overline{CD}$	4. Addition property of equality
5. $\overline{AC} \cong \overline{BD}$	5. Partition property of equality

Check Your Understanding of Section 4.3

A. Multiple-Choice

1. Which property of equality would justify the statement $\overline{ST} \cong \overline{ST}$?
   (1) segment addition
   (2) substitution
   (3) reflexive
   (4) transitive

2. If $\overline{MN} \cong \overline{PQ}$ and $\overline{PQ} \cong \overline{RS}$, what would the transitive property let you conclude?
   (1) Q is the midpoint of $\overline{MS}$
   (2) $\overline{MN} \cong \overline{RS}$
   (3) $\overline{QP} \cong \overline{NM}$
   (4) $MN = \dfrac{1}{2}PQ$

3. A, B, and C are collinear with B between A and C. Juan wants to prove $AB = AC - BC$. His statements are the following:
$$AC = AC$$
$$AB + BC = AC$$
$$AB = AC - BC$$

   Which of the following would be three properties or postulates that justify the statements?
   (1) transitive, partition, and subtraction
   (2) partition, addition, and subtraction
   (3) reflexive, addition, and subtraction
   (4) reflexive, partition, and subtraction

4. If $m\angle 1 + m\angle 2 + m\angle 3 = 180°$ and $m\angle 2 = m\angle 4$, then the substitution property of equality could be used to state which of the following?
(1) $m\angle 1 + m\angle 2 + m\angle 3 - m\angle 4 = 180°$
(2) $m\angle 1 + m\angle 4 + m\angle 3 = m\angle 4$
(3) $m\angle 1 + m\angle 2 + m\angle 3 + m\angle 4 = 180°$
(4) $m\angle 1 + m\angle 4 + m\angle 3 = 180°$

5. Right triangles $\triangle FGI$ and $\triangle GHI$ share side $\overline{GI}$.
The reflexive property of equality could be used to justify which of the following statements?

(1) the two triangles are isosceles
(2) the length of the hypotenuse is the same in each triangle
(3) $FG + FH = GH$
(4) $GI = FH$

B. *Free-Response—show all work or explain your answer completely.*

6. Given: $\overline{RD} \cong \overline{AE}$, $\overline{DE} \cong \overline{RA}$
Prove: Perimeter of $\triangle RDE$ = perimeter of $\triangle ERA$

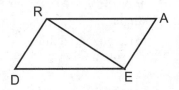

7. Given: $QS = RT$
Prove: $QR = ST$

8. Given: $\overline{MO} \perp \overline{JL}$, $\overline{KN} \perp \overline{JL}$, $\overline{KN} \cong \overline{MO}$
Prove: The area of $\triangle JLM$ equals the area of $\triangle JLK$

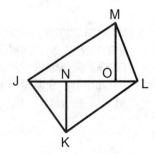

9. Given: $SI = 2HI$

Prove: $\overline{PH}$ is a median of $\triangle SIP$

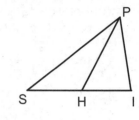

10. Given: Y is the midpoint of $\overline{WZ}$ and X is the midpoint of $\overline{WY}$

Prove: $WX = \frac{1}{4}WZ$

11. Given: $\angle ABC$ is a right angle, $\angle CBD \cong \angle EFH$, $\angle ABD \cong \angle GFH$

Prove: $\angle EFG$ is a right angle

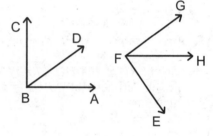

4.4 USING VERTICAL ANGLES, LINEAR PAIRS, AND COMPLEMENTARY AND SUPPLEMENTARY ANGLES

KEY IDEAS

Vertical angles, linear pairs, complementary angles, and supplementary angles are some of the angle relationships that are the building blocks of proofs. We can often infer these relationships directly from a figure without additional given information other than the existence of straight lines and right angles.

What It Looks Like	The Statement	The Reason
	$\angle 1 \cong \angle 2$	Vertical angles are congruent
	$\angle 1$ and $\angle 2$ are supplementary, or $m\angle 1 + m\angle 2 = 180°$	Linear pairs are supplementary

Linear Pairs and Supplementary Angles

Linear pairs and vertical angles are two angle relationships that can be inferred from a figure. The angles are usually not specifically stated to be linear pairs, but we can determine that they are from the figure. Linear pairs let us conclude that a pair of adjacent angles are supplementary.

Combining the definition of complementary or supplementary angles with the transitive property gives the following useful theorem:

> **Complements/Supplements of Congruent Angles Theorem**
> If two angles are supplementary or complementary to the same (or congruent) angles, then those first two angles are congruent.

Example 1

Given: $\overleftrightarrow{AB}$ and $\overleftrightarrow{CD}$ intersect at point O
Prove: $\angle AOC \cong \angle BOD$

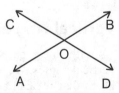

Solution: Strategy—Apply the transitive property of equality to supplementary linear pairs $\angle AOC$ and $\angle BOC$ and to $\angle BOD$ and $\angle BOC$. We are not going to use the vertical angle theorem since that is what we are trying to prove.

Statements	Reasons
1. $\overleftrightarrow{AB}$ and $\overleftrightarrow{CD}$ intersect at point O	1. Given
2. $\angle AOC$ and $\angle BOC$ are linear pairs $\angle BOC$ and $\angle BOD$ are linear pairs	2. Adjacent angles that form a straight line are linear pairs
3. $\angle AOC$ and $\angle BOC$ are supplementary $\angle BOD$ and $\angle BOC$ are supplementary	3. Linear pairs are supplementary
4. $\angle AOC \cong \angle BOD$	4. Angles supplementary to the same angle are congruent

Example 2

Given: $\overline{ABCD}$, $\angle 1 \cong \angle 3$
Prove: $\angle 2 \cong \angle 4$

Solution:

Statements	Reasons
1. $\overline{ABCD}$	1. Given
2. $\angle 1 \cong \angle 3$	2. Given
3. $\angle 1$ and $\angle 2$ are linear pairs $\angle 3$ and $\angle 4$ are linear pairs	3. Definition of linear pairs
4. $\angle 2$ and $\angle 1$ are supplementary $\angle 4$ and $\angle 3$ are supplementary	4. Linear pairs are supplementary
5. $\angle 2 \cong \angle 4$	5. Angles supplementary to congruent angles are congruent

Vertical Angles

As with linear pairs, given information in a proof will usually not include vertical angles. We may be given only the fact that two straight lines intersect. From that we can conclude that the opposite angles formed are congruent, vertical angles.

Example

Given: $\overline{ACE}$, $\angle BAC \cong \angle ECD$

Prove: $\angle BAC \cong \angle ACB$

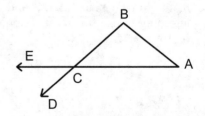

Statements	Reasons
1. $\overline{ACE}$, $\angle BAC \cong \angle ECD$	1. Given
2. $\angle ECD \cong \angle ACB$	2. Vertical angles are congruent
3. $\angle ACB \cong \angle BAC$	3. Transitive property

Complementary Angles

The measures of complementary angles sum to 90°. Therefore, any right angle partitioned into two or more angles will represent complementary angles.

Example

Given: $\overline{DPC}$, $\overline{PE} \perp \overline{APB}$
Prove: $\angle BPC$ and $\angle DPE$ are
        complementary

Solution: Strategy—show that the measures of $\angle APD$ and $\angle DPE$ sum to 90°, and substitute vertical angle $\angle BPC$ for $\angle APD$.

Statements	Reasons
1. $\overline{DPC}, \overline{PE} \perp \overline{APB}$	1. Given
2. $\angle APE$ is a right angle	2. Perpendicular lines intersect at right angles
3. m$\angle APE = 90°$	3. Right angles measure 90°
4. m$\angle APD$ + m$\angle DPE$ = m$\angle APE$	4. Partition postulate
5. m$\angle APD$ + m$\angle DPE$ = 90°	5. Transitive property
6. $\angle BPC \cong \angle APD$	6. Vertical angles are congruent
7. m$\angle BPC$ = m$\angle APD$	7. Congruent angles have equal measures
8. m$\angle BPC$ + m$\angle DPE$ = 90°	8. Substitution property
9. $\angle BPC$ and $\angle DPE$ are complementary	9. Angles that sum to 90° are complementary

| **MATH FACT** |

In proofs, we are often given the fact that three or more points are collinear. This is usually done by specifying a line, segment, or ray with multiple points, like $\overline{DPC}$ in the previous example. Without knowing $\overline{DPC}$ is a straight line, we could not conclude that $\angle BPC$ and $\angle APD$ are vertical angles.

Check Your Understanding of Section 4.4

A. *Multiple-Choice*

1. $\overrightarrow{QOT}$ intersects $\overrightarrow{ROS}$ at point O. Which reason would justify the statement $\angle QOR \cong \angle TOS$?
   (1) An angle bisector forms two congruent angles.
   (2) Vertical angles are congruent.
   (3) Supplementary angles measure 90°.
   (4) Complementary angles measure 90°.

2. If $\angle 1 \cong \angle 2$, which of the following could justify why $\angle 3 \cong \angle 4$?
   (1) $\angle 1$ and $\angle 2$ are vertical angles.
   (2) $\angle 1$ and $\angle 3$ are vertical angles.
   (3) $\angle 1$ and $\angle 2$ are supplementary, $\angle 3$ and $\angle 4$ are supplementary.
   (4) $\angle 1$ and $\angle 3$ are supplementary, and $\angle 2$ and $\angle 4$ are supplementary.

3. Which reason could be used in the following line of a proof?
   2. $m\angle 1 + m\angle 2 = 90°$ 2. _____
   (1) The measures of complementary angles sum to 90°.
   (2) The measures of supplementary angles sum to measure 90°.
   (3) $\angle 1$ and $\angle 2$ are right angles.
   (4) $\angle 1$ and $\angle 2$ are a linear pair.

4. The reason "linear pairs are supplementary" could be used in a proof when describing
   (1) the two segments formed by the midpoint of a segment
   (2) a pair of vertical angles
   (3) two adjacent right angles
   (4) the sum of the three angles in a triangle

B. *Free-Response—show all work or explain your answer completely.*

5. Given: $\overleftrightarrow{NOQ} \perp \overleftrightarrow{POM}$, m$\angle RON = 60°$.
   Prove: m$\angle QOR = 120°$

6. Given: $\overleftrightarrow{APB}$, $\overline{PD}$, $\overline{PC}$, and $\angle 1 \cong \angle 2$
   Prove: $\angle 3 \cong \angle 4$

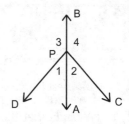

7. Given: Lines m, n, and p
   $\angle 1$ and $\angle 3$ are complementary
   Prove: m$\angle 5 = 90°$

8. Given: $\angle BOC$ and $\angle DOF$ are
   complementary
   Prove: $\angle EOC$ and $\angle AOD$ are
   complementary

9. Given: $\overrightarrow{AB}$ bisects $\angle FPC$
   Prove: $\angle DPB \cong \angle BPE$

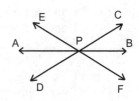

107

10. Given: $\overrightarrow{BE}$ bisects $\angle FBD$,
$\angle ABF \cong \angle CBD$
Prove: $\angle ABE \cong \angle CBE$

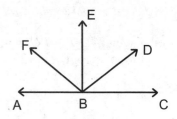

4.5 USING PARALLEL LINES

KEY IDEAS

If we are given parallel lines intersected by a transversal, we can conclude that the alternate interior angles are congruent, the corresponding angles are congruent, and the same side interior angles are congruent. We can also use the converses of these statements to prove that two lines are parallel.

What It Looks Like	The Statement	The Reason
	$\angle 1 \cong \angle 2$	The alternate interior angles formed by parallel lines are congruent
		or
	$\angle 1 \cong \angle 4$	The corresponding angles formed by parallel lines are congruent
		or
	$\angle 1$ and $\angle 3$ are supplementary	the same side interior angles formed by parallel lines are supplementary

What It Looks Like	The Statement	The Reason
	The lines are parallel	Two lines are parallel if the alternate interior angles are congruent or
	The lines are parallel	Two lines are parallel if the corresponding angles are congruent or
$m\angle 1 + m\angle 2 = 180°$	The lines are parallel	Two lines are parallel if the same side interior angles are supplementary

Proving Angle Relationships Given Parallel Lines

We can use the relationships between angles formed by parallel lines to prove angles are congruent or supplementary. Look for the transversal and pairs of alternate interior, corresponding, or same side interior angles. If there is more than one transversal, be careful that you work with pairs of angles formed by *the same* transversal.

Example 1

Given: Line m ∥ line n
Prove: $\angle 1$ and $\angle 3$ are supplementary

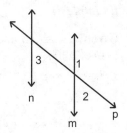

Solution: $\angle 1$ and $\angle 2$ are a linear pair, which are supplementary. $\angle 2 \cong \angle 3$ because they are corresponding angles formed by parallel lines. Substituting $\angle 3$ for $\angle 2$ in the first statement gives $\angle 1$ supplementary to $\angle 3$.

Example 2

Given: Line n || line p, $\angle 1 \cong \angle 3$
Prove: $\angle 1$ and $\angle 2$ are supplementary

Statements	Reasons
1. Line n || line p	1. Given
2. $\angle 1 \cong \angle 3$	2. Given
3. $\angle 2$ and $\angle 3$ are supplementary	3. Same side interior angles formed by parallel lines are supplementary
4. $\angle 1$ and $\angle 2$ are supplementary	4. Substitution property

Proving Lines Are Parallel

Two lines can be proven to be parallel by showing that any pair of alternate interior angles are congruent, of same side interior angles are supplementary, or of corresponding angles are congruent. You need to demonstrate the relationship for only one pair of angles.

Example 1

Given: $\angle 1 \cong \angle 3$ and $\angle 2 \cong \angle 4$
Prove: p || q

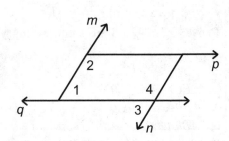

Solution: Strategy—show that the same side interior angles are supplementary.

Statements	Reasons
1. $\angle 1 \cong \angle 3$ and $\angle 2 \cong \angle 4$	1. Given
2. $\angle 3$ and $\angle 4$ are a linear pair	2. Definition of a linear pair
3. $\angle 3$ and $\angle 4$ are supplementary	3. Linear pairs are supplementary
4. $\angle 1$ and $\angle 2$ are supplementary	4. Substitution
5. $p \parallel q$	5. Two lines are parallel if the same side interior angles are supplementary

Example 2

Given: $\angle ABF = \angle DCE$, $\overrightarrow{BF} \parallel \overrightarrow{CE}$
Prove: $\overrightarrow{AB} \parallel \overrightarrow{CD}$

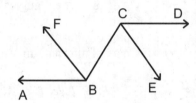

Solution:

Statements	Reasons
1. $\overrightarrow{BF} \parallel \overrightarrow{CE}$	1. Given
2. $\angle FBC \cong \angle ECB$	2. Alternate interior angles formed by parallel lines are congruent
3. $\angle ABF \cong \angle DCE$	3. Given
4. $\angle ABF + \angle FBC \cong \angle DCE + \angle ECB$	4. Addition postulate
5. $\angle ABC \cong \angle DCB$	5. Partition postulate
6. $\overrightarrow{AB} \parallel \overrightarrow{CD}$	6. If the alternate interior angles formed by two lines are congruent, then the lines are parallel

Check Your Understanding of Section 4.5

A. Multiple-Choice

1. Kevin is writing a proof to show that lines *m* and *n* are parallel. The last line in his proof might read two lines are parallel if the _____.
   (1) same side interior angles are complementary
   (2) alternate interior angles are supplementary
   (3) corresponding angles are complementary
   (4) corresponding angles are congruent

2. Lines *r* and *a* are intersected by a transversal. Which of the following could always be used to prove *m* and *n* are parallel?
   (1) A pair of vertical angles are congruent.
   (2) A pair of alternate interior angles are both right angles.
   (3) A pair of alternate interior angles are complementary.
   (4) A pair of alternate interior angles are supplementary.

B. Free-Response—show all work or explain your answer completely.

3. Given: line $t \perp$ line *s* and line $t \perp$ line *r*
   Prove: $\angle 1 \cong \angle 2$

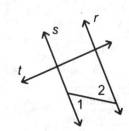

4. Given: $\overrightarrow{AD} \parallel \overline{BC}$, $\overrightarrow{AB}$ bisects $\angle CAD$
   Prove: $\angle 1 \cong \angle 2$

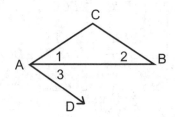

5. Given: $\overline{AC} \perp \overline{BD}$ and $\overline{AB} \parallel \overline{CD}$
   Prove: $\angle A$ and $\angle D$ are complementary

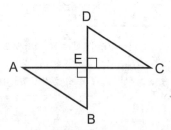

6. Given; $\overline{MSE}$, $\overline{KSO}$, and $\angle OMS \cong \angle KES$
Prove: $\angle SON \cong \angle SKY$ using a parallel line relationship

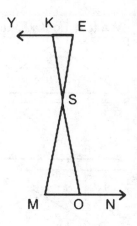

7. Given: $\overrightarrow{AB} \parallel \overrightarrow{CF}$ and $\overrightarrow{BE} \parallel \overrightarrow{CD}$
Prove: $\angle 1 \cong \angle 4$

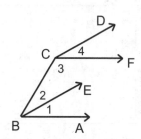

8. Given: $\overline{DFC}$, $\overline{FB}$ bisects $\angle AFC$, $\angle A \cong \angle B$, and $\angle A \cong \angle BFC$
Prove: $\overline{AB} \parallel \overline{CFD}$

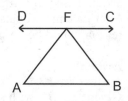

4.6 USING TRIANGLE RELATIONSHIPS

KEY IDEAS

The triangle angle sum theorem, exterior angle theorem, and isosceles triangle theorems can often be applied in proofs involving triangles to justify angle and segment relationships.

What It Looks Like	The Statement	The Reason
(triangle with angles 3 at top, 1 at bottom left, 2 at bottom right)	m∠1 + m∠2 + m∠3 = 180°	Triangle angle sum theorem
(triangle with angle 2 at top, exterior angle 1 at bottom left with extended side, angle 3 at bottom right)	m∠1 = m∠2 + m∠3	Triangle exterior angle theorem
(isosceles triangle with C at top, A at bottom left, B at bottom right, tick marks on the two equal sides)	∠A ≅ ∠B or $\overline{AC} \cong \overline{BC}$	The isosceles triangle theorem (angles opposite congruent sides in a triangle are congruent) or The converse of the isosceles triangle theorem (sides opposite congruent angles in a triangle are congruent)

Exterior Angle Theorem

When a side of a triangle is extended, we can state that the measure of the exterior angle equals the sum of the two nonadjacent interior angles. The reason would be "triangle exterior angle theorem."

Example

Given: △ABC with side $\overline{AB}$ extended
Prove: m∠1 − m∠2 = m∠4 − m∠3

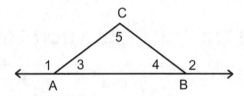

Solution:

Strategy—Apply the external triangle angle theorem to both ∠1 and ∠2. Then subtract the two relationships.

Statements	Reasons
1. $\triangle ABC$ with side $\overline{AB}$ extended	1. Given
2. m∠1 = m∠4 + m∠5	2. Triangle exterior angle theorem
3. m∠2 = m∠3 + m∠5	3. Triangle exterior angle theorem
4. m∠1 − m∠2 = m∠4 − m∠3	4. Subtraction postulate

Triangle Angle Sum Theorem

We can state that the sum of the measures of the angles of a triangle equals 180° using the "triangle angle sum theorem" for the reason.

Example 1

Given: ∠1 ≅ ∠4 and ∠3 ≅ ∠6
Prove: $\overline{CD}$ bisects ∠ACB

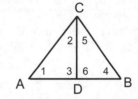

Solution:

Statements	Reasons
1. ∠1 ≅ ∠4 and ∠3 ≅ ∠6	1. Given
2. m∠1 + m∠2 + m∠3 = 180°	2. Triangle angle sum theorem
3. m∠4 + m∠2 + m∠6 = 180°	3. Substitution property
4. m∠4 + m∠5 + m∠6 = 180°	4. Triangle angle sum theorem
5. m∠2 − m∠5 = 0	5. Subtraction postulate
6. m∠2 = m∠5	6. Addition postulate
7. $\overline{CD}$ bisects ∠ACB	7. A segment that divides an angle into two congruent angles is an angle bisector

Example 2

Prove the triangle exterior angle theorem.

Given: $\triangle ABC$ and side $\overline{AB}$ extended to D
Prove: m∠3 + m∠4 = m∠1

Solution:

Strategy—∠1 and ∠2 are supplementary linear pairs. ∠2, ∠3, ∠4 are the interior angles of a triangle and sum to 180°. Combining these leads to the exterior angle theorem.

Statements	Reasons
1. $\triangle ABC$ and side $\overline{AB}$ extended to D	1. Given
2. $m\angle 2 + m\angle 3 + m\angle 4 = 180°$	2. Triangle angle sum theorem
3. $\angle 1$ and $\angle 2$ are a linear pair	3. Definition of linear pairs
4. $m\angle 2 + m\angle 1 = 180°$	4. Linear pairs are supplementary
5. $m\angle 3 + m\angle 4 - m\angle 1 = 0$	5. Subtraction property of equality
6. $m\angle 1 = m\angle 1$	6. Reflexive property of equality
7. $m\angle 3 + m\angle 4 = m\angle 1$	7. Addition property of equality

Example 3

Prove the triangle angle sum theorem
Given: $\triangle ABC$
Prove: the sum of the measures of the interior angles in $\triangle ABC$ equals 180°

   Solution: The strategy is to construct line $\overline{DE}$ through point C and parallel to $\overline{AB}$.

   We know the sum of angles 3, 4, and 5 must be 180° since they comprise a straight line. Using our angle relationships in parallel lines cut by a transversal, we find two pairs of congruent angles, $\angle 4 \cong \angle 1$ and $\angle 5 \cong \angle 2$. Substitution yields the desired relationship. The two-column proof is as follows.

Statements	Reasons
1. $\triangle ABC$	1. Given
2. Construct $\overline{DE}$ through C and parallel to $\overline{AB}$	2. There exists one and only one line through a point parallel to a line
3. $m\angle DCE = 180°$	3. A straight angle measures 180°
4. $m\angle 3 + m\angle 4 + m\angle 5 = 180°$	4. Partition postulate
5. $\angle 4 \cong \angle 1$ and $\angle 5 \cong \angle 3$	5. Alternate interior angles formed by parallel lines are congruent
6. $m\angle 1 + m\angle 2 + m\angle 3 = 180°$	6. Substitution Property

Isosceles Triangle Theorem

Whenever you have two congruent sides in a triangle, you can conclude that the opposite angles are congruent by the isosceles triangle theorem. The converse is true as well.

Example 1

Given: $\overline{RT} \cong \overline{ST}$
Prove: $\angle 1 \cong \angle 3$

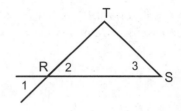

Solution:

Statements	Reasons
1. $\overline{RT} \cong \overline{ST}$	1. Given
2. $\angle 2 \cong \angle 3$	2. Angles opposite congruent sides in a triangle are congruent
3. $\angle 1 \cong \angle 2$	3. Vertical angles are congruent
4. $\angle 1 \cong \angle 3$	4. Transitive property

Example 2

Given: $\angle DAB \cong \angle ABD$, $\overline{BC} \cong \overline{AD}$
Prove: $\angle BCD \cong \angle BDC$

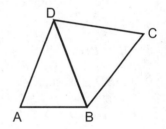

Solution:

Statements	Reasons
1. $\angle DAB \cong \angle ABD$	1. Given
2. $\overline{AD} \cong \overline{BD}$	2. Sides opposite congruent angles in a triangle are congruent
3. $\overline{BC} \cong \overline{AD}$	3. Given
4. $\overline{BC} \cong \overline{BD}$	4. Transitive property
5. $\angle BCD \cong \angle BDC$	5. Angles opposite congruent sides in a triangle are congruent

To prove a triangle is isosceles, you need to prove that two sides are congruent. Sometimes it is easier to prove two angles are congruent and then use the converse of the isosceles angle theorem to show the opposite sides are congruent.

Example

Given: $\overline{AD} \parallel \overline{BE}$, $\overline{BE}$ bisects $\angle CBD$
Prove: $\triangle ABD$ is isosceles

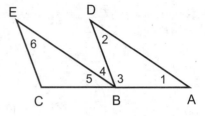

Statements	Reasons
1. $\overline{AD} \parallel \overline{BE}$	1. Given
2. $\angle 2 \cong \angle 4$	2. Alternate interior angles formed by parallel lines are congruent
3. $\overline{BE}$ bisects $\angle CBD$	3. Given
4. $\angle 4 \cong \angle 5$	4. Angle bisector divides an angle into two congruent angles
5. $\angle 2 \cong \angle 5$	5. Transitive property
6. $\angle 1 \cong \angle 5$	6. Corresponding angles formed by parallel lines are congruent
7. $\angle 1 \cong \angle 2$	7. Transitive property
8. $\overline{AB} \cong \overline{BD}$	8. In a triangle, the sides opposite congruent angles are congruent
9. $\triangle ABD$ is isosceles	9. A triangle with two congruent sides is isosceles

Check Your Understanding of Section 4.6

A. Multiple-Choice

1. One strategy for proving a triangle is isosceles is to demonstrate that
   (1) the triangle has two congruent angles
   (2) the triangle has a right angle
   (3) the triangle has a pair of complementary angles
   (4) the triangle is not obtuse

2. The triangle angle sum theorem could be used to prove directly that
   (1) an isosceles triangle has two congruent angles
   (2) two angles in a right triangle are complementary
   (3) perpendicular lines intersect at right angles
   (4) supplementary angles sum to 180°

B. *Free-Response—show all work or explain your answer completely.*

3. Given: $\triangle ABC$, $\overline{ACE}$, and $\overline{BCD}$
Prove: m$\angle DCE$ + m$\angle BAC$ + m$\angle ABC$ = 180°

4. Given: $\overline{JM} \cong \overline{JK}$, $\overline{KM}$ bisects
$\angle JKL$ and $\angle JML$
Prove: $\triangle KLM$ is isosceles

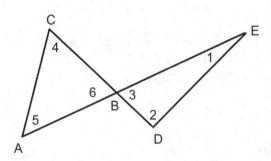

5. Given: $\triangle ABC$, $\triangle BED$,
$\overline{ABE}$, $\overline{CBD}$
Prove: m$\angle 1$ + m$\angle 2$ =
m$\angle 4$ + m$\angle 5$

6. Given: $\triangle ABE$, $\triangle EBD$, and $\triangle BCD$ are
all equilateral
Prove: B is the midpoint of $\overline{AC}$

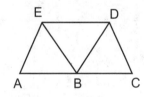

7. Given: $\triangle ABC$ with $\overline{AB}$ extended to D,
$AC = BC$
Prove: m$\angle 1$ = m$\angle 2$ + m$\angle 4$

8. Given: $\overline{HKJM}$, $\overrightarrow{IH} \parallel \overline{LM}$, and $\overline{LK} \parallel \overline{JI}$

   Prove: $\angle I \cong \angle L$

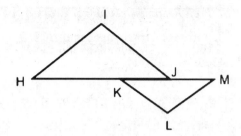

9. Given: m$\angle ADC$ = 140°, m$\angle BDC$ = 130°, and m$\angle DAB$ = 45°

   Prove: $\overline{AD} \cong \overline{BD}$

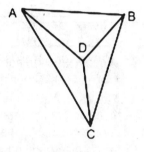

10. Given: $\overline{DABE}$, $\overline{CBF}$, m$\angle ACB$ = 50°, and m$\angle FBE$ = 40°

    Prove: $\overrightarrow{AC} \perp \overline{DABE}$

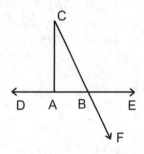

| Chapter Five | # TRANSFORMATIONS AND CONGRUENCE |

5.1 RIGID MOTION AND SIMILARITY TRANSFORMATIONS

KEY IDEAS

A geometric transformation is a function that operates on a figure and may change its position, size, shape, or orientation. The figure can be any geometric figure, picture, or even a point. The input of the function, or original figure, is called the **preimage**. The output of the function is called the **image**. Every point of the preimage is **mapped** to a corresponding point of the image by the transformation. The transformation is defined by the rule that describes the relationship between the preimage and the image. A transformation in which the distance between any two points is **preserved** is called a **rigid motion**. Rigid motions result in an image congruent to the preimage. A transformation in which the shape, but not necessarily the size, is preserved is called a **similarity** transformation. The named rigid motions studied in this course are **translations**, **rotations**, and **reflections**. The one named similarity transformation we will study is the **dilation**. The rigid motion transformation will be used to define and determine congruence between a preimage and an image.

Transformation Definitions and Notation

A transformation is a function that takes an input called the **preimage** and returns an output called the **image**. The preimage and image can be points, segments, or any geometric figure or picture. We say the transformation **maps** the preimage onto the image. Figure 5.1 shows an example of a transformation that maps $\triangle ABC$ to $\triangle A'B'C'$. $\triangle ABC$ is the preimage, and $\triangle A'B'C'$ is the image. The transformation operates on the preimage point by point so that every point in the preimage has a **corresponding point** in the image.

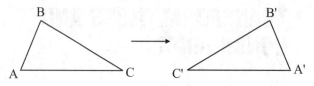

Figure 5.1 Transformation of $\triangle ABC$ to $\triangle A'B'C'$

======================== **MATH FACTS** ========================

Transformations in geometry are similar to one-to-one functions in algebra. One-to-one functions have only one possible *y*-value for each *x*-value, and only one possible *x*-value for each *y*-value. With transformations, each point in the preimage has only one corresponding point in the image. Also, given a point in the image there is only possible point in the preimage that could have led there.

The transformation rule can be specified in different ways. It can be given in terms of a geometric relationship between points in the image and preimage or as an algebraic rule that operates on the coordinates of each point in the coordinate plane.

When naming corresponding pairs of points, the same letter is used, with the single prime symbol ′ used to indicate the transformed point. An arrow, →, can also be used to indicate the operation of a transformation, as in $\triangle ABC$ → $\triangle A'B'C'$.

Rigid Motion and Invariance

A transformation in which the distance between any two points is unchanged is called a **rigid motion** or **isometry**. In other words, the lengths of any corresponding segments are equal. The image formed after a rigid motion will always be congruent to the preimage, and corresponding angles will also be congruent.

The transformation shown in Figure 5.2 is a rigid motion with congruent corresponding lengths: $AB \cong A'B'$, $BC \cong B'C'$, $CD \cong C'D'$, $DA \cong D'A'$, $AC \cong A'C'$, and $BD \cong B'D'$. Corresponding angles are also congruent: $\angle A \cong \angle A'$, $\angle B \cong \angle B'$, $\angle C \cong \angle C'$, and $\angle D \cong \angle D'$. The image is held rigid in size and shape but not necessarily in position or orientation.

Note that the distance between all corresponding points must be the same. In Figure 5.2, simply requiring the four sides of rectangle *ABCD* to maintain their length is not sufficient to ensure rigid motion. Parallelogram *ABCD* with angles not equal to 90° could still have $AB \cong A'B'$, $BC \cong B'C'$, $CD \cong C'D'$, and $DA \cong D'A'$.

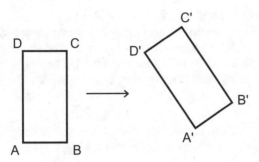

Figure 5.2 A rigid motion transformation

Quantities or properties that remain unchanged after a transformation are described as **invariant** or **preserved**. In Figure 5.2, side lengths, angle measures, and area are all invariant. Slope is not invariant since the slopes of corresponding sides $\overline{AB}$ and $\overline{A'B'}$ are not equal. Note, however, that parallelism is invariant because any pair of sides that are parallel in the preimage will still be parallel in the image.

Some properties that remain invariant under any rigid motion include:

■ Length
■ Angle
■ Parallelism
■ Area

Figure 5.3 illustrates three specific types of transformations that are rigid motions. Each will be discussed in more detail.

Line Reflection: flip   Translation: slide   Rotation: turn

Figure 5.3 Rigid motion transformations

Similarity Transformations

A transformation in which all corresponding lengths are in the same **ratio** and all corresponding angles are congruent is called a **similarity** transformation. The basic similarity transformation we will discuss is the **dilation**. Figure 5.4 illustrates a dilation with a ratio, or scale **factor** of $\frac{1}{2}$.

Dilations can be thought of as an enlargement or a reduction of the preimage. The preimage and image after a similarity transformation are not necessarily congruent; they are congruent only if the scale factor equals 1. However, corresponding angles will always be preserved. The result is an image that has the same shape as the preimage but not necessarily the same size.

Figure 5.4 Dilation with ratio equal to $\frac{1}{2}$

Orientation

Orientation of a polygon refers to the direction traveled around a figure when moving along consecutive vertices. The orientation can be clockwise or counterclockwise. In the transformation illustrated in Figure 5.1, the orientation of the preimage is clockwise since one would travel in a clockwise direction to go from vertex A to B to C. However, the orientation of the image is counterclockwise since that is the direction of travel from vertex A' to B' to C'.

The transformation shown in Figure 5.2 preserves orientation. In both the image and preimage, one travels counterclockwise from vertex A to B to C to D and also from A' to B' to C' to D'.

MATH FACTS

A rigid motion or isometry that preserves orientation is called **direct**. A rigid motion or isometry that does not preserve orientation is called **opposite**.

Translation—direct rigid motion

Rotation—direct rigid motion

Reflection—opposite rigid motion

Check Your Understanding of Section 5.1

A. Multiple-Choice

1. Which pair of figures illustrates an opposite rigid motion?

(1)

(3)

(2)

(4)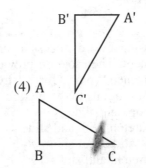

2. A triangle undergoes a rigid motion transformation. Which of the following is *not* necessarily preserved?
   (1) side length   (3) area
   (2) angle measure   (4) orientation

3. Franklin used the reduction feature on a photocopier to make smaller copies of a poster. Which best describes the copies?
   (1) dilation, image is congruent
   (2) dilation, image is not congruent
   (3) reflection, image is congruent
   (4) reflection, image is not congruent

4. Which of the following is *not* a rigid motion?
   (1) rotation   (2) translation   (3) dilation   (4) reflection

5. Which transformation is represented
   by the accompanying figure?
   (1) rotation   (3) dilation
   (2) reflection   (4) translation

6. Which of the following illustrates a transformation that could *not* be a rigid motion?
   (1) The preimage and image are both parallelograms.
   (2) The preimage and image are both rectangles, but the sides have different slopes.
   (3) The preimage and image are both triangles, but the vertices are at different locations.
   (4) The preimage and image are squares with different areas.

7. Which transformation is represented by the accompanying figure?
   (1) dilation
   (2) rotation
   (3) translation
   (4) reflection

8. Jaime is using a word processing program and has inserted a picture into his document. The program provides tools for editing the picture. The tools are named "spin," "flip," "stretch," and "slide." Which tools would apply rigid motions to the picture?
   (1) spin, flip, and slide
   (2) spin, stretch, flip, and slide
   (3) stretch, slide, and spin
   (4) flip, stretch, and slide

B. *Free-Response—show all work or explain your answer completely.*

9. After a certain transformation, the perimeter of a rectangle is unchanged. Must the transformation be an isometry? If so, explain why. If not, sketch a counterexample.

10. In the rigid motion transformation of △*LMN*, name two properties or quantities that are not preserved.

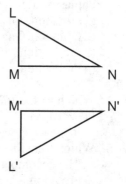

126

11. Jesse ties his dog to a stake in the ground in his backyard, allowing the dog to run around in a circular area. Fifteen minutes later, Jesse moves the stake to a different location. Is the area the dog can run in the same or different? Justify your reasoning.

5.2 PROPERTIES OF TRANSFORMATIONS

Reflections

A **line reflection** flips a figure over a line so that the figure appears as a mirror image. The line is called the **line of reflection**. The notation r_ℓ indicates a reflection over line ℓ.

The line of reflection is always the perpendicular bisector of the segment joining corresponding points in the preimage and image. As a result, every point on the preimage and its corresponding point on the image are equidistant from the line of reflection.

Figure 5.5 illustrates the reflection of point A over line ℓ. Note that line ℓ is the perpendicular bisector of $\overline{AA'}$.

Figure 5.5 Reflection of a point over a line

Figure 5.6 shows the reflection of a quadrilateral. Note that the orientation has changed from clockwise to counterclockwise. A single reflection will always change the orientation.

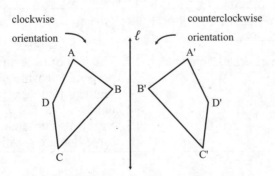

Figure 5.6 Reflection of quadrilateral *ABCD*

127

MATH FACTS

In every reflection, the segments constructed through pairs of corresponding points are parallel. In Figure 5.6, $\overline{AA'} \parallel \overline{BB'} \parallel \overline{CC'} \parallel \overline{DD'}$. The midpoints of each of these segments also lie on the line of reflection. Reflections preserve distance and angle, so they are isometries. However, orientation is not preserved.

Example 1

$\triangle F'G'H'$ is the image of $\triangle FGH$ after a reflection over line AB as shown in the figure. Classify quadrilateral $GG'H'H$. Justify your answer.

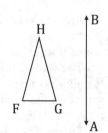

Solution: $\overline{GG'} \parallel \overline{HH'}$ because segments through corresponding pairs of points are parallel.

$\overline{GH} \cong \overline{G'H'}$ because length is preserved. Therefore $GG'H'H$ must be an isosceles trapezoid.

Example 2

Prove segments connecting corresponding points after a reflection are parallel.

Given: $\overline{CD}$ is the reflection of $\overline{AB}$ over line m

Prove: $\overline{AC} \parallel \overline{BD}$

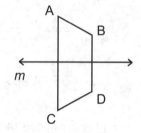

Solution:

Statements	Reasons
1. $\overline{CD}$ is the reflection of $\overline{AB}$ over line m	1. Given
2. $\overline{AC} \perp$ line m, $\overline{BD} \perp$ line m	2. The line of reflection is the perpendicular bisector of the segment joining corresponding points
3. $\overline{AC} \parallel \overline{BD}$	3. Segments perpendicular to the same line are parallel

Translations

A **translation** is a transformation that simply slides, or moves, a figure from one position to another. Translations can be specified using a **vector** with the notation $T_{\overrightarrow{FG}}$. Figure 5.7 shows pentagon *ABCDE* translated to *A'B'C'D'E'* by vector $\overrightarrow{FG}$. A vector consists of a magnitude and a direction, as illustrated by vector $\overrightarrow{FG}$. The magnitude is represented by the length of the vector. In this case, the length is *FG*. The direction is indicated by following the endpoint, point *F*, toward the arrow. Pentagon *ABCDE* is translated a distance of *FG* parallel to the direction of $\overrightarrow{FG}$.

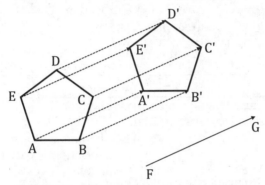

Figure 5.7 Vector definition of a translation

=== MATH FACTS ===

In every translation, the segments constructed through pairs of corresponding points are both parallel and congruent. For example in Figure 5.7, *AA'* ‖ *BB'* ‖ *CC'* ‖ *DD'* ‖ *EE'*. Also, *AA'* ≅ *BB'* ≅ *CC'* ≅ *DD'* ≅ *EE'*. Translations preserve distances and angles, so translations are an isometry. They also preserve parallelism and slope.

Example 1

Segment $\overline{MN}$ is translated by vector $\overrightarrow{FG}$ to $\overline{M'N'}$, *MN* = 4, and *FG* = 6. What can you conclude about the lengths of *M'N'*, *MM'*, and *NN'*? Explain your reasoning.

Solution: *MN* = *M'N'* = 4 because translations are isometries that preserve length. *MM'* = *NN'* = 6 because every point on $\overline{MN}$ is translated by the distance of $\overline{FG}$, which is 6.

Example 2

In example 1 above, classify quadrilateral $MNN'M'$. Justify your answer.

   Solution: $\overline{MN} \parallel \overline{M'N'}$ because translations preserve parallelism. $\overline{MM'} \parallel \overline{NN'}$ because each point on $\overline{MN}$ is translated in the same direction. Therefore both pairs of opposite sides are parallel. So $MNN'M'$ must be a parallelogram.

Example 3

Given: $\overline{CD}$ is the translation of $\overline{AB}$ by
       vector $\overrightarrow{AC}$

Prove: $m\angle BAC + m\angle ACD = 180°$

   Solution:

Statements	Reasons
1. $T_{\overrightarrow{AC}}(\overline{AB}) = \overline{CD}$	1. Given
2. $\overline{AB} \parallel \overline{CD}$	2. The image of a segment after a translation is parallel to the preimage
3. $\angle BAC$ and $\angle ACD$ are supplementary	3. Same side interior angles formed by parallel lines are supplementary
4. $m\angle BAC + m\angle ACD = 180°$	4. The measures of supplementary angles sum to 180°

Rotations

A rotation is the spinning of a figure about a pivot point called the **center of rotation**. The **angle of rotation** is measured counterclockwise unless otherwise specified. For example, Figure 5.8 shows point P rotated 60° to point P' about the center of rotation at point O. The notation $R_{C,a}$ is used to specify a rotation, where a is the angle of rotation and C is the center of rotation.

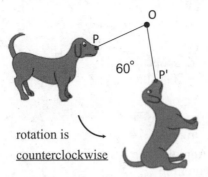

Figure 5.8 Rotation of a figure about a point

MATH FACTS

In a rotation, the angle of rotation found using any pair of corresponding points in the preimage and image will be congruent. In Figure 5.9, $\angle AOA' \cong \angle BOB' \cong \angle COC' \cong \angle DOD'$. In addition, the distances from each of a corresponding pair of points to the center of rotation are equal. In Figure 5.9, $\overline{AO} \cong \overline{A'O}$, $\overline{BO} \cong \overline{B'O}$, $\overline{CO} \cong \overline{C'O}$, and $\overline{DO} \cong \overline{D'O}$.

Figure 5.9 Rotation of trapezoid *ABCD*

Example 1

The accompanying figure could represent which transformations?

(1) translation or rotation (2) translation or reflection
(3) reflection or rotation (4) reflection or dilation

Solution: Choice (3) is correct. The figure could represent a rotation about point *A* or a reflection over line $\overleftrightarrow{AB}$.

Example 2

D'E'F'G'H' is the image of *DEFGH* after a rotation about point *O*. If m∠*DOD'* = 70°, find m∠*OGG'*.

Solution: Since the figure represents a rotation, m∠*GOG'* = 70° and $\overline{OG} \cong \overline{OG'}$. Applying the isosceles triangle theorem to △*GOG'*, m∠*OGG'* is calculated as:

$$m\angle OGG' = \frac{1}{2}(180 - 70) = 55°$$

Dilations

A dilation enlarges or reduces the size of a figure without changing its shape. All angles in the image remain congruent. However, the lengths of segments in the image are proportional to lengths in the preimage. The ratio of lengths is called the scale factor. A dilation is not a rigid motion since the image may not be congruent to the preimage. Dilations are a type of **similarity transformation** because the image is always **similar** to the preimage. Similar figures have the same shape but may differ in size. Similar figures will be discussed in detail in Chapter 8.

Dilations are specified about a center point. $D_{C,r}$ is a dilation with a center point of *C* and a scale factor of *r*. It maps point *P* onto point *P'* such that *P'* lies on the ray $\overrightarrow{CP}$ and $CP' = r \cdot (CP)$. Figure 5.10 illustrates $D_{C,2}(\overline{PQ}) = \overline{P'Q'}$.

CP' = 2CP, CQ' = 2CQ

P'Q' = 2PQ

Figure 5.10 Dilation $\overline{PQ}$ of about *C* with scale factor 2

Check Your Understanding of Section 5.2

A. Multiple-Choice

1. Which of the following transformations always preserves slope?
   (1) translation
   (2) rotation
   (3) reflection
   (4) glide reflection

2. In the accompanying figure, $\triangle A'B'C'$ is the image of $\triangle ABC$ after a translation. Which of the following is *not* necessarily true?
   (1) $\overline{BB'} \cong \overline{CC'}$
   (2) $\overline{AA'} \parallel \overline{BB'}$
   (3) $\overline{AB} \perp \overline{AA'}$
   (4) $\overline{AB} \cong \overline{A'B'}$

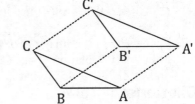

3. In the accompanying figure, $\triangle DFE$ is the image after $R_{O,80}(\triangle ABC)$. Which of the following must be true?
   (1) $m\angle A \cong 80°$
   (2) $m\angle BOF = 80°$
   (3) $m\angle A \cong m\angle C$
   (4) $m\angle ABE = 80°$

4. $\overline{GO}$ is the reflection of $\overline{EW}$ over line *m*. Which of the following is *not* necessarily true?
   (1) the midpoint of $\overline{GE}$ lies on line *m*
   (2) $\overline{GE} = \overline{OW}$
   (3) $\overline{OW} \perp$ line *m*
   (4) $\overline{GO} \cong \overline{EW}$

5. $\triangle MNP$ is the rotation of $\triangle ZXY$
through point P. Which of the
following must be true?
(1) $YM = XN$
(2) $YZ = ZX$
(3) $MN = YZ$
(4) $YZ = MP$

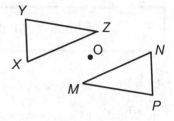

6. Pentagon $ABCDE$ is the image of pentagon $VWXYZ$ after $D_{X,3}$. Which of
the following must be true?
(1) $VW = 3AB$ (3) $CV = 3VA$
(2) $CA = 3XV$ (4) $m\angle A = 3(m\angle V)$

7. Given $\overline{PQ}$ is the image of $\overline{HG}$
after a $110°$ rotation about O,
and $m\angle HOG = 45°$, what is the
measure of $\angle GOP$?
(1) $25°$
(2) $45°$
(3) $65°$
(4) $70°$

8. Which transformation maps $\triangle ABC$
onto $\triangle A'B'C'$?
(1) $R_{A,90}$
(2) $R_{B,90}$
(3) $r_{\overline{AC}}$
(4) $T_{\overline{CC'}}$

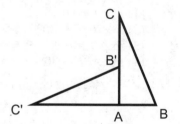

9. Given the following angle measures,
which point could be the center of
rotation that maps $\triangle ABC$ onto
$\triangle A'B'C'$?

$m\angle AXA' = 30°$ $m\angle BYB' = 40°$
$m\angle CXC' = 30°$ $m\angle BZB' = 45°$
$m\angle CYC' = 35°$ $m\angle CZC' = 48°$

(1) X
(2) Y
(3) Z
(4) none of these

10. In regular hexagon *ABCDEF*, which transformations would map △*AQF* to △*DQC*?

(1) $T_{\overline{QD}}$
(2) reflect through point *Q*
(3) $r_{\overline{BE}}$
(4) $R_{Q,60°}$

11. If $AP \cong PB \cong PC \cong PD$, which justification explains why the same rotation that maps *A* to *B* will also map *C* to *D* in the accompanying figure?

(1) points *A, B, C,* and *D* are all equidistant from *P*, and ∠*BPC* and ∠*DPA* are congruent vertical angles
(2) points *A, B, C,* and *D* are all equidistant from *P*, and ∠*APB* and ∠*CPD* are congruent vertical angles
(3) $AB \cong CD$, and ∠*APB* and ∠*BPC* are a supplementary linear pair
(4) points *A, B, C,* and *D* are all equidistant from *P*, and ∠*APB* and ∠*BPC* are a supplementary linear pair

B. *Free-response—show all work or explain your answer completely.*

12. In the figure, ∠*RPS* measures 30° and is rotated 160° about point *P*. ∠*R'PS'* is the image after the rotation. Find the measure of ∠*SPR'*.

13. Describe fully the transformation that maps $\triangle XYZ$ to $\triangle XY'Z'$.

14. When an equilateral triangle is reflected over one of its sides, what type of figure is formed? Sketch an example.

15. Given point Y is the image of point X after a reflection over $\overline{PZ}$, explain why $\angle PXZ \cong \angle PYZ$.

16. Shaina states that whenever a segment is reflected over a line, the two endpoints of the image and the two endpoints of a preimage form a trapezoid. Is she correct? Justify your reasoning.

5.3 TRANSFORMATIONS IN THE COORDINATE PLANE

KEY IDEAS

Transformations can map one point to another on the coordinate plane. Specific rules exist for finding the new coordinates given the coordinates of the preimage.

Reflections Over a Line in the Coordinate Plane

To reflect a point over any horizontal or vertical line, flip the point over the line so the distances from point to line are equal. If the line is horizontal or vertical, this can be done by counting grid units on a graph as shown in Figure 5.11. Point B is 2 units to the left of the line $x = -1$. So after $r_{x=-1}$, point B' is located 2 units to the right of $x = -1$. The same is done for point A.

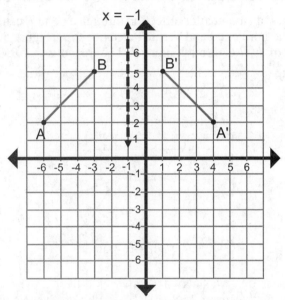

Figure 5.11 Reflection of $\overline{AB}$ over $x = -1$

Algebraically, the change in the x- or y-coordinate between the preimage and the line of reflection can be calculated and applied to locate the image on the other side of a horizontal or vertical line of reflection.

A set of rules can also be applied to the special cases of reflections in the coordinate axes, $y = x$ or $y = -x$:

- $r_{x\text{-axis}}(x, y) \rightarrow (x, -y)$
- $r_{y\text{-axis}}(x, y) \rightarrow (-x, y)$
- $r_{y=x}(x, y) \rightarrow (y, x)$
- $r_{y=-x}(x, y) \rightarrow (-y, -x)$

Example

Point $R(7, 3)$ is reflected over the line $y = x$. Find the coordinates of the image R'.

   Solution: The rule for a reflection over the line $y = x$ is $(x, y) \rightarrow (y, x)$, so we switch the two coordinates. R' has coordinates $(3, 7)$.

Reflections Through a Point in the Coordinate Plane

The coordinates of a point after reflection through a point P can be calculated by counting the horizontal and vertical distance to P. Figure 5.12 illustrates the reflection of $\triangle ABC$ through point P. The image is $\triangle A'B'C'$.

Figure 5.12 Reflection through a point

   The horizontal distance from A to P is $(0 - (-5)) = 5$. The vertical distance is $(-1 - (-4)) = 3$. Since P must be the midpoint of $\overline{AA'}$, A' must be located the same horizontal and vertical distance from P. So we add those distances to the coordinates of P. The coordinates of A' are $(0 + 5, -1 + 3)$, or $A'(5, 2)$. The same procedure is repeated to find B' and C'.

   A reflection through a point is also equivalent to a 180° rotation, which has the rule $(x, y) \rightarrow (-y, x)$. You can confirm this rule with the coordinates shown in Figure 5.12.

Translations in the Coordinate Plane

In the coordinate plane, the notation $T_{h,k}$ or $(x, y) \rightarrow (x + h, y + k)$ is used for translations. The variables h and k represent the distance the preimage is translated along the x-axis and y-axis, respectively. The signs of h and k indicate the direction of the slide:

$h > 0$ for positive x-direction	$h < 0$ for negative x-direction
$k > 0$ for positive y-direction	$k < 0$ for negative y-direction

In Figure 5.13, $\triangle A'B'C'$ is the image of $\triangle ABC$ under the translation $T_{4,-2}$. The image is translated 4 units right (positive x-direction) and 2 units down (negative y-direction).

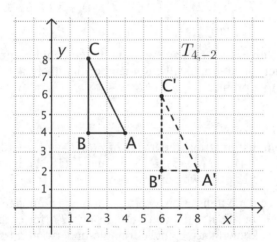

Figure 5.13 Translation of $\triangle ABC$

The coordinates of the image can be determined by counting grid units from the preimage in the appropriate x- and y-directions or can be calculated algebraically. Use the translation in Figure 5.13 as an example. The coordinates of $\triangle A'B'C'$ are calculated as follows:

$$T_{4,-2} \, A(4, 4) = A'(4 + 4, 4 - 2) = A'(8, 2)$$

$$T_{4,-2} \, B(2, 4) = B'(2 + 4, 4 - 2) = B'(6, 2)$$

$$T_{4,-2} \, C(2, 8) = B'(2 + 4, 8 - 2) = B'(6, 6)$$

When given a preimage and an image in the coordinate plane, you can determine the translation that was applied by simply counting the distance translated on the graph in each direction. You can also calculate the differences in the x-coordinates and y-coordinates between the two images.

Example 1

The vertices of $\triangle MNP$ have coordinates $M(-5, 2)$, $N(1, 2)$, $P(0, 5)$. Find the coordinates of N' and P' if vertex M is mapped to $M'(3, 6)$.

Solution: Solving this problem requires two steps. First determine the translation function. Then apply it to find N' and P'.

$$T_{h,k} \, M(-5, 2) = M'(-5 + h, 2 + k) = M'(3, 6)$$

• We know that

$$-5 + h = 3 \qquad 2 + k = 6$$
$$h = 8 \qquad k = 4$$

The translation is therefore $T_{8,4}$. N' and P' can now be found:

$$T_{8,4} \, N(1, 2) = N'(1 + 8, 2 + 4) = N'(9, 6)$$
$$T_{8,4} \, P(0, 5) = P'(0 + 8, 5 + 4) = P'(8, 9)$$

As an alternative, the preimage and the image could be graphed and the translation determined graphically. Remember that there is often more than one approach to a problem.

Example 2

After a certain transformation, the image of $\triangle BAD$ with coordinates $B(7, 1)$, $A(0, 3)$, and $D(12, 4)$ is $B'(13, 3)$, $A'(6, 5)$, and $D'(18, 1)$. Is the transformation a translation? Explain why or why not.

Solution: Under a translation, every point in the preimage undergoes the same change in coordinates. If we find the same $T_{h,k}$ for each pair of coordinates, then we know that the entire triangle underwent a translation. Calculate each h and k:

$$B'(13, 3) = B(7 + h, 1 + k) \qquad 13 = 7 + h, \, h = 6 \qquad 3 = 1 + k, \, k = 2$$
$$A'(6, 5) = A(0 + h, 3 + k) \qquad 6 = 0 + h, \, h = 6 \qquad 5 = 3 + k, \, k = 2$$
$$D'(18, 1) = D(12 + h, 4 + k) \qquad 18 = 12 + h, \, h = 6 \qquad 1 = 4 + k, \, k = -3$$

The transformation is not a translation because point D does not undergo the same translation as points B and A. An alternative approach would be to graph the two triangles and show that they are not congruent.

Rotations in the Coordinate Plane

The coordinates after rotations about the origin, $R_{\text{origin},a}(P)$ specifies a counterclockwise rotation of point P centered about the origin. If the center of rotation is not specified, it is assumed to be the origin.

The coordinates after a rotation about the origin in increments of 90° are found using the following rules:

- $R_{90}(x, y) = (-y, x)$
- $R_{180}(x, y) = (-x, -y)$
- $R_{270}(x, y) = (y, -x)$

Multiple rotations of 90° can be applied to achieve rotations of 180° and 270°. (Apply the rotation twice for R_{180} and 3 times for R_{270}.) After a rotation of 360°, the image is coincident with the preimage and the coordinates remain unchanged.

=========== **MATH FACTS** ===========

A graphical alternative to the algebraic rules for rotations is to graph the image and then rotate your paper by the specified rotation. The *x*-axis is now the *y*-axis and vice versa. Read the new coordinates, rotate the paper back to the 0° orientation, and plot the new points.

Example

Given △*MTV* with vertices *M*(1, 4), *T*(5, 6), and *V*(7, 1), graph and label triangles △*MTV*, △*M'T'V'* = R_{90}(△*MTV*), and △*M"T"V"* = R_{270}(△*MTV*).

Solution:

By applying the rule for R_{90}, we find *M'*(−4, 1), *T'*(−6, 5), and *V'*(−1, 7).

By applying the rule for R_{270}, we find *M"*(4, −1) *T"*(6, −5), and *V"*(1, −7).

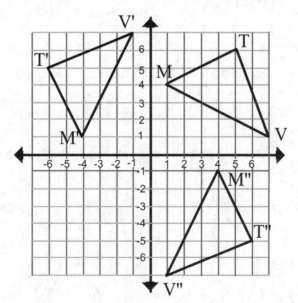

Dilations in the Coordinate Plane

To find the image of a point after a dilation about the origin, simply multiply each coordinate by the scale factor.

Example 1

Sketch $\triangle ABC$ with vertices $A(2, 1)$, $B(3, 3)$, and $C(0, 2)$. Sketch and state the coordinates of $\triangle A'B'C'$, the image after $D_3(\triangle ABC)$.

Solution: Multiply each coordinate by 3.

$A(2, 1) \rightarrow A'(6, 3)$ $\qquad$ $B(3, 3) \rightarrow B'(9, 9)$ $\qquad$ $C(0, 2) \rightarrow C'(0, 6)$

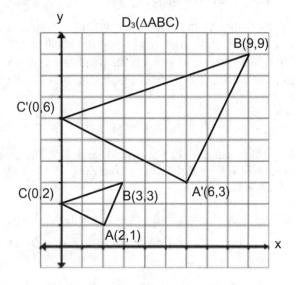

Example 2

Given $\overline{AB}$ with coordinates $A(4, 1)$ and $B(2, 3)$, find the coordinates of A' and B' after D_2 centered at $C(1, 2)$.

Solution: The horizontal distance from C to A is 3 units. The vertical distance is -1.

Scaling by a factor of 3 gives 9 units horizontally and -3 units vertically. These are counted from point C:

$$A'(1 + 9, 2 - 3)$$
$$A'(10, -1)$$

Repeat for point *B*. The horizontal distance from *C* to *B* is 1 unit. The vertical distance is 1 unit. Scaling by a factor of 3 gives 3 units horizontally and 3 units vertically.

$$B'(1 + 3, 2 + 3)$$

$$B'(4, 5)$$

The coordinates are $A'(10, -1)$ and $B'(4, 5)$.

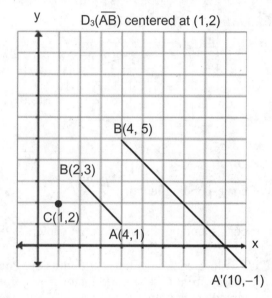

If you are given the graph of a preimage and an image, you can calculate the scale factor in one of two ways:

- Find the ratio of two corresponding lengths in the image and preimage.
- Find the ratio of the distances to the center of dilation from two corresponding points.

Example

Given $\overline{DR}$ and $\overline{D'R'}$ which is the image after a dilation about point C, write three different expressions for the scale factor of the dilation. Evaluate the ratio.

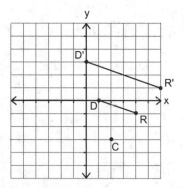

Solution:

Using the ratio of corresponding lengths: $\dfrac{D'R'}{DR}$

Using the ratio of corresponding distances to the center: $\dfrac{D'C'}{DC}$ or $\dfrac{R'C}{RC}$

From point C, D' is located 2 units left and 6 units up. Point D is located 1 unit left and 3 units up from C. The scale factor is therefore 3. The same value would be obtained using points R and C.

MATH FACT

Two examples of transformations that are neither rigid motions nor similarity transformations are the horizontal stretch and the vertical stretch. A horizontal stretch elongates a figure only in the horizontal direction. On the coordinate plane, the *x*-coordinate of every point is multiplied by a scale factor. A vertical stretch does the same thing to only the *y*-coordinate.

preimage   vertical stretch   horizontal stretch

Check Your Understanding of Section 5.3

A. Multiple-Choice

1. What is the image of point $P(5, -1)$ after a rotation of 180° about the origin?
   (1) (1, 5)   (3) (−1, 5)
   (2) (1, −5)   (4) (−5, 1)

2. Given $H(7, 2)$, what are the coordinates of H' where $T_{2,6}(H) = H'$?
   (1) (5, −4)   (2) (−5, −4)   (3) (−9, −8)   (4) (9, 8)

3. $K' = r_{y\text{-axis}}(K)$. What are the coordinates of K' if the coordinates of K are (−3, 5)?
   (1) (−3, −5)   (2) (−3, 5)   (3) (3, 5)   (4) (3, −5)

4. What is the image of (3, 12) after a dilation centered at the origin with a scale faction of 3?
   (1) (9, 36)   (2) (6, 15)   (3) (1, 4)   (4) (0, 9)

5. The image of point L after translation $(x, y) \rightarrow (x + 4, y - 3)$ is $L'(8, 1)$. What are the coordinates of point L?
   (1) (4, −2)   (2) (4, 4)   (3) (12, 2)   (4) (12, 4)

6. The coordinates of $\overline{JK}$ are $J(1, 5)$ and $K(4, 6)$. The coordinates of $\overline{J'K'}$ after a reflection through $P(2, -1)$ are
   (1) $J'(3, -7)$ and $K'(0, -8)$
   (2) $J'(0, 11)$ and $K'(3, 2)$
   (3) $J'(3, 4)$ and $K'(6, 5)$
   (4) $J'(-1, 7)$ and $K'(5, 6)$

7. D' is the image of D after a reflection over the x-axis. What are the coordinates of D' if the coordinates of D are (−8, 1)?
   (1) (−8, −1)   (2) (−8, 1)   (3) (8, −1)   (4) (8, 1)

8. What is the image of (10, −2) after a reflection over the line $y = x$?
   (1) (2, −10)   (2) (−10, 2)   (3) (−2, 10)   (4) (2, 10)

9. Which transformation would map $Z(-6, 8)$ to $Z'(-3, 4)$?
   (1) $T_{3,4}$ (2) $T_{-3,-4}$ (3) $T_{-3,4}$ (4) $D_{\frac{1}{2}}$

10. The image of $(5, 1)$ after a translation is $(2, 8)$. What is the image of $(6, 7)$ after the same translation?
    (1) $(2, 1)$ (2) $(9, 0)$ (3) $(3, 14)$ (4) $(8, 8)$

11. What transformation will map $A(1, 6)$ to $A'(6, 1)$?
    (1) $r_{y\text{-axis}}$ (2) $r_{y = x}$ (3) $R_{\text{origin},90}$ (4) $R_{\text{origin},180}$

12. In the accompanying figure, $\overline{GH}$ is the image of $\overline{MN}$ after a dilation about point Q. Which of the following represents the scale factor of the dilation?

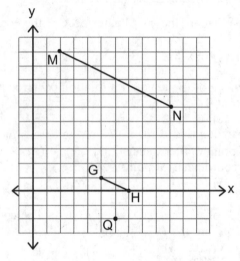

   (1) $\dfrac{MQ}{GQ}$ (2) $\dfrac{GQ}{MQ}$ (3) $\dfrac{MQ}{GH}$ (4) $\dfrac{HQ}{MQ}$

B. *Free-Response—show all work or explain your answer completely.*

13. In the accompanying figure, identify the single transformation that maps each of the preimages to its image.
    (A) $ABCD \rightarrow A'B'C'D'$ (B) $EFG \rightarrow E'F'G'$

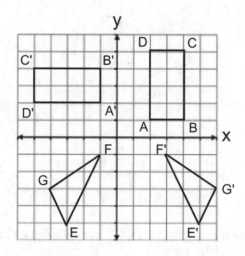

14. In the accompanying figure, identify the single transformation that maps each of the preimages to its image.
    (A) $AB \rightarrow A'B'$ (B) $CDE \rightarrow C'D'E'$ (C) $FGH \rightarrow F'G'H'$

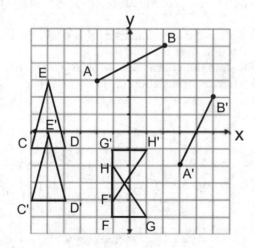

15. Identify the dilation that maps $H(6, 3)$ to $H(3, 1\frac{1}{2})$.

16. The coordinates of *GLOW* are $G(-4, 4)$, $L(-1, 3)$, $O(0, 1)$, and $W(-3, 2)$.
   (A) Graph, label, and state the coordinates of $G'L'O'W'$, the reflection of *GLOW* through the point $(1, 2)$.
   (B) Graph, label, and state the coordinates of $G''L''O''W'' = G'L'O'W'$, the image of $G'L'O'W'$ after a 270° rotation about the origin.

17. $\triangle BIG$ has coordinates $B(1, 2)$, $I(4, 5)$, and $G(6, 1)$. On graph paper:
   (A) Graph and label $\triangle BIG$.
   (B) Graph and label $\triangle B'I'G'$, the image after $R_{90°}(\triangle BIG)$.
   (C) Graph and label $\triangle B''I''G''$, the image after $r_{y=x}(\triangle B'I'G')$.

18. Rectangle *ABCD* has coordinates $A(-3, 1)$, $B(3, 3)$, $C(2, 6)$, and $D(-4, 4)$.
   (A) Graph and state the coordinates of $A'B'C'D'$, the image of *ABCD* after a reflection over the line $y = 1$.
   (B) Graph and state the coordinates of $A''B''C''D''$, the image of $A'B'C'D'$ after the translation $(x, y) \rightarrow (x + 4, y - 2)$.

5.4 SYMMETRY

KEY IDEAS

A figure has line symmetry if it can be folded in half and every point on one half maps onto a point on the second half. A figure has rotational symmetry if the figure can be rotated by an angle less than 360° and every point on the rotated image maps to a point in the preimage.

Line Symmetry

A figure has line symmetry if it can be reflected over a line that maps one half of the figure exactly onto the other half. The line is called the **line of symmetry** and divides the figure into two congruent parts. A figure can have zero, one, or more lines of symmetry. Figure 5.14 to Figure 5.17 illustrate shapes with 1, 2, 3, and 0 lines of symmetry.

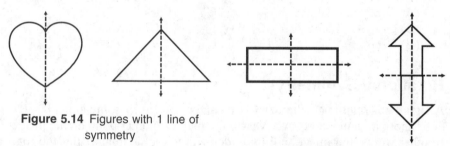

Figure 5.14 Figures with 1 line of symmetry

Figure 5.15 Figures with 2 lines of symmetry

Figure 5.16 Figure with 5 lines of symmetry

Figure 5.17 Figures with no lines of symmetry

MATH FACT

The number of lines of symmetry in a regular polygon is always the same as the number of sides.

Example 1

How many lines of symmetry are found in a regular hexagon?

(1) 1   (2) 3   (3) 6   (4) 12

Solution: (3) 6 lines of symmetry

Example 2

Which of the following letters has more than one line of symmetry?

(1)   (2)   (3)   (4)

Solution:

(3)

Rotational Symmetry

A figure has rotational symmetry if it can be rotated by some angle $0 < \theta < 360°$ about its center and have every point map to another point on the image. In other words, the image will look identical to the preimage after the rotation. Of course, the image after each of these rotations is congruent to the preimage since rotations are rigid motions.

Figure 5.18 shows rectangle *ABCD* after rotations of 0°, 90°, 180°, 270°, and 360°. Rotations of 0°, 180°, and 360° result in figures that look identical to the original. We say the figure has 180° rotational symmetry. The 0° and 360° rotations are not considered rotational symmetries. Note that the

Figure 5.18 Rectangle *ABCD* after rotations of 0°, 90°, 180°, 270°, and 360°

location of individual points has changed. After a 180° rotation, point *A* is mapped to point *C*, point *B* is mapped to point *D*, point *C* is mapped to point *A*, and point *D* is mapped to point *B*.

Some figures have more than one rotation that results in an identical figure. Figure 5.19 shows 120° and 240° rotational symmetry in an equilateral triangle.

Figure 5.19 Rotational symmetry in an equilateral triangle

Every figure will look identical after a 0° and 360° rotation about its center. This is why the rotation must be greater than 0° and less than 360° for a figure to have rotational symmetry. Rotations of 0° and 360° are always identity transformations—the image is always identical to the preimage.

MATH FACT

A regular polygon with *n* sides always has rotational symmetry, with rotations in increments equal to its central angle of 360°/*n*.

Example 1

By how many degrees must a regular octagon be rotated so that it maps onto itself? List all the rotations less than 360° that will map an octagon onto itself.

Solution: $n = 8$ for an octagon; $\dfrac{360}{n} = \dfrac{360}{8} = 45°$

Rotations of 45°, 90°, 135°, 180°, 225°, 270°, and 315°

Although the preimage and image look identical after a symmetry rotation, the position of individual points will differ.

Example 2

By how many degrees must pentagon
ABCDE be rotated about its center to map
point *A* to point *C*?

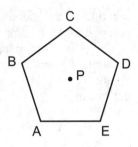

Solution: $\dfrac{360}{n} = \dfrac{360}{5} = 72°$

Since rotations are counterclockwise, the
pentagon must be rotated three increments of
72°, which equals 216°, to map point *A* to point *C*.

Check Your Understanding of Section 5.4

A. Multiple-Choice

1. Which letter has rotational symmetry?

(1)   (2)   (3)   (4)

2. What is the smallest angle of rotation that would map a hexagon onto
   itself?
   (1) 45°   (2) 60°   (3) 120°   (4) 180°

3. Which of the following figures has rotational symmetry?
   (1) equilateral triangle   (3) obtuse triangle
   (2) scalene triangle   (4) right triangle

4. Which of the following figures show exactly two lines of symmetry?

(1)   (3)

(2)   (4)

5. By how many degrees must the regular octagon be rotated to map point X to point Y?

(1) 45°   (3) 135°

(2) 90°   (4) 225°

6. By how many degrees must the regular decagon be rotated to map point R to point S?

(1) 36°   (3) 216°

(2) 144°   (4) 324°

7. Which word has a horizontal line of symmetry?

(1) **HIDE**   (2) **BOSS**   (3) **BAT**   (4) **TOOT**

8. If a regular pentagon is rotated 288° about its center P, then vertex G maps to vertex Q. What is the measure of the acute angle $\angle GPQ$?

(1) 36°   (2) 72°   (3) 84°   (4) 88°

B. Free-Response—show all work or explain your answer completely.

9. Sketch all lines of symmetry in the accompanying figure.

10. Danny and Sue are riding a Ferris wheel at a carnival. The Ferris wheel has 20 cars and spins at a rate of 6 degrees per second. Danny and Sue are in separate cars that are three positions apart. Danny's car arrives at the top of the Ferris wheel first. How many seconds later will Sue's car be at the top?

11. Sketch an example of a figure that has one line of symmetry but does not have rotational symmetry.

12. Jessie states that any triangle with line symmetry must be equilateral, but Fiona disagrees. Who is correct? Justify your answer.

5.5 COMPOSITIONS OF RIGID MOTIONS

Any of the transformations can be applied in succession, which is called a composition. The output from the first transformation becomes the input into next. For some compositions of transformations, the order in which they are applied does not matter. For some other compositions, though, the final image depends on the order in which the transformations are applied.

Compositions

A composition of transformations is a sequence of transformations with a specified order. The first transformation is applied to the preimage. The resulting image becomes the preimage of the next transformation. Multiple transformations can be chained together in this way.

Three different notations for compositions may be used. The composition symbol, ∘, may be used to separate a pair of transformations. The transformation on the right must be done first. Alternatively, transformations may be nested in parentheses. The innermost transformation must be performed first. Finally, the composition can be stated in words.

$T_{\overline{RS}} \circ R_{C,90°}\left(\overline{AB}\right)$ means rotate $\overline{AB}$ 90° about point C and then translate it by vector $\overrightarrow{RS}$.

$D_{C,2}\left(r_{\overrightarrow{PQ}}\left(\overline{AB}\right)\right)$ means reflect $\overline{AB}$ over ray $\overrightarrow{PQ}$ and then dilate through C by a scale factor of 2.

The above two compositions appear in Figure 5.20. The notations A' and B' are the result of the first transformation. The notations A'' and B'' are for the second.

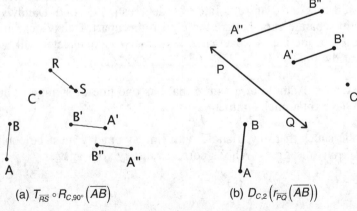

(a) $T_{\overline{RS}} \circ R_{C,90°}\left(\overline{AB}\right)$ (b) $D_{C,2}\left(r_{\overrightarrow{PQ}}\left(\overline{AB}\right)\right)$

Figure 5.20

Example 1

Which composition of transformations is represented?

(1) $R_{P,180°}(r_m(\triangle ABC))$

(2) $r_m(R_{P,180°}(\triangle ABC))$

(3) $r_m(r_{B'C'}(\triangle ABC))$

(4) $R_{P,90°}(r_m(\triangle ABC))$

Solution: (1) $\triangle A'B'C'$ is the reflection of $\triangle ABC$ over m, and $\triangle A''B''C''$ is the rotation of $\triangle A'B'C'$ by 180°.

Example 2

$\triangle RST$ has coordinates $R(6, 2)$, $S(4, -1)$, and $T(2, 3)$. Graph and label $\triangle R''S''T''$, the image of $\triangle RST$ after a reflection over the line $x = 1$ followed by a 180° rotation about the origin. State the coordinates of $\triangle R''S''T''$.

Solution:

$r_{x=1}(6, 2) \to (-4, 2)$ $R_{180°}(-4, 2) = (4, -2)$

$r_{x=1}(4, -1) \to (-2, -1)$ $R_{180°}(-2, -1) = (2, 1)$

$r_{x=1}(2, 3) \to (0, 3)$ $R_{180°}(0, 3) = (0, -3)$

• The coordinates are $R''(4, -2)$, $S''(2, 1)$, and $T''(0, -3)$.

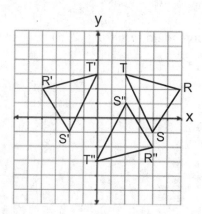

For some compositions, the order of transformations does not matter. The final image looks the same after any order of the transformations. For others, the final image will depend on the order specified.

Example 3

Point P has coordinates $(5, 1)$. Does $T_{4,2} \circ r_{y\text{-}axis}(P)$ map P to the same point as $r_{y\text{-}axis} \circ T_{4,2}(P)$?

Solution:

Using $T_{4,2} \circ r_{y\text{-}axis}(P)$: $r_{y\text{-axis}}(5, 1) \to (-5, 1)$ and $T_{4,2}(-5, 1) \to (-1, 3)$

Using $r_{y\text{-}axis} \circ T_{4,2}(P)$: $T_{4,2}(5, 1) \to (9, 3)$ and $r_{y\text{-axis}}(9, 3) \to (-9, 3)$

The compositions map point P to two different points.

MATH FACT

Compositions of transformations are similar in concept to transformations of functions you may have studied in algebra. When given two functions, $f(x) = x^2$ and $g(x) = x + 2$, we can consider the two compositions $f(g(x))$ and $g(f(x))$. In the first, the output of g becomes the input of f. In the second, the output of f becomes the input of g, If x equals 3, then $f(g(x)) = f(5) = 25$ and $g(f(x)) = g(9) = 11$. The order matters in this case. However, you can confirm for yourself that the order does not matter if $f(x) = x + 2$ and $g(x) = x + 4$.

Check Your Understanding of Section 5.5

A. *Multiple-Choice*

1. Which composition could result in $(12, 4)$ mapping to $(5, 1)$?
   (1) $D_{\frac{1}{4}}(r_{y=x})$
   (2) $D_{\frac{1}{4}}(T_{2,0})$
   (3) $R_{90°}(T_{9,-11})$
   (4) $T_{9,-11}(R_{90°})$

2. Which composition of transformations would *not* fit the two puzzle pieces together?

(1) a reflection of *HEFG* over *GF*, followed by a translation that maps *G* to *A*

(2) a reflection of *HEFG* over *GF*, followed by a translation that maps *F* to *A*

(3) a translation of *HEFG* that maps *H* to *B*, followed by a reflection over *HE*

(4) a reflection of *HEFG* over *HE*, followed by a translation that maps *H* to *B*

3. Which composition could result in the following preimage and image?

(1) dilation of $\frac{1}{2}$ followed by a 180° rotation

(2) rotation of 180° followed by a dilation of 2

(3) dilation of $\frac{1}{2}$ followed by a reflection over a vertical line

(4) translation of 5 units right followed by a dilation of $\frac{1}{2}$

4. Which composition could map (5, 1) to (1, 3)?

(1) $T_{-6,2}(r_{x\text{-axis}})$ (3) $R_{90°}(T_{2,2})$

(2) $r_{x\text{-axis}}(T_{6,2})$ (4) $T_{6,2}(r_{y\text{-axis}})$

157

5. Rayquan is designing a website logo using a computer graphics program. He uses the program's enlarge feature to triple the logo's size. Then he drags the logo horizontally 2 inches right and vertically 3 inches up. Which composition represents the change in the logo?

(1) $T_{2,3} \circ D_{\frac{1}{3}}$

(2) $T_{2,3} \circ D_3$

(3) $T_{3,2} \circ D_{\frac{1}{3}}$

(4) $D_2 \circ T_{3,2}$

6. When equilateral triangle $\triangle ABC$ undergoes a reflection over $\overline{AB}$ followed by a 240° rotation about B, which corresponding pairs of vertices are coincident?

(1) A, A'' and B, B'' only

(2) B, B'' and C, C'' only

(3) C, C'' and A, A'' only

(4) A, A'', B, B'', and C, C''

B. Free-Response—show all work or explain your answer completely.

7. The figure represents a rotation of 120° about point C followed by a reflection over line m. Sketch line m and point C in the figure.

8. The vertices of $\triangle BIG$ have coordinates $B(6, 1)$, $I(3, 2)$, and $G(7, 5)$. Graph and label $\triangle B''I''G''$, the image of $\triangle BIG$ after a 90° rotation about the origin followed by a reflection over the line $y = x$. State the coordinates.

9. The vertices of $\triangle PMN$ have coordinates $P(4, 0)$, $M(3, 3)$, and $N(6, 5)$. Graph and label $\triangle P''M''N''$, the image of $\triangle PMN$ after a reflection over the line $x = 3$ followed by the translation $T_{-4,1}$.

10. A dilation with a scale factor of $\dfrac{1}{2}$ followed by a reflection over the x-axis maps $\triangle ZED \rightarrow \triangle Z''E''D''$. If $\triangle ZED$ has coordinates $Z(0, 4)$, $E(-4, 2)$, and $D(-8, 6)$, find the coordinates of Z'', E'', and D''.

11. Rectangle $FGHK$ has coordinates $F(-7, 3)$, $G(-6, 5)$, $H(-2, 3)$, and $K(-3, 1)$. $F''G''H''K''$ is the image of $FGHK$ after the composition after a 90° rotation about the origin followed by a reflection through the origin. Graph and label $F''G''H''K''$.

Chapter Six · TRIANGLE CONGRUENCE

KEY IDEAS

When proving two triangles are congruent, we do not need to prove all three pairs of sides and all three pairs of angles congruent. There are five shortcuts commonly used—the SAS, SSS, ASA, AAS, and HL criteria. Each criterion requires only three pairs of parts to be congruent for us to conclude the triangles are congruent.

Criterion	What It Looks Like
SAS—two pairs of sides and the included angle are congruent	
SSS—three pairs of sides are congruent	
ASA—two pairs of angles and the included side are congruent	
AAS—two pairs of angles and the nonincluded side are congruent	
HL—the hypotenuses and one pair of legs are congruent in a right triangle	

6.1 THE TRIANGLE CONGRUENCE CRITERION

The SAS Criterion

Figure 6.1 shows two triangles with two pairs of corresponding sides congruent and the pair of **included angles** congruent. The included angle is the angle whose vertex is the shared endpoint of the two congruent sides. Having only these three pairs of corresponding parts congruent is sufficient to prove the triangles are congruent. We call this the SAS criterion.

- SAS Criterion

Two triangles are congruent if two pairs of corresponding sides and the pair of included angles are congruent.

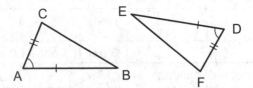

Figure 6.1 Triangles with SAS congruence

It is important to determine where the angle lies in relation to the sides when using SAS. If the angle is not the included angle, then we do not have SAS and the triangles may not be congruent. Having two sides and the nonincluded angle allows a motion that is **not rigid.** Figure 6.2 shows how a triangle with two sides congruent and the nonincluded angle congruent could move in a nonrigid manner. Side $\overline{BC}$ can pivot out, with side $\overline{AC}$ extended to complete the triangle. $\overline{AC}$ can extend because its length was not specified, and $\angle B$ can change since it was not specified either. The new configuration is now obtuse, whereas the original figure was acute.

Figure 6.2 A nonrigid triangle with parts angle-side-side congruent

Example 1

Which of the following pairs of triangles can be proven congruent using the SAS criterion?

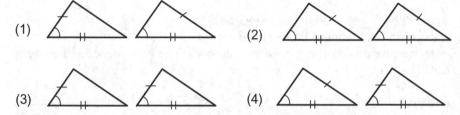

(1)

(2)

(3)

(4)

Solution: Choice (3) is the only choice in which both triangles have a pair of sides and the included angle labeled congruent.

A congruence statement between a pair of triangles will specify the pairs of corresponding parts through the order in which the vertices are listed.

Three angle congruences

$\triangle ABC \cong \triangle DEF$

Three side congruences

$\triangle ABC \cong \triangle DEF$

$\triangle ABC \cong \triangle DEF$

Example 2

Write a congruence statement for the following pair of triangles.

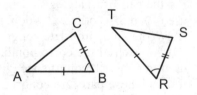

Solution: The triangles are congruent by the SAS criterion. The congruence statement $\triangle ABC \cong \triangle TRS$ will match corresponding parts in the correct order.

Often additional information is needed beyond sides and angles explicitly marked on the figure. Vertical angles, shared sides, midpoints, and parallel lines show up frequently.

Example 3

Can the two triangles be proven to be congruent? If so, state which criterion is used. Write a congruence statement.

Solution: The triangles are congruent by SAS, with $\overline{BC}$ congruent to itself by the reflexive property as the second pair of sides. The congruence statement is $\triangle ABC \cong \triangle DBC$.

Example 4

What one other piece of information must be provided before one can conclude that the two triangles are congruent by SAS?

(1) $\angle AEB \cong \angle CED$
(2) E is the midpoint of $\overline{AD}$
(3) E is the midpoint of $\overline{BC}$
(4) $\overline{AB} \cong \overline{CD}$

Solution: (3) The midpoint would allow you to conclude that $\overline{BE} \cong \overline{EC}$. Along with the vertical angles $\angle AEB \cong \angle CED$ and the marked sides $\overline{AE} \cong \overline{ED}$, there is enough information to conclude the triangles are congruent by SAS.

Example 5

For the pair of triangles shown, demonstrate the use of the SAS criteria by doing the following.
(A) Write an appropriate congruence statement.
(B) For each of the corresponding pairs of parts used to establish congruence, give a justification for why those parts are congruent.
Given: E is the midpoint of $\overline{AC}$, $\overline{AB} \| \overline{CD}$, $\overline{AB} \cong \overline{CD}$

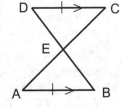

Solution:

$\triangle ABE \cong CDE$

$\overline{AB} \cong \overline{CD}$ is given information.

$\angle A \cong \angle C$ because alternate interior angles formed by parallel lines are congruent.

$\overline{AE} \cong \overline{CE}$ because a midpoint divides a segment into two congruent segments.

162

The SSS, ASA, and AAS Criterion

Besides SAS, the criteria SSS, ASA, and AAS are sufficient to prove two triangles congruent. As with SAS, the relative position of the sides and angles matters for ASA and AAS. In ASA, the side must be the included side; its endpoints must be the vertices of the two angles. In AAS, the side is not the included side. Figure 6.3 illustrates the SSS, ASA, and AAS criterion.

Figure 6.3 Triangles congruent by the SSS, ASA, and AAS criterion

---

MATH FACT

Most combinations of 3 angles and sides are sufficient to prove two triangles congruent. The two exceptions are ASS and AAA. As mentioned in the previous section, ASS allows for two different triangles to be formed from the specified parts using a nonrigid motion. AAA is not a congruent criterion because it allows for multiple triangles to be formed from the specified parts as well, actually an infinite number of possible triangles. As we will see in Chapter 8, AAA is one of the similarity criteria. Any triangle with the same shape will have all three angles congruent. However, the triangles could be enlargements or reductions of one another.

Example 1

For each pair of triangles, determine if they can be proven congruent. If so, state the criterion.

(A)   (B)   (C)   (D)

Solution:

(A) AAS   (B) SSS   (C) ASA   (D) SAS

You can apply any applicable angle or segment relationships to help show that corresponding parts of triangles are congruent.

Example 2

For each pair of triangles, identify which congruence criterion could be used to prove the triangles congruent. Explain why each of the pairs of parts used are congruent.

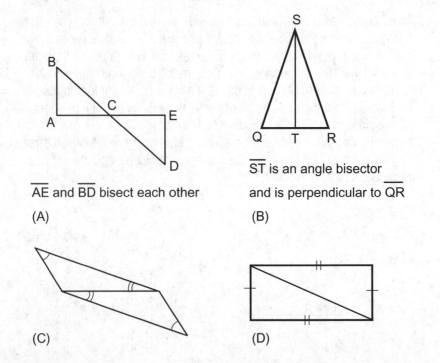

$\overline{AE}$ and $\overline{BD}$ bisect each other

(A)

$\overline{ST}$ is an angle bisector and is perpendicular to $\overline{QR}$

(B)

(C)

(D)

Solution:

(A) SAS: $\angle BCA \cong \angle DCE$ by vertical angles. $\overline{BC} \cong \overline{CD}$ and $\overline{AC} \cong \overline{CE}$ because C is a midpoint that forms two pairs of congruent segments.

(B) ASA: $\angle QST \cong \angle RST$ because an angle bisector forms two congruent angles. $\overline{ST} \cong \overline{ST}$ by the reflexive property. $\angle QTS \cong \angle RTS$ because perpendicular lines form congruent right angles.

(C) AAS: Two pairs of congruent angles are given. The included side is congruent to itself by the reflexive property.

(D) SSS: Two pairs of congruent sides are given. The third side is congruent to itself by the reflexive property.

Justification of the Triangle Congruence Criterion

The SAS criterion can be proven using rigid motions. Assuming the only three given congruent parts, we can show that there exists a composition of rigid motions that will map $\triangle DEF$ onto $\triangle ABC$, shown in Figure 6.4a. First translate the $\triangle DEF$ by vector $\overrightarrow{DA}$ so that the vertices D and A (the vertices with the congruent angles) coincide as shown in Figure 6.4b.

Next, rotate about point D by $\angle FDC$, as shown in 6.4c, so that $\overline{DF}$ coincides with $\overline{AC}$. We know this mapping is possible because $\overline{DF} \cong \overline{AC}$. Finally, reflect $\triangle DEF$ over $\overline{DF}$ so that E maps to B as shown in 6.4d. Point E must map to B because $\overline{DE} \cong \overline{AB}$. $\triangle DEF$ is mapped to $\triangle ABC$, and the triangles are congruent.

(a)   (b)

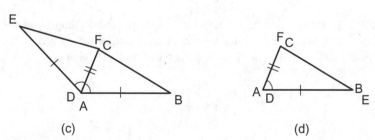

(c)   (d)

Figure 6.4 Justification of the SAS criterion for triangle congruence

The justification for the ASA criterion is similar to that for the SAS criterion. The justification for the SSS criterion differs. Given $\triangle ABC$ and $\triangle DEF$ with $\overline{AB} \cong \overline{DE}$, $\overline{BC} \cong \overline{EF}$, and $\overline{AC} \cong \overline{DF}$ as shown in Figure 6.5a, we can translate $\triangle DEF$ such that points D and A coincide. Then rotate so that $\overline{DF}$ coincides with $\overline{AC}$ as shown in Figure 6.5b. Because we have no given congruent angles, the final part of the justification is somewhat different than for SAS and ASA. Construct diagonal $\overline{BE}$ as shown in Figure 6.5c. $\triangle EAB$ is isosceles, making $\angle DEB \cong \angle DBE$. $\triangle ECB$ is also isosceles, making $\angle CEB \cong \angle CBE$. The SAS criterion now applies and the triangles are congruent.

Figure 6.5 Justification of the SSS criterion for triangle congruence

The Hypotenuse-Leg Criterion

The hypotenuse-leg (HL) criterion applies only to right triangles. When given two right triangles, if a pair of legs are congruent and the two hypotenuses are congruent, then the triangles are congruent by HL. Figure 6.6 shows two triangles congruent by HL.

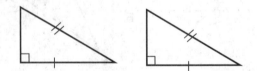

Figure 6.6 The HL criterion for triangle congruence

The HL criterion can be justified by applying the Pythagorean theorem and showing that the remaining legs in each triangle are congruent to each other. Since the two right angles are congruent, the triangles must be congruent by the SAS criterion.

Check Your Understanding of Section 6.1

A. Multiple-Choice

1. Which triangle congruence criterion can be
   used to prove the two triangles are congruent?
   (1) SAS   (3) AAS
   (2) ASA   (4) SSS

   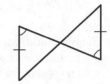

2. Which triangle congruence criterion can be
   used to prove the two triangles are
   congruent?
   (1) SAS   (3) AAS
   (2) ASA   (4) SSS

3. Which triangle congruence
   criterion can be used to prove
   the two triangles are
   congruent?
   (1) SAS   (3) AAS
   (2) ASA   (4) SSS

   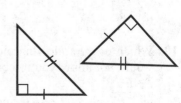

4. Which triangle
   congruence criterion can
   be used to prove the two
   triangles are congruent?
   (1) SAS   (3) AAS
   (2) ASA   (4) HL

5. Which triangle congruence
   criterion can be used to prove
   the two triangles are congruent?
   (1) SAS   (3) SSS
   (2) ASA   (4) HL

6. What would have to be true about $\overline{UN}$ to be able to prove $\triangle TUN \cong \triangle RUN$?
   (1) $\overline{UN}$ is a perpendicular bisector
   (2) $\overline{UN}$ is an angle bisector
   (3) $\overline{UN}$ is a median
   (4) $\overline{UN}$ is an altitude

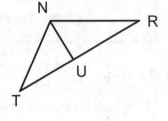

7. What additional piece of information would be needed to show that the two triangles are congruent by the HL criterion?
   (1) $\overline{GI} \cong \overline{HI}$
   (2) J is the midpoint of $\overline{GH}$
   (3) $\overline{JI}$ bisects $\angle GIH$
   (4) $\overline{IJ} \cong \overline{GI}$

8. Barbara is justifying why the SAS criterion for triangle congruence is valid. She started with two triangles, $\triangle MNP$ and $\triangle RST$, having $\angle M \cong \angle R$, $MN \cong RS$, and $MP \cong RT$. She then translated one of the triangles so that vertex M coincides with R. What would be her next step?
   (1) dilate $\triangle RST$ through center M by a scale factor of 1
   (2) translate $\triangle RST$ by vector $\overrightarrow{MP}$
   (3) reflect $\triangle RST$ through point R
   (4) rotate $\triangle RST$ about R so that $\overline{RS}$ coincides with $\overline{MN}$

9. If $\overline{WX} \parallel \overline{YZ}$, what additional piece of information would be needed to prove that $\triangle WPX \cong \triangle ZPY$ by AAS?
   (1) $\angle WPX \cong \angle ZPY$
   (2) $\overline{WP} \cong \overline{PY}$
   (3) $\overline{WP} \cong \overline{PZ}$
   (4) $\overline{WP} \cong \overline{PX}$

10. Which of the following pieces of information would *not* allow you to conclude that $\triangle ADC \cong \triangle ABC$?
    (1) $\overline{AC}$ bisects $\angle BCD$ and $\overline{CD} \cong \overline{CB}$
    (2) $\overline{AC}$ bisects $\angle BCD$ and $\overline{AD} \cong \overline{AB}$
    (3) $\overline{AC}$ bisects $\angle BCD$ and $\angle BAD$
    (4) $\overline{AC}$ bisects $\angle BAD$ and $\overline{AD} \cong \overline{AB}$

B. *Free-Response—show all work or explain your answers completely.*

11. Give a sequence of 3 rigid motions that will map $\triangle XYZ$ to $\triangle ABC$. Explain how this sequence of rigid motions demonstrates the SAS criterion for triangle congruence.

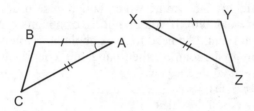

12. Gloria sketches two right triangles with one pair of corresponding legs and one pair of corresponding hypotenuses marked congruent. She knows the triangles are congruent by the HL criterion. She then concludes that SSA must also be a valid congruence criterion for all triangles because the only parts marked congruent on these triangles are two sides and the nonincluded angle. How would you explain to Gloria why her reasoning is incorrect?

6.2 PROVING TRIANGLES CONGRUENT

KEY IDEAS

Two triangles can be proven to be congruent using the congruence postulates—SSS, SAS, ASA, AAS, and HL. Each of the required pairs of corresponding parts must be shown to be congruent before the triangles can be stated to be congruent.

Using the Congruence Postulates

We can write a formal proof showing that two triangles are congruent using one of the 5 congruence postulates. Given information and information inferred from the figure are used to show that each of the two or three corresponding pairs of parts are congruent.

A good strategy will help you write a good proof. The following steps will help you consistently write a good proof.

1. *Know the material*—Be familiar with the congruence postulates and the "tools" of geometry. The tools include vocabulary as well as angle and segment relationships. The better you understand the tools of geometry, the easier you will find writing a proof.

2. *It all starts with the figure*—Sketch a figure if no figure is provided. Then use the given information and anything you can infer from the information to mark your figure with any relevant congruent markings, parallel markings, and right angle marking. These marks are crucial to helping you determine which congruence postulate to use.

3. *Plan a strategy*—Determine which congruence postulate to use. The markings should help. If you do not have enough corresponding parts marked congruent, you need to think about how you can use the given information to conclude other parts are congruent. Look for the following:
   a. Vertical angles
   b. Shared sides and angles
   c. Linear pairs
   d. Supplementary and complementary angles
   e. Parallel lines and perpendicular lines
   f. Midpoints and bisectors of segments
   g. Angle bisectors
   h. Altitudes, medians, and perpendicular bisectors in triangles
   i. Isosceles triangle theorem, exterior angle theorem, and triangle angle sum theorem

4. *If you are missing parts*—If you are still missing congruent parts, try focusing on the part or parts that would complete the postulate. Be careful not to come up with an invalid justification just because it would let you complete the proof!

5. *Write your statements and reasons*—Remember these few important rules as you write your statements and reasons:
   a. Statements are specific to the proof. This is where you state facts about named points, lines, segments, angles, triangles, and so on.
   b. Reasons never mention a point, line, segment, angle, triangle, and so forth. Reasons can only be given as postulates, theorems, and definitions.
   c. If you cannot come up with a good reason, do not use the statement.
   d. The order sometimes matters. If one statement depends on another statement, the dependent statement should come after.

6. *How to tell when you are done*—You must have one statement in your proof showing congruence for each of the corresponding pairs of parts specified by the congruence postulate. Remember that you cannot finish with the triangle congruence statement until each part of the postulate is proven. One strategy is to label with an (S) or (A) each line that proves a pair of required sides or angles congruent. So if you are using SAS, there should be two lines labeled (S) and one line

labeled (A). Another strategy is to write out your postulate and make checkboxes under each part. Check off the boxes as each required line is written.

S	A	S
✓	✓	✓

7. As a final check, keep in mind that a good triangle congruence proof should have:
   a. All required statements to demonstrate congruence
   b. No statements that cannot be justified
   c. No statement that does not help achieve the desired proof
      i. It is not incorrect to include an unnecessary statement as long as the statement is true. However, unnecessary lines in your proof will only provide an opportunity for error.

Example 1

Given: $\overline{BC}$ bisects $\overline{AD}$ at E, $\angle A \cong \angle D$
Prove: $\triangle ABE \cong \triangle DCE$

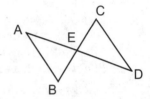

Solution: Mark the figure with the corresponding parts we know are congruent. The given bisector lets us conclude that $\overline{AE} \cong \overline{ED}$. From congruent vertical angles, we know $\angle AEB \cong \angle DEC$. Finally, we have the given congruent pair $\angle A \cong \angle D$. From the marked figure, we see that the ASA postulate applies. When writing the proof, we state that E is a midpoint before concluding that sides $\overline{AE}$ and $\overline{ED}$ are congruent.

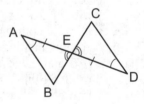

Statements	Reasons
1. $\overline{BC}$ bisects $\overline{AD}$ at E	1. Given
2. E is the midpoint of $\overline{AD}$	2. A bisector intersects a segment at its midpoint
3. $\overline{AE} \cong \overline{ED}$	3. A midpoint divides a segment into two congruent segments
4. $\angle AEB \cong \angle DEC$	4. Vertical angles are congruent
5. $\angle A \cong \angle D$	5. Given
6. $\triangle ABE \cong \triangle DCE$	6. ASA

MATH FACT

Not all correct proofs are equal. Writing proofs has its own art and style. Some proofs are more elegant than others. For example, some proofs have all the given statements written at the beginning of the proof. Others will be guided by the postulates, placing given statements in the order they are needed. Another example of a good style feature is to have any statements that indicate addition or subtraction appear in consecutive lines.

Example 2

Given: $\overline{IJ} \parallel \overline{GH}$, $\overline{FH} \parallel \overline{GJ}$, $\overline{FH} \cong \overline{GJ}$
Prove: $\triangle FHG \cong \triangle GJI$

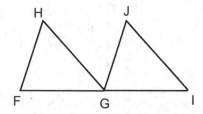

Solution:

• Strategy—Mark the two pairs of parallel lines with parallel markings. Then look for corresponding or alternate interior angles. $\angle JIG$ and $\angle HGF$ are congruent corresponding angles, as are $\angle HFG$ and $\angle JGI$. After marking all these congruent pairs, we see that the AAS postulate applies.

Statements	Reasons
1. $\overline{IJ} \parallel \overline{GH}$, $\overline{FH} \parallel \overline{GJ}$	1. Given
2. $\angle GIJ \cong \angle FGH$, $\angle IGJ \cong \angle GFH$ (A) (A)	2. Alternate interior angles formed by parallel lines are congruent
3. $\overline{FH} \cong \overline{GJ}$ (S)	3. Given
4. $\triangle FHG \cong \triangle GJI$	4. AAS

Example 3

Given: $\angle AWQ \cong \angle WAQ$, Q is the
midpoint of $\overline{PS}$, $\overline{PW} \perp \overline{SP}$,
$\overline{SA} \perp \overline{PS}$

Prove: $\triangle WPQ \cong \triangle ASQ$

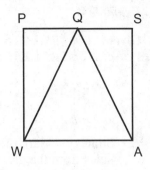

Solution:

- Strategy—Mark the figure with
  congruent markings, including
  $\overline{PQ} \cong \overline{SQ}$ because of the bisector. We
  have only a pair of congruent sides and
  right angles. However, $\triangle WQA$ has two
  congruent angles. So the triangle must
  be isosceles and $\overline{WQ} \cong \overline{AQ}$. The HL
  postulate now applies.

Statements	Reasons
1. $\angle AWQ \cong \angle WAQ$	1. Given
2. $\overline{WQ} \cong \overline{AQ}$	2. Sides opposite congruent angles in a triangle are congruent
3. Q is the midpoint of $\overline{PS}$	3. Given
4. $\overline{PQ} \cong \overline{SQ}$	4. A midpoint divides a segment into two congruent segments
5. $\overline{PW} \perp \overline{SP}$, $\overline{SA} \perp \overline{PS}$	5. Given
6. $\angle P$ and $\angle S$ are right angles	6. Perpendicular lines form right angles
7. $\triangle WPQ$ and $\triangle ASQ$ are right triangles	7. Triangles with a right angle are right triangles
8. $\triangle WPQ \cong \triangle ASQ$	8. HL

Linear pairs or complementary angles that involve a congruent pair of angles sometimes lead us to a second congruent pair of angles. We can justify the second congruence statement with "angles supplementary/complementary to congruent angles are congruent."

Example 4

Given: $\overline{AD} \cong \overline{BE}$, $\overline{AC} \cong \overline{BC}$
Prove: $\triangle DAC \cong \triangle EBC$

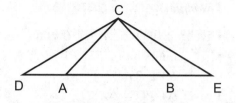

Solution:

- Strategy—Mark the figure. We can see the two pairs of congruent sides from the givens. The middle triangle is isosceles, so we can mark $\angle BAC \cong \angle ABC$. These angles are not part of the triangles we want to prove congruent. However, they form linear pairs with $\angle DAC$ and $\angle EBC$, which lets us conclude that $\angle DAC \cong \angle EBC$. The SAS postulate can be used.

Statements	Reasons
1. $\overline{AD} \cong \overline{BE}$, $\overline{AC} \cong \overline{BC}$	1. Given
2. $\angle BAC \cong \angle ABC$	2. Angles opposite congruent sides in a triangle are congruent
3. $\angle DAC$ and $\angle BAC$ are a linear pair $\angle EBC$ and $\angle ABC$ are a linear pair	3. Definition of a linear pair
4. $\angle DAC$ and $\angle BAC$ are supplementary $\angle EBC$ and $\angle ABC$ are supplementary	4. Linear pairs are supplementary
5. $\angle DAC \cong \angle EBC$	5. Angles supplementary to congruent angles are congruent
6. $\triangle DAC \cong \triangle EBC$	6. SAS

Overlapping Triangles

Certain techniques are helpful when the triangles you are trying to prove congruent overlap one another.

- Look for shared parts.
- Look for applications of the partition postulate, particularly along sides or angles where the triangles overlap.
- Sketch the figure with the triangles separated. Marking the separated figures can help you identify shared parts and the appropriate congruence postulate. If the two corresponding parts in the separated sketch have the same name, then they are shared and must be congruent. Angles, however, may be shared and congruent even if the points used to name the angle differ. The vertex must be the same. However, the second point for each ray may differ, as long as they lie on the same ray. In Figure 6.7, $\angle ABE$ and $\angle CBD$ are shared congruent angles.
- Highlight each triangle in a different color. Doing this sometimes makes it easier to formulate a strategy.

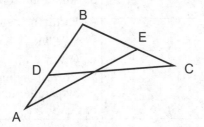

Figure 6.7 Overlapping triangles with a shared angle

Example 1

Given: $\overline{AP} \cong \overline{CQ}$, $\angle D \cong \angle R$, and $\overline{AD} \parallel \overline{PR}$

Prove: $\triangle ACD \cong \triangle PQR$

Solution:

- Strategy—Show $\overline{AC} \cong \overline{PQ}$ using the addition property and the partition postulate.

$\overline{AP} + \overline{PC} \cong \overline{PC} + \overline{CQ}$ leads to $\overline{AC} \cong \overline{PQ}$. The parallel segments gives congruent corresponding angles. We can use the AAS postulate.

Statements	Reasons
1. $\overline{AP} \cong \overline{CQ}$	1. Given
2. $\overline{PC} \cong \overline{PC}$	2. Reflexive property
3. $\overline{AP} + \overline{PC} \cong \overline{PC} + \overline{CQ}$	3. Addition property
4. $\overline{AC} \cong \overline{PQ}$ (S)	4. Partition postulate
5. $\angle D \cong \angle R$ (A)	5. Given
6. $\overline{AD} \parallel \overline{PR}$	6. Given
7. $\angle CAD \cong \angle QPR$ (A)	7. Corresponding angles formed by parallel lines are congruent
8. $\triangle ACD \cong \triangle PQR$	8. AAS

Example 2

Given: $\overline{UR} \cong \overline{VR}, \overline{TU} \cong \overline{SV}$
Prove: $\triangle RVT \cong \triangle RUS$

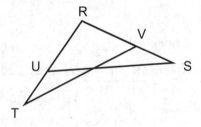

Solution: Strategy—Sketch the triangles separately to help identify the shared parts. $\angle R$ is a shared angle. Also use segment addition to show $\overline{RT} \cong \overline{RS}$. The marked congruent parts indicate SAS.

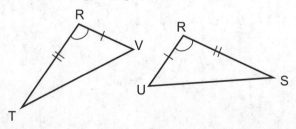

176

Statements	Reasons
1. $\overline{UR} \cong \overline{VR}$ (S)	1. Given
2. $\overline{TU} \cong \overline{SV}$	2. Given
3. $\overline{TU} + \overline{UR} \cong \overline{SV} + \overline{VR}$	3. Addition property
4. $\overline{TR} \cong \overline{SR}$ (S)	4. Partition postulate
5. $\angle R \cong \angle R$ (A)	5. Reflexive property
6. $\triangle RVT \cong \triangle RUS$	6. SAS

Overlapping angles can be handled in the same way as overlapping sides. With angles, however, a separated sketch is even more important as it can be difficult to visualize the two triangles when both angles and sides overlap. Of course, just because sides or angles overlap does not mean you will have to use segment or angle addition on those parts. In the following example, you need to use only angle addition.

=== **Math Fact** ===

Just because you sketch overlapping figures separately doesn't mean you should not mark the original figure as well. Sometimes marking both is helpful. You may not see certain relationships like isosceles triangles, vertical angles, or linear pairs in the separated figure.

Example 3

Given: $\angle RTO \cong \angle STA$, $\overline{RT} \cong \overline{ST}$
Prove: $\triangle RTA \cong \triangle STO$

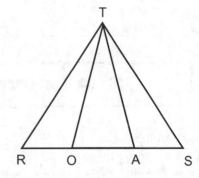

Solution:

• Strategy—Sketch the separated triangles, and mark both the original and separated figures. The markings on the original show us that △RTS is isosceles; so ∠R ≅ ∠S. Then show ∠RTA ≅ ∠STO by angle addition. The separated sketch indicates that ASA will be used.

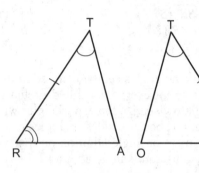

Statements	Reasons
1. ∠RTO ≅ ∠STA	1. Given
2. ∠OTA ≅ ∠OTA	2. Reflexive property
3. ∠RTO + ∠OTA ≅ ∠STA + ∠OTA	3. Addition property
4. ∠RTA ≅ ∠STO	4. Partition postulate
5. $\overline{RT} \cong \overline{ST}$	5. Given
6. ∠R ≅ ∠S	6. Angles opposite congruent sides in a triangle are congruent
7. △RTA ≅ △STO	7. ASA

Check Your Understanding of Section 6.2

A. Multiple-Choice

1. Which statement would be found in the following proof?

   Given: $\overline{LK} \cong \overline{KM}$, $\overline{JK}$ bisects ∠LKM
   Prove: △JLK ≅ △JMK
   (1) ∠LJM ≅ ∠MJK
   (2) ∠LKJ ≅ ∠MKJ
   (3) $\overline{JL} \cong \overline{JM}$
   (4) $\overline{JK}$ is a median

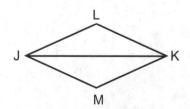

2. Anthony is trying to prove $\triangle PQR \cong \triangle PQS$. Which set of given information would allow him to justify this statement using the ASA postulate?
   (1) $\angle PQR \cong \angle PQS$ and $\angle QPR \cong \angle QPS$
   (2) $\angle R \cong \angle S$ and $\angle PQR \cong \angle PQS$
   (3) $\angle R \cong \angle S$ and $\angle QPR \cong \angle QPS$
   (4) $\overline{RQ} \cong \overline{SQ}$ and $\angle R \cong \angle S$

B. *Free-Response—show all work or explain your answers completely.*

3. Given: $\overline{NO}$ is a median of $\triangle CRN$,
   $\qquad \overline{CN} \cong \overline{RN}$
   Prove: $\triangle CON \cong \triangle RON$

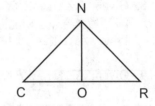

4. Given: $\overline{AU}$ and $\overline{EO}$ bisect each other at I
   Prove: $\triangle AEI \cong \triangle UOI$

5. Given: $\overline{RY} \cong \overline{YZ}, \overline{YZ} \cong \overline{ZB}, \angle E \cong \angle O$,
   $\qquad \angle B \cong \angle R$
   Prove: $\triangle REY \cong \triangle BOZ$

6. Given: $\overline{PT} \parallel \overline{QS}, \overline{TQ} \parallel \overline{SR}, \overline{PT} \cong \overline{QS}$
   Prove: $\triangle PTQ \cong \triangle QSR$

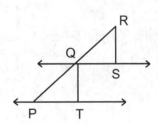

7. Given: $\overline{CD}$ bisects $\angle ACB$, $\angle B \cong \angle A$
   Prove: $\triangle ADC \cong \angle BDC$

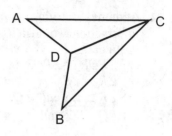

Figure for problems 8 and 9:

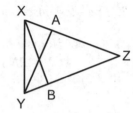

8. Given: $\triangle ZAY$, $\triangle ZBX$, $\overline{XA} \cong \overline{BY}$, $\overline{XB}$ and
   $\overline{YA}$ are altitudes
   Prove: $\triangle XBY \cong \triangle YAX$

9. Given: $\triangle ZAY$, $\triangle ZBX$, $\angle ZXB \cong \angle ZYA$,
   $\overline{AY} \cong \overline{BX}$
   Prove: $\triangle ZAY \cong \triangle ZBX$

10. Given: $\overline{OK}$ is an altitude of $\triangle DCK$, $\overline{DK} \cong \overline{CK}$
   Prove: $\triangle DOK \cong \triangle COK$

11. Given: $\overline{VS}$ and $\overline{UT}$ bisect each
   other at R
   Prove: $\triangle VRU \cong \triangle SRT$

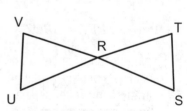

12. Given: $\overline{DY} \cong \overline{CX}$, $ABCD$ is a
   rectangle
   Prove: $\triangle BCY \cong \triangle ADX$

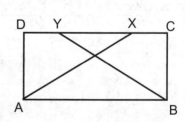

180

13. Given: $\angle A \cong \angle D$, $\overline{EC} \cong \overline{EB}$
Prove: $\triangle DBC \cong \triangle ACB$

14. Given: $\overline{ALB} \parallel \overline{DMC}$, $\overline{LN} \parallel \overline{MO}$, $\overline{AN} \cong \overline{OC}$
Prove: $\triangle ALN \cong \triangle CMO$

15. Given: $\overline{WB} \cong \overline{EB}$, $\overline{WF}$ and $\overline{EX}$ are medians
      of $\triangle WEB$
Prove: $\triangle WEX \cong \triangle EWF$

6.3 CPCTC

KEY IDEAS

Triangle congruence can be used to prove sides or angles are congruent. First identify a pair of triangles that contain the corresponding parts. Then prove the triangles are congruent. Then you can say that any pair of corresponding parts are congruent by CPCTC.

Proofs do not always ask for triangles or other polygons to be proven congruent. You may be asked to prove a pair of sides or angles are congruent or to prove a particular relationship between sides or angles. You can often accomplish this by first proving a pair of polygons or triangles are congruent. Then applying CPCPC or CPCTC. CPCPC states that all corresponding parts of

181

congruent polygons are congruent (CPCPC). The special case for triangles states that all corresponding parts of congruent triangles are congruent (CPCTC).

When writing a proof involving CPCTC or CPCPC, you will need to identify the triangles or polygons to be proven congruent and then write an appropriate congruence statement before applying CPCTC or CPCPC. Be careful to list the vertices in the correct order when writing the congruence statement—the corresponding parts must match in the correct order.

Example

Given: $\overline{ON} \cong \overline{NP}$ and $\overline{OM} \cong \overline{MP}$
Prove: $\angle O \cong \angle P$

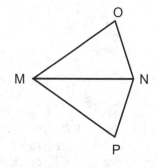

Solution:

- Strategy—first prove $\triangle MON \cong MPN$, then apply CPCTC

Statements	Reasons
1. $\overline{ON} \cong \overline{NP}$ and $\overline{OM} \cong \overline{MP}$	1. Given
2. $\overline{MN} \cong \overline{MN}$	2. Reflexive property
3. $\triangle MON \cong \triangle MPN$	3. SSS
4. $\angle O \cong \angle P$	4. CPCTC

Proving Angle and Segment Relationships

A proof may ask you to show that a segment or an angle has a certain property or is in a certain relationship with another segment or angle. CPCTC used with a congruence statement can help establish the desired relationship.

- To prove a segment is a median—prove it divides the opposite side into congruent segments.
- To prove a segment is an angle bisector—prove it divides the angle into two congruent angles.
- To prove right angles—prove a linear pair of angles are congruent since two angles equal in measure that sum to 180° must measure 90° each.
- To prove a segment is an altitude—prove it forms right angles at the opposite side.

Example 1

Given: $\overline{RQT} \parallel \overline{AP}$, $\overline{AQ} \parallel \overline{DPT}$, $\overline{QR} \cong \overline{PD}$
Prove: $\overline{AT}$ is a median of $\triangle DRT$

Solution: Strategy—a median intersects the opposite side at its midpoint. We need to show $\overline{DA} \cong \overline{AR}$. So the obvious approach would be to show $\triangle DAP \cong \triangle RAQ$. The two pairs of parallel lines give two pairs of congruent corresponding angles, and the given sides let us use AAS. CPCTC then lets us conclude $\overline{DA} \cong \overline{AR}$.

Statements	Reasons
1. $\overline{RQT} \parallel \overline{AP}$ and $\overline{DPT} \parallel \overline{AQ}$	1. Given
2. $\angle ARQ \cong \angle DAP$, $\angle RAQ \cong \angle APD$	2. Corresponding angles formed by parallel lines are congruent
3. $\overline{PD} \cong \overline{RQ}$	3. Given
4. $\triangle DAP \cong \triangle RAQ$	4. AAS
5. $\overline{AD} \cong \overline{AR}$	5. CPCTC
6. A is the midpoint of $\overline{DR}$	6. A midpoint divides a segment into two congruent segments
7. $\overline{AT}$ is a median	7. A median is a segment from a vertex to the midpoint of the opposite side of a triangle

Example 2

Given: $\overline{AP} \cong \overline{AQ}$, $\overline{PY} \cong \overline{YQ}$
Prove: $\overline{AYZ}$ is an angle bisector of $\angle XAR$

Solution:

- Strategy—prove $\triangle PAY \cong \triangle QAY$, then use CPCTC to show that $\angle PAY \cong \angle QAY$.

Statements	Reasons
1. $\overline{AP} \cong \overline{AQ}$, $\overline{PY} \cong \overline{YQ}$ (S), (S)	1. Given
2. $\overline{AY} \cong \overline{AY}$ (S)	2. Reflexive property
3. $\triangle PAY \cong \triangle QAY$	3. SSS
4. $\angle PAY \cong \angle QAY$	4. CPCTC
5. $\overline{AYZ}$ is an angle bisector of $\angle XAR$	5. An angle bisector divides an angle into two congruent angles

Double Congruence Proofs

In some proofs, we may have to prove a first pair of triangles is congruent and then use CPCTC to establish the congruence of parts needed to prove a second pair of triangles is congruent.

Example 1

Given: $\overline{DC} \cong \overline{BC}$, $\overline{AC}$ bisects $\angle DCB$
Prove: $\triangle ADE \cong \triangle ABE$

Solution:

- Strategy—prove $\triangle ADC \cong \triangle ABC$ first, and then use CPCTC to prove $\overline{AD} \cong \overline{AB}$ and $\angle DAE \cong \angle EAB$.

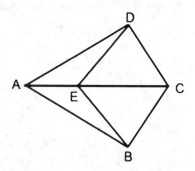

Statements	Reasons
1. $\overline{DC} \cong \overline{BC}$, $\overline{AC}$ bisects $\angle DEB$	1. Given
2. $\angle DCA \cong \angle BCA$	2. An angle bisector divides an angle into two congruent angles
3. $\overline{AC} \cong \overline{AC}$	3. Reflexive property
4. $\triangle ADC \cong \triangle ABC$	4. SAS
5. $\overline{AD} \cong \overline{AB}$	5. CPCTC
6. $\angle DAE \cong \angle BAE$	6. CPCTC
7. $\overline{AE} \cong \overline{AE}$	7. Reflexive property
8. $\triangle ADE \cong \triangle ABE$	8. SAS

Example 2

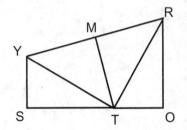

Given: $\overline{TM}$ is a perpendicular bisector of $\triangle TRY$, $\angle S$ and $\angle O$ are right angles, $\overline{YS} \cong \overline{TO}$

Prove: $\triangle YST \cong \triangle TOR$

Solution:

- Strategy—first prove $\triangle TMY \cong \triangle TMR$, and then use CPCTC to show $\overline{TY} \cong \overline{TR}$.

Statements	Reasons
1. $\overline{TM}$ is a perpendicular bisector of $\triangle TRY$	1. Given
2. $\overline{TM} \perp \overline{RY}$, $\overline{TM}$ bisects $\overline{RY}$ at M	2. Perpendicular bisectors are perpendicular to and bisect a segment
3. $\angle TMY$ and $\angle TMR$ are right angles	3. Perpendicular lines intersect at right angles
4. $\angle TMY \cong \angle TMR$	4. Right angles are congruent
5. M is the midpoint of $\overline{RY}$	5. A bisector intersects a segment at its midpoint
6. $\overline{MY} \cong \overline{MR}$	6. A midpoint divides a segment into two congruent segments
7. $\overline{TM} \cong \overline{TM}$	7. Reflexive property
8. $\triangle TMY \cong \triangle TMR$	8. SAS
9. $\overline{TY} \cong \overline{TR}$	9. CPCTC
10. $\angle S$ and $\angle O$ are right angles, $\overline{YS} \cong \overline{TO}$	10. Given

Statements	Reasons
11. $\triangle YST$ and $\triangle TOR$ are right triangles	11. Right triangles have one right angle
12. $\triangle YST \cong \triangle TOR$	12. HL

Check Your Understanding of Section 6.3

Free-Response—show all work or explain your answers completely.

1. Given: $\overline{PD} \cong \overline{FY}$, $\overline{DQ} \cong \overline{YK}$,
  $\overline{PQ} \cong \overline{FK}$
  Prove: $\angle Q \cong \angle K$

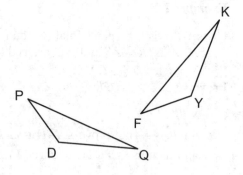

2. Given: $\overline{AC}$ and $\overline{BD}$ intersect at
  E, $\overline{AE} \cong \overline{CE}$, $\overline{BE} \cong \overline{DE}$
  Prove: $\overline{AB} \parallel \overline{CD}$

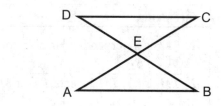

3. Given: $\angle R \cong \angle P$,
  $\angle T \cong \angle S$,
  $\overline{TI} \cong \overline{IS}$
  Prove: I is the midpoint
    of $\overline{RP}$

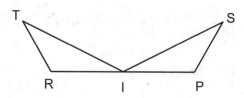

4. Given: J is the midpoint of $\overline{HK}$ and $\overline{IL}$
   Prove: $\angle I \cong \angle L$

5. Given: $\triangle PAT \cong \triangle IAR$
   Prove: $\triangle PTR \cong \triangle IRT$

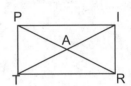

Use for problems 6 and 7:

6. Given: $\triangle PEQ$ is equilateral,
         $\angle PER \cong \angle QER$
   Prove: $\overline{RE}$ bisects $\angle PRQ$

7. Given: $\angle ERP \cong \angle ERQ$, and $\overline{PR} \cong \overline{QR}$
   Prove: $\triangle PEQ$ is isosceles

Use for problems 8 and 9:

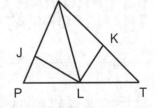

8. Given: $\overline{AL} \perp \overline{PLT}$, $\triangle PLJ \cong \triangle TLK$
   Prove: $\triangle JAL \cong \triangle KAL$

9. Given: $\overline{AL}$ bisects $\angle PAT$, $\angle JLA \cong \angle KLA$,
         $\angle JLP \cong \angle KLT$
   Prove: $\triangle PLJ \cong \triangle TLK$

Use for problems 10 and 11:

10. Given: $\overline{SP} \cong \overline{PQ}$, $\overline{SR} \cong \overline{QR}$
    Prove: $\overline{POR}$ bisects angle $\angle SPQ$

11. Given: $\overline{POR} \perp \overline{SOQ}$, $\overline{SR} \cong \overline{RQ}$
    Prove: $\overline{RO}$ is a median of triangle $\triangle SRQ$

12. Given: $\overline{LT} \parallel \overline{OV}$, $\overline{LT} \cong \overline{LE}$, $\angle E \cong \angle IVO$
Prove: $\overline{OV}$ is an angle bisector of $\triangle IVE$

13. Given: $\overline{OS}$ bisects $\angle CON$ and $\angle CSN$
Prove: $\angle OLN \cong \angle OWC$

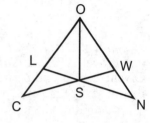

14. Given: $\overline{DMR}$ bisects $\angle LDN$,
$\angle MND \cong \angle MLD$
Prove: $\overline{TLM}$ is an altitude of $\triangle TDR$

6.4 PROVING CONGRUENCE BY TRANSFORMATIONS

KEY IDEAS

Figures can be proven congruent by identifying a sequence of rigid motions that map one onto the other. The angle and segment relationships between preimage, image, lines of reflection, center of rotation, and translation vectors can help establish specific transformations that map one point onto another.

Congruent Triangles

Two triangles are congruent if a sequence of rigid motions maps one triangle to another. Some clues to identifying rigid motions are:

- If the segments joining corresponding vertices of the two triangles are congruent and parallel, then one figure is a translation of the other.
- If a single line is the perpendicular bisector of segments formed by corresponding vertices, then one figure is the reflection of the other.
- If angles formed by corresponding vertices and a center point are all congruent, and corresponding distances to the center point are equal, then one figure is a rotation of the other.

Example 1

Given: $\triangle WXY$ and perpendicular bisector $\overline{YZ}$
Prove: $\triangle WZY \cong \triangle XZY$ using a sequence of rigid motions

Solution:

- Strategy—the perpendicular bisector suggests that $\overline{YZ}$ is a line of reflection. Show that one triangle is the image of the other after a reflection over $\overline{YZ}$.

Proof—Point Y is the image of itself after a reflection over $\overline{YZ}$ because Y lies on the line of reflection. Point Z is the image of itself for the same reason. X is the image of point W after a reflection over $\overline{YZ}$ because a line of reflection is the perpendicular bisector of the segment joining the preimage and the image after the reflection. Therefore, $r_{\overline{YZ}}(\triangle WZY) = \triangle XZY$ and the triangles are congruent because one is mapped to the other by a reflection, which is a rigid motion.

Example 2

Given: $\overline{ADBE}$, $\overline{AD} \cong \overline{BE}$, $\overline{AD} \cong \overline{CF}$, and $\overline{CF} \parallel \overline{AD}$
Prove: $\triangle ABC \cong \triangle DEF$

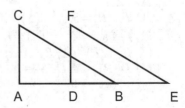

Solution: Point A translates to D by length AD. B translates to E and C translates to F by the same distances in a parallel direction. Therefore $\triangle ABC$ is a translation of $\triangle DEF$, and the triangles are congruent because translations are rigid motions.

Check Your Understanding of Section 6.4

A. Multiple-Choice

1. Fiona is given the following information: $\overline{EBDA}$, $\overline{AD} \cong \overline{BE}$, and F is the image of C after a translation by $\overrightarrow{AD}$ followed by a reflection over $\overline{EA}$. Which of the following justifications could Fiona use to explain why $\triangle ABC \cong \triangle DEF$?

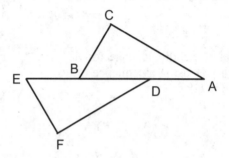

   (1) $T_{\overrightarrow{AD}}$ maps vertex A to vertex D, $T_{\overrightarrow{BE}}$ maps vertex B to vertex E, and $T_{\overrightarrow{CF}}$ maps vertex C to vertex F
   (2) each vertex in $\triangle ABC$ maps to a corresponding vertex in $\triangle DEF$ after a translation by vector $\overrightarrow{CF}$
   (3) each vertex in $\triangle ABC$ maps to a corresponding vertex in $\triangle DEF$ after a translation along vector $\overrightarrow{AD}$ followed by a reflection over $\overleftrightarrow{EA}$
   (4) each vertex in $\triangle ABC$ maps to a corresponding vertex in $\triangle DEF$ after a reflection over $\overleftrightarrow{EA}$ followed by a translation along vector $\overrightarrow{AD}$

2. Which of the following would indicate that $\triangle RST$ must be congruent to $\triangle XYZ$?
   (1) $T_{\overrightarrow{AB}}(R) = X$, $T_{\overrightarrow{BC}}(S) = Y$, $T_{\overrightarrow{CD}}(T) = Z$
   (2) $r_{\overrightarrow{AB}}(R) = X$, $r_{\overrightarrow{AB}}(S) = Y$, $r_{\overrightarrow{AB}}(T) = Z$
   (3) $R_{A,90°}(R) = X$, $R_{A,180°}(S) = Y$, $R_{A,270°}(T) = Z$
   (4) $D_2(R) = X$, $D_2(S) = Y$, $D_2(T) = Z$

B. Free-Response—show all work or explain your answers completely.

3. Given: line *m* is the perpendicular
   bisector of $\overline{FX}$, $\overline{GY}$, and $\overline{HZ}$
   Prove: $\triangle XYZ \cong \triangle FHG$

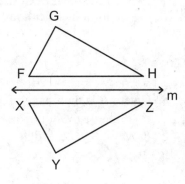

4. Given $\overline{OD}$ is an altitude of $\triangle GLD$ and $OL = OG$, explain in terms of a rigid motion why $\triangle OLD \cong \triangle OGD$.

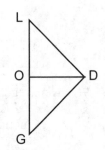

5. Given: *C* is the image of *A* after a rotation
   of 45° about *O*, *D* is the image of *B*
   after a rotation of 45° about *O*
   Prove: $\triangle AOB \cong \triangle COD$

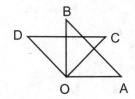

6. Given: *L* is the midpoint of $\overline{DV}$ and $\overline{UI}$
   is congruent and parallel to $\overline{LD}$
   Prove: $\triangle UDL \cong \triangle ILV$ using rigid
   motions

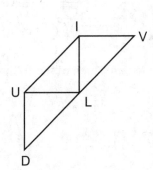

7. Given: I is the midpoint of $\overline{KIH}$ and $\overline{JIG}$
   Prove: $\triangle KIJ \cong \triangle HIG$, using a rigid motion for a justification

8. Given: B is the image of A after a reflection over $\overline{CQ}$
   Prove: $\triangle AQC \cong \triangle BQC$

9. Given: regular decagon P with center P
   Prove: $\triangle PRS \cong \triangle PUT$ using a rigid motion justification

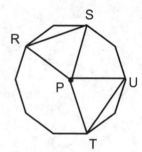

10. Given: $\angle ACO \cong \angle TCG$, $\overline{AC} \cong \overline{CO}$, $\overline{CT} \cong \overline{CG}$
    Prove: $\triangle CAT \cong \triangle COG$

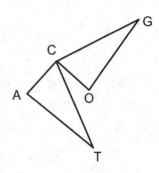

Chapter Seven

GEOMETRY IN THE COORDINATE PLANE

7.1 LENGTH, DISTANCE, AND MIDPOINT

$$\wedge$$
=========== $\overset{}{\text{Key Ideas}}$ ===========
$$\diagdown$$

Coordinate geometry takes the geometric building blocks of point and line and then maps them onto the coordinate plane by assigning each point a coordinate pair (x, y). Given the coordinates of the endpoints of a segment, (x_1, y_1) and (x_2, y_2), the following formulas allow us to calculate the length, midpoint, and point that divides a segment in a given ratio:

$$length = \sqrt{(x_1 - x_2)^2 + (y_1 - y_2)^2}$$

midpoint: $x_{MP} = \dfrac{1}{2}(x_1 + x_2)$, $y_{MP} = \dfrac{1}{2}(y_1 + y_2)$

divide segment in a specified ratio: $ratio = \dfrac{x - x_1}{x_2 - x}$, $ratio = \dfrac{y - y_1}{y_2 - y}$

Brief Review of the Coordinate Plane

Coordinate geometry associates every point with a coordinate pair (x, y), which locates the position of that point in the plane, as shown in Figure 7.1. The coordinates refer to the distance along a pair of perpendicular axes, the x-axis and the y-axis. The plane is divided into four quadrants (I, II, III, and IV) by the x-axis and y-axis. The intersection of these axes is called the origin and is assigned the coordinates $(0, 0)$. The sign of each coordinate refers to a direction from the origin as follows:

- positive x: right
- negative x: left
- positive y: up
- negative y: down

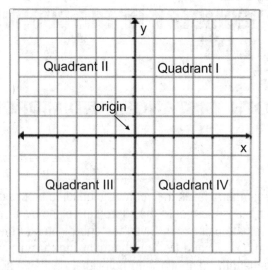

Figure 7.1 The coordinate plane

The Distance Formula

The formula for calculating the distance between any two points is derived from the Pythagorean theorem. Consider two points, A and B, with coordinates $A(x_1, y_1)$ and $B(x_2, y_2)$ as shown in Figure 7.2. To calculate the length $\overline{AB}$, we first complete a right triangle by sketching horizontal segment $\overline{AC}$ and vertical segment $\overline{BC}$. The lengths of these segments are $(x_2 - x_1)$ and $(y_2 - y_1)$. The Pythagorean theorem can now be used to find the distance, d, from A to B.

$$d^2 = a^2 + b^2$$
$$d^2 = (x_2 - x_1)^2 + (y_2 - y_1)^2$$
$$d = \sqrt{(x_1 - x_2)^2 + (y_1 - y_2)^2}$$

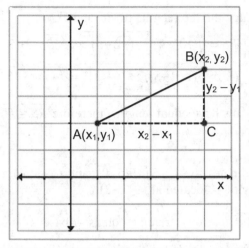

Figure 7.2 Using the Pythagorean theorem to calculate length

Distance Formula

The distance, d, between two points with coordinate (x_1, y_1) and (x_2, y_2) is

$$d = \sqrt{(x_1 - x_2)^2 + (y_1 - y_2)^2}$$

Example 1

Find the length of the segments whose endpoints have coordinates $(1, 2)$ and $(5, 4)$.

Solution:

$$d = \sqrt{(x_1 - x_2)^2 + (y_1 - y_2)^2}$$
$$d = \sqrt{(5-1)^2 + (4-2)^2} = \sqrt{20} = 2\sqrt{5}$$

MATH FACT

An alternative to remembering the distance formula is to complete the right triangle, count the lengths of the horizontal and vertical legs, and use the Pythagorean theorem.

Example 2

Are segments $\overline{RS}$ and $\overline{TU}$ congruent given the coordinates $R(4, 1)$, $S(10, -2)$, $T(-5, 3)$, $U(1, 6)$?

Solution: Calculate the lengths of $\overline{RS}$ and $\overline{TU}$.

$$RS = \sqrt{(10-4)^2 + (-2-1)^2} = \sqrt{45} = 3\sqrt{5}$$

$$TU = \sqrt{(1-(-5))^2 + (6-3)^2} = \sqrt{45} = 3\sqrt{5}$$

Congruent segments have equal lengths, therefore $\overline{RS} \cong \overline{TU}$.

Example 3

$\triangle ABC$ has coordinates $A(-4, -3)$, $B(5, 0)$, and $C(-3, 4)$. $\overline{CD}$ is an altitude of the triangle, with D located at $(-1, -2)$.

Solution: The area of a triangle equals $\dfrac{1}{2}$ base $\times$ height. The height is the length of altitude $\overline{CD}$, and the base is side AB. A sketch helps identify the appropriate side to use for the base. Apply the distance formula:

$$AB = \sqrt{(-4-5)^2 + (-3-0)^2} = \sqrt{90} = 3\sqrt{10}$$

$$CD = \sqrt{(-3-(-1))^2 + (4-(-2))^2} = \sqrt{40} = 2\sqrt{10}$$

$$\text{Area} = \frac{1}{2} \cdot 3\sqrt{10} \cdot 2\sqrt{10} = 30$$

Midpoint

The coordinates of the midpoint of a segment, (x_{MP}, y_{MP}), can be calculated from the coordinates of its endpoints, (x_1, y_1) and (x_2, y_2), using the midpoint formula. Figure 7.3 shows the midpoint of a segment and its coordinates.

Midpoint Formula

$$x_{MP} = \frac{1}{2}(x_1 + x_2), \quad y_{MP} = \frac{1}{2}(y_1 + y_2)$$

Figure 7.3 Midpoint of a segment

Example 1

Find the coordinates of the midpoint of $\overline{PQ}$, given coordinates $P(1, 8)$ and $Q(-5, 2)$.

Solution:

$$x_{MP} = \frac{1}{2}(x_1 + x_2) = \frac{1}{2}(1 + (-5)) = -2$$

$$y_{MP} = \frac{1}{2}(y_1 + y_2) = \frac{1}{2}(8 + 2) = 5$$

The coordinates of the midpoint are $(-2, 5)$.

The midpoint formula can also be used in reverse to calculate the coordinates of an endpoint using the coordinates of the other endpoint and the midpoint.

Example 2

M is the midpoint of $\overline{PQ}$. Find the coordinates of point Q given $P(3, 8)$ and $M(7, -2)$.

Solution:

$$x_{MP} = \frac{1}{2}(x_1 + x_2)$$

$$7 = \frac{1}{2}(3 + x_2)$$

$$14 = 3 + x_2$$

$$x_2 = 11$$

$$y_{MP} = \frac{1}{2}(y_1 + y_2)$$

$$-2 = \frac{1}{2}(8 + y_2)$$

$$-4 = 8 + y_2$$

$$y_2 = -12$$

The coordinates of M are $(11, -12)$.

Example 3

$\overline{DPG}$ is a diameter of circle P. The coordinates of D are $(1, 4)$, and the coordinates of P are $(5, 8)$. Find the coordinates of G.

Solution: The center of the circle, P, is the midpoint of the diameter. So we can use the midpoint formula to find the coordinates of G.

$$x_{MP} = \frac{1}{2}(x_1 + x_2)$$

$$5 = \frac{1}{2}(1 + x_2)$$

$$10 = 1 + x_2$$

$$x_2 = 9$$

$$y_{MP} = \frac{1}{2}(y_1 + y_2)$$

$$8 = \frac{1}{2}(4 + y_2)$$

$$16 = 4 + y_2$$

$$y_2 = 12$$

The coordinates of G are $(9, 12)$.

Dividing a Directed Segment in a Specified Ratio

Besides dividing a segment into two congruent parts by finding a midpoint, we can divide a **directed segment** into two parts whose lengths are in any specified ratio. A directed segment is simply a segment in which we consider the first endpoint as the starting point and the second endpoint as the ending point. To divide a directed segment in a specified ratio, we set up and solve the following proportion that relates the coordinates of the endpoints of the segment, P and Q, to the specified ratio. The proportion is determined twice, once for the x-coordinate and once for the y-coordinate.

$$\text{ratio} = \frac{x - x_1}{x_2 - x}, \text{ratio} = \frac{y - y_1}{y_2 - y}$$

For example, consider directed segment $\overline{JK}$ with coordinates $J(1, -2)$ and $K(11, 3)$. We want to find the coordinates of point L that divide it into a $2:3$ ratio. $\overline{JL}$ is the smaller segment and $\overline{LK}$ is the larger, as shown in Figure 7.4.

2x   3x

J(1, –2) L(x,y)   K(11, 3)

Figure 7.4 Directed line segment

To find the *x*-coordinate of point *L*, we use the proportion as follows:

$$\text{ratio} = \frac{x - x_1}{x_2 - x}$$

$$\frac{2}{3} = \frac{x - 1}{11 - x}$$

$$2(11 - x) = 3(x - 1)$$

$$22 - 2x = 3x - 3$$

$$25 = 5x$$

$$x = 5$$

Repeat for the *y*-coordinate:

$$\text{ratio} = \frac{y - y_1}{y_2 - y}$$

$$\frac{2}{3} = \frac{y - (-2)}{3 - y}$$

$$2(3 - y) = 3(y + 2)$$

$$6 - 2y = 3y + 6$$

$$0 = 5y$$

$$y = 0$$

The coordinates of *L* are (5, 0). Figure 7.5 shows a graph of the segment and point *L*. If desired, a check can be performed by calculating the lengths of *JL* and *LK* and then showing they are in a 2 : 3 ratio.

Figure 7.5 Segment divided in a 2 : 3 ratio

Sometimes a point is stated to be a certain fraction of the way from one end of a segment to the other. For example, Q lies $\frac{1}{3}$ of the way from A to B. In this case, the entire segment is broken up into 3 parts, with AQ equal to $\frac{1}{3}$ of the total length and QB equal to the remaining $\frac{2}{3}$. The ratio of the parts is therefore $1:2$.

$$AQ:QB = 1:2$$

Example

Points U and V have coordinates $U(-3, 7)$ and $V(9, 1)$. Point W lies on segment $\overline{UV}$ such that it is $\frac{1}{6}$ of the way from U to V. Find the coordinates of W.

Solution: The segment is broken up into 6 parts, with UW equal to $\frac{1}{6}$ the total length and WV equal to $\frac{5}{6}$ of the total length. The ratio is therefore $1:5$.

$$\frac{UW}{WV} = \frac{1}{5} = \frac{x-(-3)}{9-x}$$

$$9 - x = 5(x + 3)$$

$$9 - x = 5x + 15$$

$$-6 = 6x$$

$$x = -1$$

Repeat for the y-coordinate:

$$\frac{UW}{WV} = \frac{1}{5} = \frac{y-7}{1-y}$$

$$1 - y = 5(y - 7)$$

$$1 - y = 5y - 35$$

$$36 = 6y$$

$$y = 6$$

The coordinates of W are $(-1, 6)$.

Check Your Understanding of Section 7.1

A. Multiple-Choice

1. What is the distance between points $A(4, 1)$ and $B(11, 0)$?
   (1) $5\sqrt{2}$ (2) $5\sqrt{10}$ (3) $2\sqrt{7}$ (4) 7

2. What is the midpoint of segment $\overline{MR}$ with coordinates $M(-4, 1)$ and $R(10, 6)$?
   (1) $(7, 3\frac{1}{2})$ (2) $(7, 2\frac{1}{2})$ (3) $(3, 3\frac{1}{2})$ (4) $(3, 2\frac{1}{2})$

3. The midpoint of $\overline{CD}$ is $P(2, 6)$. If the coordinates of C are $(8, 0)$, what are the coordinates of point D?
   (1) $(5, 3)$ (2) $(3, -3)$ (3) $(14, -6)$ (4) $(-4, 12)$

4. $\overline{AB}$ is a diameter of circle O. If the coordinates of O are $(3, 7)$ and the coordinates of A are $(-1, 3)$, what are the coordinates of B?
   (1) $(2, 2)$ (2) $(7, 11)$ (3) $(-5, -1)$ (4) $(1, 5)$

5. What is the length of a segment with endpoints $(-2, 3)$ and $(10, 9)$?
   (1) $6\sqrt{5}$ (2) $4\sqrt{13}$ (3) $3\sqrt{2}$ (4) $\sqrt{108}$

6. A circle has a center at coordinates $(-2, 4)$. A point on the circle has coordinates $(6, 6)$. What is the length of the diameter of the circle?
   (1) 58 (2) 29 (3) $4\sqrt{17}$ (4) $2\sqrt{17}$

B. Free-Response—show all work or explain your answer completely.

7. Find the coordinates of the point that divides directed segment $\overline{PQ}$ in a $1:2$ ratio, given $P(-10, 2)$ and $Q(8, 11)$.

8. Directed segment $\overline{SX}$ has coordinates $S(-3, 1)$ and $X(17, 11)$. Find the coordinates of the point that is $\dfrac{3}{10}$ of the way from S to X.

9. Directed segment $\overline{AB}$ has coordinates $A(1, -6)$ and $B(9, 10)$. Find the coordinates of the point that divides the segment in a $3:5$ ratio.

10. Directed segment $\overline{MN}$ has coordinates $M(-6, -1)$ and $N(8, 13)$. Find the coordinates of the point that divides the segment in a $3:4$ ratio.

11. Jamie wants to determine if the point (5, 6) lies on the perpendicular bisector of the segments with endpoints (2, 3) and (10, 5). Explain how she can do this.

12. The coordinates of A, B, C, and D are $A(1, 6)$, $B(4, 1)$, $C(-7, -9)$, and $D(-4, -4)$. Is $\overline{AB} \cong \overline{CD}$? Justify your answer.

7.2 PERIMETER AND AREA USING COORDINATES

KEY IDEAS

The perimeter of any polygon can be calculated from the coordinates of the polygon's vertices using the distance formula. To calculate the area, divide a polygon into rectangles and right triangles, whose areas can be easily calculated.

Perimeter

The perimeter of any polygon is easily calculated from the vertices. The length of each side is calculated using the distance formula and then all the sides are added. Alternatively, the figure can be graphed and each length can be calculated using the Pythagorean theorem. The distance formula and the Pythagorean theorem often result in radical expressions. Simplify radicals and combine like radicals when adding the lengths of the sides.

Example

Find the perimeter of the triangle with vertices $A(-3, 4)$, $B(5, 0)$, and $C(3, 6)$.

Solution: Calculate the lengths of each of the three sides of $\triangle ABC$ using the distance formula:

$$AB = \sqrt{(-3-5)^2 + (4-0)^2} = \sqrt{64+16} = \sqrt{80} = 4\sqrt{5}$$
$$BC = \sqrt{(5-3)^2 + (0-6)^2} = \sqrt{4+36} = \sqrt{40} = 2\sqrt{10}$$
$$CA = \sqrt{(3-(-3))^2 + (6-4)^2} = \sqrt{36+4} = \sqrt{40} = 2\sqrt{10}$$

The perimeter of $\triangle ABC = AB + BC + CA$

$$= 4\sqrt{5} + 2\sqrt{10} + 2\sqrt{10}$$

$$= 4\sqrt{5} + 4\sqrt{10}.$$

Area

The area of any polygon can be calculated by graphing the figure and sketching the smallest rectangle that bounds the polygon. The regions of the rectangle that go beyond the polygon are all right triangles or rectangles. Their areas can be subtracted from the area of the rectangle to give the area of the polygon.

Example 1

Find the area of the hexagon whose vertices are $A(-4, 4)$, $B(1, 3)$, $C(4, -3)$, $D(1, -6)$, $E(-2, -5)$, and $F(-5, -1)$.

 Solution: The figure is graphed and bounded by the rectangle shown in the figure. The region outside the hexagon and within the rectangle is divided into the 8 regions shown.

Area of the bounding rectangle = $9 \cdot 10 = 90$

Area of region $1 = \dfrac{1}{2}(1 \cdot 5) = 2.5$

Area of region $2 = \dfrac{1}{2}(1 \cdot 5) = 2.5$

Area of region $3 = 3 \cdot 1 = 3$

Area of region $4 = \dfrac{1}{2}(3 \cdot 6) = 9$

Area of region $5 = \dfrac{1}{2}(3 \cdot 3) = 4.5$

Area of region $6 = \dfrac{1}{2}(1 \cdot 3) = 1.5$

Area of region $7 = 1 \cdot 3 = 3$

Area of region $8 = \dfrac{1}{2}(3 \cdot 4) = 6$

Area of hexagon = $90 - (2.5 + 2.5 + 3 + 9 + 4.5 + 1.5 + 3 + 6) = 58$

Calculating the area of a figure with a curved perimeter can be estimated using the reverse of the above technique. The curved figure is filled in with squares, rectangles, and triangles as close as desired to the figure's perimeter. Smaller squares, rectangles, and triangles allow you to fill in closer to the perimeter, giving a more accurate estimate.

Example 2

Estimate the area of the ellipse shown in the accompanying figure.

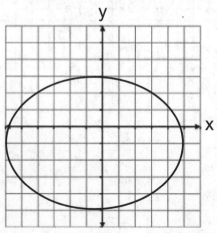

Solution: We can estimate the area by filling in the ellipse with 5 rectangles. Triangles could have been used as well to get a more accurate estimate of the area.

Calculate the areas of the rectangles.
Rectangle 1 area = 7 · 6 = 42
Rectangle 2 area = 4 · 2 = 8
Rectangle 3 area = 5 · 1 = 5
Rectangle 4 area = 4 · 2 = 8
Rectangle 5 area = 5 · 1 = 5
The approximate area of the ellipse is 42 + 8 + 5 + 8 + 5 = 68.

For comparison, the actual area of the ellipse is 69. So our estimate was within 2% of the actual area.

Check Your Understanding of Section 7.2

A. Multiple-Choice

1. What is the perimeter of a regular octagon if the coordinates of two consecutive vertices are (3, 2) and (7, 8)?
   (1) $2\sqrt{10}$ (2) $16\sqrt{10}$ (3) $16\sqrt{13}$ (4) $32\sqrt{13}$

2. $\triangle ABC$ has vertices with coordinates $A(5, -1)$, $B(1, 1)$, and $C(3, 3)$. $\overline{AF}$ is an altitude and point F has coordinates (2, 2). What is the area of the triangle?
   (1) 6   (2) 12   (3) 18   (4) 24

3. A triangle has vertices with endpoints (−1, 3), (4, 3), and (3, 6). What is the perimeter of the triangle?
   (1) $5+\sqrt{10}$　　(2) $10+\sqrt{10}$　　　　(3) 15　　　　　　(4) $\sqrt{30}$

4. Find the perimeter of parallelogram *FROG* with coordinates $F(1, 2)$, $R(4, 2)$, $O(7, 6)$, and $G(4, 6)$.
   (1) 12　　　　　(2) 14　　　　　　(3) 16　　　　　　(4) 18

B. *Free-Response—show all work or explain your answer.*

5. Calculate the area of the accompanying figure.

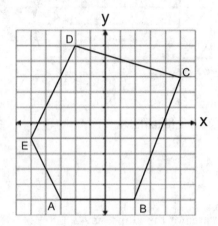

6. Find the area of the polygon whose vertices are $A(−6, −3)$, $B(5, −3)$, $C(3, 2)$, $D(−2, 5)$, and $E(−4, 4)$.

7. Calculate the area of the accompanying figure.

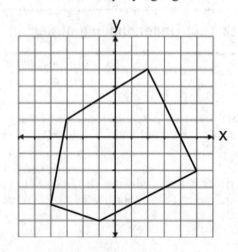

8. Estimate the area of the accompanying figure.

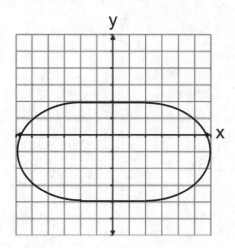

9. Estimate the area of the heart using 2 triangles and 3 rectangles.

10. Estimate the area of the accompanying figure.

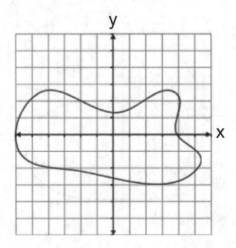

7.3 SLOPE AND EQUATIONS OF LINES

Slope

The slope of a straight line is a measure of its steepness, as shown in Figure 7.6. It is also the change in the y-coordinate divided by the change in the x-coordinate between any two points on the line. The change in the y-coordinate is called the rise, and the change in the x-coordinate is called the run. Slope is calculated from the coordinates of each point using the slope formula.

Slope Formula:

$$\text{slope} = \frac{\text{rise}}{\text{run}} = \frac{y_2 - y_1}{x_2 - x_1}$$

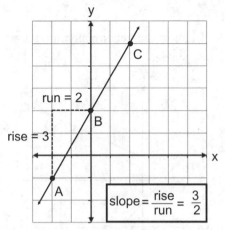

Figure 7.6 Slope of a line

In Figure 7.6, the slope is calculated between points A and B. The rise is the difference in the y-coordinates: $2 - (-1) = 3$. The run is the differences in the x-coordinates: $0 - (-2) = 2$. The result is a slope of $\frac{3}{2}$. The rise and run can be counted from a graph if you can identify 2 points on the line whose coordinates are easily determined. Keep in mind several important features of slope:

- The slopes between any pair of points on a straight line are equal.
- The slope between two points A and B on a straight line will be the same whether it is calculated from A to B or B to A.
- Slope is positive if the straight line is directed up and to the right.
- Slope is negative if the straight line is directed down and to the left.
- The slope of any horizontal line equals zero since the rise is zero.
- The slope of any vertical line is undefined since the run is zero and we cannot divide by zero.
- The slope of parallel lines are equal.
- The slope of perpendicular lines are negative reciprocals (except for horizontal and vertical lines, which are also perpendicular).

Example

Find the slope of each line in the accompanying figure.

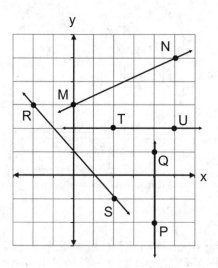

Solution: For each line, determine the coordinates of two points. Then apply slope $= \dfrac{y_2 - y_1}{x_2 - x_1}$.

$\overline{RS}$ with $R(-2, 3)$ and $S(2, -1)$. Slope $= \dfrac{(-1-3)}{(2-(-2))} = -1$

$\overline{MN}$ with $M(0, 3)$ and $N(5, 5)$. Slope $= \dfrac{(5-3)}{(5-0)} = \dfrac{2}{5}$

$\overline{TU}$ with $T(2, 2)$ and $U(5, 2)$. Slope $= \dfrac{(2-2)}{(5-2)} = 0$

$\overline{QP}$ with $Q(4, 1)$ and $P(4, -2)$. Slope $= \dfrac{(-2-1)}{(4-4)}$ is undefined

Collinearity

Slope can be used to determine the collinearity of three points. Given points A, B, and C, the three points will be collinear if the slope of $\overline{AB}$ equals the slope of $\overline{BC}$.

Example

Are points $D(1, 4)$, $E(4, 6)$, and $F(7, 8)$ collinear?

Solution: Using slope $= \dfrac{y_2 - y_1}{x_2 - x_1}$, the slope of $\overline{DE} = \dfrac{6-4}{4-1} = \dfrac{2}{3}$ and the

slope of $\overline{EF} = \dfrac{8-6}{7-4} = \dfrac{2}{3}$. The slopes are equal. So the points are collinear.

Equations of Lines

A straight line in the x-y-plane is described by a function that relates a value of the independent variable x to a value of the dependent variable y. The functions that describe straight lines are **linear**. A **linear function** is one in which the variables are raised only to the first power.

- Two common ways to write the equation of a line are the point-slope form and the slope-intercept form.

Point-Slope Form of a Line:

$$y - y_1 = m(x - x_1)$$

where m is the slope and where x_1 and y_1 are the coordinates of any point on the line.

Slope-Intercept Form of a Line:

$$y = mx + b$$

where m is the slope and b is the y-intercept of the line. The y-intercept is the y-coordinate at which the line crosses the y-axis.

Figure 7.7 shows the graph of the line whose point–slope equation is $y - 3 = 4(x - 1)$. This represents a line that passes through the point $(1, 3)$ and has a slope of 4. The line can be graphed using $(1, 3)$ as a starting point and then graphing additional points by following a rise/run of 4. The same line in slope-intercept form would be $y = 4x - 1$. The slope is 4, and the y-intercept is -1. Start at the point $(0, -1)$ and follow a rise/run of 4 to locate additional points on the line.

211

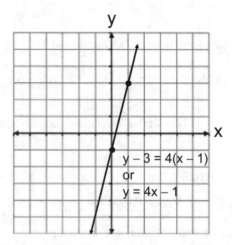

Figure 7.7 Graph of the line $y - 3 = 4(x - 1)$

MATH FACT

The x-intercept of a line occurs when the y-coordinate is zero, and the y-intercept occurs when the x-coordinate is zero. Either intercept can be found from the equation of the line by setting the other coordinate equal to zero.

Example 1

Graph the line $y = \dfrac{1}{2}x - 3$ and the line $y - 1 = -\dfrac{2}{3}(x - 1)$.

Solution: The line $y = \dfrac{1}{2}x - 3$ has a slope of $\dfrac{1}{2}$ and a y-intercept of -3. Graph a point at $(0, -3)$, and then continue the line with a slope of $\dfrac{1}{2}$. The line $y - 1 = -\dfrac{2}{3}(x - 1)$ passes through the point $(1, 1)$ and continues with a slope of $-\dfrac{2}{3}$. The graphs are shown in the accompanying figure.

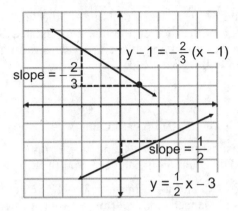

Example 2

Write the equations of the lines $\overleftrightarrow{GQ}$ and $\overleftrightarrow{WF}$ shown in the accompanying figure.

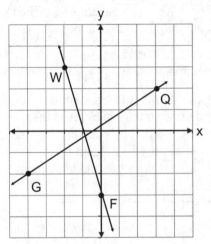

Solution: Using points G and Q, the slope of $\overleftrightarrow{GQ}$ is $\dfrac{2-(-2)}{3-(-4)} = \dfrac{4}{7}$, and the equation of $\overleftrightarrow{GQ}$ in point-slope form is $y - 2 = \dfrac{4}{7}(x-3)$.

Using points W and F, the slope of $\overleftrightarrow{WF} = \dfrac{-3-3}{0-(-2)} = -3$. The y-intercept is -3. The slope-intercept form of the equation of $\overleftrightarrow{WF}$ is $y = -3x - 3$.

When written in slope-intercept form, the equation allows you to calculate the corresponding y-coordinate for any desired x-coordinate easily. For example, consider the line given by the equation $y = 2x + 3$. When $x = 1$, $y = 2(1) + 3 = 5$. When $x = 2$, $y = 2(2) + 3 = 7$. The resulting coordinate pairs are (1, 5) and (2, 7). These points lie on the line. Think about it the other way. Every point that lies on the line must satisfy the equation of the line. This means if we substitute the x- and y-values from the coordinate pair (x, y), the equation will be balanced.

Example 3

Do the points (4, 8) and (6, 9) lie on the line whose equation is $y = 3x - 4$?

Solution: By substituting $x = 4$ and $y = 8$ into $y = 3x - 8$, we get $8 = 3(4) - 4 = 8$. The equation is balanced, and (4, 8) is on the line.

By substituting $x = 6$ and $y = 9$ into $y = 3x - 8$, we get $9 = 3(6) - 4 \neq 14$. The equation is not balanced, and (6, 9) is not on the line.

Writing the Equation of a Line Given Two Points or a Slope and Point

When given two points, the equation of the line passing through those points can be written by first calculating the slope and then substituting the slope and the coordinates of one point into the point-slope form of the line.

MATH FACT

If you are given two points, the equation can be found by graphing the two points and drawing the line between them with a straightedge. You can determine the slope by counting the rise and the run from the graph. You can find the y-intercept by following the line back to the y-axis. You now have the slope and y-intercept to use in the equation $y = mx + b$.

Example

Write the equation of the line passing through (1, 5) and (3, 11). Give both the point-slope form and the slope-intercept form.

Solution: Slope $= \dfrac{11 - 5}{3 - 1} = 3$. Substituting the slope and the first coordinate pair into $y - y_1 = m(x - x_1)$ yields the point-slope equation $y - 5 = 3(x - 1)$.

Find the slope-intercept form by solving for y:

$$y - 5 = 3(x - 1)$$
$$y = 3(x - 1) + 5 = 3x - 3 + 5$$
$$y = 3x + 2$$

Check Your Understanding of Section 7.3

A. Multiple-Choice

1. Which point is collinear with $R(-1, 3)$ and $S(1, 8)$?
   (1) $J(4, 13)$ (2) $K(0, 11)$ (3) $L(5, 10)$ (4) $M(3, 13)$

2. Which point lies on the line whose equation is $y = 5x - 3$?
   (1) $(3, 5)$ (2) $(1, 3)$ (3) $(-1, -2)$ (4) $(3, 12)$

3. What is the slope of the line passing through the points $(6, 0)$ and $(4, 6)$?
   (1) $-\dfrac{1}{3}$ (2) $\dfrac{1}{3}$ (3) -3 (4) 3

4. What are the slope and y-intercept of the line whose equation is $4x + 5y = 10$?

   (1) slope $= -\dfrac{4}{5}$, y-intercept $= 2$

   (2) slope $= \dfrac{4}{5}$, y-intercept $= 2$

   (3) slope $= -\dfrac{4}{5}$, y-intercept $= 10$

   (4) slope $= -\dfrac{5}{4}$, y-intercept $= 1$

5. What is the x-intercept of the line $y + 2 = \dfrac{1}{7}(x + 3)$?

   (1) 11   (2) $-1\dfrac{4}{7}$ (3) 2   (4) 3

6. Which of the following is the equation of a vertical line?
   (1) $y = 3$ (2) $x = 4$ (3) $x + y = 0$ (4) $y = 0$

7. The point $(8, -2)$ satisfies the equation of which line?
   (1) $y + 2 = 2(x + 8)$ (3) $y + 2 = 2(x - 8)$
   (2) $y - 2 = 2(x - 8)$ (4) $y - 2 = 2(x + 8)$

8. Which is the equation of a line with a y-intercept of 11 and a slope of 5?

   (1) $y = -5x + 11$ (3) $y = 5x - 11$

   (2) $y = -5x - 11$ (4) $y = 5x + 11$

B. *Free-Response—show all work or explain your answer completely.*

9. Are the points $(2, 5)$, $(5, 6)$, and $(11, 8)$ collinear? Justify your answer.

10. Points A, B, and C have coordinates $A(-2, -4)$, $B(3, 1)$, and $C(8, 5)$. Are the points collinear? Justify your answer.

11. Write the equation of the line passing through the points $D(-6, 7)$ and $E(2, 3)$.

12. Write the equation of the line passing through the points $D(4, -6)$ and $E(2, 8)$.

13. Write the equation of the line passing through the points $(3, 1)$ and $(7, -2)$.

14. What is the y-intercept of the line passing through the points $(-5, 2)$ and $(5, 4)$?

15. What is the x-intercept of the line passing through the points $(-3, 6)$ and $(-5, 4)$?

7.4 EQUATIONS OF PARALLEL AND PERPENDICULAR LINES

KEY IDEAS

We have two theorems regarding slopes of parallel and perpendicular lines:

- If two lines are parallel, then their slopes are equal.
- If two lines are perpendicular, then their slopes are negative reciprocals.

These can be used to write equations of a line parallel or perpendicular to a given line or to write the equation of the perpendicular bisector of a given segment.

Slopes and Equations of Parallel and Perpendicular Lines

> ## Slope of Parallel Lines Theorem
> Parallel lines have equal slopes.

When given a line, there are an infinite number of parallel lines. All have the same slope but different y-intercepts. Figure 7.8 shows a family of parallel lines and their equations. Notice that only the y-intercept changes.

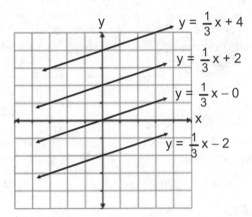

$$y = \frac{1}{3}x + 4$$

$$y = \frac{1}{3}x + 2$$

$$y = \frac{1}{3}x - 0$$

$$y = \frac{1}{3}x - 2$$

Figure 7.8 Parallel lines

You can determine if two lines are parallel by comparing their slopes. If the equations are not in either slope-intercept or point–slope forms, rewrite them in one of those forms so the slope can be easily identified.

Example 1

Are the lines $y = 4x + 2$ and $2y - 8x - 10 = 0$ parallel?

Solution: The slope of the first equation is 4. The second equation must be rearranged:

$$2y - 8x - 10 = 0$$

$$2y = 8x + 10$$

$$y = 4x + 5$$

The second line also has a slope of 4, so the lines are parallel.

Example 2

What is the slope of a line parallel to $6x + 3y = 12$?

Solution: Rearrange the equation so it is in slope-intercept form:

$$6x + 3y = 12$$
$$3y = -6x + 12$$
$$y = -2x + 4$$

The slope of the line is -2, so a parallel line would also have a slope of -2.

Slope of Perpendicular Lines Theorem
Perpendicular lines have negative reciprocal slopes unless they are horizontal and vertical.

The negative reciprocal of a slope m is $= -\dfrac{1}{m}$. There are an infinite number of lines perpendicular to a given line. They will all have the same slope but different y-intercepts. Figure 7.9 shows a family of perpendicular lines. Since all the lines perpendicular to $y = \dfrac{1}{2}x + 2$ have the same slope, they must be parallel to each other.

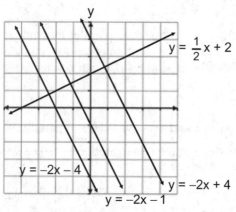

Figure 7.9 Perpendicular lines

The product of a pair of negative reciprocals is always -1. If two lines are perpendicular, the product of their slopes is -1.

Example 1

What is the slope of a line perpendicular to $x + 5y = 10$?

Solution: Rearrange into slope-intercept form:

$$x + 5y = 10$$
$$5y = -x + 10$$
$$y = -\frac{1}{5}x + 2$$

The slope of the line is $-\frac{1}{5}$. The slope of a perpendicular line is the negative reciprocal, which is 5.

We can combine what we have learned about parallel lines, perpendicular lines, and equations of lines to write the equation of a line parallel or perpendicular to a given line and through a given point. The given line is used to find the desired slope. The given point becomes the x_1 and y_1 in the point-slope equation of the line. Remember that the y-intercept of the new equation may be different from the y-intercept of the given line.

Example 2

Find the equation of the line parallel to $y = 3x - 7$ and passing through the point (4, 1).

Solution: The slope of the given line is 3. By substituting $m = 3$, $x_1 = 4$, and $y_1 = 1$ into the point-slope equation, we get:

$$y - y_1 = m(x - x_1)$$
$$y - 1 = 3(x - 4) \text{ or } y = 3x - 11$$

Example 3

Find the equation of the line perpendicular to $y = 6x + 3$ and passing through the point (3, 5).

Solution: The slope of the given line is 6, so the slope of the perpendicular line is $-\frac{1}{6}$. By substituting $m = -\frac{1}{6}$, $x_1 = 3$, and $y_1 = 5$ into the point-slope equation, we get:

$$y - y_1 = m(x - x_1)$$
$$y - 5 = -\frac{1}{6}(x - 3) \text{ or } y = -\frac{1}{6}x + \frac{11}{2}$$

Equation of a Perpendicular Bisector

Using the tools above, we can find the equation of a perpendicular bisector of a segment. Recall that the bisector of a segment passes through the segment's midpoint and that we have a formula for finding the midpoint of a segment given the two endpoints. First calculate the coordinates of the midpoint and the slope of the segment. Then substitute the negative reciprocal of the slope and the coordinates of the midpoint into the point-slope equation. Figure 7.10 shows the line $y = -x$, which is the perpendicular bisector of segment $\overline{AB}$. The line $y = -x$ passes through the midpoint M of $\overline{AB}$. The slope of $\overline{AB}$ is 1, and the slope of $y = -x$ is -1, which are negative reciprocals.

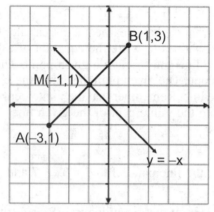

Figure 7.10 Perpendicular bisector of a segment

MATH FACT

You must find the midpoint of the segment when asked for the equation of a perpendicular bisector. What would happen if you used one of the endpoints? You would get a perpendicular line, but it would pass through the endpoint instead of the midpoint. In Figure 7.10, the line would pass through A or B instead of M.

Example

Find the equation of the perpendicular bisector of the segment with endpoints $A(-2, 1)$ and $B(4, 3)$.

Solution: First find the midpoint:

$$x_{MP} = \frac{1}{2}(x_1 + x_2) = \frac{1}{2}(-2 + 4) = 1$$

$$y_{MP} = \frac{1}{2}(y_1 + y_2) = \frac{1}{2}(1 + 3) = 2$$

The midpoint is (1, 2).

Find the slope:

$$\text{Slope} = \frac{y_2 - y_1}{x_2 - x_1} = \frac{3 - 1}{4 - (-2)} = \frac{2}{6} = \frac{1}{3}$$

Use the negative reciprocal of the slope, which is -3.

Now substitute into the point slope equation:

$$y - y_1 = m(x - x_1)$$
$$y - 2 = -3(x - 1) \text{ or } y = -3x + 5$$

Check Your Understanding of Section 7.4

A. Multiple-Choice

1. What is the slope of the line parallel to $y + 7 = 4(x - 3)$?
   (1) -3 (2) -4 (3) 4   (4) -12

2. What is the slope of the line perpendicular to $x + 3y = 12$?
   (1) 3   (2) -3 (3) 12   (4) -12

3. Which is the equation of a line parallel to $y - 4 = 2(x - 6)$?
   (1) $y - 4 = 5(x + 7)$ (3) $y - 3 = 7(x - 6)$

   (2) $y - 3 = 2(x + 1)$ (4) $y + 2 = -\frac{1}{2}(x + 5)$

4. Which line is parallel to the line given by the equation $y + 2x = 4$?

(1) $3y = 6x + 9$

(3) $y = -\dfrac{1}{2}x + 4$

(2) $y = 2x + 6$

(4) $3y + 6x = 9$

5. Which is the equation of the line parallel to $2x + 5y = 7$?

(1) $y = -\dfrac{2}{5}x + 3$

(3) $y = -\dfrac{5}{2}x + 3$

(2) $y = -5x + \dfrac{7}{2}$

(4) $y = 5x + \dfrac{7}{2}$

6. Which is the equation of a line perpendicular to $y = \dfrac{2}{9}x + 1$?

(1) $2x - 9y = 1$

(3) $9x + 2y = 1$

(2) $2x + 9y = 1$

(4) $9x - 2y = 1$

7. Which equation represents a line perpendicular to $6x + 2y = 8$?

(1) $y = -3x + 4$

(3) $y = \dfrac{1}{3}x + 4$

(2) $y = 3x + 4$

(4) $y = \dfrac{3}{4}x + 2$

8. If the equation of line m is $y + 3 = 4(x - 5)$, which of the following would result in a parallel line?
   (1) a reflection over the y-axis
   (2) a reflection over the x-axis
   (3) a rotation of 90° about the origin
   (4) a translation of 2 units vertically

B. *Free-Response—show all work or explain your answer completely.*

9. Write the equation of the line parallel to $y = -\dfrac{1}{4}x + 2$ and passing through the point (4, 4).

10. Write the equation of the line parallel to $y - 3x + 6 = 0$ and passing through the point (1, 2).

11. Write the equation of a line parallel to $5x + 2y = 4$ and passing through the point (2, 0).

12. Write the equation of the line perpendicular to $y - 4x = 2$ and passing through the point (4, 1).

13. Write the equation of the line perpendicular to $y = \dfrac{2}{7}x + 2$ and passing through the point $(2, -3)$.

14. Write the equation of the line perpendicular to $-2x + y = 4$ and passing through the point $(2, 3)$.

15. Write the equation of the perpendicular bisector of the segment with endpoints $(-7, 0)$ and $(1, 6)$.

16. Write the equation of the perpendicular bisector of the segment with endpoints $(-4, 6)$ and $(4, 4)$.

7.5 EQUATIONS OF LINES AND TRANSFORMATIONS

Finding the Line of Reflection

We know that a line of reflection is always the perpendicular bisector of the segment joining corresponding points. Therefore, when given a pair of corresponding points, you can find the line of reflection using the procedure to find the perpendicular bisector.

Example

The reflection of $T(4, 1)$ over the line $\overleftrightarrow{PQ}$ is $T'(5, 4)$. Write the equation of $\overleftrightarrow{PQ}$.

Solution: First find the slope of $\overline{TT'}$:

$$\text{Slope of } \overline{TT'} = \frac{y_2 - y_1}{x_2 - x_1}$$

$$= \frac{4 - 1}{5 - 4}$$

$$= 3$$

The slope of the line of reflection is the negative reciprocal.

$$\text{Slope of } \overleftrightarrow{PQ} = -\frac{1}{3}$$

Next find the midpoint of $\overline{TT'}$:

$$\text{Midpoint} = \frac{1}{2}(x_1 + x_2), \frac{1}{2}(y_1 + y_2)$$

Midpoint of $\overline{TT'} = \frac{1}{2}(4+5), \frac{1}{2}(1+4)$

Midpoint of $\overline{TT'} = (4.5, 2.5)$

By using the midpoint and slope, we get:

$$y - 2.5 = -\frac{1}{3}(x - 4.5) \text{ or } y = -\frac{1}{3}x + 4$$

Finding the Image of a Point After a Reflection

In chapter 5, we saw rules for finding the image of a point after a reflection over horizontal and vertical lines and over the lines $y = x$ and $y = -x$. For any other lines, the best approach is graphical. Find the slope of the given line of reflection. Then use the negative reciprocal of the slope to trace a path from the preimage to the line of reflection. Continue the same distance on the other side to locate the image.

Example

Point $A(-1, 2)$ is reflected over the line $y = 3x - 5$ to A'. Find the coordinates of A'.

  Solution: First graph A and the given line. The slope of the line is 3, and the negative reciprocal is $-\frac{1}{3}$. From A, follow a path with a slope of $-\frac{1}{3}$ to the line, which is 3 jumps right and 1 jump down in this case. Continue the same number of jumps on the other side of the line to (5, 0). The coordinates of A' are (5, 0).

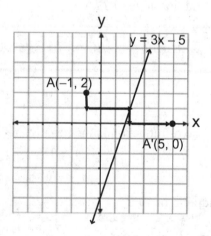

Finding the Equation of a Line After a Dilation

When given a line, the equation of the image of the line after a dilation through the origin can be found as follows:

- Find the coordinates of the y-intercept, and dilate by the specified dilation.
- The new line has the same slope as the original since slope is preserved.

Example

Write an equation for the line that represents the dilation of $2y - x + 4 = 0$ by a scale factor of 3 through the origin.

Solution: Rewrite the equation in $y = mx + b$ form

$$2y - x + 4 = 0$$

$$2y = x - 4$$

$$y = \frac{1}{2}x - 2$$

The slope of the preimage is $\frac{1}{2}$, and the y-intercept has coordinates $(0, -2)$.

Dilate the y-intercept by a factor of 3 to get the new intercept $(0, -6)$, and use the same slope.

The equation of the image is $y = \frac{1}{2}x - 6$.

Check Your Understanding of Section 7.5

A. Multiple-Choice

1. Which of the following lines represents the image of $y = 6x - 2$ after $D_{origin,4}$?

   (1) $y = \frac{3}{2}x - 2$ (3) $y = 6x + 2$

   (2) $y = 24x - 2$ (4) $y = 6x - 8$

2. Which of the following methods could always be used to find the equation of the line of reflection, given preimage A and image A'?
   (1) find the equation of a line whose slope is the negative reciprocal of the slope of $\overline{AA'}$ and that passes through A'
   (2) find the equation of a line whose slope is the negative reciprocal of the slope of $\overline{AA'}$ and that passes through the midpoint of $\overline{AA'}$
   (3) find the equation of the line that is a 90° rotation of $\overline{AA'}$ about the origin
   (4) find the equation of the line parallel to $\overline{AA'}$ and passing through the origin

3. Which of the following lines represents the image of $y = \dfrac{1}{5}x + 1$ after a rotation of 90° centered about the origin?
   (1) $y = -5x$ (3) $y = -5x - 5$
   (2) $y = -5x + 1$ (4) $y = -5x + 5$

4. Which of the following transformations could map a line with a slope of $\dfrac{3}{4}$ to a line with a slope of $-\dfrac{4}{3}$?
   (1) rotation about the origin
   (2) translation
   (3) dilation
   (4) translation in the x-direction

5. Which of the following does *not* always represent an invariant quantity?
   (1) the slope of a line after a translation of 1 unit in the x-direction and 2 units in the y-direction
   (2) the coordinates of a point on a line after a 90° rotation about that point
   (3) the coordinates of the y-intercept of a line after a dilation about the origin
   (4) the coordinates of a point A that lies on line m, after a reflection over line n, if line m and line n intersect at A

B. *Free-Response—show all work or explain your work completely.*

6. The line $y = 3x + \dfrac{1}{2}$ is dilated through the origin with a scale factor of 4. Write the equation of the resulting line.

7. The line $y = -x + 2$ is dilated by a factor of 2 through the point $(3, -1)$ by a scale factor of 2. Write the equation of the resulting line.

7.6 THE CIRCLE

KEY IDEAS

The center-radius equation of the circle is $(x - h)^2 + (y - k)^2 = r^2$, where (h, k) is the center of the circle and r is the radius. In general form, the equation is $x^2 + Cx + y^2 + Dy + E = 0$. We can convert from general form to center-radius form by completing the square.

Equation of the Circle—Center-Radius Form

Circles are one of the conic sections, along with parabolas, ellipses, and hyperbolas. The name **conic section** comes from the fact that each curve is obtained by taking a cross-sectional slice from a cone. A circle can be represented by an algebraic relationship between the x- and y-coordinates.

The equation is derived from the Pythagorean theorem. Figure 7.11 shows circle C with radius r and C located at coordinates (h, k). Let point P be any point on the circle with coordinates (x, y). Recall the theorem:

The set of points a distance r from a given point P is a circle centered at P with radius r.

No matter where on the circle we choose point P, the distance PC must be equal to r. If we complete the right triangle with r as the hypotenuse, the base has length $(x - h)$ and the height has length $(y - k)$. The Pythagorean theorem or the distance formula applied to the triangle gives:

$$a^2 + b^2 = c^2$$

$$(x - h)^2 + (y - k)^2 = r^2$$

Center Radius Form of the Equation of a Circle
$(x - h)^2 + (y - k)^2 = r^2$ is a circle with center (h, k) and radius r

---

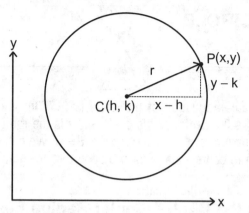

Figure 7.11 Circle in the coordinate plane

Graphing a circle from the equation in $(x - h)^2 + (y - k)^2 = r^2$ form is very straightforward. The center (h, k) and the radius r are easily identified. Remember that the constant term on the right is the square of the radius—we need to take the square root to find the radius. Also note that the h- and k-terms are subtracted, so be careful with the signs. Once the center and radius are determined, graph a point at the center. The first four points on the circle are plotted by translating the center up r units, down r units, right r units, and left r units. Finally complete the circle passing through the first four points. A careful graph can be made using a compass, or a rough sketch can be made by drawing the circle freehand.

MATH FACT

A circle is not a function, and it is not one-to-one. For every x-coordinate, there are two corresponding y-coordinates. Also there are two different x-coordinates that lead to the same y-coordinate.

Example 1

Graph the equation $(x - 2)^2 + (y + 1)^2 = 9$.

Solution: The center is located at
(2, −1). Watch out for the signs! In the
equation, 2 is subtracted from x and −1
is subtracted from y. The radius is
$\sqrt{9} = 3$. We plot the center point at
(2, −1) and then points up, down, right,
and left 3 from the center. Then
complete the circle.

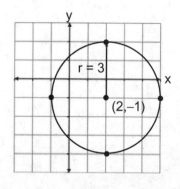

Example 2

Write the equations of the three circles shown in the accompanying figure.

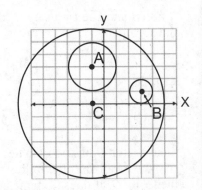

Solution:

The center of circle A is $(-1, 3)$, and the radius is 2. So the equation is $(x + 1)^2 + (y - 3)^2 = 4$.

The center of circle B is $(3, 1)$, and the radius is 1. So the equation is $(x - 3)^2 + (y - 1)^2 = 1$.

The center of circle C is $(0, -1)$, and the radius is 6. So the equation is $x^2 + (y + 1)^2 = 36$.

If the coordinates of the center and one point on the circle are given, we can apply the distance formula to find the radius. If only the two endpoints of a diameter are given, the midpoint and distance formulas can be used to find the missing center and radius.

Example 3

Find the equation of the circle with center C at $(2, 7)$ and point $P(4, 12)$ on the circle.

Solution: The distance formula used on the center C and point P will give the radius, r, of the circle:

$$r = \sqrt{(x_1 - x_2)^2 + (y_1 - y_2)^2}$$
$$r = \sqrt{(4 - 2)^2 + (12 - 7)^2} = \sqrt{29}$$

The radius squared is 29. Substituting into $(x - h)^2 + (y - k)^2 = r^2$ yields $(x - 2)^2 + (y - 7)^2 = 29$.

Example 4

$\overline{BD}$ is a diameter of circle O. The coordinates of B are $(-4, 9)$, and the coordinates of D are $(6, 3)$. Find the equation of the circle.

Solution: The first step is to find the center of the circle using the midpoint formula:

$$x_{MP} = \frac{1}{2}(x_1 + x_2) = \frac{1}{2}(-4 + 6) = 1$$

$$y_{MP} = \frac{1}{2}(y_1 + y_2) = \frac{1}{2}(9 + 3) = 6$$

The center O has coordinates (1, 6).

Now we can find the radius by calculating the distance from point O to either B or D. It doesn't matter which one we pick since any radius has the same length. When using D, we get:

$$r = \sqrt{(6-1)^2 + (3-6)^2} = \sqrt{34}$$

By substituting the coordinates of O and the radius into the equation of the circle, we get:

$$(x - 1)^2 + (y - 6)^2 = 34$$

The equation of a circle can also be used to determine if a particular point lies on the circle. As with equations describing lines, parabolas, and other curves, the equation of a circle will be satisfied by the coordinates of any point that lies on the circle.

Example 5

Does the point (−9, 5) lie on the circle described by $(x + 4)^2 + (y - 1)^2 = 41$?

Solution: By substituting the coordinates into the equation, we find:

$$(-9 + 4)^2 + (5 - 1)^2 = 41$$

$$25 + 16 = 41$$

$$41 = 41$$

The equation is balanced, so the point does lie on the circle.

Example 6

Does the point ($\sqrt{7}$, 3) lie on the circle centered at (0, 4) and having a radius of 3?

Solution: We first write the equation in center-radius form. Then substitute the coordinates $(\sqrt{7}, 3)$:

$$x^2 + (y - 4)^2 = 3^2$$
$$\left(\sqrt{7}\right)^2 + (3-4)^2 = 9$$
$$7 + 1 = 9$$
$$8 \neq 9$$

The equation is not balanced, so the point does not lie on the circle.

General Form of the Equation of the Circle

We can expand the two squared terms in the center-radius form of the circle.

General Form of the Equation of a Circle

$$x^2 + y^2 + Cx + Dy + E = 0$$

For example, the circle $(x + 2)^2 + (y + 5)^2 = 4$ in general form is:

$$(x + 2)^2 + (y + 5)^2 = 4$$
$$x^2 + 4x + 4 + y^2 + 10y + 25 = 4$$
$$x^2 + y^2 + 4x + 10y + 25 = 0$$

Going from general form to center-radius form involves completing the square. First rearrange the equation so all the x-terms and all the y-terms are grouped on one side and the constant is on the other. Using the previous equation as an example, we get:

$$x^2 + 4x + y^2 + 10y = -25$$

Let b be the coefficient of the x-term. The constant needed to complete the square for the terms in x is $(\frac{1}{2}b)^2$. In this case, $b = 4$ and the constant needed is $(\frac{1}{2} \cdot 4)^2$ or 4. To complete the square for the y-terms, we need a constant equal to $(\frac{1}{2} \cdot 10)^2$ or 25. After adding the required constants on both sides, we can factor the terms with x's and the terms with y's to get back where we started from.

$$x^2 + 4x + 4 + y^2 + 10y + 25 = -25 + 4 + 25$$
$$(x + 2)^2 + (y + 5)^2 = 4$$

Example

Find the coordinates of the center and the length of the radius of a circle whose equation is $x^2 + y^2 + 6x + 8y + 12 = 0$.

Solution: Using the procedure for completing the square:

$$x^2 + y^2 + 6x + 8y + 12 = 0$$
$$x^2 + 6x + y^2 + 8y = -12 \qquad \text{group the } x \text{ and } y \text{ terms}$$
$$\left(\frac{1}{2} \cdot 6\right)^2 = 9 \text{ and } \left(\frac{1}{2} \cdot 8\right)^2 = 16 \qquad \text{calculate the required constants}$$
$$x^2 + 6x + 9 + y^2 + 8y + 16 = -12 + 9 + 16 \qquad \text{add the constants to both sides}$$
$$(x + 3)^2 + (y + 4)^2 = 13 \qquad \text{complete the square}$$

Check Your Understanding of Section 7.6

A. *Multiple-Choice*

1. The center and radius of the circle described by $(x + 8)^2 + (y - 4)^2 = 49$ are
   (1) center (8, –4), radius 49   (3) center (–8, 4), radius 7
   (2) center (8, –4), radius 7   (4) center (–8, 4), radius 49

2. The equation of the circle with center (9, –2) and radius $\sqrt{11}$ is
   (1) $(x - 9)^2 + (y + 2)^2 = 11$ (3) $(x + 2)^2 + (y - 9)^2 = 11$
   (2) $\left(x - 9\right)^2 + \left(y + 2\right)^2 = \sqrt{11}$ (4) $(x + 9)^2 + (y - 2)^2 = 11$

3. A and B are the endpoints of diameter $\overline{AB}$ of circle O. Given $A(5, 1)$ and $B(9, -3)$, which of the following is the equation of the circle?
   (1) $(x - 5)^2 + (y - 1)^2 = 8$ (3) $(x - 7)^2 + (y + 1)^2 = 32$
   (2) $(x - 9)^2 + (y + 3)^2 = 8$ (4) $(x - 7)^2 + (y + 1)^2 = 8$

4. What is the equation of the circle shown in the accompanying figure?

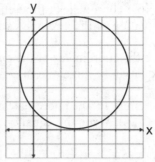

(1) $(x + 3)^2 + (y + 3)^2 = 4$

(2) $(x - 3)^2 + (y - 3)^2 = 4$

(3) $(x + 3)^2 + (y + 3)^2 = 16$

(4) $(x - 3)^2 + (y - 4)^2 = 16$

5. What is the equation of the circle shown in the accompanying figure?

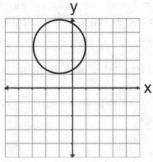

(1) $(x - 1)^2 + (y + 3)^2 = 4$

(2) $(x - 1)^2 + (y + 3)^2 = 16$

(3) $(x + 1)^2 + (y - 3)^2 = 4$

(4) $(x - 1)^2 + (y + 3)^2 = 16$

6. Which of the following points lies on the circle $(x + 1)^2 + y^2 = 12$?

(1) $(2, -\sqrt{3})$ (2) $(0, 11)$ (3) $(\sqrt{12}, 0)$ (4) $(-2, \sqrt{3})$

7. Point $Q(\sqrt{5}, \sqrt{3})$ lies on a circle whose center is at the origin. What is the radius of the circle?

(1) $\sqrt{7}$ (2) $\sqrt{15}$ (3) $\sqrt{8}$ (4) $\sqrt{13}$

8. Point $P(8, 3)$ lies on a circle whose radius is $\sqrt{50}$. Which of the following could be the coordinates of the center of the circle?

(1) $(7, 1)$ (2) $(2, 5)$ (3) $(1, 4)$ (4) $(6, 2)$

B. *Free-Response—show all work or explain your answer completely.*

9. (1, 8) and (9, −4) are the coordinates of the endpoints of a diameter of circle P. Write the equation of the circle in center-radius form.

10. (4, −1) and (−2, 7) are the coordinates of the endpoints of a diameter of circle P. Write the equation of the circle in center-radius form.

11. $\overline{BC}$ is the diameter of circle P. The coordinates of B are (−3, 2), and the coordinates of C are (5, 6). Write the equation of the circle.

12. The equation of a circle is $x^2 + y^2 - 6x + 4y - 4 = 0$. Write the equation in center-radius form.

13. Circle O has the equation $x^2 + y^2 + 2x - 8y - 9 = 0$. What is the length of the diameter of circle O?

14. The equation of a circle is $x^2 + y^2 - 4x + 12y + 15 = 0$. Find the coordinates of the center of the circle and the length of the radius.

15. The equation of a circle is $x^2 + y^2 + 2x + 14y + 20 = 0$. Find the coordinates of the center of the circle and the length of the diameter.

16. For what value of k does the point $(\sqrt{5}, 1)$ lie on the circle whose equation is $x^2 + (y - 6)^2 = k$?

Chapter
Eight

SIMILAR FIGURES AND TRIGONOMETRY

8.1 SIMILAR FIGURES

KEY IDEAS

Two figures are similar if they have the same shape. Their sizes may be different. Similar figures have the additional following properties:

- All pairs of corresponding angles are congruent.
- All pairs of corresponding lengths are proportional; the constant of proportionality is called the scale factor.
- The areas are proportional to the scale factor squared.
- The volumes are proportional to the scale factor cubed.
- A figure may be mapped to a similar figure with a similarity transformation, which will include a dilation.

Definition of Similarity and Proportional Parts

Two figures that have the shape but possibly different sizes are called similar. You can think of similar figures as a photocopy enlargement or reduction of one another. Any two similar figures must have certain properties.

Properties of Similar Figures

- All corresponding lengths are proportional.
- All corresponding angles are congruent.

The constant of proportionality is called the scale factor. Figure 8.1 shows two similar quadrilaterals, *ABCD* and *WXYZ*. We can write a similarity

Figure 8.1 Similar quadrilaterals *ABCD* and *WXYZ*

235

statement using the similarity symbol ~. When pairing the corresponding vertices, the similarity statement is $ABCD \sim WXYZ$. The four pairs of congruent corresponding angles are as follows:

$$\angle A \cong \angle W$$
$$\angle B \cong \angle X$$
$$\angle C \cong \angle Y$$
$$\angle D \cong \angle Z$$

The four corresponding pairs of proportional sides are the following:

$$\frac{AB}{WX} = \frac{BC}{XY} = \frac{CD}{YZ} = \frac{DA}{ZW}$$

Each of the ratios is equivalent to $\frac{1}{2}$, so the scale factor is $\frac{1}{2}$.

MATH FACT

It's not just corresponding sides that are proportional in similar figures. Any two segments whose endpoints are corresponding points will be proportional. For example in Figure 8.1, diagonal lengths *AC* and *WY* are also in a 1:2 ratio. In similar triangles, all corresponding lengths must be proportional.

Example 1

$\triangle PST$ has side lengths $PS = 3$, $ST = 4$, and $TP = 5$. $\triangle BLM$ has side lengths $BL = 9$, $LM = 12$, and $MB = 15$. If the two triangles are similar, what is the scale factor? Write a similarity statement for the two triangles.

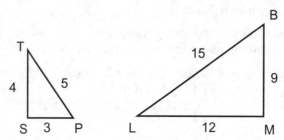

Solution: Find the scale factor by writing the ratios of corresponding sides and pairing them up so all three ratios are equal. To be sure you are forming ratios with the correct parts, pair up the smallest side from each figure, then the next biggest, and so on.

The three ratios are the following: $\dfrac{PS}{MB} = \dfrac{3}{9}$

$$\frac{ST}{LM} = \frac{4}{12}$$

$$\frac{TP}{BL} = \frac{5}{15}$$

Each ratio simplifies to $\frac{1}{3}$, which is the scale factor. The similarity statement is $\triangle PST \sim \triangle BML$. The two largest sides are TP and BL. The two medium sides are ST and LM. The two smallest sides are PS and MB. These pairs match up to the corresponding sides as specified in the similarity statement.

Example 2

Given similar triangles $\triangle ABC \sim \triangle JKL$, $m\angle A = 102°$, and $m\angle K = 43°$. Find $m\angle B$ and $m\angle C$.

Solution: The strategy here is to combine two theorems. From the similarity relationship, we know corresponding angles are congruent.

$$m\angle B = m\angle K = 43°$$

From the triangle angle sum theorem:

$$m\angle A + m\angle B + m\angle C = 180°$$
$$102° + 43° + m\angle C = 180°$$
$$145° + m\angle C = 180°$$
$$m\angle C = 35°$$

Finding a Missing Side

If you know three out of the four parts of a proportion made from corresponding parts, you can solve for the fourth. Use the fact that the cross products of a proportion are equal to write an equivalent equation.

Example 1

Given $ABCDE \sim VWXYZ$, with $AB = 18$, $BC = 30$, and $WX = 20$. Find the length of WV.

Solution: From the similarity statement, $\overline{AB}$ and $\overline{VW}$ are corresponding parts, as are $\overline{BC}$ and $\overline{WX}$.

$$\frac{AB}{VW} = \frac{BC}{WX}$$ corresponding parts are proportional

$$\frac{18}{VW} = \frac{30}{20}$$

$$30VW = 18 \cdot 20$$

$$VW = \frac{18 \cdot 20}{30}$$ cross products are equal

$$= 12$$

The equation that results from solving the proportion may be quadratic. In that case, use any technique you are most comfortable with to solve the quadratic, such as factoring, the quadratic formula, completing the square, or trial and error. Also look out for given information that may give a clue to the scale factor, such as midpoints.

Example 2

Rectangle *TRIP* is similar to rectangle *BSAP*. A is the midpoint of $\overline{PI}$, $RI = x^2$, $SA = x + 4$, and $TR = 10x$. Find BS.

Solution: Midpoint A divides $\overline{PI}$ in half, so $PA = \frac{1}{2}PI$. This gives us the scale factor between the two rectangles.

$$\frac{SA}{RI} = \frac{1}{2}$$ corresponding parts are proportional

$$\frac{x+4}{x^2} = \frac{1}{2}$$

$x^2 = 2(x + 4)$ cross products are equal

$x^2 = 2x + 8$

$x^2 - 2x - 8 = 0$

$(x - 4)(x + 2) = 0$ factor

$x - 4 = 0 \quad x + 2 = 0$

$x = 4 \quad\quad\quad x = -2$

$x = 4$ length cannot be negative, so use only $x = 4$

$TR = 10 \cdot 4$ substitute

$BS = \dfrac{1}{2} TR$

$BS = \dfrac{1}{2} \cdot 40$ apply the scale factor of $\dfrac{1}{2}$

$ = 20$

Scaled Drawing

Similar figures, scale factors, and dilations are frequently used by architects, engineers, graphic designers, and many other professions. An engineer or an architect may create a design on paper of a building, car, or any other object. The drawing is called a scaled drawing because it is similar to the actual object it represents. All lengths are reduced or enlarged by the same scale factor. The scaled drawings let you see what the final object will look like, give guidance to someone who will build the object, and help designers determine how multiple parts fit together.

When working with a scaled drawing, apply the similar figure relationships between corresponding lengths, angles, area, and volume.

Example

An interior designer is sketching a new layout for a living room that calls for installing carpet in the room. The room is rectangular in shape, and her sketch measures 12 inches by 9 inches. If the scale factor on the drawing is 1:20 and carpet costs $3.99 per square foot, how much should she budget for the carpet?

Solution:

First find the actual dimension of the room.

$$\text{length}_{\text{actual}} = \text{length}_{\text{drawing}} \cdot \text{scale factor}$$
$$= 12 \text{ in.} \cdot 20$$
$$= 240 \text{ in.}$$
$$\text{width}_{\text{actual}} = \text{width}_{\text{drawing}} \cdot \text{scale factor}$$
$$= 9 \text{ in.} \cdot 20$$
$$= 180 \text{ in.}$$

$$A_{\text{room}} = \text{length} \cdot \text{width}$$
$$= 240 \text{ in.} \cdot 180 \text{ in.}$$
$$= 43{,}200 \text{ in.}^2$$

Now convert the area to square feet so the cost can be calculated. Note that the scale factor is squared when converting in.^{2} to ft^2.

$$A_{room} = 43{,}200 \text{ in.}^2 \cdot \left(\frac{1 \text{ ft}}{12 \text{ in.}}\right)^2$$

$$= 300 \text{ ft}^2$$

The total cost will be the area multiplied by the cost per square foot.

$$\text{cost} = A_{room} \cdot \frac{\text{cost}}{\text{ft}^2}$$

$$\text{cost} = 300 \text{ ft}^2 \cdot \frac{\$3.99}{\text{ft}^2}$$

$$= \$1{,}197$$

The designer should budget $1,197 for the carpet.

Scaling Perimeter, Area, and Volume

Looking back to the carpet problem in the previous section, there is a relationship between the area of the drawing, the actual area of the room, and the scale factor. The area of the room in the drawing is 12 in. · 9 in. = 108 in.^{2}
The ratio of the two areas is

$$\frac{\text{Area}_{drawing}}{\text{Area}_{actual}} = \frac{108 \text{ in.}^2}{43{,}200 \text{ in.}^2}$$

$$= \frac{1}{400}$$

This ratio is exactly equal to the scale factor squared:

$$\left(\frac{1}{20}\right)^2 = \frac{1}{400}$$

Whenever a figure undergoes a similarity transformation, the areas of the preimage and image are in a ratio equal to the scale factor squared. This is because area will always be equal to the product of two lengths and each length is multiplied by the scale factor. Volume will always be the product of three lengths. So the volume of two similar solids is proportional to the scale factor cubed.

Effect of a Dilation on Perimeter, Area, and Volume

Quantity	Proportional To
Length	Scale factor
Area	(Scale factor)^2
Volume	(Scale factor)^3

Example 1

$\triangle SKY \sim \triangle BLU$, $SK = 12$, $BL = 3$, and the area of $\triangle BLU$ is 5 cm^2. What is the area of $\triangle SKY$?

Solution: The ratio of the areas is equal to the square of the ratio of corresponding sides. Use the corresponding lengths SK and BL.

$$\frac{A_{\triangle SKY}}{A_{\triangle BLU}} = \left(\frac{SK}{BL}\right)^2$$

$$\frac{A_{\triangle SKY}}{5 \text{ cm}^2} = \left(\frac{12 \text{ cm}}{3 \text{ cm}}\right)^2$$

$$\frac{A_{\triangle SKY}}{5 \text{ cm}^2} = (4)^2$$

$$A_{\triangle SKY} = 16 \cdot 5 \text{ cm}^2$$
$$A_{\triangle SKY} = 80 \text{ cm}^2$$

Example 2

The volume of a rectangular prism is 10 in.^{3} If the prism undergoes a dilation with a scale factor of 4, what is the volume of the dilated prism?

Solution: Every dimension will be multiplied by 4. So the ratio of any pair of corresponding lengths will be 4. Volume is proportional to scale factor cubed.

$$\frac{\text{volume}_{\text{new}}}{\text{volume}_{\text{original}}} = (\text{scale factor})^3$$

$$\frac{\text{volume}_{\text{new}}}{10 \text{ in.}^3} = (4)^3$$

$$\text{volume}_{\text{new}} = (4^3) \cdot 10 \text{ in.}^3$$
$$\text{volume}_{\text{new}} = 640 \text{ in.}^3$$

Example 3

Kate wants to buy Styrofoam spheres for a craft project. Last week, she bought 3 in. diameter spheres for $0.75 each. This week, she wants to buy 8 in. spheres. Estimate the cost for each 8 in. sphere, rounded to the nearest dollar. Justify your reasoning.

Solution: The scale factor is the ratio of the diameters, or $\frac{8}{3}$. Volume is scaled by a factor of $\left(\frac{8}{3}\right)^3$. A reasonable estimate for the cost is to scale the cost in the same way as the volume since the amount of material in the sphere is proportional to the volume.

$$\text{cost}_{8 \text{ in. sphere}} = \text{cost}_{3 \text{ in. sphere}} \cdot (\text{scale factor})^3$$

$$\text{cost}_{8 \text{ in. sphere}} = \$0.75 \cdot \left(\frac{8}{3}\right)^3$$

$$= \$14.22$$

$$= \$14 \text{ rounded to the nearest dollar}$$

Example 4

A painter uses 8 gallons of paint to paint a house. At his next job, he estimates the length and width of the house are twice that of the first house he painted. Approximately how much paint should he expect to use?

Solution: The amount of paint will be proportional to the surface area of the house, and surface area is proportional to the scale factor of the length squared. In this case, the scale factor is 2.

$$\text{amount of paint} = \text{amount}_{\text{previous}} \cdot (\text{scale factor})^2$$
$$= 8 \text{ gallons} \cdot 2^2$$
$$= 32 \text{ gallons}$$

Check Your Understanding of Section 8.1

A. *Multiple-Choice*

1. Pentagon *ABCDE* is similar to pentagon *RSTUV*. If *DE* = 18 and *UV* = 15, which of the following is *not* necessarily true?

   (1) $m\angle E = m\angle V$

   (2) $\dfrac{\text{area of } ABCDE}{\text{area of } RSTUV} = \dfrac{6}{5}$

   (3) $\dfrac{AB}{RS} = \dfrac{6}{5}$

   (4) $\dfrac{BC}{ST} = \dfrac{6}{5}$

242

2. Given $\triangle JOG \sim \triangle RUN$, $JG = 20$, $RN = 12$, and $OG = 6$, what is the length of UN?
(1) 10   (2) 8.2   (3) 6.4   (4) 3.6

3. Given $\triangle LBJ \sim \triangle JFK$, $LB = (x + 6)$, $BJ = (3x + 2)$, $JF = 4$, and $FK = 6$, find the value of x.

(1) $3\dfrac{1}{3}$ (2) $4\dfrac{2}{3}$ (3) $5\dfrac{1}{3}$ (4) $6\dfrac{1}{3}$

4. Rectangle $ABCD$ is similar to rectangle $HIJK$. If diagonal $AC = 5$, diagonal $HJ = 2$, $CD = (3x - 4)$, and $JK = (x + 5)$, what is the value of x?

(1) $\dfrac{30}{13}$ (2) 4.5   (3) 33   (4) $42\dfrac{1}{4}$

5. In the accompanying figure of similar triangles $\triangle FLO$ and $\triangle TRY$, $FL = 15$, $LO = 20$, and $FO = 30$. In $\triangle TRY$, $TR = 12$, $RY = 6$, and $TY = 8$. Which of the following is a correct similarity statement?

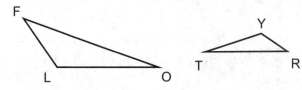

(1) $\triangle FLO \sim \triangle RYT$ (3) $\triangle LOF \sim \triangle RTY$
(2) $\triangle FLO \sim \triangle TRY$ (4) $\triangle LOF \sim \triangle YRT$

6. The area of two similar regular hexagons are in a ratio of $9:1$, and the larger has a side length of 6. What is the perimeter of the smaller hexagon?

(1) $\dfrac{2}{3}$ (2) 4   (3) 12   (4) 108

7. Frank is looking at an architect's design of a home. The scale on the drawing is 1 in. = 2 ft. If the rectangular basement playroom has dimensions 15 in. by 12 in., what is the area of the actual playroom?
(1) 54 ft²   (2) 369 ft²   (3) 525 ft²   (4) 720 ft²

8. Lianna has a drawing of her dog Bucky. In the drawing, Bucky's tail is 4 cm long. Lianna makes a photocopy of the drawing with the enlargement function set to 125%. The next day, she takes the enlarged copy and reduces it by 25%. What is the length of Bucky's tail on the final drawing?

(1) $3\dfrac{3}{4}$ cm   (2) $3\dfrac{1}{2}$ cm   (3) 3 cm   (4) $2\dfrac{3}{4}$ cm

9. *DRAG* and *DYLF* are similar isosceles
   trapezoids. *F* is the midpoint of $\overline{GD}$,
   GR = (8*x* + 2), *DL* = (3*x* + 6), and *DY*
   = 5*x*. What is the length of *DR*?
   (1) 20   (2) 50   (3) 60   (4) 80

10. Which of the following are *not* necessarily similar?
    (1) two regular octagons   (3) two isosceles triangles
    (2) two squares   (4) two segments

B. *Free-Response—show all work or explain your answer completely.*

11. In the accompanying figure,
    △*ABC* ~ △*AVW*, *AB* = 24, *AV* = 15,
    and *WC* = 18. What is the length of
    AW?

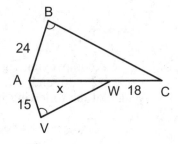

12. Given △*PUT* ~ △*HAG*, m∠*P* = 37°, and m∠*G* = 50°, what is the
    measure of angle *U*?

13. Given △*TOP* ~ △*CAB* with *TO* = (2*x* + 2), *OP* = (*x* − 3),
    CA = (*x* + 1), and *AB* = (*x* − 4), find the length of *TO*.

14. The length of a rectangle is 6 inches more than 4 times its width. A
    similar rectangle has a length of 18 inches and a width of 4 inches. What
    is the area of the first rectangle?

15. Frank is building a model car that is a scale model of an actual car. The
    diameter of the wheels of his model are 1.6 inches, and the wheels of
    the actual car are 24 inches in diameter. If the model is 9 inches long,
    how long is the actual car?

16. Thomas is building a miniature water tower for his model train set. He
    is scaling it from the dimension of an actual 50 ft water tower. His model
    is 10 inches high. If the real water tower can hold 27,000 gallons of
    water, how many gallons of water will the model hold?

17. Elaine states that if you have two parallelograms, you only need to
    show one pair of corresponding vertex angles are congruent in order to
    prove they are similar. Bobby disagrees. Who is correct? Explain your
    answer.

8.2 PROVING TRIANGLES SIMILAR AND SIMILARITY TRANSFORMATIONS

KEY IDEAS

Two triangles can be proven similar using the following postulates:

AA (angle-angle), SAS (side-angle-side), and SSS (side-side-side)

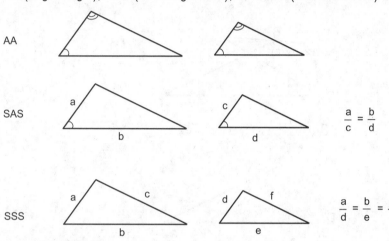

Proving Triangles Similar by the Angle-Angle Criterion

> ### The Angle-Angle (AA) Similarity Criterion
> If two pairs of corresponding angles are congruent in a pair of triangles, then the triangles are similar.

The AA criterion can be justified by showing a similarity transformation exists that maps one triangle to the other. From a transformation point of view, two figures are similar if a sequence of similarity transformations maps one figure to another. Figure 8.2a shows $\triangle ABC$ and $\triangle DEF$ with two pairs of corresponding congruent angles. A translation by vector $\overrightarrow{DA}$ will map point D to A, shown in Figure 8.2b. A rotation about point A by $\angle CAF$ will map F' onto side $\overline{AC}$ and E' on side $\overline{AB}$, as shown in Figure 8.2c. Finally, a dilation about point A by scale factor $\dfrac{AC}{DF}$ will complete the mapping. Since a dilation was used, the triangles are not necessarily congruent, but they are similar.

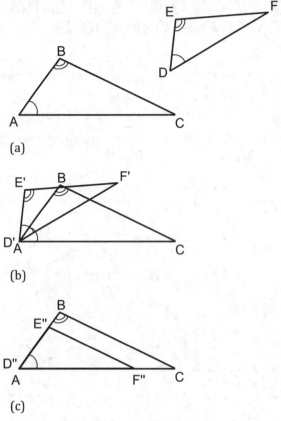

(a)

(b)

(c)

Figure 8.2 △*ABC* ~ △*DEF* by the AA criterion

| **MATH FACT** |

Having all three pairs of corresponding angles congruent in a pair of triangles is sufficient to prove them similar. (In fact, we just saw that having even two pairs is enough.) However, having all pairs of corresponding angles congruent in two polygons with more than three sides does not prove they are similar. For example, a square and a rectangle have all right angles, but they are not necessarily similar.

As with other proofs, vertical angles and parallel lines are often used when proving two triangles are similar. Also take care when writing a similarity statement. The corresponding parts must be listed in the same order.

Example 1

Given: $\overline{BER}$ and $\overline{DES}$ intersect at E,
$\overline{BER} \perp \overline{BS}$, and $\overline{BER} \perp \overline{RD}$

Prove: $\triangle RDE \sim \triangle BSE$

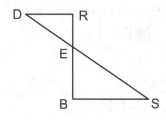

Also state a sequence of rigid motions that will map $\triangle RDE$ onto $\triangle BSE$.

Solution: Use the pair of vertical angles and pair of right angles to show the triangles are similar by AA.

Statements	Reasons
1. $\overline{BER}$ and $\overline{DES}$ intersect at E, $\overline{BER} \perp \overline{BS}$, and $\overline{BER} \perp \overline{RD}$	1. Given
2. $\angle DER \cong \angle SEB$	2. Vertical angles are congruent
3. $\angle B$ and $\angle R$ are right angles	3. Perpendicular lines intersect at right angles
4. $\angle B \cong \angle R$	4. Right angles are congruent
5. $\triangle RDE \sim \triangle BSE$	5. AA

A rotation of $180°$ about point E followed by a dilation centered at E with scale factor $\dfrac{EB}{ER}$ will map $\triangle RDE$ onto $\triangle BSE$.

Example 2

Given: $\overline{XY} \parallel \overline{PQ}$ and $\overline{XZQ}$
intersects $\overline{YZP}$ at Z

Prove: $\triangle ZXY \sim \triangle ZQP$

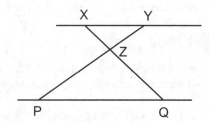

Solution: Use the congruent alternate interior angles to show two angles are congruent.

Statements	Reasons
1. $\overline{XY} \parallel \overline{PQ}$ and $\overline{XZQ}$ intersects $\overline{YZP}$ at Z	1. Given
2. $\angle PQZ \cong \angle YXZ$ and $\angle QPZ \cong \angle XYZ$	2. When two parallel lines are intersected by a transversal, the alternate interior angles are congruent.
3. $\triangle ZXY \sim \triangle ZQP$	3. AA

Example 3

Given: $\overline{BCR}$, $\overline{ACS}$, $\angle A \cong \angle S$, $AB = 8$, and $RS = 14$

Prove: $\dfrac{CR}{CB} = \dfrac{7}{4}$

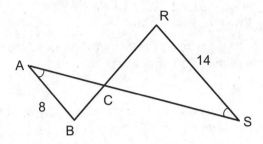

Solution: $\angle A \cong \angle S$ is given. Vertical angles $\angle ACB$ and $\angle RCS$ are congruent. Therefore, $\triangle ACB \sim \triangle SCR$ by the AA criterion. Corresponding parts of similar triangles are proportional, so $\dfrac{RS}{AB} = \dfrac{CR}{CB}$. Substituting the given lengths for RS and AB yields $\dfrac{14}{8} = \dfrac{CR}{CB}$, which simplifies to $\dfrac{CR}{CB} = \dfrac{7}{4}$.

Proving Triangles Similar by the Side-Angle-Side Criterion

If two pairs of corresponding sides are proportional in a pair of triangles and the included angles are congruent, then the triangles are similar.

Example

In $\triangle CAR$, $CA = 9$ and $AR = 12$. $\triangle TOY$ has side lengths $TO = 30$ and $TY = 40$. $m\angle A = m\angle O = 18°$. Explain why $\triangle CAR$ and $\triangle TOY$ are similar.

Solution: The two pairs of given sides are proportional.

$$\frac{CA}{TO} = \frac{9}{30}$$

$$= \frac{3}{10}$$

$$\frac{AR}{TY} = \frac{12}{40}$$

$$= \frac{3}{10}$$

The included angles in each triangle are $\angle A$ and $\angle T$. These both have the same measure, so they are congruent. The triangles are similar by the SAS criterion.

Proving Triangles Similar by the Side-Side-Side Criterion

If three pairs of corresponding sides are proportional in a pair of triangles, then the triangles are similar. Figure 8.3 shows $\triangle ABC \cong \triangle DEF$ by the SSS criterion.

$$\frac{AB}{DE} = \frac{BC}{EF} = \frac{AC}{DF}$$

Figure 8.3 $\triangle ABC \sim \triangle DEF$ by the SSS criterion

Example

In the accompanying figure, $HI = 12$, $IP = 12$, $HP = 8$, $SH = 8$, and $PS = 5\frac{1}{3}$. Prove $\triangle HOP \sim \triangle SHP$.

Solution: By matching up corresponding sides in order of length, we have:

$$\frac{HI}{SH} = \frac{12}{8}$$

$$\frac{PI}{PH} = \frac{12}{8}$$

$$\frac{PH}{PS} = \frac{8}{5\frac{1}{3}}$$

The first two ratios simplify to $\frac{3}{2}$. The third ratio also simplifies to $\frac{3}{2}$, although that one is not quite as obvious. One way to confirm that two ratios are equal is to show that the cross products are equal.

$$\frac{3}{2} = \frac{8}{5\frac{1}{3}}$$

$$3 \cdot 5\frac{1}{3} = 8 \cdot 2$$

$$16 = 16\checkmark$$

Since all three pairs of corresponding sides are proportional, the triangles are similar by SSS.

Similarity Transformations

As discussed in Section 5.2, the dilation is a similarity transformation. If given a pair of similar figures, one can always be mapped to another by a similarity transformation that will include a dilation. The concept is very much like demonstrating congruence by transformations, except that there will be a dilation to match the sizes of the figures. It is easiest to first map the two vertices that will be the center of the dilation onto each other. That way when the dilation is applied, those two vertices will still be coincident.

In the following examples, we will combine multiple aspects of similarity: proving figures similar, finding a missing side using a similarity relationship, and finding a transformation that maps one similar figure to another.

Example 1

Given: $\overline{LSI}$ and $\overline{FSP}$ intersecting at S, $FL \parallel IP$, $IP = 4$, and $FL = 6$

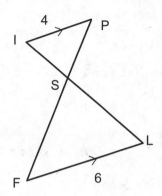

(a) Prove the two triangles formed are similar.

(b) Find the ratio $\dfrac{PS}{FS}$.

(c) State a specific similarity transformation that will map the smaller triangle onto the larger.

Solution:

(a) Given $\overline{FL} \parallel \overline{IP}$, we know $\angle F \cong \angle P$ and $\angle L \cong \angle I$ because they are the alternate interior angles formed by parallel lines and a transversal. Therefore $\triangle SIP \sim \triangle SLF$ by AA. (Note the correct order of vertices in the similarity statement.)

(b) Since corresponding parts are in the same ratio:

$$\frac{PS}{FS} = \frac{IP}{LF}$$
$$= \frac{4}{6}$$
$$= \frac{2}{3}$$

(c) A rotation of $180°$ about point S will align $\overline{SP}$ with $\overline{FS}$ and align $\overline{IS}$ with $\overline{SL}$. A dilation with center at S and a scale factor of $\dfrac{3}{2}$ will then map $\triangle SIP$ to $\triangle SLF$. Note that we have to use the reciprocal of the ratio in part (b) because we want to enlarge the smaller triangle, not reduce it. In transformation notation, the required notation is

$$R_{180°,S} \circ D_{S,\frac{3}{2}} (\triangle SIP).$$

Example 2

$ABCD$ is a rectangle, and $\overline{AE} \parallel \overline{FG}$. $AF = 15$, $FB = 3$, $BG = 4$, and $GC = 6$.

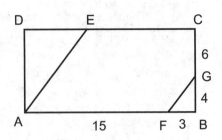

(a) Prove $\triangle ADE \sim \triangle GBF$.

(b) Find DE.

(c) What transformation maps $\triangle GBF$ to $\triangle ADE$?

Solution:

(a) Opposite angles D and B are the first pair of congruent angles. To show the second pair is congruent, we need to use the two pairs of parallel lines, $\overline{AB} \parallel \overline{CD}$ and $\overline{FG} \parallel \overline{AE}$.

Statements	Reasons
1. $ABCD$ is a rectangle and $\overline{AE} \parallel \overline{FG}$	1. Given
2. $\angle D \cong \angle B$	2. Opposite angles in a rectangle are congruent
3. $\angle BFG \cong \angle BAE$	3. Corresponding angles formed by parallel lines are congruent
4. $\overline{AB} \parallel \overline{CD}$	4. Opposite sides of a rectangle are parallel
5. $\angle BAE \cong \angle DEA$	5. Alternate interior angles formed by parallel lines are congruent
6. $\angle BFG \cong \angle DEA$	6. Transitive property
7. $\triangle ADE \sim \triangle GBF$	7. AA

(b) $\overline{BG}$ and $\overline{AD}$ are congruent corresponding sides, as are $\overline{FB}$ and $\overline{DE}$. Opposite sides of a rectangle are congruent, so $AD = BC = 10$. We can form a proportion:

$$\frac{AD}{BG} = \frac{DE}{FB}$$

$$\frac{10}{4} = \frac{DE}{3}$$

$$4 \cdot DE = 30$$

$$DE = 7.5$$

(c) Three steps are needed to map $\triangle GBF$ to $\triangle ADE$. First translate $\triangle GBF$ by vector $\overline{BD}$. This maps B to D. Next rotate 180° about point D to align BG with DA and align BF with DE. Finally a dilation is needed. The scale factor is $\dfrac{AD}{BG} = \dfrac{10}{4}$, or $\dfrac{5}{2}$, and the center is point D. The complete transformation can be expressed symbolically as $D_{D,\frac{5}{2}} \circ R_{D,180°} \circ T_{\overline{BD}}(\triangle GBF)$.

Check Your Understanding of Section 8.2

A. *Multiple-Choice*

1. Which of the following additional pieces of information would let you conclude the two triangles are similar by SSS?
   (1) $a = 30$ and $b = 18$
   (2) $a = 24$ and $b = 12$
   (3) $a = 22$ and $b = 10$
   (4) $a = 20$ and $b = 12$

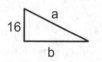

2. In the accompanying figure, $XY = 8$, $XM = 12$, $MA = 9$, $BA = 6$, and $\angle XMY \cong \angle AMB$. Which of the following statements is true?
   (1) $\triangle XYM \sim \triangle ABM$
   (2) $\triangle XYM \sim \triangle AMB$
   (3) $\triangle MYX \sim \triangle BAM$
   (4) The two triangles are not similar.

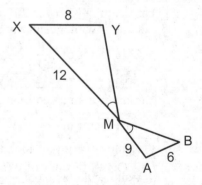

3. In the accompanying figure, $\triangle PQR \sim \triangle TSR$ and points P, R, and T are collinear. Which of the following transformations will map $\triangle TSR$ to $\triangle PQR$?

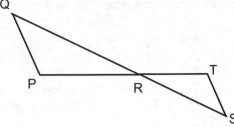

(1) reflection over line TP followed by a dilation with a scale factor of $\dfrac{QR}{TR}$ centered at Q

(2) translation by vector $\overrightarrow{TR}$ followed by a dilation with a scale factor of $\dfrac{PQ}{ST}$ centered at R

(3) rotation of $180°$ about point R followed by a dilation with a scale factor of $\dfrac{PQ}{ST}$ centered at R

(4) rotation of $180°$ about point R followed by a dilation with a scale factor of $\dfrac{QR}{TR}$ centered at R

4. If $\triangle XYZ \sim \triangle XVW$, which of the following transformations will map $\triangle XVW$ to $\triangle XYZ$?

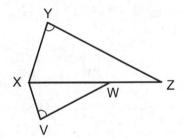

(1) reflect $\triangle XVW$ over $\overline{XW}$, and then dilate it with a center at X and scale factor of $\dfrac{XW}{WZ}$

(2) reflect $\triangle XVW$ over $\overline{XW}$, and then dilate it with a center at X and scale factor of $\dfrac{XZ}{XW}$

(3) rotate $\triangle XVW$ $180°$ about point X, and then dilate it with a center at X and scale factor of $\dfrac{XW}{WZ}$

(4) rotate $\triangle XVW$ $180°$ about point X, and then dilate it with a center at X and scale factor of $\dfrac{XZ}{XW}$

B. *Free-Response—show all work or explain your answer completely.*

5. Name a sequence of transformations that will map $\triangle RAT$ to $\triangle PIG$.

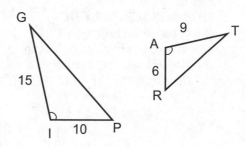

6. Given: $\overline{RK}$ and $\overline{LD}$ intersect at G, $GK = 8$, $GL = 12$, $GR = 14$, and $GD = 21$
Prove: $\triangle KGL \sim \triangle RGD$

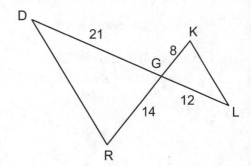

7. Given: $\triangle VLY$ and $\triangle TUX$, $VL = 8$, $LY = 10$, $VY = 4$, $TU = 4$, $UX = 5$, and $XT = 2$
Prove: $\angle V \cong \angle T$

8. Given: $\overline{ACE}$, $\overline{BC} \parallel \overline{DE}$, $\overline{AB} \parallel \overline{CD}$
Prove: $\triangle ABC \sim \triangle CDE$

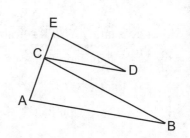

9. Square *BFED* is inscribed in △*ABC*. *EF* = 8 and *EC* = 16.
   (a) Prove △*EFA* ~ △*CDE*.
   (b) What is the ratio of *CD*:*EF*? Express your answer in simplest radical form.
   (c) Name a specific similarity transformation that will map △*EFA* to △*CDE*.

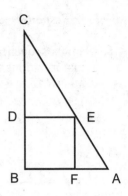

Use this figure for problems 10–11:

10. Given: parallelogram *ABCD*,
    $\overline{FEC}$ intersects $\overline{DEG}$ at *E*
    Prove: △*FEG* ~ △*CED*

11. *ABCD* is a parallelogram, $AF = \frac{1}{5}CD$, and $BG = \frac{1}{4}CD$. What is the ratio *DE*:*EF*? What transformation will map △*FEG* to △*CED*?

12. $\overline{SVT}$ bisects ∠*RSU*, *RS* = 15, *SU* = 6, *VS* = 4, and *TV* = 6
    Prove: △*RST* ~ △*USV*

13. ∠*RTS* ≅ ∠*YTS*, $\overline{TY} \cong \overline{YS}$, and $\overline{TR} \cong \overline{TS}$
    Prove: △*RTS* ~ △*TYS*

14. Given: ∠*O* and ∠*QPN* are right angles
    Prove: *NP* · *MO* = *QP* · *NO*

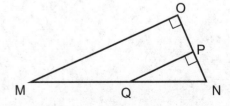

8.3 SIMILAR TRIANGLE RELATIONSHIPS

Segment Parallel to a Side of a Triangle

> ### Segment Parallel to a Side Theorem
> - A segment parallel to a side of a triangle forms a triangle similar to the original triangle.
> - If a segment intersects two sides of a triangle such that a triangle similar to the original is formed, the segment is parallel to the third side of the original triangle.

Figure 8.4 shows $\triangle ABC$ with $\overline{PQ}$ parallel to $\overline{AB}$. $\triangle PQC$ is similar to $\triangle ABC$ by the AA postulate. $\angle BAC$ and $\angle QPC$ are congruent corresponding angles formed by the parallel lines. The same is true for $\angle ABC$ and $\angle PQC$. Two pairs of corresponding angles are congruent, so the triangles are similar by AA.

The triangles can be sketched separately to help identify corresponding parts. Sometimes segment addition is needed to find a length needed to complete a proportion involving corresponding sides.

$$\overline{PQ} \parallel \overline{AB}$$
$$\triangle ABC \sim \triangle PQC$$

Figure 8.4 Segment $\overline{PQ} \parallel$ side $\overline{AB}$

Example

In $\triangle WXY$, $\overline{RS}$ is parallel to $\overline{XW}$, $RY = 8$, $XR = 12$, and $XW = 25$. Find RS.

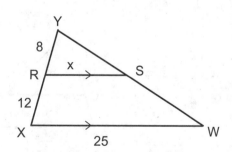

257

Solution: Start by sketching the two triangles separately and using segment addition to find XY.

$$XY = XR + RY$$
$$= 12 + 8$$
$$= 20$$

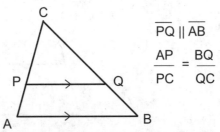

The similarity relationship is $\triangle WXY \sim \triangle SRY$. The proportion of corresponding sides is as follows:

$$\frac{XY}{RY} = \frac{XW}{RS}$$

$$\frac{20}{8} = \frac{25}{RS}$$

$$20 \cdot RS = 200$$

$$RS = 10$$

Side Splitter Theorem

A segment parallel to a side in a triangle divides the two sides it intersects proportionally.

Figure 8.5 shows $\overline{PQ}$ parallel to $\overline{AB}$ and dividing sides $\overline{AC}$ and $\overline{BC}$ proportionally. Specifically:

$$\frac{AP}{PC} = \frac{BQ}{QC}$$

$\overline{PQ} \parallel \overline{AB}$

$$\frac{AP}{PC} = \frac{BQ}{QC}$$

Figure 8.5 Parallel segment $\overline{PQ}$ divides $\overline{AC}$ and $\overline{BC}$ proportionally

The side splitter theorem is a shortcut that you can use to find a missing side when there is a line or segment parallel to a side of a triangle. You can prove the side splitter theorem by using the relationship $\dfrac{PC}{AC} = \dfrac{QC}{BC}$ and then applying the partition postulate. The important thing to remember is that it is the two sides intersected by the parallel segment that are divided proportionally. The ratio $\dfrac{AB}{PQ}$ *cannot* be combined with the ratio $\dfrac{PC}{AC}$.

Example 1

In $\triangle HRS$, $\overline{OE} \parallel \overline{RS}$, $OH = (x + 1)$, $OR = 2$, $EH = (x + 3)$, and $ES = 2.5$. Find the length of HR.

Solution: Side $\overline{OE}$ divides the two sides proportionally according to the side splitter theorem.

$$\frac{OH}{OR} = \frac{EH}{ES}$$

$$\frac{x+1}{2} = \frac{x+3}{2.5}$$

$2.5(x+1) = 2(x+3)$ set the cross products equal

$2.5x + 2.5 = 2x + 6$

$0.5x + 2.5 = 6$

$0.5x = 3.5$

$x = 7$

$HO = x + 1$
$\quad = 7 + 1$ substitute $x = 7$
$\quad = 8$

$HR = OH + OR$ segment addition
$\quad = 8 + 2$
$\quad = 10$

The side splitter theorem also applies if multiple parallel lines are intersected by two transversals. Figure 8.6 shows 4 parallel lines intersected by two transversals. Any corresponding pairs of segments will be proportional:

$$\frac{a}{b} = \frac{d}{e}, \frac{b}{c} = \frac{e}{f}, \frac{a}{c} = \frac{d}{f}$$

Example 2

In Figure 8.6, $a = 12$, $c = (x + 4)$, $d = 18$, and $f = (x + 7)$. Find c.

   Solution: Use the side splitter theorem:

$$\frac{a}{c} = \frac{d}{f}$$

$$\frac{12}{x+4} = \frac{18}{x+7}$$

$12(x+7) = 18(x+4)$ set the cross products equal

$12x + 84 = 18x + 72$

$-6x + 84 = 72$

$-6x = -12$

$x = 2$

$c = x + 4$

$c = 2 + 4$ substitute $x = 2$

$= 6$

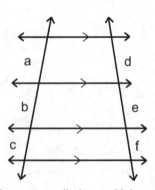

Figure 8.6 Side splitter theorem applied to multiple parallel lines intersected by two transversals

Centroid

Recall that the median of a triangle is a segment from a vertex of the triangle to the midpoint of the opposite side. Also that the medians in a triangle are all concurrent at the centroid. The centroid has the special property that it divides each median in a $1:2$ ratio.

Centroid Theorem

The centroid of a triangle divides each median in a $1:2$ ratio, with the longer segment having a vertex as one of its endpoints.

If you are given the length of either the shorter or longer piece of a median, simply multiply or divide by 2 to find the other piece. If you are given the length of the entire median, the $1:2$ ratio allows you to assign x to the shorter piece and $2x$ to the longer piece. Applying the whole equals the sum of the parts results in:

$$x + 2x = \text{length of median}$$

Example 1

In $\triangle PQR$, medians $\overline{PM}$, $\overline{QN}$ and $\overline{RO}$ intersect at G. $PM = 12$, $QG = 9$, and $OG = 2$.

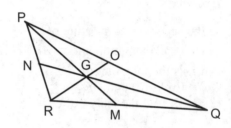

  (a) Find PG
  (b) Find QN
  (c) Find RO

   Solution: Each median is divided in a $1:2$ ratio.

(a) Let $PG = 2x$ and $GM = x$:

$$
\begin{aligned}
PG + GM &= PM \\
2x + x &= 12 \\
3x &= 12 \\
x &= 4 \\
PG &= 2x \\
&= 2(4) \\
&= 8
\end{aligned}
$$

(b) QG is the longer part of median $\overline{QN}$:

$$GN = \frac{1}{2}QG$$

$$= \frac{1}{2}(9)$$

$$= 4.5$$

$$
\begin{aligned}
QN &= QG + GN \\
&= 9 + 4.5 \\
&= 13.5
\end{aligned}
$$

261

(c) OG is the shorter part of median $\overline{RO}$:

$$GR = 2 \cdot OR$$
$$= 2(2)$$
$$= 4$$
$$RO = OG + GR$$
$$= 2 + 4$$
$$= 6$$

You may not be told explicitly that a segment is a median. However, if its endpoints are a vertex and the midpoint of the opposite side, then it must be a median. If you are given expressions for each part of the median in terms of a variable, set up a proportion using the $1:2$ ratio and then solve the proportion for the variable.

Example 2

In the accompanying figure of $\triangle ABC$, D is the midpoint of $\overline{AB}$ and E is the midpoint of $\overline{BC}$. If $GC = (6x - 12)$ and $GD = (2x + 4)$, find the length of $\overline{CD}$.

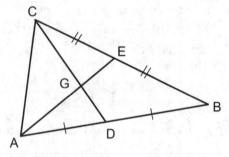

Solution: $\overline{CD}$ and $\overline{AE}$ are medians, so G is a centroid and divides $\overline{CD}$ in a $1:2$ ratio:

$$\frac{GD}{GC} = \frac{1}{2}$$
$$\frac{2x+4}{6x-12} = \frac{1}{2}$$
$$6x - 12 = 2(2x + 4)$$
$$6x - 12 = 4x + 8$$
$$2x - 12 = 8$$
$$2x = 20$$
$$x = 10$$

Now substitute $x = 10$ and solve for GC and GD:

$$GD = 6x - 12 \qquad GC = 2x + 4$$
$$= 6(10) - 12 \qquad = 2(10) + 4$$
$$= 48 \qquad\qquad = 24$$

Add the two parts to find CD:

$$CD = GD + GC$$
$$= 48 + 24$$
$$= 72$$

Midsegments

> **Definition**
> A midsegment is a segment joining the midpoints of two sides of a triangle.

> **Midsegment Theorem**
> A midsegment of a triangle is parallel to the opposite side and its length is equal to $\dfrac{1}{2}$ the length of the opposite side.

The midsegment relationships are easily proven using the theorem that states that corresponding parts of similar triangles are proportional. Figure 8.7 shows midsegment $\overline{DE}$ in $\triangle ABC$. $\triangle ABC$ can be proven similar to $\triangle ADE$ using the SAS similarity postulate. Since D and E are midpoints, $\dfrac{AD}{AB} = \dfrac{1}{2}$ and $\dfrac{AE}{AC} = \dfrac{1}{2}$, resulting in two pairs of proportional sides. The included angle, $\angle A$, is a shared angle. Therefore $\triangle ABC \sim \triangle ADE$ by SAS.

If given expressions for a midsegment and the opposite side, use the following relationship:

$$\text{opposite side} = 2 \cdot \text{midsegment}$$

Figure 8.7 Midsegment $\overline{DE}$ in $\triangle ABC$

Example 1

Points M and N are the midpoints of $\overline{DQ}$ and $\overline{RQ}$ in $\triangle DRQ$. If $MN = (x + 5)$ and $DR = (6x - 6)$, find the length of $\overline{DR}$.

Solution: First sketch the figure.

We see midsegment $\overline{MN}$ is opposite side $\overline{DR}$:

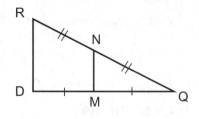

$$DR = 2MN$$
$$6x - 6 = 2(x + 5)$$
$$6x - 6 = 2x + 10$$
$$4x - 6 = 10$$
$$4x = 16$$
$$x = 4$$
$$DR = 6x - 6$$
$$= 6(4) - 6$$
$$= 18$$

Example 2

In $\triangle BIG$, the midpoint of side $\overline{BI}$ is P and the midpoint of side $\overline{IG}$ is Q. If $m\angle QPI = 38°$ and $m\angle BGI = 41°$, what is the measure of $\angle I$?

Solution: It is best to sketch the figure first.

$\overline{PQ}$ is a midsegment and is parallel to $\overline{BG}$. $\angle B$ and $\angle QPI$ are congruent corresponding angles, so $m\angle B = 38°$. We can now use the triangle angle sum theorem in $\triangle BIG$ to find $m\angle I$.

$$m\angle B + m\angle G + m\angle I = 180°$$
$$38° + 41° + m\angle I = 180°$$
$$79° + m\angle I = 180°$$
$$m\angle I = 101°$$

Similarity in Right Triangles

Altitude to a Hypotenuse Theorem

An altitude to the hypotenuse of a right triangle will divide it into two similar triangles, each of which is also similar to the original right triangle.

In Figure 8.8 of right $\triangle ABC$, altitude $\overline{CD}$ is drawn to hypotenuse $\overline{ADB}$. Three similar triangles are formed—$\triangle CBD$, $\triangle ACD$, and the original $\triangle ABC$.

When given any three parts, or sometimes two parts, any of the other parts can be found by combining the fact that corresponding parts are proportional and by using the Pythagorean theorem. Two useful strategies for dealing with these overlapping similar triangles is to sketch the triangles separately or to make a table with sides classified as "leg 1," "leg 2," and "hypotenuse." The altitude will be leg 1 of one of the interior triangles and will be leg 2 of the other.

Figure 8.8 Altitude $\overline{CD}$ drawn to hypotenuse $\overline{ADB}$ in right $\triangle ABC$

Example 1

Altitude $\overline{LE}$ is drawn to hypotenuse $\overline{BU}$ in right triangle $\overline{BLU}$. If $LU = 15$ and $LE = 9$, what is the length of BL?

Solution: Start by filling out a table of each of the sides of the three similar triangles. Let LE be leg 1 of $\triangle LUE$ and leg 2 of $\triangle BLE$. (It won't matter if you switch them.)

	$\triangle BLE$	$\triangle LUE$	$\triangle BLU$
Leg 1		9	x
Leg 2	9		15
Hypotenuse	x	15	

To make a proportion, we need two matching pairs of corresponding parts from two triangles. We don't have the necessary two pairs. However, the table shows that if we can find leg 2 of $\triangle LUE$ (side EU), we can form a proportion to find BL. We can find EU using the Pythagorean theorem:

265

$$LE^2 + EU^2 = LU^2$$
$$9^2 + EU^2 = 15^2$$
$$81 + EU^2 = 225$$
$$EU^2 = 144$$
$$EU = 12$$

We can now apply the proportional sides relationship with the following sides:

$$\frac{\text{leg } 2_{\triangle BLE}}{\text{leg } 2_{\triangle LUE}} = \frac{\text{hypotenuse}_{\triangle BLE}}{\text{hypotenuse}_{\triangle LUE}}$$

$$\frac{9}{x} = \frac{12}{15}$$

$$12x = 9 \cdot 12$$

$$12x = 135$$

$$x = 11.25$$

MATH FACT

When three numbers satisfy the proportion $\frac{a}{b} = \frac{b}{c}$, b is said to be the "mean proportional" between a and c. In geometry, the mean proportional shows up in similar right triangles. The altitude to the hypotenuse in a right triangle is the mean proportional to the two parts of the hypotenuse.

Example 2

$\overline{ON}$ is the altitude to hypotenuse $\overline{PM}$ of $\triangle PNM$. If $PO = 12$ and $MO = 16$, what is the length of $\overline{ON}$? Express your answer in simplest radical form.

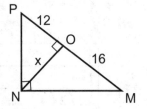

Solution: *ON* is the mean proportional between *OP* and *MO*. The proportion is

$$\frac{PO}{x} = \frac{x}{MO}$$

$$\frac{12}{x} = \frac{x}{16}$$

$$x^2 = 12 \times 16$$

$$x^2 = 192$$

$$x = \sqrt{192}$$

$$= 8\sqrt{3}$$

You would arrive at the same proportion using the table method; the proportion is equivalent to $\dfrac{\text{leg }1_{\triangle NOP}}{\text{leg }1_{\triangle MNO}} = \dfrac{\text{leg }2_{\triangle NOP}}{\text{leg }2_{\triangle MNO}}$.

Proving the Pythagorean Theorem

One of the required proofs in the curriculum is proving the Pythagorean theorem. There are many ways to prove this theorem, but the method you need to be able to use is one using similar triangles.

We just saw the theorem that an altitude to the hypotenuse of a right triangle divides the triangle into two triangles similar to the original. This theorem can be used to prove the familiar $a^2 + b^2 = c^2$.

The strategy is to write two proportions relating corresponding sides. The first proportion will use the original triangle and one interior triangle ($\triangle ABC$ and $\triangle BDC$ in the figure below). The second proportion uses the original triangle and the other interior triangle ($\triangle ABC$ and $\triangle CDA$). These can be rearranged algebraically and added to give the desired result.

Given: right triangle $\triangle ABC$ with a right angle at C, altitude $\overline{CD}$ drawn to hypotenuse $\overline{AB}$

Prove: $AC^2 + BC^2 = AB^2$

Statements	Reasons
1. Right triangle $\triangle ABC$ with a right angle at C, altitude $\overline{CD}$ drawn to hypotenuse $\overline{AB}$	1. Given
2. $\triangle BDC \sim \triangle CDA \sim \triangle BCA$	2. The altitude to the hypotenuse of a right triangle forms 3 similar right triangles
3. $\dfrac{AB}{AC} = \dfrac{AC}{AD}$ $\dfrac{AB}{BC} = \dfrac{BC}{BD}$	3. Corresponding sides in similar triangles are proportional
4. $AC^2 = AB \cdot AD$ $BC^2 = AB \cdot BD$	4. The cross products of a proportion are equal
5. $AC^2 + BC^2 = AB \cdot AD + AB \cdot BD$	5. Addition property
6. $AC^2 + BC^2 = AB(AD + BD)$	6. Distributive property
7. $AD + BD = AB$	7. Partition property
8. $AC^2 + BC^2 = AB(AB)$	8. Substitution
9. $AC^2 + BC^2 = AB^2$	9. Simplify

Check Your Understanding of Section 8.3

A. *Multiple-Choice*

Use for problems 1–3 (*figure not to scale*):

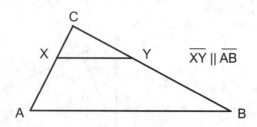

1. In $\triangle ABC$, $\overline{XY}$ is drawn parallel to $\overline{AB}$. If $AX = 9$, $XC = 3$, and $XY = 10$, what is the length of $\overline{AB}$?
   (1) 2.5   (2) 20   (3) 30   (4) 40

2. In $\triangle ABC$, $\overline{XY}$ is drawn parallel to $\overline{AB}$ such that X lies on $\overline{AC}$ and Y lies on $\overline{BC}$. If $AC = 24$, $XC = 4$, and $BC = 30$, what is the length of $\overline{YC}$?
   (1) 4.5   (2) 5   (3) 5.5   (4) 6

3. In $\triangle ABC$, $\overline{XY}$ is drawn parallel to $\overline{AB}$ such that X lies on $\overline{AC}$ and Y lies on $\overline{BC}$. If $AX = (m + 4)$, $XC = m$, $YC = (m + 3)$, and $BY = 15$, what are the two possible values of m?
   (1) 2 and 6   (2) 4 and 3   (3) 5 and 12   (4) 6 and 10

4. In $\triangle PQR$, A is a point on $\overline{PR}$ and B is a point on $\overline{QR}$ such that $\overline{AB} \parallel \overline{PQ}$. If $AR = 2$, $PA = 8$, and $BR = 3$, what is the length of $\overline{QB}$?
   (1) $\dfrac{3}{4}$ (2) 5   (3) 9   (4) 12

5. In $\triangle FOX$, Q is a point on $\overline{OF}$ and R is a point on $\overline{OX}$ such that $\overline{QR} \parallel \overline{FX}$. If $QR = 18$, $FX = 24$, and $OR = 6$, what is the length of RX?
   (1) 1   (2) 2   (3) 3   (4) 4

6. In $\triangle JAR$, S is the midpoint of $\overline{RA}$ and T is the midpoint of $\overline{JR}$. If QA = $(8x - 4)$ and $QT = 34$, what is the value of x?
   (1) 2   (3) 9
   (2) 4   (4) 12

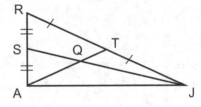

7. In the figure of $\triangle SIT$, midpoints P, O, and W are joined to form $\triangle POW$. If $TW = 8$, $SP = 7$, and $OI = 9$, what is the perimeter of $\triangle POW$?
   (1) 6   (3) 18
   (2) 12   (4) 24

Use the following figure for problems 8 and 9:

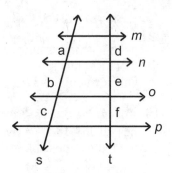

8. Lines m, n, o, and p are all parallel to each other. They are intersected by transversals s and t. If $a = 27$, $b = 30$, and $e = 25$, what is the length of d?

(1) $22\dfrac{1}{2}$ (2) 22 (3) $23\dfrac{1}{4}$ (4) $27\dfrac{7}{9}$

9. Lines m, n, o, and p are all parallel to each other. They are intersected by transversals s and t. If e is $\dfrac{3}{4}f$, which of the following is *not* *necessarily* true?

(1) b is $\dfrac{3}{4}c$ (2) $\dfrac{a}{b} = \dfrac{d}{e}$ (3) $\dfrac{b}{c} = \dfrac{e}{f}$ (4) a is $\dfrac{3}{4}b$

Use the following figure for problems 10 and 11:

10. Right triangle ABC has a right angle at C, and altitude $\overline{CD}$ is drawn to $\overline{AB}$. If $AD = 12$ and $BD = 4$, what is the length of CD?

(1) $3\sqrt{6}$ (2) $4\sqrt{3}$ (3) $4\sqrt{10}$ (4) 48

11. Right triangle ABC has a right angle at C, and altitude $\overline{CD}$ is drawn to $\overline{AB}$. If $AC = 8$ and $BC = 6$, what is the length of AD?

(1) 4.8 (2) 5.8 (3) 6.4 (4) 10

B. *Free-Response—show all work or explain your answer completely.*

12. In $\triangle LST$, medians $\overline{LD}$ and $\overline{SE}$ intersect at G. If $LG = (x^2 + x)$ and $GD = (3x + 3)$, what is the length of median $\overline{LD}$?

270

13. Jacob has a sketch of line $\overline{MN}$ and point P that is not on line $\overline{MN}$. He states that the only way to construct a line parallel to $\overline{MN}$ through P with a compass and straightedge is to use the "copy an angle" construction. Brianna says it could also be done using what she knows about the midsegment theorem. Who is correct? Explain your reasoning.

14. Given: Right triangle RST with hypotenuse $\overline{ST}$ and altitude $\overline{RU}$
Prove: $RS^2 + RT^2 = ST^2$

15. Given: D is the midpoint of $\overline{AC}$ and E is the midpoint of $\overline{BC}$, $\angle FGD$ and $\angle GFE$ are right angles
Prove: $DEFG$ is a rectangle

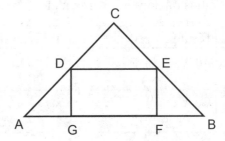

16. Prove the midsegment theorem.
Given $\triangle BAT$, G is the midpoint of $\overline{BT}$ and K is the midpoint of $\overline{AT}$.
Prove $GK = \dfrac{1}{2} AB$.

8.4 RIGHT TRIANGLE TRIGONOMETRY

Definition of Trigonometric Ratios

From the properties of similar figures, we know that the ratio of any two sides in one triangle will be equal to the ratio of two corresponding sides in a similar triangle. Right triangles occur so frequently in practical applications that mathematicians have given names to the parts of a right triangle and to the ratios formed by the sides of a right triangle.

We refer to the sides of a right triangle relative to a particular acute angle using the names **adjacent**, **opposite**, and **hypotenuse**. Figure 8.9 shows the position of the adjacent, opposite, and hypotenuse for the two acute angles A and B in $\triangle ABC$.

- Hypotenuse—the side across from the right angle
- Opposite—the side across from the specified acute angle
- Adjacent—the side included between the right angle and the specified acute angle

Figure 8.9 (a) Adjacent, opposite, and hypotenuse relative to angle *A*
(b) Adjacent, opposite, and hypotenuse relative to angle *B*

Right triangle trigonometry in this course will focus the following special ratios of sides in right triangles.

Ratio	Abbreviation	Definition
Sine	sin	$\dfrac{\text{opposite}}{\text{hypotenuse}}$
Cosine	cos	$\dfrac{\text{adjacent}}{\text{hypotenuse}}$
Tangent	tan	$\dfrac{\text{opposite}}{\text{adjacent}}$

When working with trigonometric ratios, keep the following facts in mind:

- Sine, cosine, and tangent are functions that depend on the input angle.
- Sine, cosine, and tangent will have the same value for a given angle, regardless of the side lengths of the triangle.
- The sine and cosine ratios will always be greater than 0 and less than 1. The tangent ratio can be greater that 1.

===== **MATH FACT** =====

The power of trigonometry is that these ratios will always be the same for a particular angle in a right triangle, regardless of how big or small the triangle is. This occurs because two right triangles with one pair of congruent acute angles must be similar by the AA postulate. By just knowing the angle, we know the ratio of two sides.

Example 1

Find sin(*C*), cos(*C*), and tan(*C*) in the accompanying triangle.

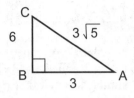

Solution:

$$\sin(C) = \frac{\text{opposite}}{\text{hypotenuse}} = \frac{3}{3\sqrt{5}} = \frac{1}{\sqrt{5}} = \frac{\sqrt{5}}{5}$$

$$\cos(C) = \frac{\text{adjacent}}{\text{hypotenuse}} = \frac{6}{3\sqrt{5}} = \frac{2}{\sqrt{5}} = \frac{2\sqrt{5}}{5}$$

$$\tan(C) = \frac{\text{opposite}}{\text{adjacent}} = \frac{3}{6} = \frac{1}{2}$$

Note that the radical expressions for sin(*C*) and cos(*C*) were rationalized by multiplying the numerator and denominator by $\sqrt{5}$ in the last step.

The Pythagorean theorem can be used to find a missing side that may be needed to form a trigonometric ratio.

Example 2

In △*ABC*, m∠*B* = 90°, *BC* = 6, and *AC* = 10. Find the value of cos(*A*).

Solution: Sketching the figure will help identify the proper parts.

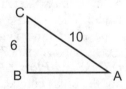

We need the adjacent side, *AB*, to form the cosine ratio. By applying the Pythagorean theorem, we get:

$$AB^2 + BC^2 = AC^2$$
$$AB^2 + 6^2 = 10^2$$
$$AB^2 + 36 = 100$$
$$AB^2 = 64$$
$$AB = 8$$

$$\cos(A) = \frac{\text{adjacent}}{\text{hypotenuse}} = \frac{8}{10} = \frac{4}{5}$$

When you are given a trigonometric ratio but no side lengths, you can assume any side lengths consistent with the given ratio. Any similar right triangle with sides in the same proportion will have the same angles and trigonometric ratios.

Example 3

In $\triangle FGH$, $m\angle H = 90°$ and $\sin(F) = \dfrac{2}{3}$. What is $\tan(F)$?

Solution: Start by sketching $\triangle FGH$. Let
$GH = 2$ and $FG = 3$. The resulting triangle will be
similar to any other right triangle with the same
angle measures.

$$GH^2 + HF^2 = GF^2 \qquad \text{Pythagorean theorem}$$
$$2^2 + HF^2 = 3^2$$
$$4 + HF^2 = 9$$
$$HF^2 = 5$$
$$HF = \sqrt{5}$$

$$\tan(F) = \frac{\text{opposite}}{\text{adjacent}}$$

$$\tan(F) = \frac{2}{\sqrt{5}} = \frac{2\sqrt{5}}{5}$$

Example 4

$\triangle PQR$ has a right angle at Q, $\sin(P) = a$, and $\cos(P) = b$. Find the values of
$\sin(R)$ and $\cos(R)$.

Solution: Sketch and label the
figure. Then choose values for the
opposite and hypotenuse that are
consistent with $\sin(P) = a$. Letting
$QR = a$ and $PR = 1$ is an obvious (but
not the only) choice. Since $\cos(P)$
must equal b and the hypotenuse
equals 1, then PQ must equal b.

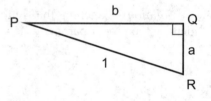

$$\sin(R) = \frac{\text{opposite}}{\text{hypotenuse}} = \frac{PQ}{PR}$$

$$\sin(R) = \frac{b}{1} = b$$

$$\cos(R) = \frac{\text{adjacent}}{\text{hypotenuse}}$$

$$\cos(R) = \frac{a}{1} = a$$

Calculator Tip

Scientific and graphing calculators can calculate the trigonometric ratios, with the **sin**, **cos**, and **tan** keys. Most calculators can be set to different units of angle measure, such as the degree, radian, or grad. It is important to know which unit the angle is given in. If the angle is followed by the degree symbol °, you are working in degrees, so the calculator should be in degree mode. For the TI-84 family of calculators, for example, you can switch between degrees and radians by pressing **mode** and using the arrow keys to select **degree** or **radian**.

Remember—Some graphing calculators are automatically set to radian mode whenever the memory is reset! You may need to set it back to degree mode. Finally, when doing trigonometric calculations using the calculator, do not round until you have reached the final answer. This is good practice for any problem involving numerical calculations.

Example 5

In $\triangle ABC$, $\angle C$ is a right angle and $m\angle A = 41°$. Find the ratios $\dfrac{AC}{AB}$, $\dfrac{BC}{AC}$, and $\dfrac{BC}{AB}$. Round to the nearest thousandth.

Solution: A sketch of the triangle shows that $\overline{BC}$ is the opposite of $\angle A$, $\overline{AC}$ is the adjacent of $\angle A$, and $\overline{AB}$ is the hypotenuse.

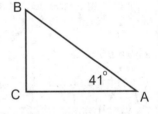

$$\frac{AC}{AB} = \cos(A)$$
$$= \cos(41°) = 0.755$$

$$\frac{BC}{AC} = \tan(A)$$
$$= \tan(41°) = 0.869$$

$$\frac{BC}{AB} = \sin(A)$$
$$= \sin(41°) = 0.656$$

Inverse Trigonometric Functions

The inverse of a function "undoes" the original function. For example, $f(x) = \sqrt{x}$ and $g(x) = x^2$ are inverse functions. If $x = 5$, then $f(g(5)) = \sqrt{5^2} = 5$. The trigonometric functions $\sin(x)$, $\cos(x)$, and $\tan(x)$ also have inverses, called the arcsin(x), arccos(x), and arctan(x). They are often abbreviated $\sin^{-1}(x)$, $\cos^{-1}(x)$, and $\tan^{-1}(x)$.

The input to the inverse trigonometric functions is a ratio, and the output is an angle. The function $\sin^{-1}(x)$, for example, returns the angle whose sine ratio is x. The same goes for $\cos^{-1}(x)$ and $\tan^{-1}(x)$.

We use the inverse trigonometric ratio to find an angle given the ratio of a pair of sides in a right triangle. On the TI-84, press the **2ND** button followed by either **sin**, **cos**, or **tan** to calculate $\sin^{-1}$, $\cos^{-1}$, or $\tan^{-1}$, respectively.

Example 1

Using the accompanying figure, find the measure of angles M and N.

Solution:

$$m\angle M = \cos^{-1}\left(\frac{3}{7}\right) = 64.6°$$

$$m\angle N = \sin^{-1}\left(\frac{3}{7}\right) = 25.4°$$

Note that as expected, the measures found for $\angle M$ and $\angle N$ add to $90°$.

Example 2

Right triangle BTP has a right angle at T, $BT = 4$, and $PT = 11$. What is the value of $\cos(P)$? Round to the nearest thousandth.

Solution: The method we saw previously was to use the Pythagorean theorem to find BP and then calculate $\cos(P)$ using $\dfrac{\text{adjacent}}{\text{hypotenuse}}$. Alternatively, we could use the inverse tangent function to find $m\angle P$ and then use the calculator to find $\cos(P)$.

$$m\angle P = \tan^{-1}\left(\frac{4}{11}\right)$$

$$= 19.9831°$$

$$\cos(P) = \cos(19.9831°)$$
$$= 0.9397$$
$$= 0.940$$

Example 3

In the accompanying figure, m∠GDO = 90°, m∠DSG = 90°, SG = 3, and SD = 5. Find the measure of angle O. Round to the nearest 0.01 degree.

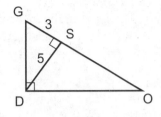

Solution: Strategy—Use the inverse tangent function to find m∠G. Then use the triangle angle sum theorem to find m∠O.

$$m\angle G = \tan^{-1}\left(\frac{5}{3}\right) = 59.0362°$$

$$m\angle D + m\angle G + m\angle O = 180°$$
$$90° + 59.0362° + m\angle O = 180°$$
$$m\angle O = 30.9637°$$
$$= 30.96°$$

Using Trigonometric Ratios to Find an Unknown Side

Given any one angle in a triangle and any one side, all the remaining sides can be determined using trigonometric ratios. Write a ratio using a ratio of the known side to the unknown side. Set that ratio equal to the appropriate trigonometric ratio. Solve the resulting proportion for the unknown side, using your calculator to find the value of the trigonometric ratio.

Example 1

Right triangle *SOX* has a right angle at *O*. The measure of angle *S* = 72°, and *XO* = 12. Find the length of *OS* to the nearest tenth.

Solution: Strategy—Relative to ∠S, the given side and the unknown side are the opposite and adjacent, which suggests using the tangent ratio.

$$\tan(S) = \frac{\text{opposite}}{\text{adjacent}} = \frac{OX}{OS}$$

$$\tan(72°) = \frac{12}{OS}$$

$$OS = \frac{12}{\tan(72°)}$$

$$OS = \frac{12}{3.07768} = 3.89903$$

$$= 3.9$$

Example 2

The perimeter of an equilateral triangle is 60 inches. What is the length of its altitude? Round your answer to the nearest 0.1 inch.

Solution: Each side of the triangle measures $\frac{1}{3}(60)$ inches, or 20 inches. By sketching the triangle, we see that relative to the 60° angle, the altitude is the opposite side and the 20-inch side is the hypotenuse. The opposite and hypotenuse form the sine ratio:

$$\sin(60°) = \frac{h}{20}$$

$$h = 20 \cdot \sin(60°)$$

$$h = 20(0.866025) = 17.3 \text{ inches}$$

Example 3

A cell phone tower is held in place with two support wires, as shown in the accompanying figure. The angle made by the wire and the ground is 55°. The distance between the base of the tower and the wire is 20 feet.

(a) How tall is the tower?
(b) What is the total length of wire used?

Round all answers to the nearest tenth.

Solution:

(a) Relative to the 55° angle, the 20-foot distance along the ground is the adjacent and the tower is the opposite. So we can use the tangent ratio.

$$\tan(55°) = \frac{\text{opposite}}{\text{adjacent}} = \frac{x}{20}$$
$$x = 20 \cdot \tan(55°)$$
$$x = 20(1.42814) = 28.6 \text{ ft}$$

The tower is 28.6 feet high.

(b) Now that we have two sides of the triangle, we can use either the Pythagorean theorem or the cosine ratio to find the length of the wire. Continuing with the trigonometry approach, we see that the 20-foot distance is the adjacent and the wire is the hypotenuse. So we use the cosine ratio.

$$\cos(55°) = \frac{20}{x}$$
$$x = \frac{20}{\cos(55°)}$$
$$x = \frac{20}{0.57357} = 34.8689 \text{ ft}$$

There are two wires, so the total length of wire is (2)(34.8689) = 69.7 ft.

Cofunctions

The two acute angles in a right triangle must be complementary. As you may have noticed, the sine of one of the acute angles is equal to the cosine of other acute angle as shown in Figure 8.10. The sine and cosine are referred to as **cofunctions**.

$$\sin(A) = \cos(B) = \cos(90 - A) = \frac{BC}{AB}$$

$$\sin(B) = \cos(A) = \sin(90 - A) = \frac{AC}{AB}$$

Figure 8.10 Cofunction relationship between sine and cosine

Cofunction Relationship

The sine of an angle is equal to the cosine of its complement: $\sin(A) = \cos(90 - A)$

   The cosine of an angle is equal to the sine of its complement: $\cos(A) = \sin(90 - A)$

Example 1

Given right triangle ABC with right angle C, $\sin(A)$ is always equal to which of the following?

(1) $\sin(90 - A)$ (3) $\cos(A)$
(2) $\cos(90 - A)$ (4) $\tan(B)$

   Solution: The correct cofunction relationship is choice (2), $\cos(90 - A)$.

Example 2

Given complementary angles M and N, if $\sin(M) = (a^2 - 2)$ and $\cos(N) = a$, find the value of a.

   Solution: Using the cofunction relationships, we know $\sin(M) = \cos(N)$. By substituting, we find:

$$a^2 - 2 = a$$
$$a^2 - a - 2 = 0$$
$$(a - 2)(a + 1) = 0$$
$$(a - 2) = 0 \qquad (a + 1) = 0$$
$$a = 2 \qquad\qquad a = -1$$

The negative value of a would lead to a negative value for cosine. The cosine ratio of any acute angle is positive so we exclude the negative value. The value of a is 2.

MATH FACT

In right triangle trigonometry, we deal with only acute positive angles. However, angles need not be acute or positive. In the study of trigonometry, the trigonometric ratios can be applied to both negative angles and obtuse angles.

Angle of Elevation and Depression

Many practical applications involve angles of elevation and angles of depression.

> **Definitions**
> *Angle of elevation:* The angle formed by the horizontal and the line directed upward to an object
> *Angle of depression:* The angle formed by the horizontal and the line directed downward to an object

An angle of elevation involves a situation where someone or something is looking upward at another object and the horizontal is often the ground. A vertical line to complete the right triangle often has to be sketched. An angle of depression involves someone or something that is at a higher elevation looking down. The horizontal may also have to be sketched. Figure 8.11 illustrates an angle of elevation and an angle of depression.

Angle of elevation and angle of depression problems involve finding a missing angle or side using a trigonometric ratio. The best strategy is to do the following:

- Start with a detailed sketch.
- Identify the triangles formed and relevant trigonometric ratios.
- Consider other parallel line and other relationships if you are still missing parts. Notice that the angle of elevation and the angle of depression shown in Figure 8.11 are congruent alternate interior angles.
- Consider working with more than one triangle, especially if the problem involves an object that moves from one position to another.

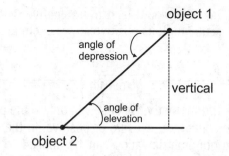

Figure 8.11 Angle of elevation and angle of depression

Example 1

An airplane is flying level and is approaching a radar station. The radar station measures a straight-line distance to the airplane of 15,000 ft when the angle of elevation is 61°.

(a) What is the altitude of the airplane?
(b) What is the horizontal distance from the airplane to the radar station?
(c) A short time later, the angle of elevation to the plane is 72°. If the altitude of the airplane remains the same, what is the horizontal distance the airplane has traveled?

Round answers to the nearest foot.

Solution:

(a) Sketch a vertical segment from the airplane to the ground to complete a right triangle. This distance, h, represents the elevation of the airplane and is opposite the angle of elevation. The 15,000 ft distance is the hypotenuse, so we should use the sine ratio.

$$\sin(61°) = \frac{h}{15,000}$$
$$h = 15,000 \cdot \sin(61°)$$
$$h = 13,119$$

The airplane is flying at an altitude of 13,119 ft.

(b) The distance along the ground is the adjacent, so we use cosine.

$$\cos(61°) = \frac{d}{15,000}$$
$$d = 15,000 \cdot \cos(61°)$$
$$d = 7,272 \text{ ft}$$

(c) The plane is now closer to the radar station. The angle of elevation is 72° and the elevation is still 13,119 ft. The altitude is opposite the angle of elevation. We are looking for the horizontal distance, d, which is adjacent to the angle of elevation. The tangent ratio applies here.

$$\tan(72°) = \frac{13,119}{d}$$
$$d = \frac{13,119}{\tan(72°)}$$
$$d = 4,263 \text{ ft}$$

Subtract to find the horizontal distance the plane has traveled.

$$7,272 - 4,263 = 3,009 \text{ ft}$$

Example 2

Frank is standing at the top of a water tower and looks down at the town below. Frank measures an angle of depression of 52° to his house. He knows the water tower is 120 ft tall. How far along the ground is the tower from his house? Round to the nearest foot.

Solution: Start with a sketch, and fill in the appropriate dimensions, horizontals, and verticals.

The angle of depression is 52°. It is easier in the problem to find the angle complementary to the angle of depression and work with that triangle. The complementary angle is 38°. The height of the tower is adjacent to the 38° angle, and the distance along the ground is adjacent to the 38° angle. So we use the tangent ratio.

$$\tan(38°) = \frac{d}{120}$$
$$d = 120 \cdot \tan(38°)$$
$$d = 94 \text{ ft}$$

The house is 94 feet from the water tower.

Example 3

A lighthouse sits on a 100 ft cliff. An observer in a boat in the harbor notes an angle of elevation to the base of the cliff of 28° and an angle of elevation to the top of the lighthouse of 44°. What is the height of the lighthouse, rounded to the nearest foot?

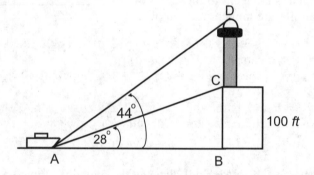

Solution: Strategy—First find the length of *AB* using △*ABC*. Then find the length of *BD* using △*ABD*.

In △*ABC*, *BC* is opposite ∠*A* and *AB* is the adjacent. So we use the tangent.

$$\tan(28°) = \frac{100}{AB}$$
$$AB = \frac{100}{\tan(28°)}$$
$$AB = 188.0726$$

In △*ABD*, *AB* is the adjacent and *BD* is the opposite. So we use the tangent again.

$$\tan(44°) = \frac{BD}{188.0726}$$
$$BD = 188.0726 \cdot \tan(44°)$$
$$BD = 181.6196$$

The height of the lighthouse, *DC*, is found by subtracting the height of the cliff from *BD*.

$$DC = BD - CB$$
$$DC = 181.6196 - 100$$
$$DC = 81.6196$$

The height of the lighthouse is 81.6 ft.

Check Your Understanding of Section 8.4

A. *Multiple-Choice*

1. Which of the following theorems best explains why sin(30°) is always equal to $\frac{1}{2}$?
   (1) the sum of the squares of the legs of a right triangle equals the square of the hypotenuse
   (2) the sum of the measures of the interior angles of a triangle equals 180°
   (3) corresponding parts of congruent triangles are congruent
   (4) corresponding parts of similar triangles are proportional

2. In $\triangle BGH$, $m\angle B = 90°$ and $m\angle G = 42°$. What is the value of $\cos(H)$?
   (1) 1   (3) 0.743
   (2) 0.804   (4) 0.669

3. The lengths of the legs of a right triangle are in a 1:3 ratio. What is the measure of the two acute angles in the triangle?
   (1) 18.4° and 71.6°   (3) 22.5° and 67.5°
   (2) 19.5° and 70.5°   (4) 30° and 60°

4. In $\triangle GFQ$, $\angle F$ is a right angle. If $GF = 12$ and $GQ = 21$, what is the measure of $\angle Q$?
   (1) 29.7°   (2) 34.8°   (3) 55.2°   (4) 60.3°

5. What is the value of the $\cos(A)$ in the following figure?
   (1) $\frac{4}{7}$ (3) $\frac{\sqrt{65}}{7}$

   (2) $\frac{3}{7}$ (4) $\frac{\sqrt{33}}{7}$

6. In $\triangle HIJ$, $m\angle I = 72°$, $m\angle J = 18°$, and $HI = 12$. What is the length of HJ?
   (1) 3.9   (2) 11.4   (3) 12.6   (4) 36.9

7. Given $\triangle ABC$ with $m\angle C = 21°$, $m\angle B = 90°$, and
   $AC = 10$. Find the length of BC.
   (1) 3.58   (3) 9.34
   (2) 6.43   (4) 9.54

8. $\triangle RST$ is a right triangle with right angle S. If
   $m\angle T = 26°$ and $RS = 8$, what is the length of ST?
   (1) 3.9   (3) 16.4
   (2) 8.9   (4) 18.2

9. Triangle JKL has a right angle at K. If $\sin(J) = c$, which expression represents $\cos(L)$?

   (1) c (2) $90 - c$ (3) $\dfrac{c}{2}$ (4) $c - 90$

10. Given acute angle L, which of the following is equivalent to $\sin(L)$ $- \cos(90 - L)$?
    (1) 1   (2) 0   (3) L (4) $-L$

11. What is the value of q if $\sin(A) = 2q$ and $\cos(90 - A) = 4q - 1$?

    (1) 2   (2) 1   (3) $\dfrac{1}{2}$ (4) 0

12. In $\triangle MNP$, $m\angle N = 90°$ and $\cos(M) = \dfrac{2}{5}$. If $MN = 8$, what is the length

    of PM?
    (1) 3.2   (2) 12   (3) 16   (4) 20

B. *Free-Response—show all work or explain your answer completely.*

13. The owner's manual for an extension ladder specifies that the angle between the ground and the ladder should be between 50° and 65°. The ladder can be extended to a maximum length of 25 feet. What is the highest point the ladder can reach on a vertical wall? Round your answer to the nearest tenth of a foot.

14. The side lengths of right $\triangle WXY$ are in a $5:12:13$ ratio. What is the measure of the larger of the two acute angles? Round your answer to the nearest tenth of a degree.

15. The altitude of an equilateral triangle is 18 inches. What is the perimeter of the triangle? Round your answer to the nearest 0.01 inch.

16. Building codes specify the angle between the ground and a wheelchair ramp to be no more than 5°. A construction company is installing a wheelchair ramp from the sidewalk to the front door of a bank. If the front door of the bank is 2.5 feet above the level of the sidewalk, what is the minimum length of the ramp? How far from the front door must the ramp start if it has no turns? Round to the nearest 0.1 foot.

17. A ladder leaning against a wall makes a 55° angle with the ground. The ladder then slips along the ground so that the height it reaches up the wall is half of what is was originally. What angle does the ladder now make with the ground? Round to the nearest 0.1 degree.

18. Given acute angle B, $\cos(B) = \left(2x - \dfrac{1}{4}\right)$, and $\sin(90 - B) = \left(x + \dfrac{1}{4}\right)$.

    Find the measure of angle B. Round your answer to the nearest tenth of a degree.

19. A firefighting helicopter is on its way to drop water on a forest fire. The pilot is flying 300 feet above the ground and sees the fire ahead at an 18° angle of depression. How much farther does he need to fly to be directly above the fire? Round your answer to the nearest foot.

20. A hawk perched at the top of a 75 ft tree sees a rabbit on the ground. The angle of depression from the hawk to the rabbit is 29°.
    (a) What is the distance from the rabbit to the tree?
    (b) What is the straight-line distance the hawk must fly to reach the rabbit?
    Round your answers to the nearest foot.

21. Jesse is standing on the ground and observes that the angle of elevation to the top of a nearby building is 25°. He then walks 40 ft toward the building and notices that the angle of elevation is now 42°. How tall is the building?

22. Scott states that the sine of an angle must be no greater than 1. Jon disagrees, saying that the sine of an angle can have any value. Who is correct? Explain your reasoning.

23. Rose is flying a kite at the park and has let out the entire 200 ft spool of string. She observes an angle of elevation of 70° from the ground to the kite.
    (a) How high is the kite above the ground?
    (b) The wind shifts, and the angle of elevation is reduced by half to 35°. By how many feet does the kite drop?

    Round your answers to the nearest foot.

24. John is jogging along a straight road at a constant speed and sees an office building that he knows is 100 feet high. At point A, John observes a 9° angle of elevation to the top of the building. At point B 30 seconds later, he finds that the angle of elevation has increased to 32°. If John is 6 ft tall, how fast is he jogging?

    Round to the nearest foot per second.

not drawn to scale

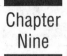

Chapter Nine PARALLELOGRAMS AND TRAPEZOIDS

9.1 PARALLELOGRAMS

=== **KEY IDEAS** ===

A parallelogram is a quadrilateral that has both pairs of opposite sides parallel. Besides having parallel sides, all parallelograms have the following properties:

- Opposite sides are congruent.
- Opposite angles are congruent.
- Adjacent angles are supplementary.
- The diagonals bisect each other.
- The two diagonals each divide the parallelogram into two congruent triangles.

Properties of Parallelograms

Definition
A parallelogram is a quadrilateral in which both pairs of opposite sides are parallel.

Besides having opposite sides parallel, parallelograms have several properties that can be proven using congruent triangles and parallel lines relationships. The current learning standards state that students should be able to prove theorems about parallelograms. Some of the theorem proofs that follow in this section are mentioned as examples. Your best strategy is to understand the basic approach and tools used in the proofs, not to memorize the proofs word for word. Many variations of these proofs are equally valid.

Theorem
Each diagonal in a parallelogram divides the parallelogram into two congruent triangles.

Theorem
Opposite sides of a parallelogram are congruent.

Theorem
Opposite angles in a parallelogram are congruent.

The proof for these theorems follows. The strategy is to use the alternate interior angles formed by the diagonal to show the two triangles are congruent by ASA. Then the rest follows by CPCTC. The second pair of opposite angles can be proven congruent using the same method with the other diagonal.

Given: parallelogram $ABCD$ and diagonal $\overline{AC}$
Prove: $\triangle ABC \; \triangle DCA$
$\qquad \overline{AB} \cong \overline{CD}, \; \overline{AD} \cong \overline{BC}$
$\qquad \angle B \cong \angle D$

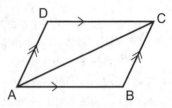

Statements	Reasons
1. Parallelogram $ABCD$	1. Given
2. $\overline{AB} \parallel \overline{CD}$ and $\overline{AD} \parallel \overline{BC}$	2. Opposite sides of a parallelogram are parallel
3. $\angle BAC \cong \angle DCA$, $\angle DAC \cong \angle BCA$	3. When parallel lines are intersected by a transversal, the alternate interior angles are congruent
4. $\overline{AC} \cong \overline{AC}$	4. Reflexive property
5. $\triangle ABC \cong \triangle CDA$	5. ASA
6. $\overline{AB} \cong \overline{CD}$, $\overline{AD} \cong \overline{BC}$	6. CPCTC
7. $\angle B \cong \angle D$	7. CPCTC

Theorem
Consecutive angles in a parallelogram are supplementary.

The proof that consecutive angles are supplementary is simply a statement of the same-side interior angle relationship for angles formed by parallel lines and a transversal.

Given: parallelogram *ABCD*
Prove: ∠*A* and ∠*B*, ∠*B* and ∠*C*, ∠*C* and ∠*D*, ∠*D* and ∠*A* are supplementary

Statements	Reasons
1. Parallelogram *ABCD*	1. Given
2. $\overline{AB} \parallel \overline{CD}$ and $\overline{AD} \parallel \overline{BC}$	2. Opposite sides of a parallelogram are parallel
3. ∠*A* and ∠*B* are supplementary ∠*B* and ∠*C* are supplementary ∠*C* and ∠*D* are supplementary ∠*D* and ∠*A* are supplementary	3. When parallel lines are intersected by a transversal, the same side interior angles are supplementary

Theorem
The diagonals of a parallelogram bisect each other.

Given: parallelogram *ABCD*, diagonals $\overline{AC}$ and $\overline{BD}$ intersect at *E*
Prove: diagonals $\overline{AC}$ and $\overline{BD}$ bisect each other

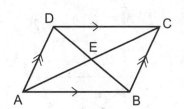

Statements	Reasons
1. Parallelogram *ABCD*	1. Given
2. $\overline{AB} \parallel \overline{CD}$ and $\overline{AD} \parallel \overline{BC}$	2. Opposite sides of a parallelogram are parallel
3. ∠*BDA* ≅ ∠*DBC*, ∠*DAC* ≅ ∠*BCA*	3. When parallel lines are intersected by a transversal, the alternate interior angles are congruent

Statements	Reasons
4. $\overline{AD} \cong \overline{BC}$	4. Opposite sides of a parallelogram are congruent
5. $\triangle DAE \cong \triangle BCE$	5. ASA
6. $\overline{DE} \cong \overline{EB}$, $\overline{AE} \cong \overline{EC}$	6. CPCTC
7. E is the midpoint of $\overline{DB}$, E is the midpoint of $\overline{AC}$	7. A midpoint divides a segment into two congruent segments
8. $\overline{AC}$ and $\overline{BD}$ bisect each other	8. A bisector intersects a segment at its midpoint

A summary of the basic parallelogram properties is shown in Figure 9.1.

- Both pairs of opposite sides of a parallelogram are parallel.
- Both pairs of opposite sides of a parallelogram are congruent.
- Both pairs of opposite angles of a parallelogram are congruent.
- All pairs of consecutive angles of a parallelogram are supplementary.
- Diagonals of a parallelogram bisect each other.
- A diagonal of a parallelogram forms two congruent triangles.

∠A ≅ ∠C, ∠B ≅ ∠D   supplementary angles
AB = CD, BC = AD   ∠A and ∠B, ∠B and ∠C
                                         ∠C and ∠D, ∠D and ∠A

$\overline{AC}$ and $\overline{BD}$ bisect each other
△ABC ≅ △CDA
△BAD ≅ △DCB

Figure 9.1 Angle and side relationships in parallelograms

If no figure is provided, be sure to sketch one. You can usually start labeling vertices anywhere on the parallelogram and can proceed in any direction as long as no vertices are skipped. The order of vertices must match the order listed in the name of the parallelogram. The exception is parallelograms with one pair of parallel sides. The parallel sides on the figure must be those specified in the problem.

Example 1

In parallelogram *QRST*, m∠*Q* is 18° more than twice m∠*R*. Find the measures of ∠*Q* and ∠*R*.

Solution:

m∠*R* = *x*°, m∠*Q* = (2*x* + 18)°   assign variables
   m∠*Q* + m∠*R* = 180°   consecutive angles of a parallelogram
                               are supplementary

$$2x + 18 + x = 180$$
$$3x = 162$$
$$x = 54$$
m∠*Q* = 2(54) + 18 = 126°
   m∠*R* = 54°

Example 2

In parallelogram *HIKE*, m∠*H* = (6*x* − 11)° and m∠*K* = (2*x* + 29)°. Find m∠*H* and m∠*K*.

Solution:

   m∠*H* = m∠*K*   opposite angles are congruent
$$6x - 11 = 2x + 29$$
$$4x = 40$$
$$x = 10$$
m∠*H* = m∠*K* = 2(10) + 29 = 49°

Example 3

Given parallelogram *ABCD*,
$\overline{FE} \perp \overline{AEB}$, m∠*EFB* = 70°,
and m∠*C* = 65°. Find
m∠*DFB* and m∠*CBF*.

Solution:

$$m\angle BEF = 90°$$

$m\angle BEF + m\angle EFB + m\angle EBF = 180°$ triangle angle sum theorem

$$90° + 70° + m\angle EBF = 180°$$

$$m\angle EBF = 20°$$

$m\angle ABC + m\angle C = 180°$ consecutive angles are supplementary

$$m\angle ABC + 65° = 180°$$

$$m\angle ABC = 115°$$

$m\angle ABF + m\angle CBF = m\angle ABC$ partition postulate

$$20° + m\angle CBF = 115°$$

$$m\angle CBF = 95°$$

$m\angle DFB + m\angle CBF = 180°$ same side interior angles are supplementary

$$m\angle DFB + 95° = 180°$$

$$m\angle DFB = 85°$$

Example 4

In parallelogram $ABCD$, diagonals AC and BD intersect at E. If $AE = (6x - 12)$ and $EC = (2x + 28)$, find AC.

Solution:

$AE = EC$ diagonals of a parallelogram bisect each other

$$6x - 12 = 2x + 28$$

$$4x = 40$$

$$x = 10$$

$$AE = EC = 6(10) - 12 = 48$$

$$AC = 96$$

Check Your Understanding of Section 9.1

A. *Multiple-Choice*

1. In parallelogram $RSTU$, $m\angle R = (8x + 12)°$ and $m\angle T = (4x + 24)°$. What is $m\angle R$?
   (1) 144°   (2) 72°   (3) 36°   (4) 18°

2. In parallelogram $ABCD$, the measures of $\angle B$ and $\angle C$ are in a $3:2$ ratio. What is the measure of $\angle B$?
   (1) 120°   (2)108°   (3) 72°   (4) 60°

3. In parallelogram *FGHI*, *FG* = $(10x - 10)$ and *HI* = $(5x + 70)$. What is the value of *x*?
   (1) 3   (2) 4   (3) 8   (4) 16

4. The perimeter of parallelogram *FGHI* is 52. If *FG* = 14, what is the length of *GH*?
   (1) 12   (2) 14   (3) 24   (4) 26

5. In parallelogram *KLMN*, m∠*L* = $(6x + 3)°$ and m∠*M* = $(3x + 15)°$. What is m∠*L*?
   (1) 21°   (2) 27°   (3) 69°   (4) 111°

6. In parallelogram *MATH*, diagonals $\overline{MT}$ and $\overline{AH}$ intersect at *P*. If *MP* = $(4x + 6)$, *PT* = $(2x + 18)$, and *AP* = $(3x + 10)$, what is the length of *MT*?
   (1) 30   (2) 40   (3) 68   (4) 60

7. In quadrilateral *RSTU*, diagonals $\overline{RT}$ and $\overline{SU}$ intersect at *O*. If *RO* = $(4x + 8)$ and if *OT* = $(14x - 22)$, for what value of *x* could *RSTU* be a parallelogram?
   (1) 2.6   (2) 3   (3) 4.75   (4) 5

B. *Free-Response—show all work or explain your answer completely.*

8. Find the measure of ∠*C* in parallelogram *ABCD*.

9.2 PROOFS WITH PARALLELOGRAMS

KEY IDEAS

A quadrilateral must be a parallelogram if any one of the following are true:

- Two pairs of opposite sides are parallel.
- Two pairs of opposite sides are congruent.
- One pair of sides is congruent and parallel.
- Two pairs of opposite angles are congruent.
- Two pairs of consecutive angles are supplementary (in parallelogram *ABCD*, angles *A* and *B*, *B* and *C*).
- The diagonals bisect each other.

Requirements for Proving a Quadrilateral Is a Parallelogram

Any one of the following properties can be used to prove a quadrilateral is a parallelogram. You usually show one of the other properties first in the process of showing both pairs of triangles are congruent. One additional useful shortcut that is added to the list is "one pair of opposite sides is congruent and parallel."

Properties Sufficient to Prove a Quadrilateral Is a Parallelogram

Property	What It Looks Like
Two pairs of opposite sides are parallel	
Two pairs of opposite sides are congruent	
One pair of sides is congruent and parallel	
Two pairs of opposite angles are congruent	
Two pairs of consecutive angles are supplementary	∠1 and ∠2, ∠2 and ∠3 are supplementary
The diagonals bisect each other	

MATH FACT

A kite is a counterexample that illustrates why one diagonal forming congruent triangles is not enough to prove that a quadrilateral is a parallelogram. Kites have two distinct pairs of consecutive sides congruent. Therefore the diagonal forms two congruent triangles. In the kite shown, $AB = BC$, $AD = CD$, and $\triangle ABD \cong \triangle BCD$. However, the figure is clearly not a parallelogram. Kites also have perpendicular diagonals. A dart is similar to a kite except one of the angles is concave. Diagonal $\overline{MO}$ in dart $MNOP$ lies outside the dart.

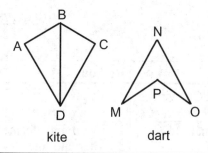

kite   dart

Parallelogram Proofs

When you are asked to prove a quadrilateral is a parallelogram, look for these common strategies:

- The diagonal forms congruent triangles: Even though on its own this is not sufficient, in most instances this can be used along with CPCTC to show opposite sides or angles are congruent.
- Congruent alternate interior angles are formed by a diagonal: This can be used to show that opposite sides are parallel. If given a parallelogram, the congruent alternate interior angles can be used later in the proof.
- Sometimes constructing a diagonal that is not explicitly shown can be helpful.

MATH FACT

Mathematicians often use the words "sufficient" and "necessary" to describe the conditions needed for a statement to be true. A sufficient condition is all that it needed—no further proof is required. A necessary condition is required, but it is not enough to prove the statement is true. For example, if given quadrilateral $ABCD$, two pairs of opposite sides being parallel is sufficient to prove that $ABCD$ is a parallelogram.

The first example is a proof of the sufficient condition "one pair of sides are congruent and parallel."

Example 1

Given: quadrilateral $ABCD$ with $\overline{AB} \parallel \overline{CD}$,
       $\overline{AB} \cong \overline{CD}$
Prove: $ABCD$ is a parallelogram

Solution:

Statements	Reasons
1. Quadrilateral $ABCD$ with $\overline{AB} \parallel \overline{CD}$, $\overline{AB} \cong \overline{CD}$	1. Given
2. Construct diagonal $\overline{AC}$	2. Two points define a segment
3. $\angle BAC \cong \angle DCA$	3. Alternate interior angles formed by parallel lines and a transversal are congruent
4. $\overline{AC} \cong \overline{AC}$	4. Reflexive property
5. $\triangle BAC \cong \triangle DCA$	5. SAS
6. $\overline{AD} \cong \overline{BC}$	6. CPCTC
7. $ABCD$ is a parallelogram	7. A quadrilateral with two pairs of opposite sides congruent is a parallelogram

Example 2

Given: $\angle E \cong \angle G$ and $\angle EDF \cong \angle GFD$
Prove: $DEFG$ is a parallelogram

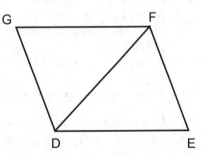

Solution:

Statements	Reasons
1. $\angle EDF \cong \angle GFD$	1. Given
2. $\overline{ED} \parallel \overline{FG}$	2. Two lines are parallel if the alternate interior angles formed are congruent
3. $\angle E \cong \angle G$	3. Given
4. $\overline{FD} \cong \overline{FD}$	4. Reflexive property
5. $\triangle EDF \cong \triangle GFD$	5. AAS
6. $\angle EFD \cong \angle GFD$	6. CPCTC
7. $\overline{GD} \parallel \overline{FE}$	7. Two lines are parallel if the alternate interior angles formed are congruent
8. $DEFG$ is a parallelogram	8. A quadrilateral with two pairs of opposite parallel sides is a parallelogram

Example 3

Given: quadrilateral $JUMP$ with $\angle J$ supplementary to $\angle P$ and $\overline{JU} \cong \overline{MP}$
Prove: $JUMP$ is a parallelogram

Solution: sketch:

$\overline{JU} \parallel \overline{PM}$ because same side interior angles $\angle J$ and $\angle P$ are supplementary.
Opposite sides $\overline{JU}$ and $\overline{MP}$ are both parallel and congruent, making $JUMP$ a parallelogram.

Example 4

Given: parallelogram $AECF$ with $\overline{AE}$ extended to B, with $\overline{FC}$ extended to D, and $\overline{DF} \cong \overline{BE}$
Prove: $ABCD$ is a parallelogram

Solution: Strategy—Prove the triangles are congruent by SAS. Then use CPCTC and alternate interior angles.

Statements	Reasons
1. Parallelogram $AEFC$ with $\overline{AE}$ extended to B and $\overline{FC}$ extended to D	1. Given
2. $\overline{AF} \cong \overline{EC}$	2. Opposite sides of a parallelogram are congruent
3. $\angle AEC \cong \angle CFA$	3. Opposite angles of a parallelogram are congruent
4. $\angle BEC$ and $\angle AEC$ are a linear pair $\angle AFD$ and $\angle CFA$ are a linear pair	4. Definition of a linear pair
5. $\angle BEC$ and $\angle AEC$ are supplementary $\angle AFD$ and $\angle CFA$ are supplementary	5. Linear pairs are supplementary
6. $\angle BEC \cong \angle AFD$	6. Angles supplementary to congruent angles are congruent
7. $\triangle ADF \cong \triangle CBE$	7. SAS
8. $\overline{AD} \cong \overline{BC}$	8. CPCTC
9. $ABCD$ is a parallelogram	9. A quadrilateral with two pairs of opposite sides congruent is a parallelogram

Check Your Understanding of Section 9.2

A. *Multiple-Choice*

1. Which of the following quadrilaterals *cannot* be proven to be a parallelogram?

(1)   (2)   (3)   (4)

2. Which of the following is sufficient to prove a quadrilateral is a parallelogram?
(1) the diagonals are congruent
(2) the diagonals bisect each other
(3) the diagonals are perpendicular
(4) one diagonal is congruent to a pair of opposite sides

B. *Free-Response—show all work or explain your answer completely.*

3. $\overline{AB} \cong \overline{BD} \cong \overline{CD}$, m∠$ABD$ = 72°, and m∠ADC = 126°. Is $ABCD$ a parallelogram? Justify your reasoning.

4. Given: Y is the midpoint of $\overline{FS}$, $\overline{HF} \parallel \overline{SI}$
Prove: $FISH$ is a parallelogram

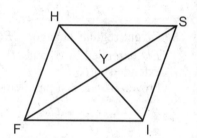

5. Given: parallelogram $ABCD$ with $\overline{AD}$ extended to E, segment $\overline{EFB}$, and F is the midpoint of $\overline{CFD}$.
Prove: D is the midpoint of $\overline{ADE}$

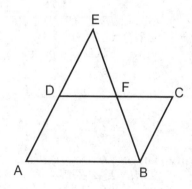

6. Given: parallelogram *RSTU*
Prove: △*UWR* ≅ △*SWT*

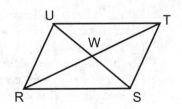

7. Given: △*ADF* ≅ △*FEC*
Prove: *DBEF* is a parallelogram

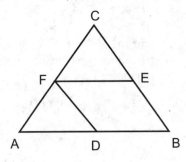

8. Given: regular hexagon *ABCDEF*,
$\overline{ED}$ extended to *G*, and $\overline{AB}$
extended to *H*
Prove: *HBGE* is a parallelogram

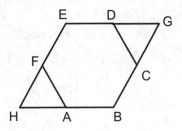

9. Given: $\overline{ID} \perp \overline{BI}$, $\overline{ID} \perp \overline{RD}$,
∠*RID* ≅ ∠*BDI*
Prove: *BIRD* is a parallelogram

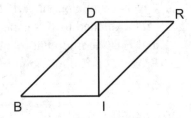

10. Given: △*FMI* ≅ △*NHG*,
∠*MIN* ≅ ∠*NGM*
Prove: *FGHI* is a
parallelogram

11. Given: $\triangle WZB \cong \triangle YXA$ and $\overline{BY} \cong \overline{WA}$
Prove: $WXYZ$ is a parallelogram

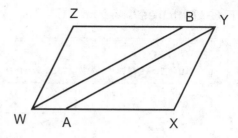

12. Given: parallelogram $QRST$, R is the midpoint of $\overline{QU}$
Prove: $RUST$ is a parallelogram

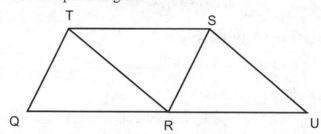

9.3 PROPERTIES OF SPECIAL PARALLELOGRAMS

KEY IDEAS

The rhombus (plural is rhombi), rectangle, and square are special parallelograms. In addition to the parallelogram properties, the special parallelograms have the following properties.

	Rectangle	Rhombus	Square
4 right angles	✓		✓
Congruent diagonals	✓		✓
4 congruent sides		✓	✓
Perpendicular diagonals		✓	✓
Diagonals bisect the angles		✓	✓

Rectangles

Definition
A rectangle is a parallelogram with a right angle, as shown in Figure 9.2.

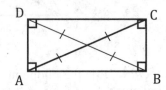

Figure 9.2 Rectangle *ABCD*

Properties of Rectangles

- All the properties of a parallelogram
- Four right angles
- Diagonals are congruent

The learning standards specify that students should be able to prove various theorems about parallelograms. As with the parallelogram property proofs, you should be familiar with the approach. However, don't try to memorize the proof word for word.

The property of congruent diagonals can be proven by showing that the two overlapping triangles formed by the diagonals are congruent.

Example 1

Given: rectangle *ABCD*
Prove: $\overline{AC} \cong \overline{BD}$

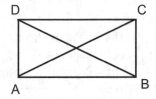

Solution:

Statements	Reasons
1. Rectangle *ABCD*	1. Given
2. $\overline{BC} \cong \overline{AD}$	2. Opposite sides of a parallelogram are congruent
3. $\overline{AB} \cong \overline{AB}$	3. Reflexive property
4. ∠*CBA* and ∠*DAB* are right angles	4. All angles in a rectangle are right angles
5. ∠*CBA* ≅ ∠*DAB*	5. All right angles are congruent
6. △*ABC* ≅ △*BAD*	6. SAS
7. $\overline{AC} \cong \overline{BD}$	7. CPCTC

Example 2

In rectangle *TRUE*, diagonals $\overline{TU}$ and $\overline{RE}$ intersect at *A*. If $TA = (10x - 12)$ and $RA = (4x + 24)$, find the length of diagonal $\overline{RE}$.

Solution: Since diagonals in a rectangle are congruent and bisect each other, $TA = AU = RA = AE$.

$$10x - 12 = 4x + 24$$
$$6x = 36$$
$$x = 6$$
$$RA = AE = 4(6) + 24 = 48$$
$$RE = 2 \cdot 48 = 96$$

Example 3

In rectangle *ABCD*, diagonals $\overline{AC}$ and $\overline{BD}$ intersect at *E*. If $m\angle DAE = 68°$, find $m\angle AEB$, $m\angle ABE$, and $m\angle CBE$.

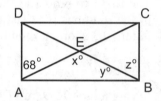

Solution:

$m\angle BAE + m\angle DAE = 90°$ angles in a rectangle measure 90°
$m\angle BAE + 68° = 90°$
$m\angle BAE = 22°$
$m\angle ABE = m\angle BAE = 22°$ △*ABE* is isosceles, with base angles
                                          $\angle BAE$ and $\angle ABE$
$m\angle AEB = 180° - 2 \cdot 32° = 136°$
$m\angle CBE + m\angle ABE = 90°$
$m\angle CBE + 32° = 90°$
$m\angle CBE = 68°$
$m\angle AEB = 116°$, $m\angle ABE = 22°$,
        $m\angle CBE = 68°$

Rhombi

Definition
A rhombus (plural rhombi) is a parallelogram with a pair of consecutive congruent sides.

Properties of a Rhombus

- All the properties of a parallelogram
- All four sides are congruent
- Diagonals are perpendicular
- Diagonals bisect the angles of the rhombus

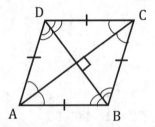

Figure 9.3 Rhombus *ABCD*

The first two properties of a rhombus are a direct result from the definition of the rhombus. The last two properties can be proven using the first two. Figure 9.3 shows a rhombus.

Example 1

In rhombus *ABCD*, m∠*DCA* = 27°.
Find m∠*BCA* and m∠*CBD*.

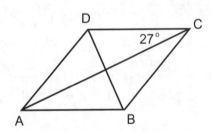

Solution:

m∠*BCA* = m∠*DCA*   diagonal of a rhombus bisects the angles
= 27°

m∠*BCD* = m∠*BCA* + m∠*DCA*   angle addition
= 27° + 27°
= 54°

m∠*ABC* + m∠*BCD* = 180°   consecutive angles are supplementary
m∠*ABC* + 54° = 180°
m∠*ABC* = 126°

m∠*CBD* = $\frac{1}{2}$m∠*ABC*   diagonals of a rhombus bisect the angles

= $\frac{1}{2}$(126°)

= 63°

306

| MATH FACT |

Since the diagonals of a rhombus are perpendicular, they will form right triangles. Be on the lookout for applications of the Pythagorean theorem when working with rhombi.

Example 2

Rhombus *ABCD* has diagonals $\overline{BD}$ and $\overline{AC}$ intersecting at *E*. If *EB* = 3 and *AE* = 9, find the perimeter of *ABCD*. Express your answer in simplest radical form.

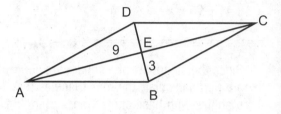

Solution: Apply the Pythagorean theorem to $\triangle ABE$

$$BE^2 + AE^2 = AB^2$$
$$3^2 + 9^2 = AB^2$$
$$90 = AB^2$$
$$AB = \sqrt{90} = 3\sqrt{10}$$
$$\text{Perimeter} = 4 \cdot AB = 12\sqrt{10}$$

Squares

Definition
A square is a parallelogram with a right angle and a pair of consecutive congruent sides.

Properties of a Square

- All the properties of a parallelogram
- All angles are right angles
- Congruent diagonals
- All four sides are congruent
- Diagonals are perpendicular
- Diagonals bisect the angles of the square

Figure 9.4 shows a square.

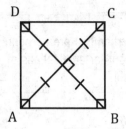

Figure 9.4 Square *ABCD*

=== **MATH FACT** ===

Since the diagonals of a square are congruent, bisect each other, and are perpendicular, the four triangles they form are all isosceles right triangles with base angles measuring 45°.

Example 1

RSTU is a square with diagonals $\overline{RT}$ and $\overline{SU}$ intersecting at *W*. Find m∠*WUR* and m∠*TWS*.

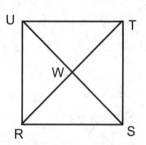

Solution:

m∠*RUT* = 90°   the angles of squares are right angles

m∠*WUR* = $\frac{1}{2}$ m∠*RUT*   diagonals of a square bisect the angles

$= \frac{1}{2}(90°)$

$= 45°$

m∠*TWS* = 90°   diagonals of a square are perpendicular

m∠*WUR* = 45°, m∠*TWS* = 90°

Example 2

The diagonal of a square measures 12 cm. What is the area of the square in cm^2?

Solution: Use the Pythagorean theorem with right triangle $\triangle ABC$ to find the side length of the square.

Let $AB = BC = x$ all sides of a square are congruent

$$AB^2 + BC^2 = AC^2$$
$$x^2 + x^2 = 12^2$$
$$2x^2 = 144$$
$$x^2 = 72$$
$$x = \sqrt{72}$$

$$\text{area} = s^2$$
$$= \left(\sqrt{72}\right)^2$$
$$= 72 \text{ cm}^2$$

Check Your Understanding of Section 9.3

A. *Multiple-Choice*

1. A parallelogram is known to be a rectangle. What must be true about the rectangle?
   (1) its diagonals divide each other in a $1:2$ ratio
   (2) its diagonals are equal in measure to the side length
   (3) its diagonals are congruent and bisect each other
   (4) its diagonals are congruent and perpendicular

2. In rectangle *MNOP*, diagonals $\overline{MO}$ and $\overline{PN}$ intersect at A. If $MA = (2x + 31)$ and $OA = (9x + 3)$, what is the length of diagonal $\overline{PN}$?
   (1) 78   (2) 39   (3) 8   (4) 4

3. In rectangle $ABCD$, $AB = 6$ and $BC = 10$. What is the length of diagonal $\overline{AC}$?
   (1) 4   (2) $2\sqrt{34}$ (3) 8   (4) $4\sqrt{11}$

4. The two diagonals of rectangle $ABCD$ are $\overline{AC}$ and $\overline{BD}$, and they intersect at E. If $AE = (4x + 12)$ and $BE = (6x - 12)$, what is the length of each diagonal?
   (1) 12   (2) 15   (3) 60   (4) 120

5. Diagonal $\overline{QS}$ is drawn in rectangle $QRST$. If m$\angle RQS = (4x + 2)°$ and m$\angle QSR = (3x + 11)°$, what is m$\angle RQS$?
   (1) 11°   (2) 44°   (3) 46°   (4) 90°

6. Rectangle $ABCD$ has side $\overline{AB}$ extended to E. If $BC = BE$ and m$\angle BAC = 32°$, what is m$\angle ACE$?
   (1) 135°
   (2) 122°
   (3) 112°
   (4) 103°

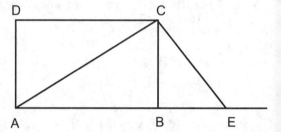

7. Quadrilateral $WXYZ$ is a rhombus. Which of the following are not necessarily true about $WXYZ$?
   (1) $\angle WYZ \cong \angle XZY$
   (2) $\angle WYZ \cong \angle WYX$
   (3) $\overline{WY}$ is the perpendicular bisector of $\overline{XZ}$
   (4) $\overline{WX} \cong \overline{XY}$

8. In rhombus $MNOP$, m$\angle MNO = 24°$. What is the measure of $\angle PMO$?
   (1) 48°   (2) 78°   (3) 96°   (4) 102°

9. In rhombus $MOTH$, m$\angle HTM = (5x + 6)°$. If m$\angle HMO = (12x - 6)°$, what is m$\angle HMO$?
   (1) 45°   (2) 51°   (3) 90°   (4) 102°

10. The perimeter of a rhombus is 40. The length of one diagonal is 12. What is the length of the other diagonal?
    (1) 6   (2) 16   (3) 24   (4) $2\sqrt{61}$

11. *ABCD* is a square, and △*AEB* is equilateral. What is the measure of ∠*DEC*?

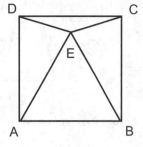

(1) 150°
(2) 140°
(3) 135°
(4) 120°

B. *Free-Response—show all work or explain your answer completely.*

12. Rhombus *EFGH* has diagonals $\overline{EG}$ and $\overline{FG}$ intersecting at *Q*. Segments $\overline{EQ}$ and $\overline{FQ}$ have lengths that are in a 2 : 1 ratio, and *EF* = 20. What is the length of $\overline{EG}$?

13. If the lengths of the diagonals of a rhombus are *a* and *b*, explain why the area of the rhombus is equal to $\dfrac{1}{2}ab$.

14. Square *GOAL* has diagonals $\overline{GA}$ and $\overline{OL}$ that intersect at *X*. If *GX* = 4, what is the perimeter of the square?

15. In square *MNOP* with diagonal $\overline{MO}$, m∠*NMO* = (4*x* + 17)°. What is the value of *x*?

16. Given square *WXYZ*, *D* is the midpoint of $\overline{YZ}$, and *WZ* = 6, find the perimeter of △*WXD*.

9.4 TRAPEZOIDS

Properties and Definitions

A trapezoid is a quadrilateral with at least one pair of parallel sides. A trapezoid with one pair of parallel sides is shown in Figure 9.5. The parallel sides are called the bases and the non-parallel sides are called the legs. The same side interior angles form when the parallel bases are supplementary.

Figure 9.5 Trapezoid with parallel bases $\overline{AB}$ and $\overline{CD}$ and with legs $\overline{BC}$ and $\overline{DA}$

Definition
Isosceles trapezoid—a trapezoid with congruent legs.

Properties of an Isosceles Trapezoid

- Same side interior angles are congruent
- Diagonals are congruent
- Base angles are congruent

Figure 9.6 shows isosceles trapezoid *ABCD* with congruent legs $\overline{BC}$ and $\overline{DA}$.

Figure 9.6 Isosceles trapezoid *ABCD*

Example 1

In isosceles trapezoid *ABCD,* m∠*A* = 72° and m∠*CDB* = 28°. Find m∠1, m∠2, m∠3, and m∠4.

Solution:

$m\angle 1 = m\angle CDB = 28°$ congruent alternate interior angles

$m\angle 1 + m\angle A + m\angle 2 = 180°$ triangle angle sum theorem

$m\angle 2 = 180° - m\angle 1 - m\angle A$
$\quad\quad = 180° - 28° - 72°$
$\quad\quad = 80°$

$m\angle 3 = m\angle CDA$ congruent base angles
$\quad\quad = 28° + 80°$
$\quad\quad = 108°$

$m\angle 1 + m\angle 4 = 72°$ congruent base angles
$\quad\quad m\angle 4 = 72° - m\angle 1$
$\quad\quad m\angle 4 = 72° - 28°$
$\quad\quad\quad\quad = 44°$

Example 2

In isosceles trapezoid $ABCD$, $\overline{AB} \parallel \overline{DC}$, $AC = (5x - 10)$, $BD = (2x + 2)$, $m\angle BAD = (3y - 12)$, and $m\angle ABC = (2y + 15)$. Find AC, $m\angle BAD$, and $m\angle ADC$.

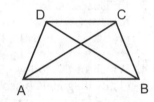

Solution:

$AC = BD$ diagonals are congruent
$5x - 10 = 2x + 2$
$\quad\quad 3x = 12$
$\quad\quad\; x = 4$
$\quad AC = 5(4) - 10$
$\quad\quad\; = 10$

$3y - 12 = 2y + 15$ base angles are congruent
$\quad\quad\; y = 27$
$m\angle BAD = 3(27) - 12$
$\quad\quad\quad = 69°$

$m\angle BAD + m\angle ADC = 180°$ same side interior angles are
$\quad\quad\quad\quad\quad\quad\quad\quad\quad$ supplementary

$m\angle ADC = 180° - 69°$
$\quad\quad\quad = 111°$

Check Your Understanding of Section 9.4

A. Multiple-Choice

1. Trapezoid *PQRS* has parallel sides $\overline{PQ}$ and $\overline{RS}$. If m∠S = $(6x - 30)°$ and m∠P = $(3x + 12)°$, what is m∠P?
   (1) 44°   (2) 54°   (3) 78°   (4) 92°

2. Isosceles trapezoid *WXYZ* has parallel sides $\overline{WX}$ and $\overline{YZ}$ and congruent legs $\overline{XY} \cong \overline{WZ}$. If m∠XYZ = 149°, what is m∠XWZ?
   (1) 15.5°   (2) 31°   (3) 90°   (4) 149°

3. In isosceles trapezoid *RSTU*, $\overline{ST} \parallel \overline{RU}$. If m∠S = $(12x + 6)°$ and m∠T = $(3x + 24)°$, what is m∠S?
   (1) 30°   (2) 54°   (3) 126°   (4) 150°

4. Which of the following is true about an isosceles trapezoid?
   (1) its diagonals bisect each other
   (2) a diagonal forms two congruent triangles
   (3) its diagonals are congruent
   (4) opposite angles are congruent

5. What are the measures of ∠1 and ∠2 in the given trapezoid?
   (1) m∠1 = 24° and m∠2 = 24°
   (2) m∠1 = 66° and m∠2 = 24°
   (3) m∠1 = 66° and m∠2 = 66°
   (4) m∠1 = 24° and m∠2 = 66°

6. Trapezoid *FGHI* has $\overline{FG} \parallel \overline{HI}$. If m∠F = $(2x + 12)°$ and m∠G = $(4x - 18)°$, for what value of x is the trapezoid isosceles?
   (1) 3   (2) 15   (3) 24   (4) 31

B. Free-Response—show all work or explain your answer completely.

7. *ABCD* is a rhombus, m∠ABD = 31°, and m∠EDA = 58°. Is *EBCD* an isosceles trapezoid? Justify your answer.

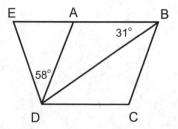

8. The height of isosceles trapezoid *MNOP* is 8, and the two
   bases measure 15 and 27. Find
   the length of leg *PM*.

9. *JKLM* is a square, and *QNOP* is an isosceles
   trapezoid. If m∠*QPJ* = 51° and m∠*POM* =
   24°, find m∠*KQN* and m∠*LNO*.

9.5 CLASSIFYING QUADRILATERALS AND PROOFS INVOLVING SPECIAL QUADRILATERALS

All quadrilaterals can be classified according to their properties. The classifications we have seen are the general quadrilateral, trapezoid, parallelogram, rectangle, rhombus, and square. It is often helpful to consider a Venn
diagram showing these classifications, as shown in Figure 9.7. Squares have
all the properties of both rectangles and rhombi, so they are considered to
be rectangles and rhombi as well as squares.

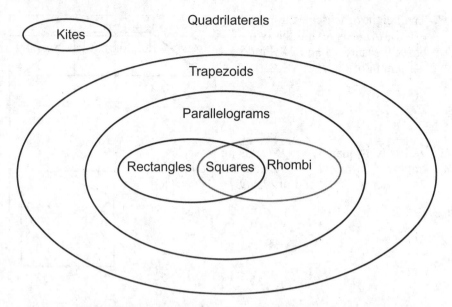

Figure 9.7 Venn diagram of quadrilateral classifications

For the special parallelograms, once you have shown that it is a parallelogram, any one of the additional properties is sufficient. The following table shows a summary of the properties needed to prove each of the parallelogram and trapezoid types.

Classification	What's Needed to Classify the Figure
Trapezoid	One pair of parallel sides
Isosceles trapezoid	One pair of parallel bases and one pair of congruent legs
Parallelogram	Any *one* of the following • Two pairs of opposite sides parallel • Two pairs of opposite sides congruent • Two pairs of opposite angles congruent • Consecutive angles supplementary • Diagonals bisect each other • One pair of sides congruent *and* parallel
Rectangle	Any one property from the parallelogram list *plus* any one of the following: • One right angle • Diagonals are congruent

Classification	What's Needed to Classify the Figure
Rhombus	Any one property from the parallelogram list *plus* any one of the following: • Diagonals are perpendicular • One pair of consecutive sides is congruent • A diagonal bisects one of the angles of the rhombus

Square	Any one property from the parallelogram list
	    plus
	Any one property from the rectangle list
	    plus
	Any one property from the rhombus list

Example 1

A quadrilateral has perpendicular diagonals that bisect each other. The quadrilateral must be which of the following figures?

  (1) rectangle   (2) rhombus   (3) square   (4) isosceles trapezoid

  Solution: Choice (2). The figure is a rhombus because it has one of the parallelogram properties (diagonals bisect each other) and one of the rhombus properties (perpendicular diagonals).

Example 2

Which combination of properties could be used to prove a quadrilateral is a square?

  (1) all four sides are congruent, and the diagonals bisect each other

  (2) the diagonals are perpendicular and congruent

  (3) the diagonals are congruent, perpendicular, and bisect each other

  (4) both pairs of opposite sides are parallel, all angles are right angles, and the diagonals are congruent

  Solution: Choice (3). Diagonals that bisect each other implies a parallelogram. Perpendicular diagonals makes it a rhombus. Congruent diagonals makes it a rectangle. Therefore the figure is a square.

Example 3

Quadrilateral *JKLM* has diagonals $\overline{JL}$ and $\overline{KM}$ that intersect at *O*. If *JO* = *OL* = *KO* = *OM*, what is the most specific classification that can be assigned to *JKLM*?

(1) rectangle   (2) rhombus   (3) square   (4) trapezoid

Solution: Choice (1). A figure with congruent diagonals that bisect each other is a rectangle.

Example 4

In parallelogram *QRST*, $\angle QSR \cong \angle QST$, $\overline{QT} \parallel \overline{RS}$, and $\overline{QT} \cong \overline{RS}$. The figure must be what type of quadrilateral?

(1) rectangle   (2) rhombus   (3) square   (4) kite

Solution: Choice (2). Sides parallel and congruent is a parallel property, and diagonals that bisect the angles is a rhombus property.

It is a good idea to write out the list of properties when doing a proof that involves parallelograms or special parallelograms. If you need to prove the figure is a particular type of parallelogram, make a plan by choosing one property from each of the required categories. When proving a figure is a rectangle, rhombus, or square, first prove the figure is a parallelogram. Then show that one of the special properties is true.

Example 5

Given: $\overline{AEBF}$, $\overline{BC} \parallel \overline{AD}$,
        $\triangle BEC \cong \triangle BFC$,
        $\angle BEC \cong \angle DCE$
Prove: *ABCD* is a rectangle

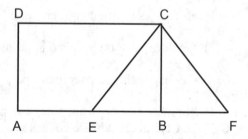

318

Solution:

- Strategy—Mark the figure. Then prove *ABCD* is a parallelogram with opposite sides parallel. Then prove it is a rectangle with a right angle.

Statements	Reasons
1. $\overline{AEBF}$, $\overline{BC} \parallel \overline{AD}$	1. Given
2. $\angle BEC \cong \angle DCE$	2. Given
3. $\overline{AB} \parallel \overline{CD}$	3. Two lines are parallel if the corresponding angles formed by a transversal are congruent
4. *ABCD* is a parallelogram	4. A quadrilateral with two pairs of opposite sides is a parallelogram
5. $\triangle BEC \cong \triangle BFC$	5. Given
6. $\angle EBC \cong \angle FBC$	6. CPCTC
7. $\angle EBC$ and $\angle FBC$ are a linear pair	7. Definition of linear pair
8. $\angle EBC$ and $\angle FBC$ are supplementary	8. Linear pairs are supplementary
9. $\angle EBC$ and $\angle FBC$ are right angles	9. Angles that are congruent and supplementary are right angles
10. *ABCD* is a rectangle	10. A parallelogram with a right angle is a rectangle

Example 6

Given: $\angle BCA \cong \angle DAC$, $\angle BAC \cong \angle BCA$,
  $\overline{BC} \cong \overline{AD}$
Prove: *ABCD* is a rhombus

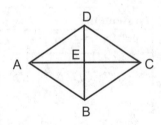

Solution:

Statements	Reasons
1. $\angle BCA \cong \angle DAC$	1. Given
2. $\overline{BC} \parallel \overline{AD}$	2. Two lines are parallel if the alternate interior angles formed by a transversal are congruent
3. $\overline{BC} \cong \overline{AD}$	3. Given

Statements	Reasons
4. *ABCD* is a parallelogram	4. A quadrilateral is a parallelogram if one pair of sides is parallel and congruent
5. ∠*BAC* ≅ ∠*BCA*	5. Given
6. △*ABC* is isosceles	6. A triangle with congruent base angles is isosceles
7. $\overline{AB} \cong \overline{BC}$	7. Sides opposite congruent angles in a triangle are congruent
8. *ABCD* is a rhombus	8. A parallelogram with a pair of consecutive congruent sides is a rhombus

Example 7

Given: ∠*BAC* ≅ ∠*ACD*, ∠*CBD* ≅ ∠*ADB*,
△*AED* ≅ △*CED*, $\overline{AB} \perp \overline{BC}$
Prove: *ABCD* is a square

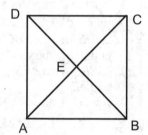

Solution:

Statements	Reasons
1. ∠*BAC* ≅ ∠*CDA*, ∠*CBD* ≅ ∠*ADB*	1. Given
2. $\overline{AB} \parallel \overline{CD}$, $\overline{BC} \parallel \overline{AD}$	2. Two lines are parallel if the alternate interior angles formed by a transversal are congruent
3. *ABCD* is a parallelogram	3. A quadrilateral whose opposite sides are parallel is a parallelogram
4. △*AED* ≅ △*CED*	4. Given
5. $\overline{AD} \cong \overline{CD}$	5. CPCTC
6. *ABCD* is a rhombus	6. A parallelogram with a pair of consecutive congruent sides is a rhombus
7. $\overline{AB} \perp \overline{BC}$	7. Given
8. ∠*ABC* is a right angle	8. Perpendicular lines intersect at right angles

Statements	Reasons
9. *ABCD* is a rectangle	9. A parallelogram with a right angle is a rectangle
10. *ABCD* is a square	10. A parallelogram that is a rectangle and a rhombus is a square

Check Your Understanding of Section 9.5

A. Multiple-Choice

1. Quadrilateral *GHIJ* has perpendicular diagonals. What other piece of information would be sufficient to prove it is a rhombus?
   (1) $\angle G \cong \angle I$ and $\angle H \cong \angle J$ (3) $\overline{GH} \cong \overline{HI}$
   (2) $\overline{GI} \cong \overline{HJ}$ (4) $\angle G \cong \angle H$

2. Quadrilateral *JKLM* has diagonals $\overline{JL}$ and $\overline{KM}$ that intersect at *A*. If $JA = AL = KA = AM$, which of the following must be true?
   (1) $\overline{KL} \cong \overline{LM}$ (3) $\angle JKM \cong \angle LKM$
   (2) $\overline{JL} \perp \overline{KM}$ (4) $\overline{KL} \perp \overline{LM}$

3. Quadrilateral *ABCD* has diagonals that form four congruent triangles. What is the most specific classification of *ABCD*?
   (1) parallelogram   (3) rhombus
   (2) rectangle   (4) square

4. Emily wants to prove that a quadrilateral is a parallelogram. Which of the following properties would *not* be sufficient for her to do so?
   (1) the diagonals bisect each other
   (2) the diagonals are congruent and perpendicular
   (3) one pair of opposite sides is congruent and parallel
   (4) two pairs of opposite angles are congruent

5. Which of the following statements is true?
   (1) all rectangles are rhombi
   (2) all parallelograms are rectangles
   (3) all squares are rectangles
   (4) all quadrilaterals are trapezoids

6. Which of the following statements is true?
   (1) all quadrilaterals with perpendicular diagonals are rectangles
   (2) all quadrilaterals with four congruent angles are squares
   (3) all quadrilaterals with four congruent sides are rhombi
   (4) all quadrilaterals with one pair of congruent sides are parallelograms

B. *Free-Response—show all work or explain your answer completely.*

7. Given: $\overline{AB} \parallel \overline{CD}$, $\overline{AB} \cong \overline{CD}$, $\overline{AB} \perp \overline{BC}$, and $\angle DEC \cong \angle DEA$
   Prove: *ABCD* is a square

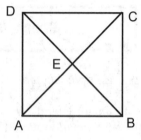

8. Given: rectangle *ABCD*, *E* is the midpoint of $\overline{AB}$, *F* is the midpoint of $\overline{BC}$, *G* is the midpoint of $\overline{CD}$, and *H* is the midpoint of $\overline{DA}$
   Prove: *EFGH* is a rhombus

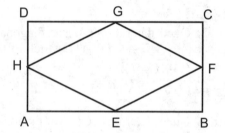

9. Given: $\angle BAC \cong \angle DCA \cong \angle ABD$, $\overline{AE} \cong \overline{EC}$
   Prove: *ABCD* is a rectangle

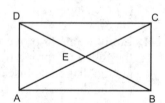

10. Given: rhombus *ABED*
    Prove: $\angle ECD \cong \angle ECB$

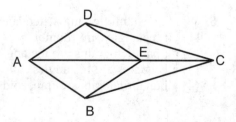

322

11. Given: $\overline{RT} \cong \overline{ST}$, Q is the midpoint of $\overline{RT}$, and P is the midpoint of $\overline{ST}$

Prove: $PQRS$ is an isosceles trapezoid

12. Given: parallelogram $EFGH$, $\overline{HJ} \cong \overline{IG}$, and $\overline{JE} \cong \overline{IF}$

Prove: $EFGH$ is a rectangle

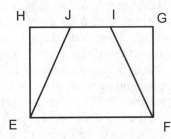

13. Gina states that a shortcut for proving a quadrilateral is a parallelogram is to show that the diagonals are perpendicular and one pair of opposite sides is congruent. Kevin does not believe her statement is true. Who is correct? Justify your reasoning.

14. $ABCDE$ is a regular pentagon and $m\angle AFC = 108°$. Is $ABCF$ a rhombus? Justify your answer.

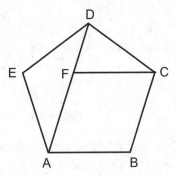

9.6 PARALLELOGRAMS AND TRANSFORMATIONS

Rigid motions can be applied to map a parallelogram onto itself.

Mapping a Quadrilateral onto Itself

The properties of parallelograms and trapezoids can be used to identify rigid motions that map the figures onto themselves. Right angles, congruent angles, congruent sides, bisected segments, and parallel lines are not only properties of the various parallelograms described in the chapter; they are also defining features of the rigid motion transformations. When given a particular parallelogram, it is possible to identify a variety of transformations that map the figure onto itself.

Figure 9.8 shows rectangle $ABCD$ with diagonals intersecting at point E. A rotation of 180° about point E would map $ABCD$ onto itself. The justification is based on the parallelogram properties. The diagonals bisect each other, so $AE = EC$. Points A and C are equidistant from the center of rotation and lie on a straight line through the center. Therefore A and B map to one another. The same justification explains why B and D map to one another.

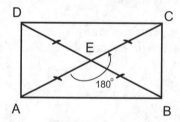

Figure 9.8 Rectangle *ABCD* under a 180° rotation

A rectangle also maps to itself after a reflection over a line through the midpoints of two opposite sides. In Figure 9.9, E and F are the midpoints of $\overline{AD}$ and $\overline{BF}$. So $AE = ED = BF = FC$. $\overline{EF}$ is the perpendicular bisector of $\overline{AD}$ and $\overline{BC}$. $\overline{EF}$ acts as the line of reflection that maps A to D and maps B to C. $r_{\overline{EF}}(ABCD) = DCBA$.

Figure 9.9 Rectangle *ABCD* reflected over a line through the midpoints of $\overline{BC}$ and $\overline{AD}$

By applying the same reasoning as with the rectangle, we find that any parallelogram can be mapped onto itself with a 180° rotation. However, a reflection through the midpoints will not necessarily work.

When a particular mapping is specified, keep corresponding parts in mind. We may need to have each point in the preimage map to a specific point in the image. A 180° rotation may not yield the desired mapping. In fact, a composition of two or more transformations may be needed. Often a translation is combined with a reflection.

Example 1

Given rectangle *ABCD*, find a rigid motion or composition of rigid motions that will map rectangle *ABCD* onto itself, such that:

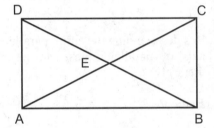

(a) the image of point *A* after the transformation is point *B*.
(b) the image of point *A* after the transformation is point *C*.
(c) the image of point *A* after the transformation is point *D*.

Solution:

A translation that maps *A* to *B* followed by a reflection over $\overline{BC}$ maps *ABCD* onto itself with point *A* mapped to point *B*.

A rotation of 180° about Point *E* maps *ABCD* onto itself with point *A* mapped to point *C*.

A translation that maps *A* to *D* followed by a reflection over $\overline{CD}$ maps *ABCD* onto itself with point *A* mapped to point *B*.

Example 2

Name 2 different rigid motions, or compositions of rigid motions, that will map parallelogram $ABCD$ onto itself such that point A maps to point C?

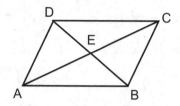

Solution:

(a) Rotate $180°$ about point E.
(b) Translate by $\overrightarrow{AC}$ followed by a rotation of $180°$ about C.

MATH FACT

When mapping a parallelogram onto itself, a vertex can only be mapped to the opposite vertex unless the figure is a rectangle or square. The angles at the two vertices must be congruent to map onto each other. Opposite angles are congruent in a parallelogram, but consecutive angles are only congruent in rectangles and squares.

Besides mapping the entire parallelogram, the triangles formed by the diagonals of parallelograms can be mapped from one to another or onto themselves.

Example 3

Using the figure from the previous example, what rigid motion will map $\triangle AED$ to $\triangle CEB$?

Solution: A rotation of $180°$ about point E maps A to C and maps B to D. So it will map $\triangle AED$ to $\triangle CEB$.

Example 4

Using the figure from Example 2, explain why a reflection across diagonal $\overline{AC}$ will *not* map $\triangle AED$ to $\triangle CEB$ unless the parallelogram is also a rhombus.

Solution: If the parallelogram is not a rhombus, the diagonals are not perpendicular. Since $DE = EB$ but $\overrightarrow{AEC}$ is not the perpendicular bisector of $\overline{DB}$, reflecting over that line cannot map D to B.

326

Check Your Understanding of Section 9.6

A. *Multiple-Choice*

1. $\overline{AB}$ is reflected through point Q such that its image is $\overline{A'B'}$. If $ABA'B'$ is a rectangle, which of the following must be true?
   (1) Q is the midpoint of $\overline{AB}$
   (2) $QA = QB$
   (3) $QA = AB$
   (4) $m\angle ABQ = 90°$

2. $\triangle ABC$ is a right triangle with a right angle at A. If $R_{B,180°}(\triangle ABC) \rightarrow \triangle A'BC'$, how can quadrilateral $ACA'C'$ be classified?
   (1) parallelogram and rectangle only
   (2) parallelogram and rhombus only
   (3) parallelogram, rectangle, rhombus, and square
   (4) parallelogram only

3. In parallelogram $ABCD$, every point on diagonal $\overline{AC}$ is equidistant from points B and D and every point on diagonal $\overline{BD}$ is equidistant from points A and C. Parallelogram $ABCD$ must be a
   (1) rectangle
   (2) rhombus
   (3) square
   (4) kite

4. A segment is rotated 90° about its midpoint. The segment and its image form the diagonals of
   (1) a square
   (2) a quadrilateral that is not a parallelogram
   (3) a rhombus but not necessarily a square
   (4) a rectangle but not necessarily a square

5. $\overline{AB}$ is translated by vector $\overrightarrow{XY}$ to $\overline{A'B'}$. If $\overrightarrow{XY} \perp \overline{AB}$ and the magnitude of $\overrightarrow{XY}$ is twice that of $\overline{AB}$, what type of figure is $ABB'A'$?
   (1) rhombus
   (2) square
   (3) rectangle
   (4) parallelogram but not a rectangle or rhombus

6. Which of the following rigid motions will map parallelogram *RSTU* onto itself such that *R* maps to *T*?
   (1) $R_{U,180°}$
   (2) $r_{\overline{TU}} \circ T_{\overline{RU}}$
   (3) $r_{\overline{RS}} \circ T_{\overline{RU}}$
   (4) $R_{O,180°}$ where *O* is the point of intersection of the diagonals

7. Segment $\overline{AB}$ is rotated 61° about point *P*, which lies on $\overline{AB}$. Which type of figure is formed by *A*, *B*, and the images of *A* and *B*?
   (1) parallelogram
   (2) isosceles trapezoid with one only pair of parallel sides
   (3) isosceles triangle
   (4) rhombus

8. Right triangle $\triangle ABC$ has a right angle at *B*. If $\triangle A'B'C'$ is the image of $\triangle ABC$ after a reflection through *B*, then the figure *ACA'C'* must always be what type of figure?
   (1) rhombus
   (2) square
   (3) rectangle
   (4) parallelogram but not a rectangle or rhombus

9. The measure of ∠*A* in rhombus *ABCD* = 85°. Which of the following will map $\triangle DEA$ to $\triangle CEB$?
   (1) reflection over $\overline{AB}$
   (2) rotation of 180° about point *E*
   (3) reflection through point *E*
   (4) translation by vector $\overrightarrow{AB}$ followed by a reflection over $\overline{BC}$

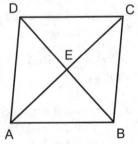

B. *Free-Response—show all work or explain your answer completely.*

10. $\overline{RS}$ is translated by vector $\overrightarrow{MN}$ such that its image is $\overline{R'S'}$. If *MN* = *RS* and $\overrightarrow{MN} \perp \overline{RS}$, classify *RSS'R'* as specifically as possible.

11. The base of isosceles $\triangle FGH$ is $\overline{FH}$. If $r_{\overline{FH}}(\triangle FGH) \rightarrow \triangle(FG'H)$, what type of parallelogram is *FGHG'*? Justify your reasoning.

12. $\triangle JKL$ is reflected through point *K* such that its image after the reflection is $\triangle J'KL'$. Explain why quadrilateral *JLJ'L'* must be a parallelogram.

13. $\overline{AC}$ is rotated 90° about its midpoint E such that its image is $\overline{BD}$. Find m$\angle BAC$.

14. Using point A and vector $\overrightarrow{RS}$, sketch the following and classify quadrilateral $ANOM$ where points M, N, and O are defined by the rigid motions shown below. Justify your classification.

A •

R ⟶ S

M is the image of Point A after a translation by vector $\overrightarrow{RS}$

N is the image of Point M after a rotation of 90° about Point A

O is the image of Point N after a translation by vector $\overrightarrow{RS}$

COORDINATE GEOMETRY PROOFS

Chapter
Ten

10.1 TOOLS AND STRATEGIES OF COORDINATE GEOMETRY PROOFS

KEY IDEAS

Special properties of triangles, quadrilaterals, and other figures can be proven using coordinate geometry. Certain tools are used in coordinate geometry:

- Distance—to show segments congruent
- Midpoint—to show a segment is bisected
- Slope—to show that segments and lines are parallel (same slope) or perpendicular (negative reciprocal slopes)

Tools of Coordinate Geometry Proofs

In a coordinate geometry proof, the coordinates of the vertices of a figure are provided as givens, and we are asked to prove some feature of the figure using coordinate geometry. The coordinate geometry tools at our disposal are distance, midpoint, and slope. Equal distances indicate congruence. Slope can be used to demonstrate parallel or perpendicular lines and segments. Midpoints show if a segment is bisected.

Quantity	Formula	Use
Slope	$m = \dfrac{y_2 - y_1}{x_2 - x_1}$ or $\dfrac{\text{rise}}{\text{run}}$	Prove segments or lines are parallel (slopes are equal) or perpendicular (slopes are negative reciprocals)
Midpoint	$x_{MP} = \dfrac{1}{2}(x_1 + x_2)$ $y_{MP} = \dfrac{1}{2}(y_1 + y_2)$	Prove segments bisect each other (midpoints are concurrent)
Length	$d = \sqrt{(x_2 - x_1)^2 + (y_2 - y_1)^2}$	Segments are congruent (distances between endpoints are equal)

Follow these steps when writing a coordinate geometry proof:

- Graph the points. The graph will help you plan a strategy, check your work, and help with some calculations.
- Plan a strategy. What property will you demonstrate, and what calculations are needed?
- Perform the calculations. Write down the general equations you are using, and clearly label which segments correspond to which calculations.
- Write a summary statement. You must state in words a justification that explains why your calculations justify the proof.

10.2 PARALLELOGRAM PROOFS

KEY IDEAS

The midpoint, distance, and slope formulas can be used to prove quadrilaterals are parallelograms, rectangles, rhombi, or squares. The following tools are used:

- Midpoint can show that diagonals bisect each other.
- Distance can show sides or diagonals are congruent.
- Slope can show sides are parallel, sides are perpendicular, or diagonals are perpendicular.

Proving Parallelograms

A given quadrilateral can be proven to be a parallelogram by demonstrating one of the parallelogram properties. The three properties that can be easily proven are parallel sides, congruent sides, and diagonals bisecting each other. The calculations required and suggested summary statements are shown in the following table:

Property	Calculations	Summary Statement
2 pairs of opposite sides are parallel	Slope of each side (4 slope calculations)	The slopes of opposite sides are equal, therefore both pairs of opposite sides are parallel. A quadrilateral with two pairs of opposite sides parallel is a parallelogram.

Property	Calculations	Summary Statement
2 pairs of opposite sides are congruent	Length of each side (4 distance calculations)	The lengths of opposite sides are equal, therefore both pairs of opposite sides are congruent. A quadrilateral with two pairs of opposite congruent sides is a parallelogram.
The diagonals bisect each other	Midpoints of the diagonals (2 midpoint calculations)	The diagonals have the same midpoint, therefore they bisect each other. A quadrilateral whose diagonals bisect each other is a parallelogram.

When writing a coordinate geometry proof, be sure to show all formulas used. Show all substitutions and calculations with labels to make it clear which parts are being used in the proof. You must conclude with a summary paragraph that states what was proven and how it was done. You can think of the summary as a short paragraph proof.

Example

The coordinates of quadrilateral $ABCD$ are $A(-2, 1)$, $B(4, 4)$, $C(6, 5)$, and $D(0, 2)$. Prove $ABCD$ is a parallelogram.

   Solution:

- Strategy—find the midpoint of diagonals $\overline{AC}$ and $\overline{BD}$ to show the diagonals bisect each other.

Midpoint of $\overline{AC}$:

$$x_{MP} = \frac{1}{2}(x_1 + x_2),\ y_{MP} = \frac{1}{2}(y_1 + y_2)$$

$$x_{MP} = \frac{1}{2}(-2+6),\ y_{MP} = \frac{1}{2}(1+5)$$

$$x_{MP} = 2,\ y_{MP} = 3$$

The midpoint of $\overline{AC}$ is $(2, 3)$

Midpoint of $\overline{BD}$:

$$x_{MP} = \frac{1}{2}(x_1 + x_2),\ y_{MP} = \frac{1}{2}(y_1 + y_2)$$

$$x_{MP} = \frac{1}{2}(4+0), \, y_{MP} = \frac{1}{2}(4+2)$$

$$x_{MP} = 2, \, y_{MP} = 3$$

The midpoint of $\overline{BC}$ is (2, 3)
* Summary—The midpoints of $\overline{AC}$ and $\overline{BC}$ are the same, so the diagonals bisect each other. *ABCD* is a parallelogram because a quadrilateral whose diagonals bisect each other is a parallelogram.

Proving Special Parallelograms

To prove a quadrilateral is a special parallelogram, first prove the figure is a parallelogram. Then demonstrate one of the special properties shown in the table that follows. For squares, you need to show that the figure is both a rectangle and a rhombus.

Figure	Property	Calculation	Example of Summary Statement
Rhombus	Diagonals are perpendicular	Slope of each diagonal	The slope of the diagonals are negative reciprocals, therefore they are ⊥. A parallelogram with ⊥ diagonals is a rhombus.
	Two consecutive sides are congruent	Length of two consecutive sides	The lengths of 2 consecutive sides are equal, so they are ≅. A parallelogram with consecutive ≅ sides is a rhombus.
Rectangle	A right angle	Slope of two consecutive sides	The slopes of two consecutive sides are negative reciprocals, so the sides are ⊥ and form a right angle. A parallelogram with a right angle is a rectangle.
	Congruent diagonals	Length of each diagonal	The lengths of the diagonals are ≅, so they are congruent. A parallelogram with congruent diagonals is a rectangle.

Figure	Property	Calculation	Example of Summary Statement
Square	Show the figure is both a rectangle and a rhombus		A figure that is both a rectangle and a rhombus is a square.

The method you pick depends on what calculations you are most comfortable with. Midpoint and slope calculations are usually faster to do than distance calculations. However, finding the distance may make more sense if you need to follow up by showing the figure is a special parallelogram.

Some suggested strategies are:

Parallelogram—midpoints of the diagonals
Rectangle—slope of each side
Rhombus—length of each side or midpoint and the slope of the diagonals

Example 1

Given coordinates $A(0, 2)$, $B(4, 8)$, $C(7, 6)$, and $D(3, 0)$, prove $ABCD$ is a rectangle.

Solution:

- Strategy—Calculate the slopes of the four sides.

$$\text{Slope} = \frac{y_2 - y_1}{x_2 - x_1}$$

$$\text{Slope of } \overline{AB} = \frac{8-2}{4-0} = \frac{3}{2}$$

$$\text{Slope of } \overline{BC} = \frac{6-8}{7-4} = -\frac{2}{3}$$

$$\text{Slope of } \overline{CD} = \frac{0-6}{3-7} = \frac{3}{2}$$

$$\text{Slope of } \overline{DA} = \frac{0-2}{3-0} = -\frac{2}{3}$$

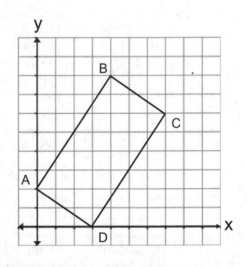

The slopes of opposite sides are equal, so $ABCD$ is a parallelogram. The slopes of consecutive sides are negative reciprocals, making the consecutive sides perpendicular and the angles right angles. $ABCD$ is a rectangle because a parallelogram with right angles is a rectangle.

334

Example 2

Given coordinates $R(-3, 0)$, $S(1, 7)$, $T(9, 6)$, and $U(5, -1)$, prove $RSTU$ is a rhombus.

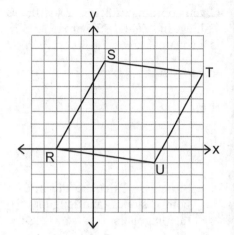

Solution:

• Strategy—Calculate the midpoints and slopes of the diagonals.

$$\text{Midpoint} = \frac{1}{2}(x_1 + x_2), \frac{1}{2}(y_1 + y_2)$$

$$\text{Midpoint of } \overline{SU} = \frac{1}{2}(1+5), \frac{1}{2}(7+(-1))$$
$$= (3, 3)$$

$$\text{Midpoint of } \overline{RT} = \frac{1}{2}(-3+9), \frac{1}{2}(0+6)$$
$$= (3, 3)$$

$$\text{Slope} = \frac{y_2 - y_1}{x_2 - x_1}$$

$$\text{Slope of } \overline{SU} = \frac{-1-7}{5-1} = -2$$

$$\text{Slope of } \overline{RT} = \frac{6-0}{9-(-3)} = \frac{1}{2}$$

The diagonals of $RSTU$ have the same midpoint, therefore they bisect each other, making $RSTU$ a parallelogram. The diagonals of $RSTU$ have negative reciprocal slopes, making them perpendicular. $RSTU$ is a rhombus because a parallelogram with perpendicular diagonals is a rhombus.

Example 3

Given coordinates $M(-4, 2)$, $A(0, 5)$, $T(3, 1)$, and $H(-1, -2)$, prove $MATH$ is a square.

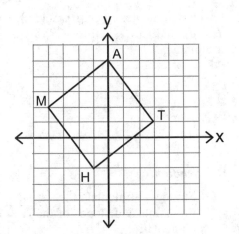

Solution:

- Strategy—Calculate the midpoints, slopes, and lengths of the diagonals.

$$\text{Midpoint} = \frac{1}{2}(x_1 + x_2), \frac{1}{2}(y_1 + y_2)$$

$$\text{Midpoint of } \overline{MT} = \frac{1}{2}(-4+3), \frac{1}{2}(2+1)$$

$$= (-\frac{1}{2}, \frac{3}{2})$$

$$\text{Midpoint of } \overline{AH} = \frac{1}{2}(0+(-1)), \frac{1}{2}(5+(-2)) = (-\frac{1}{2}, \frac{3}{2})$$

$$\text{Slope} = \frac{y_2 - y_1}{x_2 - x_1}$$

$$\text{Slope of } \overline{MT} = \frac{1-2}{3-(-4)} = -\frac{1}{7}$$

$$\text{Slope of } \overline{AH} = \frac{-2-5}{-1-0} = 7$$

$$\text{Length} = \sqrt{(x_2 - x_1)^2 + (y_2 - y_1)^2}$$

$$\text{Length of } \overline{MT} = \sqrt{(3-(-4))^2 + (1-2)^2}$$

$$= \sqrt{7^2 + (-1)^2}$$

$$= \sqrt{50}$$

Length of $\overline{AH} = \sqrt{(-1-0)^2 + (-2-5)^2}$

$$= \sqrt{(-1)^2 + (-7)^2}$$

$$= \sqrt{50}$$

The midpoints of the diagonals are the same. So the diagonals bisect each other, making $RSTU$ a parallelogram. The slopes of the diagonals are negative reciprocals, making them perpendicular. $RSTU$ is a rhombus because it is a parallelogram with perpendicular diagonals. The lengths of the diagonals are equal, making them congruent. $RSTU$ is a rectangle because it is a parallelogram with congruent diagonals. $RSTU$ is a square because it is both a rectangle and a rhombus.

Check Your Understanding of Section 10.2

A. Multiple-Choice

1. Claire wants to prove that quadrilateral $SRBD$ is a rectangle. Which combination of calculations would allow her to accomplish this?
   (1) the slope and length of one pair of opposite sides
   (2) the midpoints and slopes of the two diagonals
   (3) the slopes of the four sides
   (4) the lengths of the four sides

2. Scarlet is planning to prove a quadrilateral is a rhombus. Which of the following calculations would allow her to do so?
   (1) the slopes and midpoints of the diagonals
   (2) the slopes and lengths of the diagonals
   (3) the midpoints and lengths of the diagonals
   (4) the slopes of all four sides

3. Points $F(1, 2)$, $G(4, 7)$, and $H(9, 4)$ are the vertices of quadrilateral $FGHI$. Which of the following coordinates for point I would make $FGHI$ a parallelogram?
   (1) $(5, -2)$ (2) $(6, -1)$ (3) $(7, 0)$ (4) $(8, -2)$

B. Free-Response—show all work or explain your answer completely.

4. Given quadrilateral $ABCD$ with vertices $A(-6, -3)$, $B(-4, 3)$, $C(6, 7)$, and $D(4, 1)$, prove $ABCD$ is a parallelogram.

5. Given quadrilateral *FROG* with vertices $F(-6, 1)$, $R(1, 5)$, $O(8, 6)$, and $G(1, 2)$, prove *FROG* is a parallelogram.

6. Quadrilateral *BLUE* has vertices $B(-5, 0)$, $L(-6, 6)$, $U(6, 8)$, and $E(7, 2)$. Prove *BLUE* is a rectangle.

7. The coordinates of the vertices of quadrilateral *CRAB* are $C(-2, -1)$, $R(0, 7)$, $A(4, 6)$, and $B(2, -2)$. Prove *CRAB* is a rectangle.

8. Given points $A(2, -1)$, $B(1, 5)$, $C(7, 4)$, and $D(8, -2)$, prove quadrilateral *ABCD* is a rhombus.

9. Quadrilateral *CLUE* has vertices with coordinates $C(-5, -4)$, $L(-3, 4)$, $U(5, 6)$, and $E(3, -2)$. Prove *CLUE* is a rhombus.

10. The coordinates of *JUMP* are $J(3, -2)$, $U(1, 3)$, $M(6, 5)$, and $P(8, 0)$. Prove *JUMP* is a square.

11. Given quadrilateral *PQRS* with vertices $P(-1, 4)$, $Q(4, 7)$, $R(7, 2)$, and $S(2, -1)$, prove *PQRS* is a square.

12. Four lines have the equations $5y = 2x + 30$, $3x + 2y = 20$, $3x + 2y = 1$, and $y = \frac{2}{5}x + 12$. Do the intersections of these four lines form the vertices of a parallelogram? Justify your answer.

13. The coordinates of three vertices of a parallelogram are $A(8, 2)$, $B(2, 1)$, and $C(6, 6)$. Using the midpoint formula, find the coordinates of D, the fourth vertex of the parallelogram.

10.3 TRIANGLE PROOFS

KEY IDEAS

Coordinate geometry tools can be used to show the following:

- A triangle is scalene, isosceles, or equilateral.
- A triangle is a right triangle.
- Segments in triangles are altitudes or medians.
- Figures are congruent.

Triangles

Prove a triangle is scalene, isosceles, or equilateral by finding the length of each side. Equal lengths indicate congruent sides.

Example 1

$\triangle TBX$ has vertices with coordinates $T(1, 2)$, $B(5, 10)$, and $X(9, 2)$. Prove $\triangle TBX$ is isosceles but not equilateral.

Solution:

- Strategy—Find the length of each side, and show that two sides are congruent.

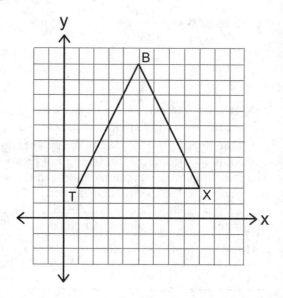

$$\text{Length} = \sqrt{(x_2 - x_1)^2 + (y_2 - y_1)^2}$$

$$TB = \sqrt{(5-1)^2 + (10-2)^2} = \sqrt{80}$$

$$BX = \sqrt{(9-5)^2 + (2-10)^2} = \sqrt{80}$$

$$TX = \sqrt{(9-1)^2 + (2-2)^2} = 8$$

$TB = BX$, but they are not equal to TX. Therefore $\triangle TBX$ has two congruent sides and is isosceles but not equilateral.

Use slope to determine if sides are perpendicular to prove a triangle is a right triangle.

Example 2

Given points $D(-2, 0)$, $J(2, 3)$, and $R(-1, 7)$, prove $\triangle DJR$ is a right triangle and identify which side is the hypotenuse.

Solution:

- Strategy—Find the slope of each side.

$$\text{Slope} = \frac{y_2 - y_1}{x_2 - x_1}$$

$$\text{Slope } \overline{DJ} = \frac{3-0}{2-(-2)} = \frac{3}{4}$$

$$\text{Slope } \overline{DR} = \frac{7-0}{-1-(-2)} = 7$$

$$\text{Slope } \overline{JR} = \frac{7-3}{-1-2} = -\frac{4}{3}$$

The slopes of $\overline{DJ}$ and $\overline{JR}$ are negative reciprocals, so they are perpendicular. $\angle J$ is therefore a right angle, and $\triangle DJR$ is a right triangle. The side opposite $\angle J$ is the hypotenuse, so $\overline{DR}$ is the hypotenuse.

Coordinate geometry can be used to prove special segments in triangles. The following table summarizes some of the segments and points of concurrency and also the calculations needed.

Segment	Calculation Needed	Summary Statement
Median	Midpoint	The segment has endpoints at a vertex and midpoint, so the segment is a median.
Altitude	Two slopes	The slopes are negative reciprocals. So the segment is ⊥ to the opposite side, making the segment an altitude.
Perpendicular bisector	Slope and midpoint	The slopes are negative reciprocals, making the segments ⊥. The segment passes through the midpoint, making it a bisector. Therefore the segment is a perpendicular bisector.
Midsegment	Two midpoints	The segment has endpoints at two midpoints, therefore the segment is a midsegment.

Segment	Calculation Needed	Summary Statement
Circumcenter	Three lengths	The distances from the point to the three vertices are equal, so the point is equidistant from each vertex. Therefore the point is the circumcenter.
Centroid	Two lengths	The distances from the point to each vertex are in a $1:2$ ratio, therefore the point is a centroid.

Example 3

The vertices of $\triangle QRS$ are $Q(0, 9)$, $R(8, 5)$, and $S(3, 0)$. Point $T(2, 3)$ lies on $\overline{QS}$. Prove $\overline{RT}$ is an altitude.

Solution:

$$\text{Slope} = \frac{y_2 - y_1}{x_2 - x_1}$$

$$\text{Slope } \overline{QS} = \frac{0 - 9}{3 - 0} = -3$$

$$\text{Slope } \overline{RT} = \frac{3 - 5}{2 - 8} = \frac{1}{3}$$

$\overline{RT}$ and $\overline{QS}$ have negative reciprocal slopes, so they are perpendicular.
Therefore $\overline{RT}$ is an altitude of $\triangle QRS$.

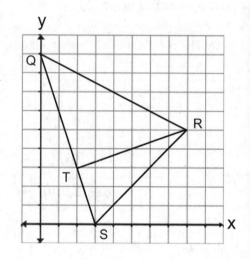

Example 4

$\triangle RST$ has vertices with coordinates $R(1, 1)$, $S(7, 3)$, and $T(2, 7)$. If point V lies on $\overline{RS}$ and has coordinates $(4, 2)$, prove $\overline{TV}$ is a median.

Solution:

$$\text{Midpoint} = \frac{1}{2}(x_1 + x_2), \frac{1}{2}(y_1 + y_2)$$

$$\text{Midpoint of } \overline{RS} = \frac{1}{2}(1+7), \frac{1}{2}(1+3) = (4, 2)$$

Since V is the midpoint of $\overline{RS}$, $\overline{TV}$ is a median.

If you need to find the centroid of a triangle, you can take either a graphical or an algebraic approach. You can graph the three medians and identify the point of concurrency, or you can find the point algebraically that divides the median in a $1:2$ ratio.

Example 5

The vertices of $\triangle ABC$ are $A(0, 0)$, $B(2, 6)$, and $C(10, 6)$. Find the centroid of $\triangle ABC$ graphically. Then demonstrate algebraically that the centroid divides a median in a $1:2$ ratio.

Solution: Let D, E, and F be the midpoints of $\overline{AB}$, $\overline{BC}$, and $\overline{AC}$, respectively. The coordinates of the midpoints are found with the midpoint formula.

$$x_{MP}, y_{MP} = \frac{1}{2}(x_1 + x_2), \frac{1}{2}(y_1 + y_2)$$

$$D: \frac{1}{2}(0+2), \frac{1}{2}(0+6)$$

$$D(1, 3)$$

$$E: \frac{1}{2}(2+10), \frac{1}{2}(6+6)$$

$$E(6, 6)$$

$$F: \frac{1}{2}(0+10), \frac{1}{2}(0+6)$$

$$F(5, 3)$$

Graph the medians to locate the centroid. The centroid is located at $G(4, 4)$.

Use the distance formula to confirm the ratio.

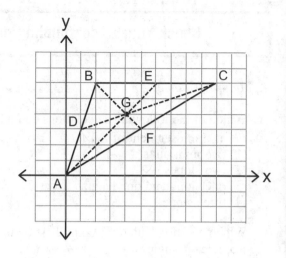

$$FG = \sqrt{(4-5)^2 + (4-3)^2}$$
$$= \sqrt{2}$$

$$GB = \sqrt{(4-2)^2 + (4-6)^2}$$
$$= \sqrt{8}$$
$$= 2\sqrt{2}$$

$$\frac{FG}{GB} = \frac{\sqrt{2}}{2\sqrt{2}}$$
$$= \frac{1}{2}$$

Point G does divide the segment in a $1:2$ ratio.

Alternatively, find the point that divides the median into a $1:2$ ratio and show its coordinates are $(4, 4)$.

Example 6

Prove algebraically that $D(2, 3)$ is the circumcenter of $\triangle ABC$, whose vertices have coordinates $A(2, -2)$, $B(-1, 7)$, and $C(5, 7)$.

Solution: D will be the circumcenter if it is equidistant from points A, B, and C.

Use the distance formula $\sqrt{(x_2 - x_1)^2 + (y_2 - y_1)^2}$.

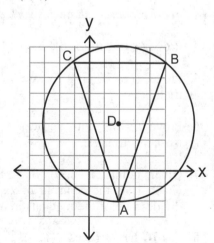

$$DA = \sqrt{(2-2)^2 + (3-(-2))^2} = 5$$

$$DB = \sqrt{(2-(-1))^2 + (3-7)^2} = 5$$

$$DC = \sqrt{(2-5)^2 + (3-7)^2} = 5$$

D is equidistant from each of the vertices of the triangle, so it must be the circumcenter of the triangle.

Check Your Understanding of Section 10.3

A. Multiple-Choice

1. Which of the following calculations could be used to prove that segment $\overline{JK}$ is a midsegment of $\triangle YAG$?
   (1) two midpoints
   (2) two slopes
   (3) one slope and one length
   (4) three slopes

2. Which set of calculations could best be used to prove $\triangle LMN$ is a right isosceles triangle?
   (1) 2 slopes and 2 midpoints
   (2) 3 midpoints
   (3) 2 slopes and 2 lengths
   (4) 2 midpoints and 2 lengths

3. In $\triangle GBH$, point D lies on $\overline{GB}$. Which of the following calculations could be used to prove $\overline{HD}$ is an altitude?
   (1) slope of $\overline{GB}$ and slope of $\overline{GH}$
   (2) slope of $\overline{GB}$ and slope of $\overline{HD}$
   (3) length of $\overline{GD}$ and slope of $\overline{DH}$
   (4) midpoint of $\overline{HD}$ and slope of $\overline{HD}$

B. Free-Response—show all work or explain your answer completely.

4. Prove $Q(1, 4)$, $B(3, 6)$, and $C(4, 1)$ are the vertices of a right triangle.

5. Prove $T(1, 5)$, $G(6, 5)$, and $H(4, 1)$ are the vertices of an isosceles triangle.

6. Prove $\triangle FGH$ with vertices $F(-2, 5)$, $G(5, 4)$, and $H(1, 1)$ is a right isosceles triangle.

7. Given points $A(-6, 0)$, $B(-4, 8)$, and $C(4, 4)$, prove graphically that $G(-2, 4)$ is the centroid of $\triangle ABC$. Demonstrate algebraically that G divides one of the medians in a $1:2$ ratio.

8. $\triangle ABC$ has vertices $A(1, 3)$, $B(5, 5)$, and $C(4, 0)$. Point $D(2, 2)$ lies on $\overline{AC}$. Prove that $\overline{BD}$ is an altitude.

9. Given points $M(-3, 2)$, $N(6, 3)$, and $O(2, -2)$, prove $m\angle M = 45°$.

10. Prove that $D(5, 5)$ is the circumcenter of $\triangle ABC$, whose vertices have coordinates $A(3, 2)$, $B(3, 8)$, and $C(7, 2)$.

11. Prove $P(4, -1)$ is the center of the circle passing through coordinates $F(9, -4)$, $G(1, 4)$, and $H(1, 4)$.

12. Prove $D(2, 8)$ is the orthocenter of $\triangle JKL$ with coordinates $J(1, 1)$, $K(3, 5)$, and $L(10, 4)$.

13. Find the values of a if $A(-4, 7)$, $B(a, -1)$, and $C(6, 2)$ are the coordinates of a right triangle with a right angle at B.

CIRCLES

Chapter
Eleven

11.1 DEFINITIONS, ARCS, AND ANGLES IN CIRCLES

KEY IDEAS

The following are the basic circle definitions and relationships:

- The measure of a central angle equals the measure of the intercepted arc.
- The measure of an inscribed angle equals $\frac{1}{2}$ the measure of the intercepted arc.
- The sum of the arc measures around a circle equals 360°.
- All radii of a circle are congruent, and congruent circles have congruent radii.
- All circles are similar, and one can be mapped to another with only a translation and a dilation.

Definitions

A **circle** is the set of points a fixed distance from a point. The point is the center of the circle, and the fixed distance is the radius.

Segments and Lines Definitions in Circles

Segment or Line	Definition
Radius	A segment with one endpoint at the center of the circle and one endpoint on the circle
Chord	A segment with both endpoints on the circle
Diameter	A chord that passes through the center of the circle
Secant	A line that intersects a circle at exactly two points
Tangent	A line that intersects a circle at exactly one point
Point of tangency	The point at which a tangent intersects a circle

Figure 11.1 Circle *P* with radius, chord, diameter, secant, and tangent

---

MATH FACT

When we say "on" the circle, we mean points on the circumference of the circle. Points in the interior of the circle are not on the circle. Consider the equation of the circle from Chapter 6, $(x - h)^2 + (y - k)^2 = r^2$. It is satisfied only by the coordinates of points that lie on the circumference. When we talk about the area of a circle, we really mean the area enclosed by the circle.

Radii, Congruence, and Similarity

Every circle has an infinite number of radii. Figure 11.1 shows only one radius. The diameter is always equal to twice the radius.

Radius and Diameter Theorems
The length of the diameter of a circle equals twice the length of its radius.
All radii of a circle are congruent.
All diameters of a circle are congruent.
Radii of congruent circles are congruent.
Diameters of congruent circles are congruent.

We can determine if two circles are congruent by comparing the radii. If the radii are congruent, the circles are congruent.

Congruent Circles Theorem
Two circles are congruent if their radii or if their diameters are congruent.

Since two circles with congruent radii are congruent, we can determine a set of rigid motions that will map one onto the other. A circle has an infinite degree of rotational symmetry (a rotation of any angle will map it onto itself). A reflection about any diameter will also map a circle to itself. Therefore, we only need to translate the centers to map one congruent circle onto another. In Figure 11.2, $\odot P \cong \odot Q$. We can map $\odot Q$ onto $\odot P$ with the translation $T_{\overrightarrow{PQ}}(\odot Q)$.

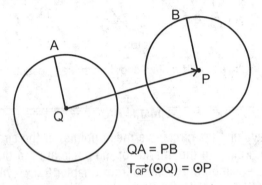

$$QA = PB$$
$$T_{\overrightarrow{QP}}(\odot Q) = \odot P$$

Figure 11.2 Mapping one congruent circle onto another

We saw from Chapter 8 that similar figures have the same shape but may differ in size. All circles by definition have the same shape. The size of any circle is determined only by its radius. When given any two circles, we can scale one circle in size so that it is congruent to the other by applying a dilation with a scale factor equal to the ratio of the radii. If the center of the dilation is the center of the circle, then the original center and its image are coincident and the circles are concentric. The congruent circles can then be mapped to one another by applying the appropriate translation as explained above.

Definition

Concentric circles—Circles having the same centers.

Similar Circles Theorems

All circles are similar.
Any circle can be mapped to another circle. First dilate by the ratio of the radii. Then translate along a vector determined by the centers of the circles.

Figure 11.3 shows $\odot Q$ being mapped to $\odot P$ with a different radius. The transformations used are as follows:

- Dilation of $\odot Q$ with scale factor equal to $\dfrac{QA}{PB}$ about center Q. The new radius, $Q'A'$, is equal to PB.
- Translation of $\odot Q'$ by $\overrightarrow{Q'P}$.

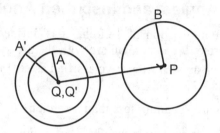

$$D_{Q,PB/QA} \circ T_{\overrightarrow{QP}}(\odot Q) = \odot P$$

Figure 11.3 Mapping of one circle to another circle of a different radius

Example 1

State a specific sequence of transformations that could be used to demonstrate that circle H is similar to circle K.

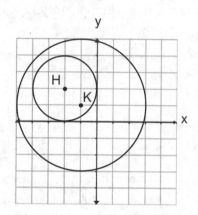

Solution: One possible transformation that would map the two circles would be a translation that maps H to K, followed by a dilation to match the radii. H needs to be translated 1 unit right and 1 unit down. A scale factor of 2 applied to circle H would make the radii congruent: $D_{K,2} \circ T_{1,-1}$ (circle H).

Example 2

Circle P has a radius of 6, and circle Q has a radius of 10. State a sequence of transformations that could be used to demonstrate the two circles are similar.

Solution: A translation that maps point P to point Q followed by a dilation centered at Q with a scale factor of $\dfrac{10}{6}$, or $\dfrac{5}{3}$, would map circle P onto circle Q. Since this represents a similarity transformation, the circles must be similar. $D_{Q,\frac{5}{3}} \circ T_{P \rightarrow Q}$ (circle P).

Arcs, Central Angles, and Inscribed Angles

An **arc** is a continuous section of a circle. It is defined by its endpoints, which are on the circle. We denote an arc with the arc symbol and the endpoints of the arc, such as $\overset{\frown}{AB}$. Arcs are classified as either major arcs, minor arcs, or semicircles:

- Minor arc—an arc spanning less than a semicircle
- Semicircle—an arc spanning exactly $\dfrac{1}{2}$ of a circle
- Major arc—an arc spanning more than a semicircle

When naming a semicircle or major arc, we add an additional point to show which semicircle is being specified. Figure 11.4 shows minor arc $\overset{\frown}{AB}$, semicircle $\overset{\frown}{ABC}$, and major arc $\overset{\frown}{ABD}$.

Figure 11.4 Minor arc, semicircle, and major arc

We can refer to two types of angles within a circle—the central angle and the inscribed angle.

Definition
Central angle—An angle whose vertex is the center of a circle and whose rays intersect the circle.

Definition
Inscribed angle—An angle whose vertex is on a circle and whose rays intersect the circle

---

Figure 11.5 shows central angle $\angle APB$ and inscribed angle $\angle CMD$. Each central angle and inscribed angle cuts off, or intercepts, an arc on the circle. Central $\angle APB$ intercepts $\overgroup{AB}$. Inscribed angle $\angle CMD$ intercepts $\overgroup{CD}$.

$$m\,\overgroup{AB} = m\angle APB \qquad m\,\overgroup{CD} = 2\,m\angle CMD$$

Figure 11.5 Central angle $\angle APB$ and inscribed angle $\angle CMD$

An arc has two different dimensions that we consider—its angle measure and its length. The angle measure of an arc is equal to the central angle that intercepts the arc. In Figure 11.5, the measure of $\overgroup{AB}$ is equal to the measure of $\angle APB$. There is also a relationship between an arc and the inscribed angle. The angle measure of an arc is equal to 2 times the angle measure of the inscribed angle that intercepts the arc.

Arc Measure Relationships

Arc—Central Angle Measures
- The angle measure of an arc equals the measure of the central angle that intercepts the arc.

Arc—Inscribed Angle Measures
- The angle measure of an arc equals twice the measure of the inscribed angle that intercepts the arc.
- The measure of an inscribed angle equals half the angle measure of the arc it intercepts.

Example 1

In circle P, $\angle ACB$ is an inscribed angle and measures $28°$. What is the measure of $\overgroup{AB}$?

Solution: $\quad m\overgroup{AB} = 2m\angle ACB = 2(28°) = 56°$

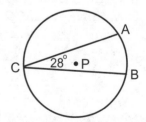

Example 2

In $\odot P$, $\angle APB$ is a central angle and $\angle ADB$ is an inscribed angle. If m$\angle APB = 110°$, find m$\angle ADB$.

Solution: The strategy here is to use the central angle to find $\text{m}\overset{\frown}{AB}$ and then use the arc to find the inscribed angle.

$$\text{m}\overset{\frown}{AB} = \text{m}\angle APB = 110°$$

$$\text{m}\angle ADB = \frac{1}{2}\text{m}\overset{\frown}{AB} = 55°$$

The definition of a central angle requires the vertex to be at the center of the circle. Therefore given any arc on a circle, there is only one possible central angle that intercepts the arc. However, there are an infinite number of possible inscribed angles that intercept an arc. They must all have the same angle measure.

Congruent Inscribed Angles Theorem
- Inscribed angles that intercept congruent arcs are congruent.
- Arcs intercepted by congruent inscribed angles are congruent.

This theorem can help us find measures of inscribed angles that intercept a common arc. We can also use this theorem to help prove that two triangles are similar.

Example 3

$\angle C$ and $\angle D$ are inscribed angles in $\odot P$.
If m$\angle C = 21°$, what is m$\angle D$?
Prove $\triangle ADQ \sim \triangle BCQ$.

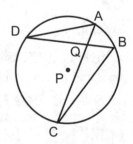

Solution: $\angle D \cong \angle C$ because they both intercept $\overset{\frown}{AB}$. So $m\angle D = m\angle C = 21°$. $\angle A \cong \angle B$ because they both intercept $\overset{\frown}{CD}$. Therefore $\triangle ADQ \sim \triangle BCQ$ by the AA postulate.

Because all radii in a circle are congruent, be on the lookout for isosceles triangles involving two radii in circle problems.

Example 4

In $\odot P$, $m\overset{\frown}{AB} = 108°$. What is the measure of $\angle A$?

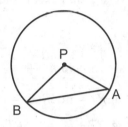

Solution: $\angle APB$ is a central angle, so $m\angle APB = m\overset{\frown}{AB} = 108°$. $\triangle PBA$ is isosceles, with $\overline{PB} \cong \overline{PA}$

$$m\angle A = \frac{1}{2}(180° - 108°) = 36°$$

Arc Addition

Consecutive arcs can be added in the same way as angles and segments to get the sum of the larger arc. Three useful arc-angle relationships that can be used in problems involving arc measures are defined.

> ### Circle—Arc Sum Theorem
> The sum of the measures of consecutive arcs around a circle equals 360°.

> ### Semicircle—Arc Sum Theorem
> The sum of the measures of consecutive arcs around a semicircle equals 180°.

Combining this last theorem with the fact that an inscribed angle measures $\frac{1}{2}$ the measure of its intercepted arc leads to the following theorem.

Inscribed Angle—Semicircle Theorem
An inscribed right angle always intercepts a semicircle.

Example 1

Diameter $\overline{SOT}$ is drawn in circle O. If $m\angle ROT =$ 110°, find $m\overarc{RS}$ and $m\overarc{STR}$.

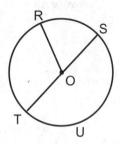

Solution:

$m\overarc{RT} = m\angle ROT = 110°$ central angle theorem

$m\overarc{RT} + \overarc{RS} = m\overarc{SRT}$ arc sum theorem

$110° + \overarc{RS} = 180°$

$\overarc{RS} = 70°$

$m\overarc{STR} + \overarc{RS} = 360°$ circle-arc sum theorem

$m\overarc{STR} + 70° = 360°$

$m\overarc{STR} = 290°$

Example 2

Diameter $\overline{APC}$ is drawn in $\odot P$. If $m\angle BPC = 31°$, what is $m\angle ARB$?

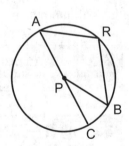

Solution: Since $\angle ARB$ is an inscribed angle, the strategy is to find the measure of the arc that the angle intercepts.

$$m\overset{\frown}{CB} = m\angle BPC = 31° \qquad \text{central angle theorem}$$

$$m\overset{\frown}{ACB} = m\overset{\frown}{AC} + m\overset{\frown}{CB} \qquad \text{arc addition}$$

$$m\overset{\frown}{ACB} = 180° + 31°$$

$$= 211°$$

$$m\angle ARB = \frac{1}{2}m\overset{\frown}{ACB} = \frac{1}{2}(211°) \qquad \text{inscribed angle theorem}$$

$$= 105.5°$$

Example 3

Points D, J, and K are located on circle O. The three points divide the circle into arcs that are in a $3:4:5$ ratio. What is the measure of the three arcs?

Solution: The sum of the arc measures must equal 360°. We can express the three arcs as $3x$, $4x$, and $5x$.

$$3x + 4x + 5x = 360$$
$$12x = 360$$
$$x = 30$$

The arc measures are $3(30)°$, $4(30)°$, and $5(30)°$
$$= 90°, 120°, \text{ and } 150°$$

Example 4

In $\odot O$, $\overline{TOS}$ is a diameter. If $RS = 4$ and $RT = 10$, find the length of $\overline{TOS}$. Express your answer in simplest radical form.

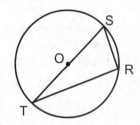

Solution: Since TOS is a diameter, inscribed $\angle R$ measures 90° and $\triangle RST$ is a right triangle. Apply the Pythagorean theorem to find TOS.

$$RS^2 + RT^2 = TOS^2$$
$$4^2 + 10^2 = TOS^2$$
$$TOS^2 = 116$$
$$TOS = \sqrt{116} = \sqrt{4 \cdot 29} = 2\sqrt{29}$$

====== **MATH FACT** ======

Quadrilaterals that are inscribed in a circle are called cyclic quadrilaterals. Cyclic quadrilaterals have opposite angles that are supplementary. This property is easily proven by looking at a sketch of a cyclic quadrilateral. Together, arcs $\overset{\frown}{ADC}$ and $\overset{\frown}{ABC}$ comprise the entire circle, so $m\overset{\frown}{ADC} + m\overset{\frown}{ABC} = 360°$. Since the intercepted angles measure $\frac{1}{2}$ the arc measures, $m\angle ABC + m\angle ADC = 180°$. So the angles are supplementary. The same justification can be used to show $\angle A$ and $\angle C$ are supplementary.

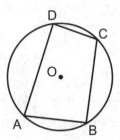

Check Your Understanding of Section 11.1

A. *Multiple-Choice*

1. Circle J has a radius of 12 and a center located at (3, 5). Circle M has a radius of 2 and a center located at (1, 8). Which of the following sequences of transformations would demonstrate that the two circles are similar?

   (1) $D_{J,\frac{1}{6}} \circ T_{2,-3}$ (circle M)   (3) $D_{M,\frac{1}{6}} \circ T_{2,-3}$ (circle J)

   (2) $D_{J,6} \circ T_{-2,3}$ (circle M)   (4) $D_{M,\frac{1}{6}} \circ T_{-2,3}$ (circle J)

2. In $\odot C$, $m\angle PCQ = 82°$ and $\overset{\frown}{PR} \cong \overset{\frown}{QR}$. What is the measure of $\overset{\frown}{PR}$?

   (1) 92°
   (2) 98°
   (3) 118°
   (4) 139°

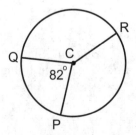

3. A circle is divided into 4 arcs that are in a 2:3:4:6 ratio. What is the measure of the smallest arc?
   (1) 12°   (2) 24°   (3) 48°   (4) 72°

4. What is the value of *x* in the following figure?
   (1) 22.5°
   (2) 45°
   (3) 51°
   (4) 67.5°

5. In ⊙*P*, ∠*APB* ≅ ∠*BPC*, and m$\overarc{ADC}$ = 208°. What is the measure of $\overarc{AB}$?
   (1) 38°
   (2) 76°
   (3) 104°
   (4) 152°

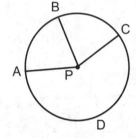

6. $\overline{RGT}$ is a diameter of ⊙*G*, and *P* lies on the circle such that m∠*PRT* = 18°. What is the measure of arc $\overarc{PR}$?
   (1) 144°   (2) 81°   (3) 72°   (4) 36°

7. △*TRY* is inscribed in ⊙*O*, with $\overline{RY}$ passing through the center *O*, *TR* = 9, and *RT* = 3. What is the length of $\overline{RY}$?
   (1) 30
   (2) $3\sqrt{10}$
   (3) $15\sqrt{3}$
   (4) $2\sqrt{5}$

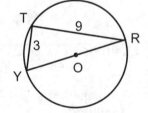

8. In ⊙*O*, m$\overarc{BC}$ = 142°. What is the measure of arc $\overarc{AB}$?
   (1) 38°
   (2) 46°
   (3) 76°
   (4) 142°

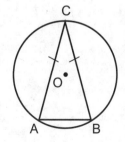

357

9. In ⊙O, $\overline{DOF}$ is a diameter, $m\widehat{DE} = 72°$, and $m\widehat{DG} = 52°$. What are the measures of ∠FGO and ∠ODE?
(1) m∠FGO = 54°, m∠ODE = 26°
(2) m∠FGO = 26°, m∠ODE = 54°
(3) m∠FGO = 64°, m∠ODE = 52°
(4) m∠FGO = 45°, m∠ODE = 26°

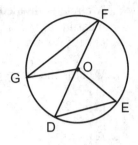

Use the accompanying figure for problems 10 and 11:

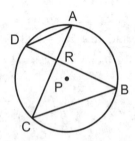

10. In ⊙P, $\overline{BRD}$ and $\overline{ARC}$ are chords, $m\widehat{AB} = 84°$, and $m\widehat{CD} = 88°$. What is the measure of ∠BRC?
(1) 86°   (2) 90°   (3) 92°   (4) 94°

11. In ⊙P, $\overline{BRD}$ and $\overline{ARC}$ are chords, m∠ARD = 85°, and m∠CBR = 54°. What is the measure of ∠C and ∠D?
(1) m∠C = 41°, m∠D = 41°   (3) m∠C = 126°, m∠D = 95°
(2) m∠C = 41°, m∠D = 47.5°   (4) m∠C = 63°, m∠D = 47.5°

12. If $AB:BC:CD:DA = 2:1:3:4$, what is the measure of ∠D?
(1) 27°
(2) 36°
(3) 48°
(4) 54°

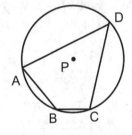

13. What is the ratio of m∠AOB : m∠ACB?
    (1) 5 : 1
    (2) 3 : 1
    (3) 2 : 1
    (4) 1 : 2

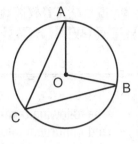

B. *Free-Response—show all work or explain your answer completely.*

14. In ⊙P, m∠N = 81°, m$\overarc{MN}$ = 100°, and
    m$\overarc{RS}$ = 74°. Find m∠R, m∠S, and m∠M.

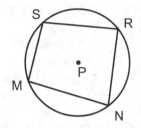

15. An equilateral triangle is inscribed in a circle. Each side of the triangle intercepts an arc of the circle. What is the measure of each of the arcs?

16. Carly states that whenever a rectangle is inscribed in a circle, the diagonals of the rectangle must lie on a diameter of the circle. Aidan doesn't think that Carly's statement is always true. Who is correct? Explain your reasoning.

17. Lisa states that any parallelogram can be inscribed in a circle. Rayshawn disagrees, saying that the only parallelograms that can be inscribed in a circle are rectangles and squares. Who is correct? Justify your reasoning.

18. Given: $\overarc{AR} \cong \overarc{IN}$
    Prove: △RAI ≅ △NIA

11.2 CONGRUENT, PARALLEL, AND PERPENDICULAR CHORDS

KEY IDEAS

The following theorems relate arc measures to parallel, perpendicular, and congruent chords:

- If two chords in a circle are congruent, they intercept congruent arcs.
- If two chords in a circle are parallel, the arcs between them are congruent.
- The perpendicular bisector of a chord passes through the center of the circle.
- A radius to a point of tangency is perpendicular to the tangent.

Congruent Chords

Congruent Chord Theorem
Congruent chords intercept congruent arcs on a circle.
Congruent arcs on a circle are intercepted by congruent chords.

The congruent chord theorem is illustrated in Figure 11.6. $\overline{AB} \cong \overline{CD}$ and $\overparen{AB} \cong \overparen{CD}$.

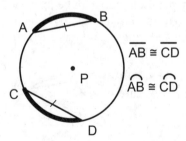

Figure 11.6 Congruent chords intercept congruent arcs

This theorem is easily proved by completing the central angle for each arc and then showing the resulting two triangles are congruent. Figure 11.7 shows the congruent chords with the central angles. $\overline{PA} \cong \overline{PB} \cong \overline{PC} \cong \overline{PD}$ because they are all radii of the same circle. $\triangle APB \cong \triangle CPD$ by SSS, and $\angle APB \cong \angle CPB$ by CPCTC. Since arcs $\overparen{AB}$ and $\overparen{CD}$ are intercepted by congruent central angles, the arcs must be congruent.

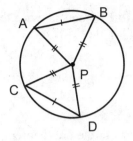

Figure 11.7 Congruent chords forming congruent central angles in a circle

The congruent chord theorem can be used to help find the measures of arcs and angles. In addition to congruent chords, look out for inscribed angles, central angles, and circle-arc sum relationships around a circle or semicircle. There's often more than one relationship involved in circle problems.

Example 1

In $\odot A$, $MN = JK$. If the measure of $\overset{\frown}{MN}$ is equal to $(3x + 8)°$ and the measure of $\overset{\frown}{JK}$ is equal to $(4x - 33)°$, what is the measure of each arc?

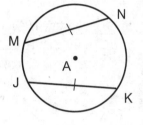

Solution: The two arcs are intercepted by congruent chords, so their angle measures must be equal.

$$4x - 33 = 3x + 8$$
$$x = 41$$

$$\text{m}\overset{\frown}{MN} = \text{m}\overset{\frown}{JK} = 3(41) + 8 = 131°$$

Example 2

$\angle ABC$ is inscribed in $\odot O$ and measures $35°$. If $\overline{AB} \cong \overline{BC}$, what is the measure of $\overset{\frown}{AB}$?

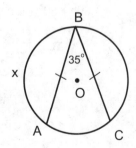

Solution: The measure of $\overset{\frown}{AC}$ equals twice the measure of the inscribed angle.

$$m\overset{\frown}{AC} = 2m\angle B$$
$$= 2 \cdot 35°$$
$$= 70°$$

$$m\overset{\frown}{AB} + m\overset{\frown}{BC} + 70° = 360° \quad \text{circle-arc theorem}$$
$$x + x + 70° = 360° \quad \text{congruent chord theroem}$$
$$2x + 70° = 360°$$
$$2x = 290°$$
$$x = 145°$$

$$m\overset{\frown}{AB} = 145°$$

Example 3

In $\odot C$, chords $\overline{AB}$ and $\overline{CD}$ are congruent, $m\overset{\frown}{BC} = 60°$, and $m\overset{\frown}{CD} = 80°$. Find $m\angle ABD$.

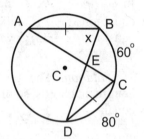

Solution: $\angle ABD$ is an inscribed angle, so its measure will be $\dfrac{1}{2} m\overset{\frown}{AD}$.

$$m\overset{\frown}{AB} + m\overset{\frown}{BC} + m\overset{\frown}{CD} + m\overset{\frown}{AD} = 360° \quad \text{circle-arc sum theorem}$$
$$m\overset{\frown}{AB} = m\overset{\frown}{CD} = 80° \quad \text{congruent chord theorem}$$
$$80° + 60° + 80° + m\overset{\frown}{AD} = 360°$$
$$220° + m\overset{\frown}{AD} = 360°$$
$$m\overset{\frown}{AD} = 140°$$
$$m\angle ABD = \frac{1}{2} m\overset{\frown}{AD}$$
$$= 70°$$

Parallel Chords

Parallel Chords Theorem
The two arcs formed *between* a pair of parallel chords are congruent.
If the two arcs formed *between* a pair of chords are congruent, the chords are parallel.

The parallel chord theorem is illustrated in Figure 11.8. $\overline{AB} \parallel \overline{CD}$ and the arcs between these chords, *AC* and *BD*, are congruent. Notice that the location of these two arcs is different from the location of congruent arcs formed by congruent chords.

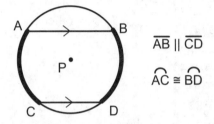

$$\overline{AB} \parallel \overline{CD}$$

$$\overset{\frown}{AC} \cong \overset{\frown}{BD}$$

Figure 11.8 Congruent arcs between a pair of parallel chords

The parallel chord theorem can be proved by constructing a transversal intersecting the two parallel chords, as shown in Figure 11.9. ∠*CDA* ≅ ∠*BAD* because they are alternate interior angles formed by the parallel chords. *AC* and *BD* are intercepted by congruent inscribed angles, so the arcs themselves must be congruent.

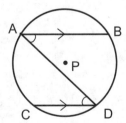

Figure 11.9 Parallel chords with transversal

We can apply this theorem to find arc and angle measures in the same way as with the congruent chord theorem. Just do not confuse the two different pairs of congruent arcs.

Example 1

In $\odot P$, chords $\overline{AB}$ and $\overline{CD}$ are parallel.
If $m\overparen{AB} = 118°$ and $m\overparen{CD} = 78°$, find $m\overparen{AC}$.

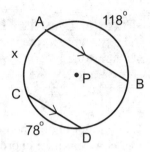

Solution: $\overparen{AC}$ and $\overparen{DB}$ are the congruent arcs between a pair of parallel chords.

$$m\overparen{AB} + m\overparen{BD} + m\overparen{CD} + m\overparen{AC} = 360° \qquad \text{circle-arc sum theorem}$$
$$m\overparen{AC} = m\overparen{DB} \qquad\qquad\qquad \text{parallel chord theorem}$$
$$118° + x + 78° + x = 360°$$
$$196° + 2x = 360°$$
$$2x = 164°$$
$$x = 82°$$
$$m\overparen{AC} = 82°$$

If the chords are both parallel and congruent, then there will be two pairs of congruent arcs, one from the congruent chord relationship and one from the parallel chord relationship.

Example 2

In circle $\odot P$, $\overline{AB} \cong \overline{CD}$ and $\overline{AB} \parallel \overline{CD}$.
If $m\overparen{BD} = 75°$, find $m\overparen{CD}$.

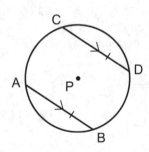

Solution:

$$m\overparen{AB} + m\overparen{BD} + m\overparen{CD} + m\overparen{AC} = 360° \qquad \text{circle-arc sum theorem}$$
$$m\overparen{AC} = m\overparen{BD} \qquad\qquad\qquad \text{parallel chord theorem}$$
$$m\overparen{AB} = m\overparen{CD} \qquad\qquad\qquad \text{congruent chord theorem}$$
$$x + 75° + x + 75° = 360°$$
$$2x + 150° = 360°$$
$$2x = 210°$$
$$x = 105°$$
$$m\overparen{CD} = 105°$$

Example 3

Trapezoid *RSTU* has parallel bases $\overline{RS}$ and $\overline{TU}$ and is inscribed in circle *P*.
If m$\angle T$ = 109° and m$\overset{\frown}{RU}$ = 92°, find m$\overset{\frown}{TU}$.

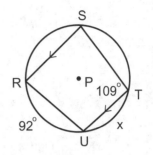

Solution:

$$m\overset{\frown}{SRU} = 2m\angle T \qquad\qquad \text{inscribed angle } \angle T \text{ intercepts } \overset{\frown}{SRU}$$
$$= 2(109°)$$
$$= 218°$$
$$m\overset{\frown}{ST} = m\overset{\frown}{RU} \qquad\qquad \text{parallel chords theorem}$$
$$= 92°$$
$$m\overset{\frown}{SRU} + m\overset{\frown}{ST} + m\overset{\frown}{TU} = 360° \qquad \text{circle-arc theorem}$$
$$218° + 92° + m\overset{\frown}{TU} = 360°$$
$$m\overset{\frown}{TU} = 50°$$

Perpendicular Chords

You should be familiar with two perpendicular chord relationships, the chord–perpendicular bisector theorem and the radius-tangent theorem.

Chord–Perpendicular Bisector Theorem
The perpendicular bisector of any chord passes through the center of the circle.
A diameter or radius that is perpendicular to a chord bisects the chord.
A diameter or radius that bisects a chord is perpendicular to the chord.

You can think of this theorem as linking three properties of a chord:

- Bisecting another chord
- Perpendicular to another chord
- Is a diameter

When any two of the above are true, the third is implied to be true.

Figure 11.10 illustrates the theorem. $\overline{BED}$ is a perpendicular bisector of chord $\overline{AC}$. Therefore $\overline{BED}$ passes through center *P* and is a diameter.

Figure 11.10 A perpendicular bisector of a chord

Example

$\overline{BDP}$ is a radius of $\odot P$ and is perpendicular to $\overline{AC}$. If $AC = 24$ and $PC = 13$, find PD and DB.

Solution: Radius $\overline{BDP}$ is perpendicular to $\overline{AC}$, so it must also bisect $\overline{AC}$.

$$DC = \frac{1}{2}AC$$

$$DC = \frac{1}{2}(24)$$

$$DC = 12$$

$PD^2 + DC^2 = PC^2$	Pythagorean theorem
$PD^2 + 12^2 = 13^2$	
$PD^2 + 144 = 169$	
$PD^2 = 25$	
$PD = 5$	
$PB = PC = 13$	All radii of a circle are congruent.
$PD + DB = PB$	partition postulate
$5 + DB = 13$	
$DB = 8$	

Check Your Understanding of Section 11.2

A. *Multiple-Choice*

1. $\overline{GH}$ and $\overline{MN}$ are chords in $\odot O$ and are parallel. Which of the following must be true?
   (1) $\overline{GH} \cong \overline{MN}$
   (2) $\overparen{GM} \cong \overparen{HN}$
   (3) $\overline{MH}$ is a diameter
   (4) $\angle GMN$ is a right angle

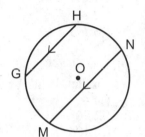

2. A diameter of circle A intersects a chord of the circle at its midpoint. Which of the following must be true?
   (1) the chord must be congruent to the radius of the circle
   (2) the diameter must be congruent to the chord
   (3) the diameter must be $\perp$ to the chord
   (4) the chord's length must $\dfrac{1}{3}$ that of the diameter

3. In $\odot C$, $\overline{QR} \cong \overline{ST}$, $m\overparen{QR} = 90°$, and $m\overparen{QT} = 160°$. What is the measure of $\overparen{RS}$?
   (1) $10°$
   (2) $20°$
   (3) $35°$
   (4) $45°$

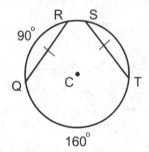

4. $\overline{OSCT}$ is a diameter of $\odot C$. $JS = 6$, $SB = 6$, and $CT = 9$. What is the length SC?
   (1) $3\sqrt{5}$
   (2) $5\sqrt{3}$
   (3) 3
   (4) $\sqrt{117}$

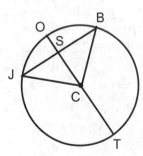

5. In ⊙*P*, chord $\overline{FG}$ is parallel to chord $\overline{HI}$. Which of the following is not necessarily true?

(1) $\angle G \cong \angle H$
(2) $\widehat{FH} \cong \widehat{GI}$
(3) $\angle H \cong \angle F$
(4) $\widehat{FG} \cong \widehat{HI}$

6. In ⊙*O*, $\overline{UV}$ and $\overline{XY}$ are both congruent and parallel. If $m\widehat{UX} : m\widehat{UV} = 1:4$, what is $m\widehat{XY}$?

(1) 45°
(2) 90°
(3) 125°
(4) 144°

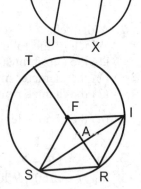

7. In circle *F*, diameter $\overline{TFR}$ is perpendicular to chord $\overline{IAS}$. Which of the following is *not* necessarily true?

(1) $FS = TF$
(2) $FA = AR$
(3) $SR = RI$
(4) $SA = AI$

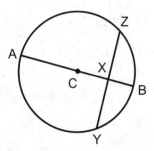

8. Rectangle *ABCD* is inscribed in ⊙*P*, and $\widehat{AB}$ measures 30°. What are the measures of $\widehat{BC}$ and $\widehat{CD}$?

(1) $m\widehat{BC} = 60°$, and $m\widehat{CD} = 120°$
(2) $m\widehat{BC} = 120°$, and $m\widehat{CD} = 60°$
(3) $m\widehat{BC} = 150°$, and $m\widehat{CD} = 30°$
(4) $m\widehat{BC} = 30°$, and $m\widehat{CD} = 150°$

9. In ⊙*C*, diameter $\overline{ACXB}$ is perpendicular to $\overline{ZXY}$. If $AB = 24$ and $ZY = 18$, find *CX*.

(1) $14\sqrt{3}$
(2) $3\sqrt{7}$
(3) $12\sqrt{7}$
(4) $7\sqrt{3}$

B. *Free-Response—show all work or explain your answer completely.*

10. In $\odot P$, $m\overparen{KNM} = 220°$ and $m\overparen{KL} = m\overparen{LM}$. What is the measure of $\angle K$?

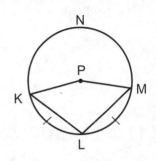

11. A regular pentagon is inscribed in a circle. What is the measure of each of the arcs formed by the vertices of the pentagon?

12. Given: Chords $\overline{MN} \cong \overline{PQ}$ and $\overline{MN} \parallel \overline{PQ}$
Prove: $m\overparen{NQ} + m\overparen{QP} = 180°$

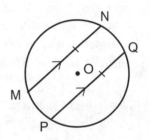

13. Jason states that if you construct a circle with two congruent chords $\overline{AB}$ and $\overline{CD}$ such that arc $\overparen{ABDC}$ is formed, then *ABDC* must be an isosceles trapezoid. Robin says that *ABDC* does not have to be an isosceles trapezoid. Who is correct? Explain your reasoning.

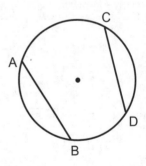

14. Prove the arcs intercepted by vertical angles are congruent if the chords are congruent.
Given: $\overline{AC} \cong \overline{BD}$
Prove: $\overparen{AB} \cong \overparen{CD}$

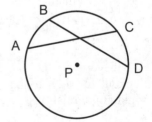

15. Given: $\overline{HO} \cong \overline{HE}$, $\overline{OS} \cong \overline{ES}$
Prove: $\overline{HS}$ is a diameter

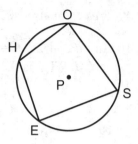

11.3 TANGENTS

Tangent to a Radius

From any point outside a circle, there are exactly two lines that can be drawn tangent to a circle. We call those lines tangents. There will be only one radius of the circle that can be drawn to the point of tangency, and that radius will be perpendicular to the tangent. In Figure 11.11, diameter $\overline{CPB}$ is perpendicular to tangent $\overline{AB}$ at the point of tangency, B.

Tangent-Radius Theorem
A diameter or radius to a point of tangency is perpendicular to the tangent. A line perpendicular to a tangent at the point of tangency passes through the center of the circle.

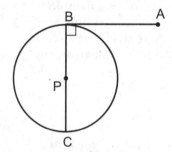

Figure 11.11 Tangent perpendicular to a diameter

Whenever you see a radius intersecting a tangent, you should be on the lookout for right triangles, complementary angles, or other applications of right angle relationships. The Pythagorean theorem is often involved.

Example 1

Radius $\overline{DG}$ and tangent $\overline{OG}$ of $\odot D$ intersect at point G. If $DG = 4$ and $OG = 6$, what is the length of OD in simplest radical form?

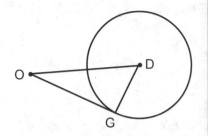

Solution: The radius and tangent are perpendicular, so $\triangle GDO$ is a right triangle. We can apply the Pythagorean theorem.

$$OG^2 + DG^2 = OD^2$$
$$6^2 + 4^2 = OD^2$$
$$36 + 16 = OD^2$$
$$52 = OD^2$$
$$OD = \sqrt{52}$$
$$OD = \sqrt{4 \cdot 13} = 2\sqrt{13}$$

The next example is a good exercise in working your way angle by angle toward the desired angle. If the path to the solution is not clear at first, then try to identify any applicable relationship and find any unknown angle. You will usually find that eventually, with a little perseverance, you will get to the goal.

Example 2

In $\odot P$, $\overline{NPM}$ is a diameter, $\overline{MG}$ is a tangent, $\overline{GM} \cong \overline{GKL}$, and $\text{m}\overset{\frown}{MK} = 44°$. What is the measure of $\angle LMG$?

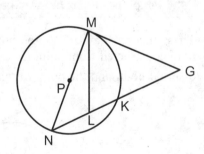

Solution:

$$m\angle N = \frac{1}{2}m\widehat{MK}$$ inscribed angle theorem

$$= \frac{1}{2}(44°)$$

$$= 22°$$

$$m\angle N + m\angle NMG + m\angle G = 180°$$ triangle angle sum theorem
$$22° + 90° + m\angle G = 180°$$
$$m\angle G = 68°$$

$$m\angle LMG = \frac{1}{2}(180° - m\angle G)$$ $\triangle LMG$ is isosceles

$$= \frac{1}{2}(180° - 68°)$$

$$= 56°$$

Tangents from a Common Point

Tangents from a Common Point Theorem
Given a circle and external point P, there are exactly two tangents from P to the circle.

Congruent Tangent Theorem
Given a circle and external point P, segments between the external point and the point of tangency are congruent.

These theorems are illustrated in Figure 11.12. Tangents $\overrightarrow{QA}$ and $\overrightarrow{QB}$ are both constructed from point Q. The portion of the two tangents between the external point and the point of tangency are congruent, $\overline{QA} \cong \overline{QB}$.

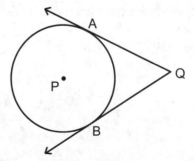

Figure 11.12 Common tangents to circle P

The congruence of the two tangents from a common point can be proved by completing the two triangles shown in Figure 11.13. Radii $\overline{OL} \cong \overline{OM}$, $\overline{KNO} \cong \overline{KNO}$, and $\angle KLO$ and $\angle KMO$ are right angles because they are formed by a radius and tangent. $\triangle KMO \cong \triangle KLO$ by the HL postulate, so $\overline{KM} \cong \overline{KL}$ by CPCTC.

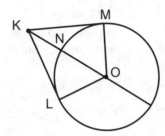

Figure 11.13 Congruent tangents $\overline{KL}$ and $\overline{KM}$

A triangle circumscribed about a circle is comprised of a series of tangents from common points—the vertices of the triangles. We can use the congruent tangents to help find the perimeter of the triangle.

Example 1

$\triangle RST$ is circumscribed about a circle. If $SX = 6$, $XR = 7$, and $TZ = 8$, what is the perimeter of $\triangle RST$?

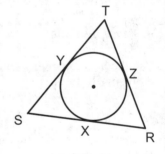

Solution: $SY = SX$, $TY = TZ$, and $ZR = XY$ because each pair are tangents from the same point. The perimeter is therefore $2 \cdot 6 + 2 \cdot 7 + 2 \cdot 8 = 42$.

The next example combines several of the circle theorems we have seen so far.

Example 2

Tangents $\overline{TR}$ and $\overline{TQ}$ are drawn from point T to circle A. $\overline{RAS}$ is a diameter of circle A.

If $m\angle T = 46°$, what is the measure of $\overset{\frown}{SQ}$?

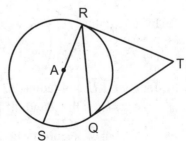

Solution:

$\overline{TR} \cong \overline{TQ}$ congruent tangent theorem

$m\angle TRQ = \dfrac{1}{2}(180° - m\angle T)$ $\triangle TRQ$ is isosceles

$\qquad\qquad = 67°$

$\qquad\qquad m\angle SRT = 90°$ tangent-radius theorem
$m\angle SRQ + m\angle TRQ = 90°$ partition postulate
$\qquad m\angle SRQ + 67° = 90°$
$\qquad\qquad m\angle SRQ = 23°$

$m\overset{\frown}{SQ} = 2m\angle SRQ$ inscribed angle theorem

$\qquad = 46°$

In the next section, we will see a set of angle theorems involving tangents, secants, and chords that could also have been applied to this example.

Check Your Understanding of Section 11.3

A. Multiple-Choice

Use the accompanying figure for problems 1–3:

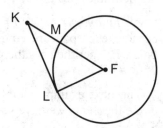

1. In $\odot F$, radius $\overline{FL}$ is drawn to tangent $\overline{KL}$. If $LF = 8$ and $KL = 12$, what is the length of $\overline{KMF}$?
(1) $4\sqrt{13}$ (2) $4\sqrt{5}$ (3) $2\sqrt{5}$ (4) 4

2. In $\odot F$, radius $\overline{FL}$ is drawn to tangent $\overline{KL}$. If $m\angle K = 29°$, what is the measure of $\angle F$?
(1) $25.5°$ (2) $29°$ (3) $48°$ (4) $61°$

3. Tangent $\overline{KL}$ and radius $\overline{FL}$ are drawn in circle F. If $KM = 8$ and $KL = 24$, what is the length of $\overline{FM}$?
(1) 6 (2) 32 (3) $4\sqrt{6}$ (4) $6\sqrt{4}$

4. In circle *O*, radius $\overline{OA}$, tangent $\overline{AC}$, and
   diameter $\overline{MON}$ are drawn. Which of the
   following is *not* necessarily true?
   (1) $OA \cong AC$
   (2) $OA = OM$
   (3) $m\angle AOM \cong m\angle BON$
   (4) the length of $\overparen{CBO}$ is greater
       than the length of $\overline{AC}$

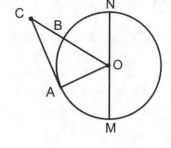

5. △*ABC* is circumscribed about
   circle *P*. Which of the following
   must be true?
   (1) $m\angle C = m\angle B$
   (2) the perimeter of △*ABC* is equal
       to one-half the circumference
       of circle *P*
   (3) $AC = AB$
   (4) $BD = BE$

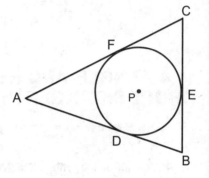

Use the accompanying figure for
problems 6 and 7:

6. In circle *O*, tangents $\overline{KL}$ and $\overline{KM}$ are
   drawn from point *K*. $\overline{OL}$ and $\overline{OM}$ are
   radii. If $OL = 5$ and $KNO = 15$, what is
   the perimeter of *KMOL*?
   (1) $5+10\sqrt{10}$ (2) $10+20\sqrt{2}$ (3) $10+10\sqrt{10}$ (4) 40

7. In circle *O*, tangents $\overline{KL}$ and $\overline{KM}$ are drawn from point *K*. $\overline{OL}$ and
   $\overline{OM}$ are radii. If $m\overparen{MN} = 60°$, what is $m\angle LKO$?
   (1) 12°   (2) 15°   (3) 24°   (4) 30°

B. *Free-Response—show all work or explain your answer completely.*

8. Equilateral triangle *WXY* is circumscribed about circle *P*, and radius $\overline{QP}$ is extended to *Y*. Prove $\overline{QPY}$ bisects ∠*WYX*.

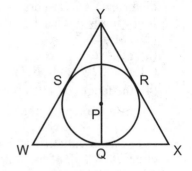

11.4 ANGLE-ARC RELATIONSHIPS WITH CHORDS, TANGENTS, AND SECANTS

Angles formed by intersecting chords, secants, and tangents have the following relationships with the intercepted arcs:

Vertex inside the circle	Angle measure = $\frac{1}{2}$ the *sum* of the intercepted arcs
![C,B,A,D circle diagram with angles 1,2,3,4]	$m\angle 1 = m\angle 2 = \frac{1}{2}\left(m\widehat{AC} + m\widehat{DB}\right)$ $m\angle 3 = m\angle 4 = \frac{1}{2}\left(m\widehat{BC} + m\widehat{AD}\right)$
Vertex on the circle	Angle measure = $\frac{1}{2}$ of the intercepted arc
![F,C,D,E,B,A circle diagram]	$m\angle ABC = \frac{1}{2}m\widehat{BC}$ $m\angle DEF = \frac{1}{2}m\widehat{DF}$

Vertex outside the circle	Angle measure $= \dfrac{1}{2}$ the *difference* of the intercepted arcs
	$$\mathrm{m}\angle P = \frac{1}{2}\left(\mathrm{m}\widehat{ACB} - \mathrm{m}\widehat{AB}\right)$$
	$$\mathrm{m}\angle P = \frac{1}{2}\left(\mathrm{m}\widehat{CD} - \mathrm{m}\widehat{AB}\right)$$
	$$\mathrm{m}\angle P = \frac{1}{2}\left(\mathrm{m}\widehat{AC} - \mathrm{m}\widehat{AB}\right)$$

Angles with Vertex Inside the Circle

When two chords intersect inside a circle, they form a pair of congruent vertical angles that intercept two arcs.

Angle-Arc Theorem for Intersecting Chords (*Vertex Inside the Circle*)

The measures of a pair of vertical angles formed by two intersecting chords are equal to one-half the sum of the measures of the intercepted arcs.

Figure 11.14 shows chords $\overline{AB}$ and $\overline{CD}$ intersecting inside a circle. Vertical angles $\angle 1$ and $\angle 2$ intercept arcs $\overset{\frown}{AC}$ and $\overset{\frown}{DB}$. Vertical angles $\angle 2$ and $\angle 3$ intercept arcs $\overset{\frown}{BC}$ and $\overset{\frown}{AD}$. The angle measures are calculated from the arcs as follows:

$$m\angle 1 = m\angle 2 = \frac{1}{2}\left(m\overset{\frown}{AC} + m\overset{\frown}{DB}\right)$$

$$m\angle 3 = m\angle 4 = \frac{1}{2}\left(m\overset{\frown}{BC} + m\overset{\frown}{AD}\right)$$

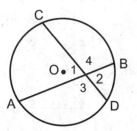

Figure 11.14 Angles formed by intersecting chords

This theorem can be proved by sketching $\overline{AD}$ to complete triangle $\triangle ADE$ within the circle as shown in Figure 11.15. $\angle 1$ is an exterior angle of $\triangle ADE$. By the exterior angle theorem, $m\angle 1 = m\angle 4 + m\angle 5$. $\angle 5$ is an inscribed angle, equal in measure to $\frac{1}{2}m\overset{\frown}{AC}$. $\angle 4$ is an inscribed angle equal in measure to $\frac{1}{2}m\overset{\frown}{BD}$. Combining these equations gives

$$m\angle 1 = \frac{1}{2}m\overset{\frown}{AC} + \frac{1}{2}m\overset{\frown}{BD} = \frac{1}{2}\left(\overset{\frown}{AC} + \overset{\frown}{BD}\right).$$

The same procedure can be used to show the matching relationship with other pairs of vertical angles.

Figure 11.15 Proving the angle-arc theorem for intersecting chords

Example 1

Chords $\overline{AGB}$ and $\overline{CGD}$ intersect at G in circle P. $m\overarc{AC} = 41°$ and $m\overarc{BD} = 133°$. Find $m\angle AGC$, $m\angle BGD$, $m\angle AGD$, and $m\angle BGC$.

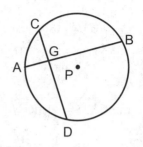

Solution:

$m\angle AGC = m\angle BGD = \dfrac{1}{2}\left(m\overarc{AC} + m\overarc{BD}\right)$ angle $= \dfrac{1}{2}$ sum of intercepted arcs

$m\angle AGC = m\angle BGD = \dfrac{1}{2}(41° + 133°)$

$m\angle AGC = m\angle BGD = 87°$

$m\angle AGC + m\angle BGC = 180°$ linear pairs are supplementary

$\qquad 87° + m\angle BGC = 180°$

$\qquad\qquad m\angle BGC = 93°$

$m\angle AGD = m\angle BGC = 93°$ vertical angles are congruent

Example 2

In $\odot A$, $\overline{MAKN}$ is a diameter, $m\overarc{GN} = 105°$, and $m\angle GKM = 64°$. Find $m\overarc{MF}$ and $m\overarc{NF}$.

Solution:

$$\widehat{GM} = 180° - m\widehat{GN} \qquad \text{semicircle-arc sum theorem}$$
$$\widehat{GM} = 180° - 105°$$
$$= 75°$$

$$m\angle GKM = \frac{1}{2}\left(m\widehat{GM} + m\widehat{NF}\right) \qquad \text{Angle measures } \frac{1}{2} \text{ the sum of}$$
$$\qquad\qquad\qquad\qquad\qquad\qquad \text{the intercepted arcs.}$$
$$64° = \frac{1}{2}\left(75° + m\widehat{NF}\right)$$
$$128° = 75° + m\widehat{NF}$$
$$m\widehat{NF} = 53°$$

We now have three out of four arcs around the circle and can solve for the fourth.

$$m\widehat{GM} + m\widehat{GN} + m\widehat{NF} + m\widehat{MF} = 360° \qquad \text{circle-arc sum theorem}$$
$$75° + 105° + 53° + m\widehat{MF} = 360°$$
$$m\widehat{MF} = 127°$$

Confusing central angles with angles formed by two chords is a common error. Be careful not to assume the angle formed by the chords is equal to the measure of the intercepted arc—that is the relationship for central angles only.

MATH FACT

The theorem that states that the measure of a central angle equals the measure of the intercepted arc is just a special case of the intersecting chords theorem. When the chords are diameters, the vertical angles formed are central angles.

Angles with Vertex on the Circle

Angle-Arc Theorem for Chords and Tangents (*Vertex on the Circle*)

The measure of the angle formed by two chords, or a chord and tangent, with the vertex on the circle is equal to $\frac{1}{2}$ the measure of the intercepted arc.

Figure 11.16 illustrates the three cases of the theorem. Figure 11.16a shows ∠*ABC* formed by tangent $\overline{AB}$ and chord $\overline{BC}$. The measure of ∠*B* equals $\frac{1}{2}$ the measure of arc $\overparen{BC}$. Figure 11.16b shows the special case where the chord is a diameter. The angle formed is equal to $\frac{1}{2}(180°)$ or 90°. When the angle is formed by two chords, shown in Figure 11.16c, we get the familiar theorem that the measure of an inscribed angle is equal to $\frac{1}{2}$ the measure of the intercepted arc.

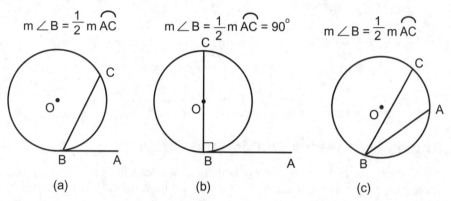

Figure 11.16 Angle with vertex on the circle

Example 1

In circle *P*, chord $\overline{ST}$ intersects tangent $\overline{TR}$ at point *T*. m$\overparen{ST}$ = 152°. What is the measure of ∠*RTS*?

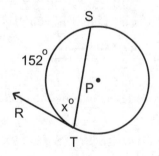

Solution:

$$m\angle RTS = \frac{1}{2}m\overparen{RS} \qquad \text{angle measures } \frac{1}{2} \text{ the intercepted arc}$$

$$m\angle RTS = \frac{1}{2}(152°)$$

$$= 76°$$

Example 2

In circle P, tangent $\overrightarrow{TR}$ intersects chord $\overline{TV}$ at T. If $m\overset{\frown}{TV} = 116°$, what is the measure of $\angle RTV$?

Solution: The intercepted arc is $\overset{\frown}{TSV}$, so we begin by finding the measure of $\overset{\frown}{TSV}$.

$$m\overset{\frown}{TV} + m\overset{\frown}{TSV} = 360° \qquad \text{circle-arc sum theorem}$$

$$116° + m\overset{\frown}{TSV} = 360°$$

$$m\overset{\frown}{TSV} = 244°$$

$$m\angle RTV = \frac{1}{2}m\overset{\frown}{TSV} \qquad \text{angle measures } \frac{1}{2} \text{ the intercepted arc}$$

$$= \frac{1}{2}(244°)$$

$$= 122°$$

Angles with Vertex Outside the Circle

When an angle is formed by two tangents, two secants, or a tangent and secant, the vertex lies outside the circle and we have the following theorem:

> **Angle-Arc Theorem for Tangents and Secants (*Vertex Outside*)**
> The measure of the angle formed by two secants, two tangents, or a secant and tangent is equal to $\frac{1}{2}$ the difference of the measures of the two intercepted arcs.

Figure 11.17 shows the angle-arc relationship for the three cases where the vertex is outside the circle.

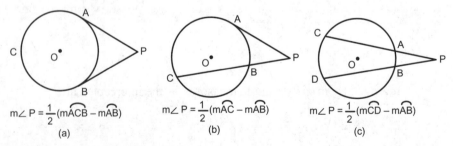

$m\angle P = \frac{1}{2}(m\overset{\frown}{ACB} - m\overset{\frown}{AB})$

(a)

$m\angle P = \frac{1}{2}(m\overset{\frown}{AC} - m\overset{\frown}{AB})$

(b)

$m\angle P = \frac{1}{2}(m\overset{\frown}{CD} - m\overset{\frown}{AB})$

(c)

Figure 11.17 Angle-arc relationships, (a) two tangents, (b) a tangent and a secant, and (c) two secants

382

Example 1

In $\odot P$, secants $\overline{RST}$ and $\overline{RUV}$ intersect at R.

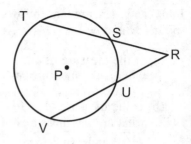

(a) If the measure of $\overset{\frown}{TV}$ = 140° and the measure of arc $\overset{\frown}{SU}$ is 42°, what is the measure of $\angle R$?

(b) If m$\angle R$ = 36° and m$\overset{\frown}{SU}$ = 48°, find m$\overset{\frown}{TV}$.

Solution: (a) The angle measure equals the difference between the measures of the intercepted arcs.

$$m\angle R = \frac{1}{2}\left(m\overset{\frown}{TV} - m\overset{\frown}{SU}\right)$$

$$= \frac{1}{2}(140° - 42°)$$

$$= \frac{1}{2}(98°)$$

$$= 49°$$

(b) Here we work backward from the relationship to find the angle.

$$m\angle R = \frac{1}{2}\left(m\overset{\frown}{TV} - m\overset{\frown}{SU}\right)$$

$$36° = \frac{1}{2}\left(m\overset{\frown}{TV} - 48°\right)$$

$$72° = m\overset{\frown}{TV} - 48°$$

$$m\overset{\frown}{TV} = 120°$$

When the angle is formed by two tangents, only one of the arc measures is needed to find the angle. We can find the other arc measure using the fact that the two measures must sum to 360°.

Example 2

$\overline{AM}$ and $\overline{AN}$ are tangent to $\odot P$ at points M and N.

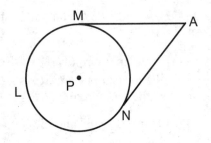

(a) If the measure of $\widehat{MN} = 115°$, what is the measure of $\angle A$?

(b) If the measure of $\angle A$ is 32°, what are the measures of arcs $\widehat{MLN}$ and $\widehat{MN}$?

Solution:

(a) First we need to find $m\widehat{MLN}$.

$$m\widehat{MLN} + m\widehat{MN} = 360°$$
$$m\widehat{MLN} + 115° = 360°$$
$$m\widehat{MLN} = 245°$$

Now apply the tangent-arc relationship.

$$m\angle A = \frac{1}{2}\left(m\widehat{MLN} - m\widehat{MN}\right)$$
$$= \frac{1}{2}(245° - 115°)$$
$$= \frac{1}{2}(130°)$$
$$= 65°$$

(b) Here both arc measures are unknown. However, they are related by the fact that their sum equals 360°. So we really have only one variable. Let $m\widehat{MLN} = x°$ and $m\widehat{MN} = (360 - x)°$. It doesn't matter which arc we define as x. We will still come up with the same answer in the end. Either way, watch out for the signs—there will be multiple negative signs in the equation.

$$m\angle A = \frac{1}{2}\left(m\widehat{MLN} - m\widehat{MN}\right)$$

$$32° = \frac{1}{2}(x - (360 - x))°$$

$$32° = \frac{1}{2}(2x - 360)°$$

$$64° = 2x - 360°$$

$$424° = 2x$$

$$x = 212°$$

Now substitute to find each arc.

$$m\widehat{MLN} = x°$$
$$= 212°$$

$$m\widehat{MN} = (360 - x)°$$
$$= (360 - 212)°$$
$$= 148°$$

The next example brings together several of the circle theorems we have seen.

Example 4

In circle P, $m\angle R = 47°$ and $m\widehat{SU} = 86°$. Are points T, P, and U collinear? Is $ST = SU$? Justify your reasoning.

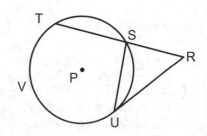

Solution: The strategy here is to find all the arc measures. For T, P, and U to be collinear, they must form a straight line through the center and lie on a diameter. The measure of arc $\widehat{TVU}$ would have to be 180°. If $ST = SU$, the two chords would have to intercept congruent arcs and $m\widehat{ST} = m\widehat{SU}$. Using the tangent-secant relationship, we can find $\widehat{TVU}$.

$$m\angle R = \frac{1}{2}\left(m\widehat{TVU} - m\widehat{SU}\right)$$

$$47° = \frac{1}{2}\left(m\widehat{TVU} - 86°\right)$$

$$94° = m\widehat{TVU} - 86°$$

$$m\widehat{TVU} = 180°$$

So points T, P, and U must be collinear.
We can now find the measure of $\widehat{ST}$ by setting the sum of the arc measures equal to 360°.

$$m\widehat{TVU} + m\widehat{SU} + m\widehat{ST} = 360°$$

$$180° + 86° + m\widehat{ST} = 360°$$

$$m\widehat{ST} = 94°$$

Since arcs $\widehat{SU}$ and $\widehat{ST}$ are not congruent, we know that $\overline{ST}$ and $\overline{SU}$ are not congruent and $ST \neq SU$.

Check Your Understanding of Section 11.4

A. Multiple-Choice

Use the accompanying figure for problems 1 and 2:

1. In $\odot A$, $m\widehat{LF} = 48°$ and $m\widehat{OP} = 52°$. What is the measure of $\angle 1$?
   (1) 42°   (3) 50°
   (2) 48°   (4) 52°

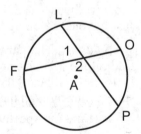

2. In $\odot A$, $m\widehat{LO} = 59°$ and $m\widehat{FP} = 151°$.
   What is the measure of $\angle 2$?
   (1) 104.5°
   (2) 105°
   (3) 121°
   (4) 150.5°

Use the accompanying figure for problems 3 and 4:

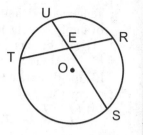

3. In ⊙O, chords $\overline{UES}$ and $\overline{TER}$ intersect at
   point E. If m$\widehat{UT}$ = 38° and m$\widehat{RS}$ = 74°, what
   is the measure of ∠TES?
   (1) 56°   (3) 106°
   (2) 71°   (4) 124°

4. In ⊙O, chords $\overline{UES}$ and $\overline{TER}$ intersect at point E. Which of the following
   is *not* true?
   (1) m∠UET = m$\widehat{UT}$
   (2) m∠UET + m∠TES = 180°
   (3) m∠UET = m∠RES
   (4) m$\widehat{UT}$ + m$\widehat{RS}$ = twice the measure of ∠RES

Use the accompanying figure for problems 5 and 6:

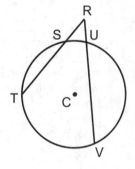

5. Secant $\overline{RST}$ intersects ⊙C at S and T. Secant
   $\overline{RUV}$ intersects ⊙C at U and V. If m$\widehat{TV}$ =
   117° and m$\widehat{SU}$ = 21°, what is the measure of
   angle R?
   (1) 10.5°   (3) 48°
   (2) 24°   (4) 63°

6. Secant $\overline{RST}$ intersects ⊙C at S and T. Secant $\overline{RUV}$ intersects ⊙C at
   U and V. If m∠R = 39° and m$\widehat{SU}$ = 29°, what is the measure of $\widehat{TV}$?
   (1) 51°   (2) 87°   (3) 107°   (4) 136°

Use the accompanying figure for problems
7 and 8:

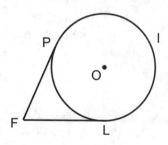

7. Tangents $\overline{FP}$ and $\overline{FL}$ intersect circle O
   at P and L respectively. If m∠F = 62°,
   what is the measure of $\widehat{PL}$?
   (1) 136°   (3) 124°
   (2) 128°   (4) 118°

8. Tangents $\overline{FP}$ and $\overline{FL}$ intersect circle O at P and L respectively. Which of the following is *not necessarily* true?
   (1) FO bisects $\angle F$
   (2) $\angle FLO$ is a right angle
   (3) the difference of the measures of arcs $\overarc{LIP}$ and $\overarc{LP}$ equals twice the measure of $\angle F$
   (4) $FP = PO$

9. Given $\odot P$ with diameter $\overline{BPD}$ intersecting $\overline{AEC}$ at E, $m\overarc{AB} = 112°$, and $m\overarc{BC} = 52°$, what is the measure of $\angle AED$?
   (1) $5°$
   (2) $56°$
   (3) $60°$
   (4) $82°$

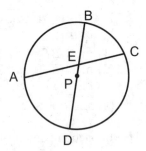

10. In $\odot A$, chords $\overline{EF}$ and $\overline{CD}$ are parallel. $m\overarc{EF} = 80°$, $m\overarc{CGD} = 110°$, and $m\angle E = 54°$. What is the measure of $\angle 1$?
    (1) $54°$
    (2) $56°$
    (3) $67°$
    (4) $82°$

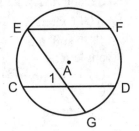

Use the accompanying figure for problems 11 and 12:

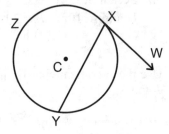

11. In $\odot C$, tangent $\overrightarrow{XW}$ intersects chord $\overline{XY}$ at X. If $m\overarc{XY} = 142°$, what is the measure of $\angle WXY$?
    (1) $18°$ (3) $81°$
    (2) $71°$ (4) $88°$

12. In $\odot C$, tangent $\overarc{XW}$ intersects chord $\overline{XY}$ at X. If $m\overarc{XZY} = 212°$, what is the measure of $\angle WXY$?
    (1) $74°$ (2) $80°$ (3) $82°$ (4) $106°$

Use the accompanying figure for problems 13 and 14:

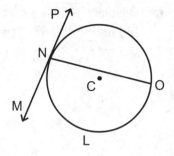

13. $\overleftrightarrow{PNM}$ is tangent to circle C at N. If m$\angle ONP = 85°$, what is the measure of $\overarc{NLO}$?
 (1) 150° (3) 170°
 (2) 160° (4) 190°

14. $\overleftrightarrow{PNM}$ is tangent to circle C and intersects chord $\overline{ON}$ at N. If m$\overarc{ON}$: m$\overarc{OLN}$ = 2:3, what is the measure of $\angle PNO$?
 (1) 60°   (2) 72°   (3) 80°   (4) 88°

15. In $\odot Q$, tangent $\overleftrightarrow{BA}$ intersects chord $\overline{BC}$ and diameter $\overline{BQD}$ at B. If m$\overarc{CD}$ = 48°, what is the measure of $\angle ABC$?
 (1) 66°
 (2) 52°
 (3) 48°
 (4) 42°

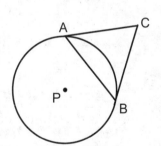

B. Free-Response—show all work or explain your answer completely.

16. In $\odot P$, tangents $\overline{AC}$ and $\overline{BC}$ intersect at C. If m$\angle C = 46°$, what is the measure of $\overarc{AB}$?

17. In $\odot P$, chords $\overline{CEA}$ and $\overline{BED}$ intersect at E. If m$\angle D = 48°$, m$\angle C = 38°$, and m$\overarc{CD}$ = 152°, what is the measure of $\overarc{BA}$?

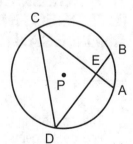

18. Diameter $\overline{ACM}$ is extended to point M and intersects $\overline{SRM}$ at M. If m∠M = 21° and m$\overset{\frown}{SR}$ = 85°, what is the measure of $\overset{\frown}{AS}$?

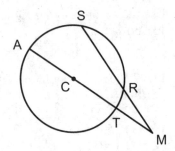

19. In ⊙C, secant $\overline{DHGE}$ intersects tangent $\overline{DB}$ at D. m$\overset{\frown}{EF}$ = 38°, m$\overset{\frown}{FH}$ = 72°, and m∠EGF = 58°. Find m∠D.

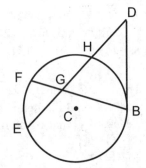

20. Given chord $\overline{XAZ}$, tangent $\overline{WX}$, tangent $\overline{WY}$, m∠W = 64°, and m$\overset{\frown}{YZ}$ = 64°, determine if $\overline{XAZ}$ is a diameter. Justify your reasoning.

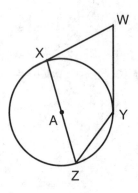

11.5 SEGMENT RELATIONSHIPS IN INTERSECTING CHORDS, TANGENTS, AND SECANTS

KEY IDEAS

The relationships between segment lengths formed by intersecting chords, tangents, and secants are shown in the following table:

Intersecting chords	Products of the parts are equal $a \cdot b = c \cdot d$
Intersecting secants	outside · whole = outside · whole $PA \cdot PC = PB \cdot PD$
Intersecting tangent and secant	outside · whole = outside · whole $PA^2 = PB \cdot PC$

Segments Formed by Intersecting Chords

Theorem
When two chords intersect in a circle, the products of the parts of the two chords are equal.

Figure 11.18 shows the two segments formed within each of two intersecting chords. The product of the parts are equal, giving the relationship $a \cdot b = c \cdot d$.

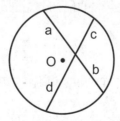

Figure 11.18 Product of the parts of intersecting chords are equal

The theorem is based on the similar triangles formed by completing the two triangles, which is illustrated in Figure 11.19. $\angle C \cong \angle A$ because they intercept the same arc. $\angle B \cong \angle D$ for the same reason. $\triangle ADE \cong \triangle CBE$ by AA. Corresponding parts are in the same ratio, which gives:

$$\frac{DE}{BE} = \frac{AE}{CE}$$

$$DE \cdot CE = BE \cdot AE$$

$$a \cdot b = c \cdot d$$

Figure 11.19 Proof of the intersecting chord theorem

Example 1

Chords $\overline{BA}$ and $\overline{CD}$ intersect at E in $\odot O$.

(a) If $BE = 6$, $AE = 12$, and $DE = 8$, find EC.
(b) If $DC = 12$, $BE = 4$, and $AE = 9$, find DE and EC.

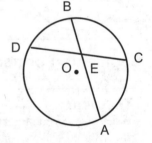

Solution: (a) Apply the product of the parts relationship.

$$DE \cdot EC = BE \cdot AE$$
$$8 \cdot EC = 6 \cdot 12$$
$$EC = 9$$

(b) We are given the entire length of DE instead of the length of the parts. However, we can use the fact that the sum of the parts equals the whole.

Let $DE = x$
$EC = DC - x$
$ = 12 - x$
$DE \cdot EC = BE \cdot AE$ product of the parts are equal
$x \cdot (12 - x) = 4 \cdot 9$
$12x - x^2 = 36$
$-x^2 + 12x - 36 = 0$ The equation is quadratic, so bring all terms to one side.

$$x^2 - 12x + 36 = 0$$ Multiply by -1 so the leading coefficient equals 1.

$$(x - 6)\,(x - 6) = 0$$ factor

$$x - 6 = 0 \quad x - 6 = 0$$ zero product property

$$x = 6,\, x = 6$$

$$DE = 6$$

$$EC = 12 - 6$$ substitute for x

$$EC = 6$$

$$EC = DE = 6$$

Example 2

In $\odot O$, chords $\overline{AD}$ and $\overline{BC}$ intersect at E. If $AE = 4$, $AB = 12$, $BE = 15$, and $CD = 8$, what are the lengths of $\overline{DE}$ and $\overline{CE}$?

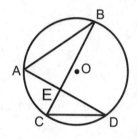

Solution: $\triangle ABE \sim \triangle CDE$ by AA because $\angle A$ and $\angle C$ both intercept $\overarc{BD}$ and $\angle D$ and $\angle B$ both intercept $\overarc{AC}$.

$$\frac{AB}{CD} = \frac{BE}{DE}$$ corresponding parts are proportional

$$DE \cdot AB = CD \cdot BE$$

$$DE = \frac{CD \cdot BE}{AB}$$

$$DE = \frac{8 \cdot 15}{12}$$

$$= 10$$

$$AE \cdot DE = CE \cdot BE$$ product of the parts of intersecting chords

$$4 \cdot 10 = CE \cdot 15$$ are congruent

$$CE = \frac{4 \cdot 10}{15}$$

$$= 2\frac{2}{3}$$

Example 3

$\overline{BGPD}$ is a diameter of $\odot P$ and bisects $\overline{AC}$. If $GD = 12$ and $AD = 15$, find BG and BC.

Solution: We know $\triangle AGD$ and $\triangle CGB$ are similar triangles. Vertical angles $\angle AGD$ and $\angle BGC$ are congruent. $\angle C \cong \angle D$ because they are inscribed angles intercepting the same arc. So $\triangle AGD \sim \triangle BGC$. The diameter bisects $\overline{AC}$, so it must be perpendicular to $\overline{AC}$. We can use the Pythagorean theorem to find AG:

$$AG^2 + GD^2 = AD^2$$
$$AG^2 + 12^2 = 15^2$$
$$AG^2 + 144 = 225$$
$$AG^2 = 81$$
$$AG = 9$$

Since $\overline{AC}$ is bisected at G,

$$GC = AG = 9$$

We now have enough information to apply a similarity relationship.

$$\frac{AG}{BG} = \frac{GD}{GC}$$
$$\frac{9}{BG} = \frac{12}{9}$$
$$12BG = 81$$
$$BG = 6.75$$

$$\frac{AG}{BG} = \frac{AD}{BC}$$
$$\frac{9}{6.75} = \frac{15}{BC}$$
$$9BC = 101.25$$
$$BC = 11.25$$

$$BG = 6.75, \ BC = 11.25$$

Segments Formed by Tangents and Secants

> ## Theorem
> - When two secants from a common point external to a circle intersect that circle, the products of the outer segments and the entire secants are equal.
> - When a secant and tangent from a common point external to a circle intersect that circle, the products of the outer segments and the entire secant or tangent are equal.

$$PA \cdot PC = PB \cdot PD$$

(a)

$$PA^2 = PB \cdot PC$$

(b)

$$PA = PB$$

(c)

Figure 11.20 Segment relationship between secants and tangent intersecting a circle

When a secant intersects a circle, it is divided into a segment outside the circle and a segment within the circle. Figure 11.20a shows two segments intersecting circle O from point P. The outer part of secant $\overline{PAC}$ is $\overline{PA}$, and the outer part of secant $\overline{PBD}$ is $\overline{PB}$. The secant-segment theorem states:

$$PA \cdot PC = PB \cdot PD$$

The relationship is similar when a tangent and secant meet at a common point outside the circle. The only difference is that the entire length is outside the circle as shown in Figure 11.20b. So the outer length and whole length are the same, giving the relationship

$$PA \cdot PA = PB \cdot PC$$
$$PA^2 = PB \cdot PC$$

The theorem also holds for two tangents as shown in Figure 11.20c. The relationship is

$$PA^2 = PB^2 \text{ or}$$
$$PA = PB$$

Example 1

Secants $\overline{ABC}$ and $\overline{ADE}$ intersect $\odot P$ at B and D.

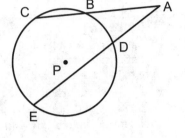

(a) If $AB = 9$, $BC = 3$, and $AD = 6$, find DE.

(b) If $BC = 9$, $AD = 12$, and $AE = 30$, find AB.

Solution:

(a) $AB \cdot AC = AD \cdot AE$ outside · whole = outside · whole

$9(9 + 3) = 6(AE)$

$108 = 6AE$

$AE = 18$

$DE = AE - AD$

$= 18 - 6$

$= 12$

(b) Let $AB = x$ and $AC = x + BC$

$AB \cdot AC = AD \cdot AE$ outside · whole = outside · whole

$x(x + 9) = 12(30)$

$x^2 + 9x = 360$

$x^2 + 9x - 360 = 0$

$(x - 15)(x + 24) = 0$ factor

$x - 15 = 0,\ x + 24 = 0$ zero product rule

$x = 15$ or $x = -24$

$AB = 15$ use the positive value

Example 2

Secant $\overline{KNG}$ intersects $\odot P$ at N and G. Tangent $\overline{KI}$ intersects $\odot P$ at I. If $KN = 5$ and $NG = 7$, find IK. Express your answer in simplest radical form.

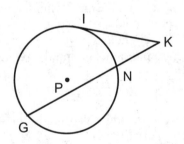

Solution:

$$IK^2 = KN \cdot KG$$

$$IK^2 = 5(5+7)$$

$$IK^2 = 60$$

$$IK = \sqrt{60}$$

$$IK = \sqrt{4 \cdot 15}$$

$$IK = 2\sqrt{15}$$

Check Your Understanding of Section 11.5

A. *Multiple-Choice*

1. In circle P, chords $\overline{CD}$ and $\overline{EF}$ intersect at point X. Which of the following theorems could be used to justify the statement $CX \cdot XD = EX \cdot XF$?
   (1) corresponding parts of similar triangles are in the same ratio
   (2) congruent chords in a circle intercept congruent arcs
   (3) the arcs between parallel chords in a circle are congruent
   (4) the area of a sector of a circle is proportional to the square of the radius of the circle

Use the accompanying figure for problems 2 and 3:

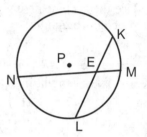

2. In circle P, chords $\overline{KL}$ and $\overline{MN}$ intersect at E. If $KE = 8$, $EL = 11$, and $EM = 5$, what is the length of $\overline{EN}$?
   (1) 13.8   (3) 17.6
   (2) 14   (4) 24

3. Chords $\overline{KL}$ and $\overline{MN}$ intersect at E in $\odot P$. $LE = 56$, $NE = 112$, and $ME = 24$. What is the length of KE?
   (1) 16
   (2) 32
   (3) 40
   (4) 48

Use the accompanying figure for problems 4 and 5:

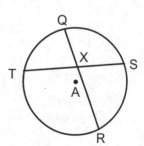

4. Chords $\overline{ST}$ and $\overline{QR}$ intersect at X in $\odot A$. If $XT = 14$, $XS = 16$, and $QX = 12$, what is the length of XR?
   (1) $18\frac{2}{3}$ (3) $19\frac{1}{2}$
   (2) $18\frac{3}{4}$ (4) $20\frac{3}{4}$

5. In circle A, $XR = 15$, $QX = 8$, and $TX = 10$. What is the length of XS?
   (1) 12   (2) 14   (3) 15   (4) 16

397

Use the accompanying figure for problems 6 and 7:

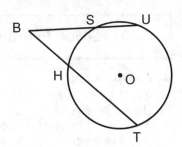

6. Secants $\overline{BSU}$ and $\overline{BHT}$ intersect at B. If $BS = 8$, $SU = 2$, and $BH = 5$, what is the length of $\overline{HT}$?

(1) $5\dfrac{1}{5}$ (2) 7   (3) 11   (4) 15

7. Secant $\overline{BSU}$ intersects circle O at S and U. Secant $\overline{BHT}$ intersects circle O at H and T. $BS = 6$, $BH = 3$, and $BT = 15$. What is the length of SU?

(1) 6   (2) 4   (3) 2   (4) 1.5

Use the accompanying figure for problems 8 and 9:

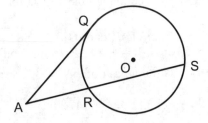

8. Secant $\overline{ARS}$ intersects ⊙O at R and S. If $AR = 7$ and $RS = 9$, what is the length of $\overline{AQ}$?

(1) $3\sqrt{7}$ (3) $7\sqrt{3}$
(2) $4\sqrt{7}$ (4) 16

9. Secant $\overline{ARS}$ intersects ⊙O at R and S. Tangent $\overline{AQ}$ intersects ⊙O at Q. If $AR = 3$ and $AQ = 6$, what is the length of $\overline{RS}$?

(1) 9   (2) 10   (3) 12   (4) 14

10. In circle O, $AB = 20$, $KI = 5$, and $IJ = 20$. What is the length of AI?

(1) 8
(2) 10
(3) 12
(4) 15

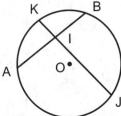

11. Points A, B, C, and D lie on circle O. If $AE = 8$, $ED = 10$, and $CE = 4$, which transformation will map the smaller triangle onto the larger?

(1) dilation by a scale factor of 2.5, followed by a reflection through point E
(2) dilation by a scale factor of 2.0, followed by a reflection through point E
(3) rotation by $\angle COA$ about point O, followed by a dilation by a scale factor of 2.5 centered at A
(4) translation from B to D, followed by a reflection over $\overline{AD}$, followed by a dilation by a scale factor of 2 centered at D

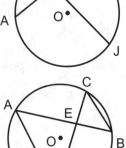

B. *Free-Response—show all work or explain your answer completely.*

12. Given: circle C with chords $\overline{JK}$ and $\overline{LM}$
    intersecting at point A
    Prove: $JA \cdot KA = LA \cdot MA$

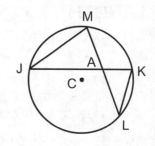

13. In $\odot O$, $ED = 10$, $AE = 4$, $AD = 12$, and
    $CB = 15$. What are the lengths of $\overline{CE}$
    and $\overline{EB}$?

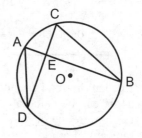

14. In $\odot O$, $\overline{AC}$ is tangent at point C, $AF = 4$,
    $AC = 8$, and $BE = 4$. Find EC and DE.

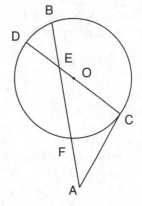

11.6 AREA, CIRCUMFERENCE, AND ARC LENGTH

The radian is a unit of angle measure. 2π radians is equal to one complete revolution around a circle, or 360°. You can convert between units using the following formulas:

- radians $= \dfrac{\pi}{180°} \cdot$ degrees

- degrees $= \dfrac{180°}{\pi} \cdot$ radians

The length of an arc and the area of a sector formed by that arc and its central angle can be calculated from the measure of the central angle. Multiply the circumference or area of the circle by the fraction of the circle spanned by the central angle.

Unit of Angle Measure	Arc Length	Sector Area
Degrees	$\dfrac{\theta}{360°} c$	$\dfrac{\theta}{360°} A$
Radians	$r \cdot \theta$	$\dfrac{1}{2} r^2 \cdot \theta$

θ is the central angle measure, r is the radius of the circle, c is the circumference of the circle, and A is the area of the circle.

Area and Circumference

The circumference, C, of a circle is the length around the circle. You can think of it as the length of a piece of string that would wrap exactly once around the circle. The area, A, of a circle is the area enclosed by the circle or the area of the interior region.

Formulas
In a circle of radius r and diameter d,
Circumference of a circle: $C = 2\pi r = \pi d$
Area of a circle: $A = \pi r^2 = \dfrac{1}{4} \pi d^2$

When working with area and circumference, be sure to read the question carefully to determine if answers should be rounded or be in terms of π.

Example 1

Find the area and circumference of a circle whose diameter is 12 in. Express your answer in terms of π.

Solution: The radius of the circle equals $\frac{1}{2}$ the diameter, or 6 in.

$$C = 2\pi r = 12\pi \text{ in.}$$
$$A = \pi r^2 = \pi \,(6 \text{ in.})^2 = 36\pi \text{ in.}^2$$

We can also work backward to find the radius or diameter from the circumference or area.

MATH FACT

When working with the quantity π, you may be asked to express your answer either in terms of π or as a rounded value. When giving a rounded value, you should always use the π button on your calculator instead of 3.14. This way you will be certain to retain enough precision in your answer.

Example 2

Anna has a round table for which she would like to buy a tablecloth. She wants to get an accurate measurement of the table's diameter. So she wraps a tape measure around the table and measures the length around the table to be $188\frac{1}{2}$ inches. What is the diameter of the table in feet? Round to the nearest foot.

Solution: We are given a circumference of $188\frac{1}{2}$ inches. So we can use the circumference formula to solve for the diameter.

$$C = \pi d$$
$$188.5 \text{ in.} = \pi d$$
$$d = 60.001 \text{ in.}$$

Multiply by $\dfrac{1\,\text{ft}}{12\,\text{in.}}$ to find the diameter in feet.

$$d = 60.001\,\text{in.} \cdot \dfrac{1\,\text{ft}}{12\,\text{in.}} = 5\,\text{ft}$$

Example 3

An athletic field is in the shape of a square with semicircular ends. If the distance between the two semicircles is 30 meters, what is the area of the athletic field in terms of π?

Solution: All sides of a square are congruent, so the sides of the square that form the diameter of the semicircles are also 30 meters. The radius of the two semicircles is 15 meters. The area of each semicircle will be half of the area of a full circle.

$$\text{Area}_{\text{semicircle}} = \frac{1}{2}\pi r^2$$

$$= \frac{1}{2} \cdot 15^2\,\pi$$

$$= 112.5\pi$$

$$\text{Area}_{\text{square}} = s^2$$

$$= 30^2$$

$$= 900$$

$$\text{Area}_{\text{total}} = 2 \cdot \text{Area}_{\text{semicircle}} + \text{Area}_{\text{square}}$$

$$= 2(112.5\pi) + 900$$

$$= 225\pi + 900$$

Arc Length and Radian Measure

Arcs have length, which equals the length of a string wrapped around a circle starting and ending at the two endpoints of the arc. We will refer to the arc length with the variable s. Angle measures are often represented with the Greek letter θ (theta). Figure 11.21 shows $\overset{\frown}{MN}$, angle measure θ, and arc length s.

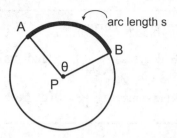

Figure 11.21 Arc length *s* versus arc angle measure θ

The circumference is the length of an arc spanning the entire circle. If the arc spans less than the full circle, the arc length is reduced proportionally. The fraction of the circle spanned by a central angle measuring θ degrees is θ/360°. This fact makes the intercepted arc length equal to θ/360° · circumference.

Degrees are not the only unit of measure for angle. The other unit of angle measure you need to know is the **radian**.

Definition

Radian—A unit of angle measure. In a circle, a central angle measuring 1 radian intercepts an arc with length equal to the radius of the circle.

Figure 11.22 illustrates the concept of the radian. ∠*APB* is a central angle whose measure is 1 radian and intercepts $\overset{\frown}{AB}$. The length of $\overset{\frown}{AB}$ equals *s*, which equals *r* in this case. 2π radians is equivalent to 1 full revolution.

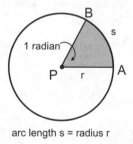

arc length s = radius r

Figure 11.22 An arc intercepted by a central angle measuring 1 radian

The following formulas are used to convert between radians and degrees. These are found on the reference sheet.

Degree—Radian Conversion Formulas

$$\theta(\text{degrees}) \cdot \frac{\pi}{180°} = \theta(\text{radians})$$

$$\theta(\text{radians}) \cdot \frac{180°}{\pi} = \theta(\text{degrees})$$

Arc Length Formula

$s = \dfrac{\theta}{360°}$ circumference   where θ is the measure of the central angle in degrees

$s = r\theta$ where θ is the measure of the central angle in radians

---

Example 1

The radius of a circle is 4 inches. Find the length of the arc intercepted by a central angle measuring 1.2 radians.

Solution:

$$s = r\theta = 4(1.2) = 4.8 \text{ inches}$$

Example 2

The radius of a circle is 3 cm. What is the length of the arc intercepted by a central angle measuring 42°? Round your answer to the nearest tenth of a centimeter.

Solution: The angle is given in degrees, so use the formula

$$s = \frac{\pi}{180°} r\theta$$

$$s = \frac{\pi}{180°} \cdot 3 \cdot 42° = 2.19911$$

$$s = 2.2 \text{ cm}$$

Example 3

$\angle FKG$ is an inscribed angle measuring $\dfrac{\pi}{6}$ radians. What is the length of $\overset{\frown}{FG}$ if the radius of the circle is 10 meters? Express your answer in terms of π.

Solution: Sketching the figure will help here:

The central angle that intercepts the same arc will have a measure twice that of the intercepted angle, or $2 \cdot \dfrac{\pi}{6} = \dfrac{\pi}{3}$ radians.

$$s = r\theta$$
$$s = \frac{\pi}{3} \cdot 10$$
$$= \frac{10\pi}{3} \text{ m}$$

Example 4

Frannie is planting a flower bed in the area around a corner of her backyard fence. The two sections of fence come together at an angle of $\dfrac{4\pi}{3}$ radians. She wants the flower bed to have a radius of 6 feet. Frannie wants to put a brick border around the flower bed. What will be the length of the border to the nearest inch?

Solution: The region is bounded by a circular arc with the angle given in radians. So we use the formula

$$s = r \cdot \theta$$
$$s = 6 \cdot \frac{4\pi}{3}$$
$$s = \frac{24\pi}{3}$$
$$s = 25.1327 \text{ ft}$$

We convert to inches using the ratio $\dfrac{12 \text{ in.}}{1 \text{ ft}}$:

$$12.1327 \text{ in.} \cdot \frac{12 \text{ in.}}{1 \text{ ft}} = 301.592 \text{ in.}$$

The border will have a length of 302 in.

Sectors

A sector is the region bounded by a circle and a central angle. Figure 11.23 shows sector *APB*, the shaded region, in circle *P*.

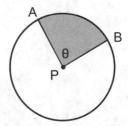

Figure 11.23 Sector *APB*

The area of a sector is equal to the fraction of the circle spanned by the central angle multiplied by the area of the circle.

Formula for Area of a Sector

$A_{\text{sector}} = \dfrac{\theta}{360}\pi r^2$ where θ is the measure of the central angle in degrees

$A_{\text{sector}} = \dfrac{1}{2}\theta r^2$ where θ is the measure of the central angle in radians

---

Example 1

Find the area of a sector of a circle if the radius is 12 cm and the sector is bounded by a central angle measuring 0.25 radians.

Solution:

$$A_{\text{sector}} = \frac{1}{2}\theta r^2$$

$$A_{\text{sector}} = \frac{1}{2}(0.25)(12)^2 = 18 \text{ cm}^2$$

Example 2

In a circle of radius 8 in., a sector has an area of 40 in.² What is the measure of the central angle that bounds the sector? Express your answer in radians.

Solution:

$$A_{sector} = \frac{1}{2}\theta r^2$$

$$40 = \frac{1}{2}\theta(8)^2$$

$$40 = 32\theta$$

$$\theta = 1.25 \text{ radians}$$

Example 3

A spinner for a board game has a radius of 9 inches and is divided into 5 regions. The central angles that bound regions 2 through 5 each measure 60°. What is the area of region 1? Round your answer to the nearest 0.1 square inch.

Solution: The central angle that bounds region 1 is equal to 360° minus the sum of the remaining central angles.

$$\theta_1 = 360° - \theta_2 - \theta_3 - \theta_4 - \theta_5$$
$$\theta_1 = 360 - 4(60°) = 120°$$

$$A_{sector1} = \frac{1}{360}\theta\pi r^2 = \frac{1}{360}(120)(\pi)(9^2)$$

$$A_{sector1} = 84.8 \text{ in.}^2$$

Example 4

The discus-throwing area at a track and field match is in the shape of a sector of a circle. The length along each side is 80 meters, and the angle is 34.9°.
  (a) What is the area of the sector? Round your answer to the nearest tenth of a square meter.
  (b) Officials want to mark off the entire area with a chalk-line boundary. What is the total length of the chalk-line? Round your answer to the nearest tenth of a meter.

Solution: (a) The region is a circular sector with the angle given in degrees, so we can use the formula

$$A_{sector} = \frac{1}{360}\theta\pi r^2$$

$$A_{sector} = \frac{1}{360}34.9\pi(80^2)$$

$$A_{sector} = 1949.2 \text{ m}^2$$

(b) The curved part of the sector is an arc, whose length is given by

$$s = \frac{\pi}{180°}r\theta$$

$$s = \frac{\pi}{180°} \cdot 80 \cdot 34.9$$

$$s = 48.7 \text{ m}$$

Now add the two straight 80-meter segments.
Total length = 48.7 + 80 + 80 = 208.7 m.

Arc Length and Sector Area in Similar Circles

All circles are similar and differ only in the length of their radii. If two circles have radii in a ratio of $r_1/r_2 = k$, we can determine the ratio of arc lengths and sector areas given the same central angle.
The ratio of arc lengths is

$$\frac{s_1}{s_2} = \frac{\theta_1 r_1}{\theta_2 r_2}$$

Since the central angle measures are congruent, $\frac{\theta_1}{\theta_2} = 1$. The ratio $\frac{r_1}{r_2}$ is the constant of proportionality between the two circles, or the dilation factor k. By substituting we get

$$\frac{s_1}{s_2} = k \text{ or}$$

$$s_1 = ks_2$$

Arc Length Theorem for Similar Circles

The ratio of arc lengths, given congruent central angles, is equal to the ratio of the radii of the circles. The result is the same whether the central angles are measured in radians or degrees.

By applying the same process to sector areas we find

$$\frac{A_{sector1}}{A_{sector2}} = k^2$$

$$A_{sector1} = k^2 \cdot A_{sector2}$$

Sector Area Theorem for Similar Circles

The ratio of sector areas, given congruent central angles, is equal to the ratio of the radii squared. As with arc length, the result is the same whether central angles are measured in radians or degrees.

Check Your Understanding of Section 11.6

A. Multiple-Choice

1. What is the degree measure of an angle measuring $\frac{5\pi}{6}$ radians?

   (1) 2.6°   (2) 30°   (3) 150°   (4) 300°

2. What is the degree measure of an angle measuring $\frac{\pi}{9}$ radians?

   (1) 90°   (2) 60°   (3) 40°   (4) 20°

3. What is the measure in radians of an angle whose measure is 45°?

   (1) $\frac{\pi}{6}$ (2) $\frac{\pi}{4}$ (3) $\frac{\pi}{3}$ (4) $\frac{\pi}{2}$

4. What is the measure in radians of an angle whose measure is 270°?

   (1) $\frac{3\pi}{2}$ (2) π (3) $\frac{3\pi}{4}$ (4) $\frac{3\pi}{8}$

Use the accompanying figure for problems 5 and 6:

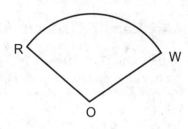

5. Arc $\overset{\frown}{RW}$ is centered at point O. If m$\angle ROW = 115°$ and $RO = 9$, what is the length of $\overset{\frown}{RW}$?

(1) $\dfrac{23\pi}{4}$　　(2) $\dfrac{23\pi}{8}$　　　(3) $\dfrac{207\pi}{8}$　　　(4) $\dfrac{46\pi}{4}$

6. In circular sector ROW, $OW = 12$ and m$\angle ROW = \dfrac{7\pi}{6}$ radians. What is the area of sector ROW?

(1) 7π　　　(2) 84π　　　(3) 120π　　　(4) 168π

Use the accompanying figure for problems 7 and 8:

7. What is the length of arc s?

(1) 1.6π　　(3) $5\dfrac{1}{3}\pi$

(2) $\dfrac{3}{4}\pi$　　(4) 64π

8. What is the area of the sector?

(1) 16π　　　(2) 32π　　　(3) 64π　　　(4) 128π

Use the accompanying figure for problems 9 and 10:

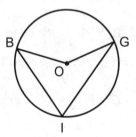

9. Inscribed angle $\angle I$ measures $\dfrac{2\pi}{5}$ radians, and radius $BO = 5$. What is the length of arc $\overset{\frown}{BG}$?

(1) $\dfrac{20\pi}{5}$　　(3) 4π

(2) $\dfrac{10\pi}{5}$　　(4) 9π

10. Inscribed angle $\angle I$ measures $71°$, and radius $OG = 10$. What is the area of sector BOG?

(1) 12.4　　　(2) 61.9　　　(3) 93.7　　　(4) 123.9

11. $\angle AOB$ and $\angle CPD$ are central angles in circles O and P, respectively. If $\angle AOB \cong \angle CPD$, $OA = 6$, and $PC = 2$, the ratio $\dfrac{\text{Area(sector } AOB)}{\text{Area(sector } CPD)}$ equals

(1) $\dfrac{1}{3}$ (2) 3   (3) 6   (4) 9

12. The image of $\odot X$ after D_4 is $\odot Y$. What is the ratio of an arc length on $\odot X$ to an arc length on $\odot Y$ if the arcs are intercepted by congruent central angles?

(1) $\dfrac{1}{4}$ (2) $\dfrac{1}{16}$ (3) 4   (4) 16

B. *Free-Response—show all work or explain your answer completely.*

13. Two arc lengths, s_1 and s_2, are defined by the accompanying figure. Which arc is longer? Justify your answer.

7 s_1 0.28 radians   3 s_2 0.6 radians

14. The beam of the searchlight on a Coast Guard rescue ship is in the shape of a circular sector. The central angle measures $33°$. The crew has found that the intensity of the light is sufficient to search the ocean up to 150 feet from the searchlight. What is the area of the region that can be searched without moving the light? Round to the nearest square foot.

15. $\triangle ABC$ is inscribed in $\odot P$. $AB = BC$, $m\angle B = 54°$, and $PB = 4$. What is the length of arc $\overset{\frown}{BC}$? Round to the nearest tenth.

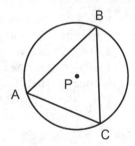

16. The accompanying figure is comprised of a right triangle and a semicircle. What is the total area of the figure? Round your answer to the nearest tenth.

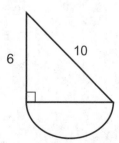

6   10

SOLIDS AND MODELING

Chapter
Twelve

12.1 PRISMS AND CYLINDERS

A prism is a solid with two polygons for bases that are parallel and congruent. A circular cylinder is a solid with two congruent, parallel circles for bases. Both prisms and cylinders may be oblique or right.

	Prism	Circular Cylinder
Volume (V)	$B \cdot h$	$\pi r^2 h$

(B is the area of the base, h is the height, r is the radius)

Definitions

The following definitions are used in our discussion of solid figures:

- A *solid* is any 3-D figure that is fully enclosed.
- A *face* of a solid is any of the surfaces that bound the solid.
- A *polyhedron* is any solid whose faces are polygons.
- An *edge* is the intersection of two faces in a polyhedron.
- A *lateral face* is any face of a solid other than its bases.
- *Volume* refers to the amount of space contained within a solid figure. We measure volume in units of length^3, such as in.^{3}, cm^3, and ft^3. These units refer to the amount of space occupied by a cube 1 unit in length on each side. For example, 1 in.^{3} is the volume of a cube 1 in. × 1 in. × 1 in. A cylinder with a volume of 50 in.^{3} would be able to contain 50 of these 1 in. × 1 in. × 1 in. cubes.

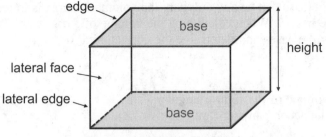

Figure 12.1 Parts of a prism

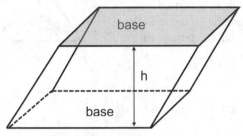

Figure 12.2 Oblique prism

Prisms

A prism is a polyhedron with:

- Two bases that are parallel and congruent polygons
- Rectangular lateral faces

The prism shown in Figure 12.1 above is a rectangular prism. The edges between the lateral faces are called lateral edges. The height of the prism is the perpendicular distance between the two bases. Prisms are named based on the type of polygons that make up the bases.

A prism can also be classified as right or oblique. A right prism, such as the one in Figure 12.1, has its lateral edges perpendicular to the bases and the bases are aligned one over the other. An oblique prism has lateral edges that are not perpendicular to the bases as shown in Figure 12.2 above. Note that the height in an oblique prism is still the perpendicular distance between the bases.

Volume of a Prism
Volume (V) $= B \cdot h$
where B is the area of the base and h is the height.

Since the base can be any polygon, the volume formula is in terms of the area of the base B. When calculating the volume of a prism, the first step is to identify the bases and determine the correct area formula.

Example

Find the volume of the right triangular prism in the accompanying figure.

Solution: The bases of the prism are the two triangles. So use the formula for the area of a triangle to find B. The base of the triangle is the 6 cm dimension, and the height of the triangle is 8 cm.

$$B = \frac{1}{2} b \cdot h$$

$$= \frac{1}{2}(6)(8)$$

$$= 24 \text{ cm}^2$$

Now apply the volume formula for the cylinder. The height of the prism is the 12 cm distance between the two triangles.

$$V = B \cdot h$$
$$= 24(12)$$
$$= 288 \text{ cm}^2$$

MATH FACT

When working with prisms be careful not to confuse the height of the polygonal base with the height of the prism. For example, a prism with a 10 cm height may have a triangular base whose height is 4 cm. The 4 cm height of the triangle is used in the area formula $\frac{1}{2}bh$ to calculate the area of the triangular base. The 10 cm height of the prism is used in $V = Bh$ to calculate the volume of the prism.

Circular Cylinders

A circular cylinder is a solid figure with two parallel and congruent circular bases and a curved lateral surface. The height is the perpendicular distance between the bases. As with the prisms, cylinders can be right or oblique. Figure 12.3 shows each type of cylinder.

right circular cylinder   oblique circular cylinder

Figure 12.3 Cylinder

Volume of a Circular Cylinder
Volume (V) $= \pi r^2 h$
where r is the radius of the base and h is the height.

Example 1

A right circular cylinder has a diameter of 10 cm and a height of 5 cm. Find the volume. Express your answers in terms of π.

Solution:

$$\text{Radius} = \frac{1}{2}\text{diameter}$$

$$r = \frac{1}{2}(10) = 5\ \text{cm}$$

$$V = \pi r^2 h$$
$$= \pi(5\ \text{cm})^2 \cdot 5\ \text{cm}$$
$$= 125\pi\ \text{cm}^3$$

We can work backward to find the height or the radius given the volume of a cylinder.

Example 2

A right circular cylinder has a volume of 200 in.^{3} If the height of the cylinder is 6 in., what is the radius of the cylinder? Round your answer to the nearest hundredth.

Solution: Set up the formula for volume. Substitute the known values. Then solve for the radius.

$$V = \pi r^2 h$$
$$200\ \text{in.}^3 = \pi(r^2)(6\ \text{in.})$$
$$r^2 = 10.61032\ \text{in.}^2$$
$$r = \sqrt{10.61032}\ \text{in.}$$
$$= 3.26\ \text{in.}$$

Check Your Understanding of Section 12.1

A. *Multiple-Choice*

1. What is the volume of a prism with a square base if the side length of the base is 8 inches and the height is 5 inches?
   (1) 40 in.^{3} (2) 200 in.^{3} (3) 300 in.^{3} (4) 320 in.^{3}

2. The bases of a prism are rectangles. The rectangle has a perimeter of 24 ft and the ratio of its length to width is 2 : 1. The height of the prism is 2 ft. What is the volume of the prism?
(1) 48 ft³   (2) 54 ft³   (3) 64 ft³   (4) 553 ft³

3. A right circular cylinder has a diameter of 6 in. and a height of 13 in. What is the volume of the cylinder?
(1) 78π in.³   (2) 117π in.³   (3) 468π in.³   (4) 1,014π in.³

4. A right circular cylinder has a volume of 432 cm³. If the height of the cylinder is 4.5 cm, what is the radius of the cylinder?
(1) 5.5 cm   (2) 30.56 cm   (3) 9.24 cm   (4) 1,944 cm

5. A cylinder has a volume of 52π in.³ and a radius of 4 in. What is the height of the cylinder?
(1) 3 in.   (2) $3\frac{1}{4}$ in.   (3) $3\frac{1}{2}$ in.   (4) $3\frac{3}{4}$ in.

6. The volume of the prism shown in the accompanying figure is 12.5 m³. What is the length of the dimension labeled x?
(1) 2.5 m   (3) 4.5 m
(2) 4 m   (4) 5 m

2.5 m

2 m

x

7. A prism has bases in the shape of a regular pentagon. If the volume of the prism is 60 in.³ and its height is 5 in., what is the area of each base?
(1) 6 in.²   (2) 9 in.²   (3) 12 in.²   (4) 24 in.²

8. A 1-meter-long length of pipe has an inner radius of 2.5 cm and an outer radius of 2.8 cm. The pipe is made of a metal with a density of 7.8 grams/cm³. What is the weight of the pipe?
(1) 3.2 kg   (2) 3.4 kg   (3) 3.9 kg   (4) 4.2 kg

9. A prism with a hexagonal base has a height of 10 in. and a volume of 60 in.³ Which of the following is *not necessarily* true about the prism?
(1) the lateral faces are all rectangles
(2) the area of each base is 3 in.²
(3) the perimeter of the hexagon equals the area of the lateral faces divided by 10 in.
(4) the two hexagonal bases are congruent

10. A prism with a square base and a cylinder have the same volume and the same height. What is the ratio of the side length of the square to the radius of the cylinder?

(1) $\sqrt{\pi}$ (2) π (3) π^2 (4) $\dfrac{1}{\sqrt{\pi}}$

B. *Free-Response—show all work or explain your answer completely.*

11. A cylindrical pot is 8 inches in diameter and 7 inches high. Tom wants to pour cans of soup into the pot. The cylindrical soup cans are 5 inches high and 3 inches in diameter. How many cans of soup can Tom pour in if he wants to leave at least 3 inches from the top of the pot?

12. A prism has bases in the shape of regular octagons. If the volume of the prism is 400 ft³ and the height of the prism is 8 ft, what is the area of each base?

13. A cylindrical water tower holds 100,000 gallons of water. A new water tower is being built that is twice the radius and 1.5 times the height. How many gallons of water will the new tower hold?

14. An 8-m diameter round swimming pool has a uniform depth of 1.5 m. The pump for the pool's filter can move water at a rate of 3 m³ per hour. How long will it take to pump a volume equal to the entire pool through the filter? Round to the nearest hour.

12.2 CONES, PYRAMIDS, AND SPHERES

KEY IDEAS

A circular cone is a solid with one circular base and an apex. A pyramid is a solid with one polygonal base and an apex. A sphere is the solid that is the set of all points a fixed distance from a center point.

	Circular Cone	Pyramid	Sphere
Volume (V)	$\dfrac{1}{3}\pi r^2 \cdot h$	$\dfrac{1}{3}B \cdot h$	$\dfrac{4}{3}\pi r^3$

B is the area of the base, h is the height

Circular Cones

A circular cone is a solid with one circular base and a curved lateral surface that comes to a point at the apex. The height of the cone is the length of the segment from the apex perpendicular to the base. In a right circular cone, the height will intersect at the center of the circular base.

Figure 12.4 Right circular cone

Volume of a Circular Cone

Volume $(V) = \dfrac{1}{3}\pi r^2 h$

where r is the radius of the base and h is the height.

These formulas are applied in a manner similar to those for the cylinder. The Pythagorean theorem can be used to relate slant height, height, and radius. So if you know any two, you can find the third.

Example

Jenna is planting a garden and has a pile of topsoil delivered to her house. A dump truck delivers the topsoil and dumps it in a pile that is cone shaped. The radius of the pile is 18 ft and the height is 12 ft. What is the volume of the pile? Round to the nearest tenth.

Solution:

$$V = \frac{1}{3}\pi r^2 h$$
$$= \frac{1}{3}\pi (18 \text{ ft})^2 (12 \text{ ft})$$
$$= 4,071.5 \text{ ft}^3$$

Pyramids

A **pyramid** is a solid with one polygonal base and triangular lateral faces that meet at an apex. The height of the pyramid is the length of the segment from the apex perpendicular to the base. In a right pyramid, the height intersects at the center of the base. Figure 12.5 shows right pyramids with bases in the shape of a triangle and a rectangle.

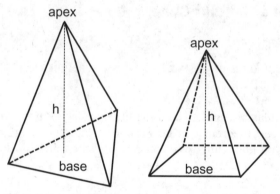

Figure 12.5 Triangular and square-based pyramids with height *h*

Volume of a Pyramid

Volume $(V) = \frac{1}{3}B \cdot h$

where B is the area of the base and h is the height.

Example

Find the volume of a 5 ft high right pyramid with a square base whose side length is 4 ft.

Solution: First find the area of the base, which is a square.

$$B = s^2$$
$$= 4^2$$
$$= 16$$

Now apply the volume formula

$$V = \frac{1}{3} B \cdot h$$

$$= \frac{1}{3} 16 \cdot 5$$

$$= \frac{80}{3} \text{ ft}^3$$

MATH FACT

If the base of a pyramid is a regular polygon, then the lateral faces are all isosceles triangles.

Spheres

A **sphere**, as shown in Figure 12.6, is the solid figure comprised of the set of all points a fixed distance (the radius) from a center point. The only dimension needed to define the volume and area of a sphere is its radius.

Figure 12.6 Sphere with radius *r*

Area and Volume of a Sphere

Volume $(V) = \frac{4}{3} \pi r^3$

Example

(a) Find the volume of a sphere whose diameter is 14 inches. Round to the nearest hundredth.

(b) Find the radius of a sphere whose volume equals 432.5 cm³. Round to the nearest tenth.

Solution:

(a) $V = \dfrac{4}{3}\pi r^3$

$\quad = \dfrac{4}{3}\pi(7)^3$ \qquad Divide the diameter by 2 to get the radius.

$\quad = 1,436.76$ in.³

(b) Working backward, write the volume equation and fill in the known quantities.

$$V = \frac{4}{3}\pi r^3$$

$$432.5 = \frac{4}{3}\pi r^3$$

$$r^3 = 103.2517$$

$$r = \sqrt[3]{103.2517} \qquad \text{Use the cube root function on your calculator.}$$

$$= 4.7 \text{ cm}$$

Check Your Understanding of Section 12.2

A. Multiple-Choice

1. What is the volume of a sphere whose radius is 16 ft?
   (1) 64π ft³   (2) $341\dfrac{1}{3}\pi$ ft³   (3) $4{,}096\pi$ ft³   (4) $5{,}461\dfrac{1}{3}\pi$ ft³

2. What is the radius of a sphere whose volume is 130 in.³?
   (1) 3.14 in.   (2) 5.57 in.   (3) 11.14 in.   (4) 12.56 in.

3. What is the volume of a pyramid with a square base if the perimeter of the base is 60 inches and the height is 20 inches?
(1) 124 in.^{3} (2) 520 in.^{3} (3) 1,500 in.^{3} (4) 129,600 in.^{3}

4. What is the volume of a cone with a diameter of 4 ft and a height of 6 ft?
(1) 8π ft^3 (2) 16π ft^3 (3) 24π ft^3 (4) 32π ft^3

5. What is the radius of a cone whose volume is 455 in.^{3} and height is 12 in.?
(1) 3.16 in.   (2) 6.02 in.   (3) 6.16 in.   (4) 10.4 in.

B. Free-Response—show all work or explain your answer completely.

6. Jack and Jill built a snowman composed of three snowballs. The bottom snowball has a diameter of 75 cm. Each snowball above that is a dilation of the one below with a scale factor of $\frac{1}{2}$. What is the total volume, in cubic meters, of the snowman? Round to the nearest hundredth.

7. The Everswift tennis ball company packages three tennis balls in a cylindrical tube with the balls stacked one above the other. Assuming no gaps at the top, bottom, or sides between the balls and the tube, what fraction of the tube is occupied by the tennis balls?

8. The office of Dr. Barnett has a water cooler for the patients. The cooler is cylindrical with a 16-inch diameter and a 24-inch height. Paper cups are provided that are cone shaped, 4 inches high, and 5 inches in diameter. How many servings of water can one full cooler provide if, on average, people fill their cups 80% full?

9. A cone and a cylinder have the same height. The volumes of the two solids are equal. Find the ratio of the radius of the cone to the radius of the cylinder.

10. Ryan has 30 identical marbles. He wants to measure the diameter but cannot get an accurate measurement with his ruler. He has a 1-liter measuring cup that can hold all the marbles. He fills the measuring cup with the marbles and then fills the cup to the 1-liter line with water. When Ryan removes the marbles, the water comes up to the 750 mL line. What is the diameter of each marble? Round to the nearest 0.1 cm.

11. Paxton works in an ice cream parlor, where she scoops ice cream cones. The ice cream comes in cylindrical containers 12 inches in diameter and 18 inches high. She scoops the ice cream in 3-inch diameter scoops. How many scoops can she get out of one container?

12. The base of a cone is described by the curve $x^2 + y^2 = 16$. If the altitude of the cone intersects the center of its base and the volume of the cone is 72π, what is the height of the cone?

12.3 CROSS SECTIONS AND SOLIDS OF REVOLUTION

KEY IDEAS

The intersection of a plane and a solid is an area called a cross section.

- Cross sections parallel to the base of a prism and cylinder are congruent to the base.
- Cross sections parallel to the base of a cone or pyramid are similar to the base.
- Cross sections of spheres are similar circles.
- Spheres are generated by rotating circles.

Cones, cylinders, and spheres are solids of revolution. This means they can be generated by rotating a particular cross section around a line.

- Cones are generated by rotating triangles.
- Cylinders are generated by rotating rectangles.
- Spheres are generated by rotating circles.

Cross Sections

A **cross section** of a solid is the 2-dimensional figure created when a plane intercepts a solid. Cross sections are often made parallel or perpendicular to the base of a figure. Figure 12.7 shows a cross section of a cone. The base lies in plane P, and plane Q is parallel to plane P. Plane Q intercepts a cross section in the shape of a circle.

Prisms and cylinders, which have two parallel bases, have cross sections that are congruent to the bases when the intercepting plane is parallel to the base. The cylinder in Figure 12.8 has circular cross sections all congruent to the bases. Cones and pyramids have only one base. Their cross sections are all similar to the base when the plane is parallel to the base. The cross sections of the pyramid in Figure 12.8 are all rectangles similar to the base. As the cross sections approach the apex, their size approaches zero until the cross section is just a point. The cross sections of a sphere are all circles, which get smaller as the distance from the center increases.

Figure 12.7 Cross sections of a cone

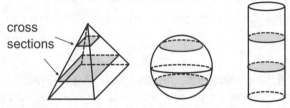

Figure 12.8 Cross sections of a pyramid, sphere, and cylinder

Example 1

A right circular cylinder has a radius of 4 and a height of 10. What is the area of a cross section taken parallel to the bases? Write your answer in terms of π.

   Solution: The cross section is congruent to the base. So it is a circle with radius of 4. The area is πr^2, or 16π.

   Cross sections formed by planes perpendicular to a base may have specific shapes as well. Consider the following cross sections of a cone, pyramid, cylinder, and prism.

Example 2

Plane Q intersects a prism with a square base perpendicular to the base and passes through the apex. What shape represents the intersection of the plane and the pyramid?

   Solution: The cross section perpendicular to the base is a triangle. Since the base is a regular polygon, the triangle must be isosceles.

   Finally, cross sections can be taken at an angle that is neither parallel nor perpendicular to the base of a solid. Some examples of these types of cross sections are shown in Figure 12.9. The circle, ellipse, parabola, and hyperbola are known as the conic sections because they represent the possible cross sections formed by a plane intersecting a cone.

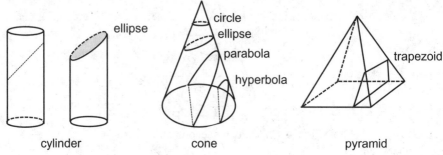

Figure 12.9 Some cross sections of solids

Example

Planes *A* and *B* both intersect a cylinder. If the cross section formed by plane *A* is a rectangle and the cross section formed by plane *B* is an ellipse, which of the following could be true?

(1) plane *A* is parallel to the base of the cylinder, and plane *B* forms a 45° angle with the base

(2) plane *A* is perpendicular to the base of the cylinder, and plane *B* is parallel to the base

(3) plane *A* is perpendicular to the base of the cylinder, and plane *B* forms a 45° angle with the base

(4) plane *A* forms a 45° angle with the base of the cylinder, and plane *B* is perpendicular to the base

Solution: The correct choice is (3). The two cross sections are shown in the accompanying figures.

Solids of Revolution

Certain solids can be generated by rotating a planar figure around a line. We call these *solids of revolution*. Some examples of solids of revolution and the figures that generate them are shown in the following table.

426

Figure Rotated	Solid
Right triangle, rotate 360° about leg $\overline{AB}$	Cone
Rectangle *ABCD*, rotate 360° around side $\overline{CD}$	Cylinder
Circle *P*, rotate 180° about diameter $\overline{AB}$	Sphere

The dimensions of the planar figure that is being rotated will determine the dimensions of the resulting solid. For example, when a cone is generated by rotating a right triangle, the lengths of the two legs will be equal to the radius and height of the cone.

Some planar figures can be rotated about more than one side. The right triangle in the table could have been rotated about leg $\overline{AC}$ to generate a short, wide cone with height *AC* and radius *AB*. Rectangle *ABCD* could have been rotated about $\overline{AD}$ to generate a cylinder with height *AD* and radius *CD*.

Example

Right triangle *JGH* has a right angle at *G* and legs with lengths *GH* = 7 cm and *GJ* = 6 cm. What solid figure is generated when △*GHJ* is rotated 360° about *JG*? What is the volume in terms of π?

Solution: The solid figure is a cone. The radius equals *GH*, and the height equals *GJ*.

Apply the formula for the volume of a cone with the radius = 7 and height = 6.

$$V = \frac{1}{3}\pi r^2 h$$

$$= \frac{1}{3}\pi \cdot 7^2 \cdot 6$$

$$= 98\pi \ \text{cm}^3$$

MATH FACT

The largest cross section of a sphere is called a "great circle." A great circle is also the circle with the largest circumference that can be traced on the sphere. The center of a great circle is always the center of the sphere. On a flat surface, the shortest distance between two points is a straight line. The corresponding theorem on the curved surface of a sphere is that the shortest distance between two points is along the great circle that contains those points.

Check Your Understanding of Section 12.3

A. Multiple-Choice

1. The bases of a prism are right triangles. Which of the following describes the cross sections of the prism taken parallel to the base?
   (1) right triangles
   (2) isosceles triangles
   (3) rectangles
   (4) trapezoids

2. A cross section of a cylinder is in the shape of a square. What must be true about the cylinder?
(1) the cylinder is oblique
(2) the height of the cylinder is twice the length of its radius
(3) the cross section was formed by the intersection of the cylinder and a plane parallel to one of its bases
(4) the area of the square is equal to the area of one of the bases of the cylinder

3. Which of the following statements is *not* true about the pentagonal prism shown in the accompanying figure?
(1) all cross sections taken parallel to the bases are congruent
(2) all cross sections taken perpendicular to the bases are congruent
(3) all cross sections taken parallel to the bases are pentagons
(4) all cross sections perpendicular to the bases are rectangles

4. Sarah is designing a plumbing system for a new building using a computer-aided design program. She wants to create a model of a hollow pipe with the program by sketching a cross section and then rotating it 360° around a line. Which of the following cross sections and lines would result in a hollow cylindrical pipe?

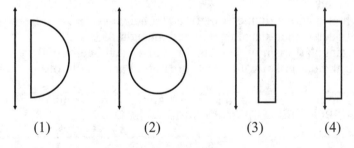

(1)   (2)   (3)   (4)

5. Which of the following is *not* true about the cross sections of a sphere?
(1) all the cross sections have the same radius
(2) all the cross sections are circles
(3) all the cross sections are similar
(4) the cross sections with the largest diameter pass through the center of the sphere

6. A plane intercepts a cube. Which of the following is true?
   (1) the cross section could be a circle
   (2) the cross section could be a rectangle
   (3) the cross section is always a square
   (4) the cross section never has an area larger than one of the faces of the cube

7. Three planes intercept a solid, forming cross sections in the shape of a rectangle, a circle, and an ellipse. Which of the following could represent the solid?
   (1) sphere
   (2) cone
   (3) cylinder
   (4) pyramid

8. $\triangle JKL$ is a right triangle with a right angle at K, $JK = 6$, and $KL = 10$. Which of the following solids is generated when $\triangle JKL$ is rotated $360°$ about KL?
   (1) a cylinder with a height of 10 and a diameter of 12
   (2) a cylinder with a height of 6 and a diameter of 20
   (3) a right circular cone with a height of 10 and diameter of 12
   (4) a right circular cone with a height of 6 and diameter of 20

9. A quadrilateral with coordinates $A(0, 3)$, $B(5, 3)$, $C(5, 0)$, and $D(0, 0)$ is rotated $360°$ about the x-axis. Which of the following describes the solid that is formed?
   (1) a cylinder with a diameter of 10 and a height of 3
   (2) a cylinder with a diameter of 6 and a height of 5
   (3) a rectangular prism with a base area of 15 and height of 3
   (4) a rectangular prism with a base area of 30 and height of 6

10. In which of the following solids are all cross sections similar?
    (1) cube
    (2) sphere
    (2) cylinder
    (4) pyramid with a triangular base

11. A sphere has a radius of r. What is the area of a cross section taken at a distance of $\frac{1}{2}r$ from the center of the sphere?

    (1) $\frac{1}{4}r^2$ (2) $\frac{1}{2}r^2$ (3) $\frac{1}{2}\pi r^2$ (4) $\frac{3}{4}\pi r^2$

B. *Free-Response—show all work or explain your answer completely.*

12. A triangle with vertices $A(0, 0)$, $B(4, 0)$, and $C(0, 6)$ is rotated about the y-axis. Completely describe the solid formed. Find its volume in terms of π.

13. Rectangle $MNOP$ has lengths $MN = 10$ and $NO = 6$. $MNOP$ is rotated $360°$ about $\overline{OP}$. What solid figure is generated, and what is its volume? Round to the nearest hundredth.

14. A semicircle with a radius of 12 cm is rotated $180°$ about the line that contains the radius. What is the volume of the solid that is generated? Express your answer in terms of π.

15. An I-beam is a solid piece of steel with a cross section in the shape of an "I." What is the volume of a 12-foot-long I-beam with the cross section shown in the accompanying figure? Express your answer in ft^3 and round to the nearest 0.1 ft^3.

12.4 PROVING VOLUME AND AREA BY DISSECTION, LIMITS, AND CAVALIERI'S PRINCIPLE

KEY IDEAS

The formulas for circumference may depend on some parameter, such as the number of sides and area of a circle. The formulas for the volume of a cylinder, cone, pyramid, and sphere can all be derived using dissection, limits, and Cavalieri's principle and the set of properties.

Dissection—Any region or solid can be divided into smaller, nonoverlapping regions or solids such that the sum of the smaller areas or volumes equals the whole area or volume.

Limit—A figure's properties may depend on some parameter, such as the number of sides. A limit in geometry describes the set of properties that are approached as the parameter approaches some new value. We sometimes let the parameter approach **infinity**. For example, as the number of sides in a polygon approaches infinity, the shape of the polygon approaches that of a circle.

Cavalieri's principle—If two solids are bounded by the same two parallel planes *and* if any plane parallel to those two planes intercepts regions of equal area, then the two solids have the same volume. Alternatively, if two parallel planes intercept solids of equal volume, the original solids have equal volumes.

Dissection

The proofs of the area and volume formulas we will look at in this section rely on the mathematical concepts that we will define here—dissection, limits, and Cavalieri's principle.

The term "dissection" describes dividing a figure or solid into multiple parts. It is really just another way of saying "the whole equals the sum of the parts." Figure 12.10a shows the dissection of a regular hexagon into 6 equilateral triangles. The area of the hexagon is equal to the sum of the areas of the 6 triangles. This type of dissection was the basis for deriving the formula for the area of a regular hexagon, $A = \frac{1}{2} \cdot \text{perimeter} \cdot \text{apothem}$. Figure 12.10b shows the dissection of a right circular cylinder into three smaller cylinders. The volume of the large cylinder is equal to the volume of the three smaller cylinders.

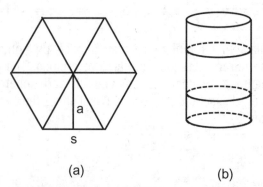

(a)   (b)

Figure 12.10 (a) Dissection of a hexagon into 6 triangles
(b) Dissection of a cylinder into smaller cylinders

432

Limits

An algebraic limit is the value that an expression approaches as a variable gets closer and closer to some specified value. A simple example is the limit of the value of $\left(4+\dfrac{1}{x}\right)$ as x approaches infinity. We can make a table showing values of the expression as a function of x.

x	$4+\dfrac{1}{x}$
1	5
2	4.5
10	4.1
100	4.01
1,000	4.001

You can see from the pattern that the limit of $\left(4+\dfrac{1}{x}\right)$ approaches 4 as x approaches infinity. Of course, the value of x cannot actually equal infinity since infinity is not a real number.

We can consider limits in the context of geometry. Figure 12.11 shows $\triangle ABC$ for three values of the measure of angle A. As $m\angle A$ approaches $90°$, $\triangle ABC$ approaches a right triangle. In this case, we can actually reach the limit and $\triangle ABC$ will become a right triangle. Figure 12.12 shows a geometric example in which we do not actually reach the limit. Consider $\overleftrightarrow{AB}$ and $\overleftrightarrow{CD}$, which are not parallel. Now translate point D in the direction of vector $\overrightarrow{AB}$. The slope of $\overleftrightarrow{CD}$ will begin to approach the slope of $\overleftrightarrow{AB}$. The farther we translate D in the direction of $\overrightarrow{AB}$, the closer the two slopes will be to each other. The limit of the slope of $\overleftrightarrow{CD}$ as the translation distance approaches infinity is the slope of $\overleftrightarrow{AB}$.

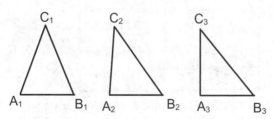

Figure 12.11 $\triangle ABC$ as $m\angle A$ approaches $90°$

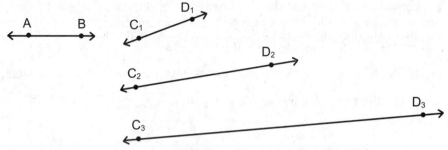

Figure 12.12 The slope of $\overline{CD}$ approaches the slope of $\overline{AB}$ as D translates away from C in the direction of $\overline{AB}$

The final concept we will use in proving area and volume formulas is Cavalieri's principle.

Cavalieri's Principle

If two solids are contained between two parallel planes and if every parallel plane between these two planes intercepts regions of equal area, then the solids have equal volume. Also, any two parallel planes intercept two solids of equal volume.

Both the triangular prism and the rectangular prism shown in Figure 12.13 are bounded by planes R and T. "Bounded" means that the solids have at least one point contained in each of planes R and T and that the solids do not pass through to the other sides of the planes. As plane S sweeps upward from R toward T, corresponding cross sections are formed in the triangular prism and the rectangular prism. If every pair of corresponding cross sections are equal in area (though not necessarily congruent), then the two solids have the same volume.

Figure 12.13 Cavalieri's principle applied to a prism with a triangular base and to a prism with a rectangular base

The second part of Cavalieri's principle is illustrated in Figure 12.14. If the conditions required for Cavalieri's principle are true, then any pair of parallel planes R and S will intercept regions of equal volume.

Figure 12.14 Equal volumes contained between two parallel planes

---

MATH FACT

The part of the cone and the part of the pyramid intercepted by planes R and S in Figure 12.14 are called a truncated cone and truncated pyramid, respectively. Another term used to describe these shapes is "frustum."

We can use Cavalieri's principle to infer information about cross sections of solids with equal volume or about the volume of solids with corresponding cross sections of the same area.

Example 1

Two stacks of quarters are shown in the accompanying figure. If each stack contains 7 identical quarters, what must be true about the volume of the two stacks? Explain your answer in terms of Cavalieri's principle.

Solution: Each of the quarters has a uniform and congruent cross section, and any plane parallel to the bases of the two stacks must intercept cross sections of equal area. Cavalieri's principle states that the volumes of the stacks must be equal.

Example 2

A right circle cylinder has a radius of 3 in. and a height of 4 in. A prism has bases in the shape of a square with side lengths $3\sqrt{\pi}$ in. and a height of 4 in. In terms of Cavalieri's principle, explain why the two solids have the same volume.

Solution: Every cross section of a cylinder parallel to its base is congruent, and the base of the given cylinder has an area of $A = \pi r^2 = 9\pi$ in.[2] Every cross section of a prism parallel to its base is congruent, and its base has an area of $A = s^2 = 9\pi$ in.[2] Since the solids have the same height, they are contained by the same two parallel planes. Every cross section of the two solids are equal in area. Therefore, Cavalieri's principle states the two solids have the same volume.

Circumference and Area of a Circle

Circumference of a Circle

The formula for the area of a circle can be derived using a limit argument by considering a polygon inscribed in a circle and letting the number of sides increase towards infinity.

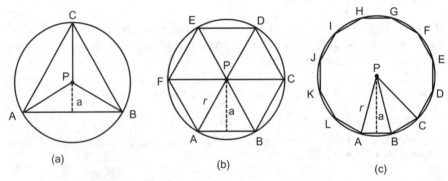

(a)

(b)

(c)

Figure 12.15 Proving the area of a circle using inscribed polygons

Consider the three polygons inscribed in circle P, as shown in Figure 12.15. The perimeter of the polygon will depend on both the radius of the circle and the number of sides of the polygon. We can write the equation

$$\text{perimeter of polygon with } N \text{ sides} = k_N \cdot r$$

where r is the radius of the circle and k_N is a number that depends on N. As N gets larger, the perimeter of the polygon approaches the circumference of the circle. The value of k_N also gets larger, but will never exceed a value of

436

2π. This fact can be shown by using trigonometry to write an expression for k_N in terms of N, but it is outside the scope of this course.

We can say that in the limit as the number of sides of an inscribed polygon approaches infinity, the figure will approach a circle and its perimeter will approach $2\pi r$. In other words, the circumference of a circle equals $2\pi r$.

Area of a Circle

The area of a circle can also be derived from a limit argument and inscribed polygons, as shown in Figure 12.15. The formula for the area of a polygon is $A = \dfrac{1}{2}$ perimeter $\cdot$ apothem. The apothem is the length a shown in the figure. As the number of sides increases towards infinity, the perimeter approaches the circumference of the circle and the apothem approaches the radius of the circle.

$$A = \frac{1}{2}\text{perimeter}\cdot\text{apothem} \to \frac{1}{2}(2\pi r)r$$

$$A_{\text{circle}} = \frac{1}{2}(2\pi r)r = \pi r^2$$

Volume of a Cylinder

The formula for the volume of a cylinder can be derived starting with the formula for a prism. Figure 12.16a shows a prism with a triangular base inscribed inside a right circular cylinder of the same height. The volume of the cylinder is larger than the volume of the prism.

Figure 12.16b shows the same cylinder with a 6-sided prism inscribed inside it. The volume of the cylinder is still larger but by not as much. As the number of sides of the prism approaches infinity, the area of the base of the prism approaches the area of the circular base of the cylinder. Therefore the prism and cylinder must have the volume formula:

$\text{Volume}_{\text{cylinder}} = \text{Volume}_{\text{prism}} = B \cdot h$, where B is the area of the base

The formula for the area of a circle can be substituted for B:

$$V_{\text{cylinder}} = \pi r^2 h$$

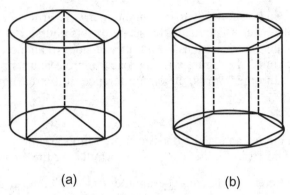

Figure 12.16 Prisms with a triangular base and with a hexagonal base inscribed inside a right circular cylinder

Volume of a Pyramid

The volume of a pyramid can be derived by the dissection of a cube with side length s as shown in Figure 12.17. Let point P be the center of the cube. The four bottom vertices and point P form a pyramid with a square base as shown in Figure 12.17a. Since P is the center of the cube, a congruent pyramid is formed by the top four vertices and point P as shown in Figure 12.17b. Four more congruent pyramids can be made using point P and four vertices at the front, back, left, and right. None of the pyramids overlap. So the volume of the cube must be equal to the volume of the 6 pyramids:

$$6 \, \text{Volume}_{\text{pyramid}} = \text{Volume}_{\text{cube}}$$
$$6 \, \text{Volume}_{\text{pyramid}} = s^3$$

$$\text{Volume}_{\text{pyramid}} = \frac{1}{6}s^3$$

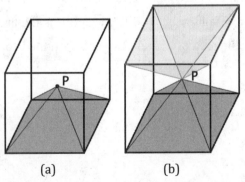

Figure 12.17 Dissection of a cube into 6 pyramids

The height of each pyramid, shown in Figure 12.17b, is equal to $\frac{1}{2}s$, or $s = 2h$. We can separate the s^3 term into $s^2 \cdot s$ and then substitute $s = 2h$ to get the final formula:

$$\text{Volume}_{\text{pyramid}} = \frac{1}{6}s^3$$

$$= \frac{1}{6}s^2 \cdot s$$

$$= \frac{1}{6}\left(s^2\right)2h$$

$$= \frac{1}{3}\left(s^2\right)h$$

The area of the base of each pyramid is the same as the area of the base of the cube, or s^2.

$$\text{Volume}_{\text{pyramid}} = \frac{1}{3}Bh$$

Volume of a Cone

The same approach that was used for the volume of a cylinder can be applied to derive the formula for the volume of a cone. Starting with a pyramid whose volume is $\frac{1}{3}Bh$, let the number of sides in the polygonal base approach infinity. The shape of the base will approach a circle, but the apex will be unchanged. So the formula remains the same:

$$V_{\text{cone}} = \frac{1}{3}Bh$$

Check Your Understanding of Section 12.4

A. Multiple-Choice

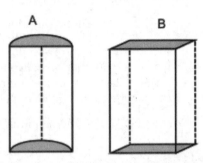

1. Solids *A* and *B*, shown in the accompanying figure, each have uniform cross sections perpendicular to the shaded bases.

   If the heights of the two solids are equal, which of the following statements represents an application of Cavalieri's principal?
   - (1) the surface areas of solid *A* and solid *B* may be different even though their volumes are equal
   - (2) solid *A* can be generated by rotating a rectangle, while solid *B* is not a solid of revolution
   - (3) if the area of base *A* equals the area of base *B*, the volumes of the solids are equal
   - (4) if the volume and height of solids *A* and *B* are equal, their surface areas must be equal

2. Two right circular cones each have a height of 4 in., a volume of 30 in.^{3}, and coplanar bases. Which of the following is *not* an application of Cavalieri's principle?
   - (1) the bases of each cone are equal in area
   - (2) the two cones have the same surface area
   - (3) A cross section parallel to and $\frac{3}{4}$ in. above the base of one cone must be equal in area to a cross section $\frac{3}{4}$ in. above and parallel to the base of the other cone
   - (4) plane *A* is parallel to and 1 in. above the base of each cone; plane *B* is parallel to and $\frac{3}{4}$ in. above plane *A*. The two solids intercepted by planes *A* and *B* are equal in volume.

3. Which of the following could be used to explain why the formula for the volume of a pyramid is $\frac{1}{3}Bh$?

   (1) as the area of the base approaches zero, the slant height of a pyramid approaches its height

   (2) a point $\frac{1}{3}$ of the way along its altitude, upward from the base, divides a pyramid into two pyramids of equal volume

   (3) as the number of sides in the base of a pyramid approaches infinity, the base approaches a circle

   (4) a cube can be divided into 6 congruent pyramids, each with a height equal to $\frac{1}{2}$ the height of the cube

4. Polygon A is inscribed in circle P, and polygon B is circumscribed about circle P. Which of the following statements is *not* true?

   (1) perimeter$_{\text{polygon } A}$ > circumference$_{\text{circle } P}$ > perimeter$_{\text{polygon } B}$

   (2) perimeter$_{\text{polygon } B}$ > circumference$_{\text{circle } P}$ > perimeter$_{\text{polygon } A}$

   (3) as the number of sides of polygon A approaches infinity, the perimeter of polygon A approaches the circumference of circle P

   (4) as the number of sides of polygon B approaches infinity, the perimeter of polygon B approaches the circumference of circle P

5. Paul wants to estimate the value of π using a hexagon inscribed inside a circle. He knows the radius of the circle is 10 inches. Which of the following would give an estimate as to the value of π?

   (1) the area of the hexagon divided by twice the radius of the circle

   (2) the area of the hexagon divided by the radius of the circle

   (3) the perimeter of the hexagon divided by twice the radius of the circle

   (4) the perimeter of the hexagon divided by the radius of the circle

6. Which of the following could be used to explain why the formulas for the volume of a cylinder and the volume of a prism are both equal to the area of the base multiplied by the height?

   (1) exactly one cylinder can fit inside any prism of the same height

   (2) both cylinders and prisms have two bases

   (3) given a prism inscribed in a cylinder of the same height, as the number of sides of the prism's bases increase to infinity, the prism's base approaches circles congruent to the cylinder's bases

   (4) given a prism inscribed in a cylinder of the same height, any plane parallel to the base of each solid intercepts corresponding congruent figures

7. In the limit that the number of sides in the base of a pyramid approaches infinity, the shape of the pyramid will approach that of a
   (1) cylinder
   (2) cone
   (3) sphere
   (4) cube

8. Two cones share bases with a right circular cylinder and have common apex P as shown in the accompanying diagram. The cylinder has height h and radius r. What is the volume of the region within the cylinder and outside the two cones?

   (1) $\frac{1}{3}\pi r^2 h$ (3) $\frac{1}{2}\pi r^2 h$

   (2) $\frac{2}{3}\pi r^2 h$ (4) $\frac{3}{4}\pi r^2 h$

9. A cylinder and a prism have the same height and the same volume. Their bases lie in the same planes. Which of the following is *not necessarily* true?
   (1) the area of their bases are equal
   (2) a plane parallel to the bases of both the cylinder and prism intercepts two corresponding figures with equal areas
   (3) two planes parallel to the bases of each solid intercept corresponding solids with equal volumes
   (4) the cylinder and prism are congruent

10. Caitlyn is estimating the value of π by comparing the area of a circle to the area of a polygon inscribed inside a circle. Which of the following would lead to a more accurate estimate?
    (1) increasing the number of sides of the polygon
    (2) decreasing the number of sides of the polygon
    (3) increasing the radius of the circle
    (4) decreasing the radius of the circle

B. Free-Response—show all work or explain your answer completely.

11. A stack of identical 10 playing cards is arranged as shown in the figure on the left. Anne picks up the cards and straightens the stack as shown in the figure on the right. Explain why the two stacks have the same volume.

12. A right rectangular prism and an oblique rectangular prism each have bases that are 5 cm by 2 cm and heights of 10 cm. Explain, in terms of Cavalieri's principle, why do the two solids have the same volume?

13. Gordon knows the formula for the circumference of a circle is equal to $2\pi r$, where r is the radius of the circle. He also knows the formula for the area of a polygon is $A = \frac{1}{2}$ perimeter $\cdot$ apothem. Explain how Gordon can use this information to derive the formula for the area of the circle. Refer to the accompanying figure.

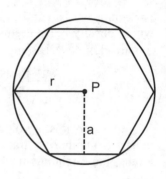

12.5 MODELING AND DESIGN

KEY IDEAS

Physical objects can be modeled using the basic solids—prisms, cylinders, cones, pyramids, and spheres. Volume and area consider-ations can be used to design physical objects or to meet a condition. Dimensions can be found to maximize or minimize area, volume, cost, and so on. Density can be used to find mass, population, or any other quantity specified on a per-area or per-volume basis. Numerical and graphical methods may be needed to find solutions to problems that do not result in equations that are easily solved.

Modeling with Volume

Many physical objects can be modeled using the basic solids discussed in this chapter. Some examples include:

Cylinders—trees, cans, barrels, people

Spheres—balls, balloons, planets

Prisms—bricks, boxes, aquariums, swimming pools, rooms, books

The dimensions needed for a solid figure can be found in the text of the problem. An added layer to these problems is that volumes or areas are used to calculate some other quantity. Some common relationships to look for include:

- Mass = density × volume
- Total cost = cost per unit volume × volume
- Energy contained in a material = heat content × volume

For example, if a can has a volume of 100 cm^3 and it is filled with a liquid whose density is 1.2 grams/cm^3, the mass of the liquid inside the can can be found:

$$\text{mass} = \text{density} \times \text{volume}$$
$$\text{mass} = 100 \text{ cm}^3 \cdot 1.2 \text{ grams/cm}^3$$
$$= 120 \text{ grams}$$

In some problems, a volume of material may be transferred from one container to another or transformed into another shape. In these cases, the total volume remains unchanged and the volumes before and after must be equal.

Example 1

A cylindrical piece of metal has a radius of 15 in. and a height of 2 in. It goes through a hot-press machine that reduces the height of the cylinder to $\frac{1}{4}$ in. What is the new radius of the cylinder, assuming no material is lost? Round to the nearest thousandth.

Solution: Strategy—Find the volume of the cylinder before pressing, and set it equal to the expression for the volume after pressing. Use this equation to solve for the radius after pressing.

$$V = \pi r^2 h$$
$$V_{\text{before}} = \pi \cdot 15^2 \cdot 2$$
$$= 450\pi \text{ in.}^3$$
$$V_{\text{after}} = \pi \cdot r^2 \cdot \frac{1}{4} \qquad \text{we don't know the new radius}$$
$$V_{\text{before}} = V_{\text{after}}$$

$$\pi \cdot r^2 \cdot \frac{1}{4} = 450\pi \qquad \text{volumes must be equal}$$

$$\frac{r^2}{4} = 450$$

$$r^2 = 1{,}800$$

$$r = \sqrt{1{,}800}$$

$$= 42.426$$

Example 2

An engineer for a construction company needs to calculate the total volume of a home that is under construction so the proper-sized heating and air conditioning equipment can be installed. The house is modeled as a rectangular prism with a triangular prism for the roof as shown in the accomanying figure. The roof is symmetric with $EF = ED$. The dimensions are $AB = 30$ ft, $BC = 50$ ft, and $BD = 25$ ft, and the measure of $\angle EDF = 40°$. What is the total volume of the house? Round to the nearest cubic foot.

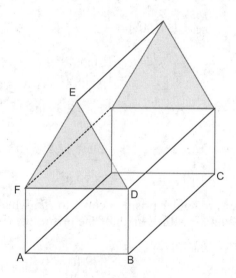

Solution: Calculate the volume of each section.

Rectangular prism:

length = BC width = AB height = BD
length = 50 ft   width = 30 ft   height = 25 ft

$$\text{Volume}_{\text{rectangular prism}} = \text{length} \cdot \text{width} \cdot \text{height}$$
$$= (50\ \text{ft})\,(30\ \text{ft})\,(25\ \text{ft})$$
$$= 37{,}500\ \text{ft}^3$$

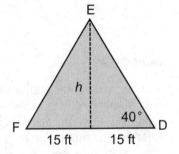

Triangular prism:
The first step is to find the area of the triangular base, $\triangle FED$. FD has a length equal to AB, or 30 ft. The height of the isosceles triangle divides the base into two 15 ft lengths. The tangent ratio can be used to find the height of $\triangle FED$.

$$\tan = \frac{\text{opposite}}{\text{adjacent}}$$

$$\tan(40°) = \frac{h}{15}$$

$$h = 15\tan(40°)$$
$$= 12.586\ \text{ft}$$

$$\text{Area}_{\text{triangle}} = \frac{1}{2}\,\text{base} \cdot \text{height}$$

$$= \frac{1}{2}\,30\ \text{ft} \cdot 12.586\ \text{ft}$$

$$= 188.79\ \text{ft}^2$$

The volume of a triangular prism is equal to Bh, where B is the area of the triangular base and h is the length of the prism, which is equal to BC.

$$V_{\text{triangular prism}} = Bh$$
$$= 188.79\ \text{ft}^2 \cdot 50\ \text{ft}$$
$$= 9{,}439.5\ \text{ft}^3$$

Add the two volumes to get the total volume of the house.

$$\text{Volume} = 37{,}500\ \text{ft}^3 + 9{,}439.5\ \text{ft}^3$$
$$= 46{,}939.5\ \text{ft}^3$$
$$= 46{,}940\ \text{ft}^3$$

Example 3

A monument consists of a pyramid with a square base on top of a prism with a square base. A central cylindrical shaft in the center houses a spiral staircase. Point *H* indicates the center of the circular base of the shaft. The dimensions of the monument are $AB = 24$ ft, $BE = 84$ ft, $GE = 30$ ft, and m$\angle GEH = 60°$. The central shaft has a diameter of 6 ft. The monument is solid stone except for the shaft. What was the weight of the stone used to construct the monument, to the nearest ton? The stone has a density of 120 lb/ft^3.

Solution: The volume of the monument is the sum of the volumes of the prism and pyramid minus the volume of the shaft. Find each volume separately.

Prism:

length = *AB*	width = *BC*	height = *BE*
length = 24 ft	width = 24 ft	height = 84 ft

$$\text{Volume}_{\text{square prism}} = \text{length} \cdot \text{width} \cdot \text{height}$$
$$= 24 \text{ ft} \cdot 24 \text{ ft} \cdot 84 \text{ ft}$$
$$= 48,384 \text{ ft}^3$$

Pyramid:
First calculate the height of the pyramid, *GH*, using $\triangle GHE$ and the sine ratio.

$$\sin(60°) = \frac{GH}{GE}$$

$$\sin(60°) = \frac{GH}{30}$$

$$GH = \sin(60°) \cdot 30 \text{ ft}$$
$$= 25.980 \text{ ft}$$

The base of the pyramid is a square with side length 24 ft.

$$B = s^2$$
$$= (24\ \text{ft})^2$$
$$= 576\ \text{ft}^2$$

$$\text{Volume}_{\text{pyramid}} = \frac{1}{3}Bh$$

$$= \frac{1}{3}(576\ \text{ft}^2)(25.980\ \text{ft})$$

$$= 4,988.16\ \text{ft}^3$$

Cylinder:
The cylindrical shaft has a diameter of 6 ft, a radius of 3 ft, and a height of 84 ft.

$$\text{Volume}_{\text{cylinder}} = \pi r^2 h$$
$$= \pi\ (3\ \text{ft})^2\ (84\ \text{ft})$$
$$= 2,375.04\ \text{ft}^3$$

$$\text{Volume of monument} = \text{Volume}_{\text{prism}} + \text{Volume}_{\text{pyramid}} - \text{Volume}_{\text{cylinder}}$$
$$= 48,384\ \text{ft}^3 + 4,988.16\ \text{ft}^3 - 2,375.04\ \text{ft}^3$$
$$= 50,997.12\ \text{ft}^3$$

Calculate the weight of the stone

$$\text{weight} = \text{volume} \cdot \text{density}$$

$$= (50,997.12\ \text{ft}^3)\left(120\frac{\text{lb}}{\text{ft}^3}\right)$$

$$= 6,119,654.4\ \text{lb}$$

Finally, convert the weight to tons using the conversion factor found on the reference sheet.

$$1\ \text{ton} = 2,000\ \text{lb}$$

$$\text{weight} = 6,119,654.4\ \text{lb} \cdot \frac{1\ \text{ton}}{2,000\ \text{lb}}$$

$$= 3,059.8\ \text{tons}$$
$$= 3,060\ \text{tons}$$

The stone has a weight of 3,060 tons.

Density and Area

Density is usually defined as the amount of some quantity per unit volume. However, it can also be calculated on a per unit area basis. Some example of densities per unit area that appear in real-world applications include the following:

- *Population density*—The number of people living in a region divided by the area of the region is the population density. High population density is typical of an urban area, while low population density may be found in a rural area.
- *Power density*—Solar power panels produce power that is proportional to their area. The power produced per unit area is an important measure of the quality and efficiency of a solar panel.
- *Computer applications*—Storage devices such as hard drives may be described in terms of number of gigabytes of data per unit area.

Example

The city of Deerfield sits along the bank of the Fox Run River. A map of the city is shown. The downtown region, indicated by region A, is $\frac{1}{2}$ mile wide and 1 mile long. The population density of Deerfield, except for the city center, is 800 people per square mile. The population density of the downtown region is 4,000 people per square mile. What is the population of Deerfield?

Fox Run River

Deersfield

A

4 miles

2 miles

Solution: We can model the town as a semicircle and the downtown region as a rectangle. To calculate the population, we first find the area of each region. Let the area of the downtown region be represented by Area$_A$ and the area of the remainder of the town be represented by Area$_B$.

$$\text{Area}_A = \text{length} \cdot \text{width}$$
$$= \frac{1}{2} \cdot 1$$
$$= 0.5 \text{ mi}^2$$

The area of the remainder of the town is a semicircle minus the area of the downtown region.

$$\text{Area}_B = \frac{1}{2}\pi r^2 - 0.5 \qquad \text{area of semicircle circle} - \text{area of rectangle.}$$

$$= \frac{1}{2}\pi(2)^2 - 0.5$$

$$= 5.78318 \text{ mi}^2$$

The population of each region is the product of the area and the population density.

$$\begin{aligned}\text{Population} &= \text{Area}_A \cdot 800 \text{ people/mi}^2 + \text{Area}_B \cdot 4{,}000 \text{ people/mi}^2 \\ &= 5.78318 \text{ mi}^2 \cdot 800 \text{ people/mi}^2 + 0.5 \text{ mi}^2 \cdot 4{,}000 \text{ people/mi}^2 \\ &= 6{,}626 \text{ people}\end{aligned}$$

The population of Deerfield is 6,626 people.

Check Your Understanding of Section 12.5

A. Multiple-Choice

1. An orange juice processing plant fills 375 mL bottles with orange juice. The average orange has a diameter of 2.5 in. and is 10% juice by volume. How many bottles are needed to process 15,000 oranges?
   (1) 3,875   (2) 4,290   (3) 536   (4) 687

2. Jack is making a scale drawing of the floor plan of his home. The scale factor is 1 in. : 4 ft. The drawing of his living room is a rectangle measuring 5 inches by 3 inches. He is planning to purchase new carpet for the living room that costs $4 per square foot. How much will the carpet cost?
   (1) $720   (2) $840   (3) $960   (4) $1,040

3. A construction worker is building a brick wall using bricks that are 8 in. × 4 in. × 2 in. Each brick weighs 4 lb. How many pounds of bricks are needed to build a wall 1 ft wide, 4 ft high, and 8 ft long?
   (1) 2,834   (2) 3,024   (3) 3,200   (4) 3,456

B. *Free-Response—show all work or explain your answer completely.*

4. A farmer has a roll of 600 ft of wire fence. He wants to fence in a rectangular area for his sheep to graze. One side of the rectangle is against a river and doesn't need to be fenced. However, the other three sides need to be fenced. What is the area in square feet of the largest rectangular region that can be enclosed?

5. Carrie is designing a new patio for her backyard. The patio is to be 24 ft by 18 ft, which includes a 4 in. border made using 8 in. by 4 in. red bricks. How many red bricks will she need for the border if no bricks are to be cut?

6. Patrick uses heating oil to heat his house in the winter. He has an oil storage tank in his basement that is 2 m long and 1 m in diameter. The energy content of heating oil is 0.034 million BTU/liter, and he is charged $1,300 to fill the tank completely. Patrick is considering switching to natural gas, which costs $10 per million BTU. Which fuel would be less expensive to use? Justify your answer.

7. The Metro Zoo is designing a new enclosure for their 6 polar bears. The diameter of the island is determined by the requirement that the population density of polar bears on the island be equal to 0.004 polar bears per square foot. What will be the volume of water in the moat if the moat must be 6 feet deep?

5 ft

8. Kevin wants to install solar panels capable of providing 3,000 watts of electrical power to his home. The supplier he selected sells panels that are rectangular with a length of 60 in. and a width of 40 in. The panels cost $275 each. Kevin has determined that these solar panels installed on his roof could produce electrical power with an energy density of 14 watts/ft^2. How much money will Kevin have to spend on solar panels?

9. Trayvon has two beakers to use in his chemistry lab. Beaker A has a diameter of 2 cm and a height of 12 cm. Beaker B has a diameter of 3 cm and a height of 9 cm.
   (a) Which beaker has the greater volume?
   (b) The lab instructions say to add 50 grams of a liquid into one of the beakers. If the liquid has a density of 0.89 g/cm^3, will it fit into one of his beakers? Explain your answer.

451

10. An Egyptian pyramid has a square base 220 ft on each side and a height of 80 ft. The average size of the stone blocks used in the construction is 50 in. × 50 in. × 30 in. and have a density of 228 lb/ft³. If the pyramid is 90% solid, how many blocks were used and how many tons did they weigh? Round to the nearest ton.

11. Jessie and Jamie go to the movies and buy a large tub of popcorn. The popcorn container is cylindrical with a diameter of 9 inches and a height of 9 inches. Popcorn has a density of 34 ounces per cubic foot. If the tub of popcorn costs $6, what is the price per ounce? Round to the nearest $0.01/ounce.

12. Amanda purchases a fish tank that is in the shape of a hexagonal prism. The base, shown in the accompanying figure, has an apothem of 30 cm and a side length of 52 cm. The height of the tank is 60 cm. The aquarium shop owner tells her that a good rule for determining the maximum number of fish that can be put into a tank is 6 cm of fish per 10 liters (1 liter = 1,000 cm³).

30 cm

52 cm

   If Amanda wants to purchase clownfish, angelfish, and butterfly fish in a 4 : 2 : 1 ratio, what is the total number of fish she can buy?

Clownfish	8 cm
Angelfish	16 cm
Butterfly fish	20 cm

13. A winery ages its wine in barrels that have a diameter of 25 in. and a height of 35 in. There are currently 125 full barrels of wine in the cellar of the winery. If the winery can sell a 750 mL bottle of wine for $14, what is the value of wine in the cellar? Round to the nearest dollar.

14. The towns of Marionville and Springfield are shown on the map. Marionville has a population of 8,100 people. If Springfield has the same population density as Marionville, estimate the total population of Springfield, rounding to the nearest hundred.

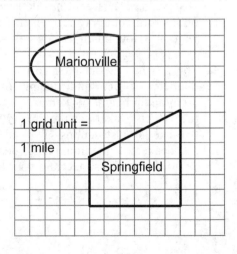

Marionville

1 grid unit =
1 mile

Springfield

15. A hot air balloon can lift a weight that is equal to the difference between the weight of the hot air inside the balloon and the weight of an equal volume of air at the outside temperature. If the air inside the balloon is heated to 200°F and the outside air is at a temperature of 60°F, what must be the diameter of the balloon in order to lift a total of 550 lb? Round to the nearest 0.1 ft, and use the density of air given in the table:

Temperature	Density
60°F	0.060 lb/ft^3
200°F	0.076 lb/ft^3

16. A cylindrical piece of aluminum has a diameter of 30 cm and a length of 75 cm. The cylinder will be rolled into aluminum foil, and the foil will be packaged in 1-pound boxes to be sold in a grocery store. The density of aluminum is 2.7 g/cm^3. How many boxes of aluminum foil can be produced from the piece of aluminum?

17. A construction company is working on plans that call for digging a straight 0.25-mile tunnel through a mountain. The cross section of the tunnel is shown in the accompanying figure.

10ft
20ft

As the workers dig through the mountain, the rubble is brought to a gravel company to be processed into gravel. The cargo beds of the dump trucks are 18 ft long by 10 ft wide by 6 ft high, and the trucking company charges $450 per load delivered to the gravel company. Approximately how much money should the engineer budget into the project for trucking costs?

18. The Red Ribbon Orchard sorts the apples it harvests by size. The three size categories are shown in the accompanying table.

	Grade B	Grade A	Grade AA
Average diameter	4 inches	5 inches	6 inches

After one day's harvest, there were 3,000 pounds of grade B apples, 4,200 pounds of grade A apples, and 3,200 pounds of grade AA apples. If the average density of the apples is 28.76 lb/ft^3, what is the total number of apples that were harvested? Round to the nearest 1,000 apples.

19. An engineer is designing a
    hinge to be used on the
    landing gear of a new
    airplane. She sketches the
    cross section of the hinge
    on a computer, which is
    shown in the accompanying
    figure. The circle represents
    a hole for a hinge pin.

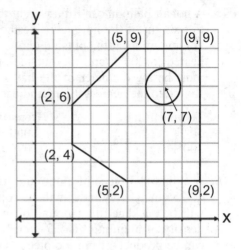

   The engineer next uses a
function on the computer
that creates a solid by
translating the cross section
in a direction perpendicular
to the sketch. The engineer
enters a translation distance
of 10 units. In the final step, the engineer enters a scale factor of 1
unit = 2 centimeters. The resulting computer image of the solid is then
sent to a factory to be manufactured out of a high-strength metal. If
the density of the metal is 8.44 g/cm^3 and the metal costs \$32 per
kilogram, what is the expected cost for the metal to make one hinge?
Round to the nearest dollar.

20. An oil pipeline is being built that uses pipe 4 ft in diameter with a 0.5 in.
    thickness. The sections used are 40 ft in length. The pipe sections are
    delivered by a flatbed truck with a maximum capacity of 48,000 lb. How
    many truckloads of pipe are needed to construct 1 mile of pipe? The
    density of the steel used is 490 lb/ft^3.

ANSWERS AND SOLUTIONS
TO PRACTICE EXERCISES

CHAPTER 1

Section 1.1

A.

1. (1)	**3.** (2)	**5.** (1)	**7.** (4)	**9.** (3)	**11.** (4)
2. (3)	**4.** (4)	**6.** (1)	**8.** (2)	**10.** (4)	

B.

12.

13. $\overline{AB}, \overline{AC}, \overline{AD}, \overline{BC}, \overline{BD}, \overline{CD}$

14. $\angle MOA$, $\angle MOT$, $\angle MOH$, $\angle AOT$, $\angle AOH$, $\angle TOH$

Section 1.2

A.

1. (2)	**3.** (4)	**5.** (4)	**7.** (1)	**9.** (4)
2. (2)	**4.** (1)	**6.** (2)	**8.** (3)	**10.** (1)

B.

11. (a) $\overline{AB}$ or $\overline{CD}$

(b) plane ABE and plane ABC

(c) plane $BAE \parallel$ plane CDF

12. Juan is correct. The intersection of two planes is a line, so M, N, and P must lie on that line.

13. All three legs will touch the floor. Any three points define a plane. The stool will tilt until the plane defined by the tips of the three legs lies on the plane defined by the floor.

Section 1.3

A.

1. (2)	**3.** (4)	**5.** (4)	**7.** (3)	**9.** (3)
2. (4)	**4.** (3)	**6.** (2)	**8.** (1)	**10.** (1)

B.

11. $x = \pm 3$

12. $x = -3 \pm \sqrt{17}$

13. $3\sqrt{75} = 15\sqrt{3}$ and $7\sqrt{27} = 21\sqrt{3}$
    $15\sqrt{3} < 21\sqrt{3}$
    So $3\sqrt{75}$ is less than $7\sqrt{27}$.
14. $CD = 6$, $AB = 27$
15. $m\angle A = 66°$, $m\angle B = 114°$
16. 12
17. 36° and 54°
18. $x = 7$
19. $x = -14$ or $x = 4$

CHAPTER 2

Section 2.1

A.
 1. (2)   3. (3)   5. (2)   7. (1)
 2. (2)   4. (1)   6. (1)   8. (4)
 9. $m\angle KOG = 59°$ and $m\angle IOH = 149°$
10. 90°
11. $m\angle A = 65°$ and $m\angle B = 25°$
12. $\angle SRT$ increases by 15°, and $\angle SRU$ decreases by 15°.

Section 2.2

A.
 1. (2)   3. (3)   5. (4)   7. (4)   9. (4)
 2. (4)   4. (1)   6. (3)   8. (2)
B.
10. 102
11. $x = 5$ and $y = 12$
12. $m\angle TRA + m\angle ARP = m\angle PRT$. Substitute to get $62 + (5x + 1) = 10x + 3$. Solve to find, $x = 12$. Substitute to find $m\angle ARP = 5(12) + 1 = 61°$. The angles are not congruent, so RA is not an angle bisector.

Section 2.3

A.
 1. (3)   3. (3)   5. (1)   7. (3)
 2. (4)   4. (1)   6. (4)   8. (2)
B.
 9. 135°   10. 90°   11. 60°   12. 117°

Section 2.4

A.
 1. (3)   **3.** (2)   **5.** (4)   **7.** (3)   **9.** (1)
 2. (3)   **4.** (4)   **6.** (4)   **8.** (3)   **10.** (3)

B.
11. m∠*BDE* = (3*x* − 6°) by alternate interior angles. Around point *D*,
  (4*x* + 10) + (4*x*)+ (3*x* − 6) = 180°, so *x* = 16. m∠*ADB* = 64° and
  m∠*BDE* = 42°. m∠*ADE* = 64° + 42° = 106°. m∠*DEB* = 74°. ∠*ADE*
  and ∠*DEB* are same side interior angles and are supplementary, so
  AD ∥ *BE*.
12. 117°
13. 115°
14. 72°
15. 150° and 30°
16. The rays are not parallel because the alternate interior angles are not
  congruent. They measure 32° and 29°.

Section 2.5

A.
 1. (2)   **3.** (1)   **5.** (2)   **7.** (4)   **9.** (4)
 2. (4)   **4.** (3)   **6.** (3)   **8.** (4)   **10.** (3)

B.
11. 24°
12. $a^2 + b^2 = c^2$, $5^2 + 6^2 = 8^2$, $61 \neq 64$. Since the Pythagorean theorem is
  not satisfied, the triangle is not a right triangle.
13. 24°
14. $4\sqrt{5}$
15. $a^2 + b^2 = c^2$, $5^2 + 12^2 = c^2$, $c = 13$. Since $\frac{13}{20} < \frac{3}{4}$, Stansha does not
  qualify for a free bus pass.
16. $5\sqrt{2}$
17. 8
18. $20\sqrt{2}$
19. Measure the diagonal length. If its length squared equals $24^2 + 10^2$,
  the Pythagorean theorem is satisfied and the posts form right
  angles.
20. 18°
21. 30°
22. 90°

CHAPTER 3

Section 3.1

A.
1. (3)　　**2.** (3)　　**3.** (4)
B.
4. Use the construction for copying a segment.
5. Construct a segment by copying segment $\overline{AR}$. Copy twice more, using the endpoint of the previous segment for the next starting point. Use a straightedge to extend the previous segment to the new endpoint.
6. Copy $\overline{MA}$, and then copy $\overline{TH}$ onto the end of the first copy. Use a straightedge to extend the copy of $\overline{MA}$ to the new endpoint.
7. Use the construction for copying an angle.
8. Use the construction for copying an angle. Then copy it a second time onto one of the rays of the first copy.
9. Use the angle bisector construction.
10. Use the construction for perpendicular bisector.
11. Use the construction for perpendicular bisector, and then bisect it.
12. Use the construction for an equilateral triangle.
13. Use the construction for an equilateral triangle. Then bisect one of its angles.
14. Use the construction for perpendicular bisector.

Section 3.2

A.
1. (3)　　**2.** (3)　　**3.** (1)　　**4.** (2)
B.
5. Use the construction for a perpendicular line through a point not on the line.
6. Use the construction for a perpendicular line through a point on the line.
7. Use the construction for a parallel line.
8. Use the construction for a regular hexagon.
9. Use the construction for inscribing an equilateral trangle.
10. Use the construction for inscribing a square.
11. Construct equilateral triangle ABC. Construct a line perpendicular to $\overline{AB}$ and passing through point C. Label the intersection of $\overline{AB}$ and the perpendicular as point D. $\triangle ADC$ has angles measuring 30°, 60°, and 90°.
12. Construct a square. Bisect one pair of opposite sides. The two midpoints and two of the vertices of the square will form the rectangle.

Section 3.3

1. (1) **3.** (3) **5.** (3) **7.** (2) **9.** (1)
2. (4) **4.** (4) **6.** (1) **8.** (3) **10.** (3)

CHAPTER 4

Section 4.2

A.
1. (1) **2.** (3) **3.** (1) **4.** (4) **5.** (2)
B.

6. $\overline{XD} \cong \overline{DY}$; given. D is a midpoint; a midpoint divides a segment into two congruent segments. $\overline{ZD}$ is a median; a median is a segment from a vertex to the midpoint of the opposite side.

7. $m\angle ZDY = 90°$; given. $\angle ZDY$ is a right angle; angles measuring 90° are right angles. $\overline{ZD} \perp \overline{XY}$; segments intersecting at right angles are $\perp$. $\overline{ZD}$ is an altitude; a segment from a vertex of a triangle $\perp$ to the opposite side is an altitude.

8. $AZ = \frac{1}{2}(12) = 6$ and $AX = \frac{1}{2}(10) = 5$ because a midpoint divides a segment in half. The perimeter of $\triangle AXZ = 6 + 5 + 4 = 15$ because the perimeter equals the sum of the sides of a triangle.

Section 4.3

A.
1. (3) **2.** (2) **3.** (4) **4.** (4) **5.** (2)
B.
6. $RD = AE$ and $DE = RA$ are given. $RE = RE$ by the reflexive postulate. By the addition postulate, $RD + DE + RE = AE + RA + RE$. Perimeter = sum of the side lengths, so the perimeters are equal.

7. $QS = RT$; given. $QS = QR + RS$ and $RT = RS + ST$; partition postulate. $QR + RS = RS + ST$; substitution property. $RS = RS$; reflexive property. $QR = ST$; subtraction property.

8. $\overline{MO}$ and $\overline{NK}$ are congruent altitudes. $\triangle JLM$ and $\triangle JLK$ share base JL, which is congruent to itself. The area of each triangle is $\frac{1}{2}$ base · height, which must therefore be equal for the two triangles.

9. $SI = 2HI$ is given, so $HI = \frac{1}{2}SI$ and H is a midpoint. $\overline{PH}$ is a segment from a vertex to the opposite midpoint, so it is a median.

10. Y is the midpoint of $\overline{WZ}$ and X is the midpoint of $\overline{WY}$; given. $WY = \frac{1}{2}WZ$ and $WX = \frac{1}{2}WY$; a midpoint divides a segment in half. $\frac{1}{2}WY = \frac{1}{4}WZ$; multiplication property. $WX = \frac{1}{4}WZ$; substitution property.

11. $\angle ABC$ is a right angle, $\angle CBD = \angle EFH$, and $\angle ABD \cong \angle GFH$; given. $m\angle ABC = m\angle ABD + m\angle CBD$ and $m\angle EFG = m\angle EFH + m\angle GFH$; partition postulate. $m\angle EFG = m\angle CBD + m\angle ABD$; substitution property. $m\angle ABC = 90°$; definition of a right angle. $m\angle ABD + m\angle CBD = 90°$; transitive property. $m\angle EFG = 90°$; transitive property.

Section 4.4

A.

1. (2)   2. (4)   3. (1)   4. (3)

B.

5. $\overrightarrow{NOQ} \perp \overrightarrow{POM}$ and $m\angle RON = 60°$; given. $\angle QOR$ and $\angle RON$ are a linear pair; definition of linear pair. $m\angle QOR + m\angle RON = 180°$; linear pairs are supplementary. $m\angle QOR + 60° = 180°$; substitution property. $m\angle QOR = 120°$; subtraction property.

6. $\angle 1$ is supplementary to $\angle 3$ and $\angle 2$ is supplementary to $\angle 4$; definition of a linear pair. $\angle 1 \cong \angle 2$; given. $\angle 3 \cong \angle 4$; angles supplementary to congruent angles are congruent.

7. $m\angle 1 + m\angle 2 + m\angle 3 = 180°$; sum of the adjacent angle measures around a straight line equals 180°. $\angle 1$ and $\angle 3$ are complementary; given. $m\angle 1 + m\angle 3 = 90°$; definition of complementary angles. $m\angle 2 = 90°$; subtraction property. $m\angle 2 = m\angle 5$; vertical angles are congruent. $m\angle 5 = 90°$; transitive property.

8. $\angle BOC$ and $\angle DOF$ are complementary; given. $\angle DOF \cong \angle EOC$ and $\angle BOC \cong \angle AOD$; vertical angles are congruent. $\angle EOC$ and $\angle AOD$ are complementary; substitution property.

9. $\overrightarrow{AB}$ bisects $\angle FPC$; given. $\angle FPB \cong \angle CPB$; an angle bisector forms two congruent angles. $\angle DPF \cong \angle EPC$; vertical angles are congruent. $\angle DPF + \angle FPB \cong \angle EPC + \angle CPB$; addition property. $\angle DPB \cong \angle BPE$; partition postulate.

10. $\overrightarrow{BE}$ bisects $\angle FBD$; given. $\angle FBE \cong \angle DBE$; definition of angle bisector. $\angle ABF \cong \angle CBD$; given. $\angle FBE + \angle ABF \cong \angle DBE + \angle CBD$; addition. $\angle ABE \cong \angle CBE$; partition.

Section 4.5

A.
  1. (4)   **2.** (2)
B.
  3. Line $t \perp$ line s and line $t \perp$ line r; given. Line $s \parallel$ line r; lines perpendicular to the same line are parallel. $\angle 1 \cong \angle 2$; alternate interior angles formed by parallel lines are congruent.

  4. $\overline{AD} \parallel \overline{BC}$ and $\overline{AB}$ bisects $\angle CAD$; given. $\angle 3 \cong \angle 2$; alternate interior angles formed by parallel lines are congruent. $\angle 1 \cong \angle 3$; an angle bisector divides an angle into two congruent angles. $\angle 1 \cong \angle 2$; transitive property.

  5. $\overline{AC} \perp \overline{BD}$ and $\overline{AB} \parallel \overline{CD}$; given. $\angle AEB$ is a right angle; perpendicular lines form right angles. $m\angle AEB = 90°$; right angles measure 90°. $m\angle B = m\angle D$; alternate interior angles formed by parallel lines are congruent. $m\angle A + m\angle B + m\angle AEB = 180°$; triangle angle sum theorem. $m\angle A + m\angle B + 90° = 180°$; substitution property. $m\angle A + m\angle D = 90°$; subtraction postulate. $\angle A$ and $\angle D$ are complementary; angles whose measures sum to 90° are complementary.

  6. $\angle OMS \cong \angle KES$; given. $\overline{MON} \parallel \overline{EKY}$; two lines are parallel if their alternate interior angles are congruent. $\angle MOS \cong \angle EKS$; alternate interior angles formed by parallel lines are congruent. $\angle SON$ and $\angle MOS$ are a linear pair and $\angle EKS$ and $\angle SKY$ are a linear pair; definition of a linear pair. $\angle SON \cong \angle SKY$; angles supplementary to congruent angles are congruent.

  7. $\overrightarrow{AB} \parallel \overrightarrow{CF}$ and $\overrightarrow{BE} \parallel \overrightarrow{CD}$; given. $\angle ABC$ and $\angle 3$ are supplementary; same side interior angles formed by parallel lines are supplementary. $m\angle ABC + m\angle 3 = 180°$; definition of supplementary angles. $m\angle ABC = m\angle 1 + m\angle 2$; partition postulate. $m\angle 1 + m\angle 2 + m\angle 3 = 180°$; substitution property. $m\angle 2 + m\angle 3 + m\angle 4 = 180°$; same reasons as previous. $m\angle 1 - m\angle 4 = 0$; subtraction postulate. $m\angle 1 \cong m\angle 4$, subtraction postulate. $\angle 1 \cong \angle 4$; definition of congruent.

  8. $\overline{FB}$ bisects $\angle AFC$; given. $\angle AFB \cong \angle BFC$, an angle bisector divides an angle into two congruent angles. $\angle A \cong \angle B$; given. $\angle B \cong \angle BFC$; transitive property; $\overline{AB} \parallel \overline{CFD}$; two lines are parallel if their alternate interior angles are congruent.

Section 4.6

A.
  1. (1)   **2.** (2)
B.
  3. $\angle DCE \cong \angle ACB$; vertical angles are congruent. $m\angle ACB + m\angle BAC + m\angle ABC = 180°$; triangle angle sum theorem. $\angle DCE + m\angle BAC + m\angle ABC = 180°$; substitution property.

4. $\overline{JM} \cong \overline{JK}$, $\overline{KM}$ bisects $\angle JKL$, and $\angle JML$; given. $\angle JKM \cong \angle JMK$; angles opposite congruent sides in a triangle are congruent. $\angle JKM \cong \angle LKM$ and $\angle JMK \cong \angle LMK$; an angle bisector divides an angle into two congruent angles. $\angle JMK \cong \angle LKM$; transitive property. $\angle LKM \cong \angle LMK$; transitive property. $\triangle LMK$ is isosceles; a triangle with congruent base angles is isosceles.

5. $m\angle 1 + m\angle 2 + m\angle 3 = 180°$; triangle angle sum theorem. $m\angle 4 + m\angle 5 + m\angle 6 = 180°$; triangle angle sum theorem. $m\angle 1 + m\angle 2 + m\angle 3 = m\angle 4 + m5 + m\angle 6$; transitive property. $\angle 3 = m\angle 6$; vertical angles are congruent. $m\angle 1 + m\angle 2 = m\angle 4 + m\angle 5$; subtraction property.

6. $\triangle ABE$, $\triangle EBD$, and $\triangle BCD$ are all equilateral; given. $AB = BE$, $BE = BD$, and $BD = BC$; definition of equilateral triangle. $AB = BD$ and $AB = BC$; transitive property. B is a midpoint; a midpoint divides a segment into two congruent segments.

7. $AC = BC$; given. $m\angle 2 = m\angle 3$; angles opposite congruent sides in a triangle are congruent. $m\angle 1 = m\angle 3 + m\angle 4$; exterior angle theorem. $m\angle 1 = m\angle 2 + m\angle 4$; substitution property.

8. $\overline{IH} \parallel \overline{LM}$ and $\overline{LK} \parallel \overline{JI}$; given. $m\angle H = m\angle M$ and $m\angle IJK = m\angle LKJ$; alternate interior angles formed by parallel lines are congruent. $m\angle H + m\angle I + m\angle IJK = 180°$ and $m\angle L + m\angle M + m\angle LKJ = 180°$; triangle angle sum theorem. $m\angle I - \angle L = 0$; subtraction property. $m\angle I = m\angle L$; addition property. $\angle I \cong \angle L$; definition of congruent.

9. $m\angle ADC = 140°$, $m\angle BDC = 130°$, and $m\angle DAB = 45°$; given. $m\angle ADB + m\angle ADC + m\angle BDC = 360°$; sum of angle measures around a point equals 360°. $m\angle ADB + 140° + 130° = 360°$; substitution property. $m\angle ADB = 90°$; subtraction postulate. $m\angle DBA + m\angle ADB + m\angle DAB = 180°$; triangle angle sum theorem. $m\angle DBA + 90° + 45° = 180°$; substitution property. $m\angle DBA = 45°$; subtraction postulate. $\overline{AD} \cong \overline{BD}$; sides opposite congruent angles in a triangle are congruent.

10. $m\angle ACB = 50°$ and $m\angle FBE = 40°$; given. $m\angle FBE \cong m\angle ABC$; vertical angles are congruent. $m\angle ABC = 40°$; transitive property. $m\angle BAC + m\angle ACB + m\angle ABC = 180°$; triangle angle sum theorem. $m\angle BAC + 50° + 40° = 180°$; substitution property. $m\angle BAC = 90°$; subtraction postulate. $\angle BAC$ is a right angle; right angles measure 90°. $\overline{AC} \perp \overline{DABE}$; lines that intersect at right angles are perpendicular.

CHAPTER 5

Section 5.1

A.

1. (2) 3. (2) 5. (1) 7. (4)
2. (4) 4. (3) 6. (4) 8. (1)

B.

9. The transformation may not be an isometry. A counterexample is a rectangle of length 10 and width 4 being transformed to a rectangle of length 9 and width 5. The perimeters are equal, but the rectangles are not congruent.
10. Orientation and slope of $\overline{LN}$
11. The area is the same because the circular region was translated, which is a rigid motion. Area is preserved by rigid motions.

Section 5.2

A.

1. (1)	**3.** (2)	**5.** (4)	**7.** (3)	**9.** (1)	**11.** (2)
2. (3)	**4.** (2)	**6.** (2)	**8.** (1)	**10.** (2)	

B.
12. $130°$
13. A rotation of $90°$ about point X
14. A rhombus.

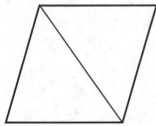

15. P is the image of itself and Z is the image if itself after the same reflection over $\overline{PZ}$. Every point is mapped by the same reflection, which is a rigid motion, so the images are congruent.
16. In a reflection, segments joining corresponding points are congruent, so the figure will have at least two parallel sides.

Section 5.3

A.

1. (4)	**3.** (3)	**5.** (2)	**7.** (1)	**9.** (4)	**11.** (2)
2. (4)	**4.** (1)	**6.** (1)	**8.** (3)	**10.** (3)	**12.** (2)

B.
13. (A) $R_{origin,90°}$
    (B) $r_{x=1}$
14. (A) $r_{y=x}$
    (B) $T_{0,-3}$
    (C) $r_{y=-3}$
15. $D_{\frac{1}{2}}$

16. (A) $G'(6, 0)$, $L'(3, 1)$, $O'(2, 3)$, $W'(5, 2)$
  (B) $G''(0, -6)$, $L''(1, -3)$, $O''(3, -2)$, $W''(2, -5)$
17. (B) $B'(-2, 1)$, $I'(-5, 4)$, $G'(-1, 6)$
  (C) $B''(1, -2)$, $I''(4, -5)$, $G''(6, -1)$
18. (A) $A'(-3, 1)$, $B'(3, -1)$, $C'(2, -4)$, $D'(-4, -2)$
  (B) $A''(1, -1)$, $B''(7, -3)$, $C''(6, -6)$, $D''(0, -4)$

Section 5.4

A.

1. (3)	**3.** (1)	**5.** (3)	**7.** (1)
2. (2)	**4.** (4)	**6.** (3)	**8.** (2)

B.
  9. 5 lines of symmetry

10. 9 seconds
11. An example is a trapezoid.

12. Fiona is correct. An isosceles triangle with only two congruent sides has one line of symmetry.

Section 5.5

A.

1. (4)	**2.** (2)	**3.** (3)	**4.** (4)	**5.** (2)	**6.** (2)

B.

7.

8. $B'(-1, 6)$, $I'(-2, 3)$, $G'(-5, 7)$, and $B''(6, -1)$, $I''(3, -2)$, $G''(7, -5)$

9. $P'(2, 0)$, $M'(3, 3)$, $N'(0, 5)$ and $P''(-2, 1)$, $M''(-1, 4)$, $N''(-4, 6)$

10. $Z'(0, 2)$, $E'(-2, 1)$, $D'(-4, 3)$ and $Z''(0, -2)$, $E''(-2, -1)$, $D''(-4, -3)$

11. $F'(-3, -7)$, $G'(-5, -6)$, $H'(-3, -2)$, $K'(-1, -3)$ and $F''(3, 7)$, $G''(5, 6)$, $H''(3, 2)$, $K''(1, 3)$

CHAPTER 6

Section 6.1

A.

1. (3)	**3.** (2)	**5.** (4)	**7.** (1)	**9.** (2)
2. (4)	**4.** (3)	**6.** (1)	**8.** (4)	**10.** (2)

B.

11. Translate $\triangle XYZ$ so X maps to A. Rotate $\triangle XYZ$ about A so $\overline{XZ}$ maps to $\overline{AC}$. Reflect $\triangle XYZ$ over $\overline{AC}$.

12. Because the triangles are right triangles with two pairs of corresponding congruent sides, the Pythagorean theorem can be used to show the third pair of corresponding sides are also congruent, allowing congruence by SSS or SAS. The Pythagorean theorem applies only to right triangles, so SSA cannot be used for other types of triangles.

Section 6.2

A.

1. (2)   **2.** (1)

B.

3. $\overline{NO}$ is a median of $\triangle CRN$; given. O is the midpoint of $\overline{CR}$; a median is a segment from a vertex to the midpoint of the opposite side. $\overline{CO} \cong \overline{RO}$; a midpoint divides a segment into two congruent segments. $\overline{ON} \cong \overline{ON}$; reflexive property. $\overline{CN} \cong \overline{RN}$; given. $\triangle CON \cong \triangle RON$ by SSS.

4. $\overline{AU}$ and $\overline{EO}$ bisect each other at I; given. I is the midpoint of $\overline{AU}$ and $\overline{EO}$; bisectors intersect at the midpoint. $\overline{EI} \cong \overline{IO}$ and $\overline{AI} \cong \overline{UI}$; a midpoint divides a segment into two congruent segments. $\angle AIE \cong \angle OIU$; vertical angles are congruent. $\triangle AEI \cong \triangle UOI$ by SAS.

5. $\overline{RY} \cong \overline{YZ}$ and $\overline{YZ} \cong \overline{ZB}$; given. $\overline{RY} \cong \overline{ZB}$; transitive property. $\angle E \cong \angle O$ and $\angle B \cong R$; given. $\triangle REY \cong \triangle BOZ$ by AAS.

6. $\overline{PT} \parallel \overline{QS}$ and $\overline{TQ} \parallel \overline{SR}$; given. $\angle PQT \cong \angle QRS$ and $\angle PTQ \cong \angle QSR$; corresponding angles formed by parallel lines are congruent. $\overline{PT} \cong \overline{QS}$; given. $\triangle PTQ \cong \triangle QSR$ by AAS.

7. $\overline{CD}$ bisects $\angle ACB$; given. $\angle ACD \cong \angle BCD$; an angle bisector divides an angle into two congruent angles. $\overline{DC} \cong \overline{DC}$; reflexive property. $\angle A \cong \angle B$; given. $\triangle ADC \cong \triangle BDC$ by AAS.

8. $\overline{XA} \cong \overline{BY}$; given. $\overline{XY} \cong \overline{XY}$; reflexive property. $\overline{AY} \perp \overline{XA}$ and $\overline{XB} \perp \overline{BY}$; altitudes are perpendicular to the opposite side of a triangle. $\angle XAY$ and $\angle YBX$ are right angles; perpendicular lines intersect at right angles; $\angle XAY \cong \angle YBX$; all right angles are congruent; $\triangle XBY$ and $\triangle YAX$ are right triangles; triangles with a right angle are right triangles. $\triangle XBY \cong \triangle YAX$ by HL.

9. $\angle ZXB \cong \angle ZYA$ and $\overline{AY} \cong \overline{BX}$; given. $\angle Z \cong \angle Z$; reflexive property. $\triangle ZAY \cong \triangle ZBX$ by AAS.

10. Altitude $\overline{OK}$; given. $\overline{OK} \perp \overline{DC}$; an altitude is perpendicular to the opposite side of a triangle. $\angle DOK$ and $\angle COK$ are right angles; perpendicular lines form right angles. $\angle DOK \cong \angle COK$; all right angles are congruent; $\triangle DOK$ and $\triangle COK$ are right triangles; triangles with a right angle are right triangles. $\overline{OK} \cong \overline{OK}$; reflexive property. $\overline{DK} \cong \overline{CK}$; given. $\triangle DOK \cong \triangle COK$ by HL.

11. $\overline{VS}$ and $\overline{UT}$ bisect each other at R; given. R is the midpoint of $\overline{VS}$ and $\overline{UT}$; bisectors intersect at the midpoint. $\overline{VR} \cong \overline{RS}$ and $\overline{UR} \cong \overline{RT}$; a midpoint divides a segment into two congruent segments. $\angle URV \cong \angle TRS$; vertical angles are congruent. $\triangle VRU \cong \triangle SRT$ by SAS.

12. $\overline{DY} = \overline{CX}$ and $ABCD$ is a rectangle, given. $\overline{YX} \cong \overline{YX}$; reflexive property. $\overline{DY} + \overline{YX} \cong \overline{CX} + \overline{YX}$; addition property. $\overline{DX} \cong \overline{CY}$; partition postulate. $\overline{AD} \cong \overline{BC}$; opposite sides of a rectangle are congruent. $\angle C$ and $\angle D$; all angles of a rectangle are congruent. $\triangle ADX = \triangle BCY$ by SAS.

13. $\angle A \cong \angle D$ and $\overline{EC} \cong \overline{EB}$; given. $\angle EBC \cong \angle ECB$; angles opposite congruent sides in a triangle are congruent. $\overline{BC} \cong \overline{BC}$; reflexive property. $\triangle DBC \cong \triangle ACB$ by AAS.

14. $\overline{ALB} \parallel \overline{DMC}$, $\overline{LN} \parallel \overline{MO}$; given. $\angle LAN \cong \angle MCO$; alternate interior angles formed by parallel lines are congruent. $\angle LNA \cong \angle MOC$; alternate exterior angles formed by parallel lines are congruent. $\overline{AN} \cong \overline{OC}$; given. $\triangle ALN \cong \triangle CMO$ by ASA.

15. $\overline{WB} \cong \overline{EB}$; given. $\angle EWX \cong WEB$; angles opposite congruent sides in a triangle are congruent. $\overline{WF}$ and $\overline{EX}$ are medians; given. F and X are midpoints of $\overline{EB}$ and $\overline{WB}$, respectively; definition of a median. $FE = \frac{1}{2}EB$ and $WX = \frac{1}{2}WB$; a midpoint divides a segment in half. $WB = EB$; definition of congruence. $FE = WX$; halves of equal quantities are equal. $\overline{EF} \cong \overline{WX}$; definition of congruence. $\overline{WE} \cong \overline{WE}$; reflexive property. $\triangle WEX \cong \triangle EWF$ by SAS.

Section 6.3

B.

1. $\overline{PD} \cong \overline{FY}$, $\overline{DQ} \cong \overline{YK}$, and $\overline{PQ} \cong \overline{FK}$; given. $\triangle PDQ \cong \triangle FYK$; SSS. $\angle Q \cong \angle K$; CPCTC.

2. $\overline{AC}$ and $\overline{BD}$ intersect at E, $\overline{AE} \cong \overline{CE}$, and $\overline{BE} \cong \overline{DE}$; given. $\angle AEB \cong \angle CED$; vertical angles are congruent. $\triangle AEB \cong \triangle CED$; SAS. $\angle A \cong \angle C$; CPCTC. $\overline{AB} \parallel \overline{CD}$; two segments are parallel if the alternate interior angles formed by a transversal are congruent.

3. $\angle R \cong \angle P$, $\angle T \cong \angle S$, and $\overline{TI} \cong \overline{IS}$; given. $\triangle TRI \cong \triangle SPI$; AAS; $\overline{RI} \cong \overline{PI}$; CPCTC. I is the midpoint of $\overline{RP}$; a point that divides a segment into two congruent segments is a midpoint.

4. J is the midpoint of $\overline{HK}$ and $\overline{IL}$; given. $\overline{HJ} \cong \overline{KJ}$ and $\overline{IJ} \cong \overline{LJ}$; a midpoint divides a segment into two congruent segments. $\angle HJI \cong \angle LJK$; vertical angles are congruent. $\triangle HJI \cong \triangle KJL$; SAS. $\angle I \cong \angle L$; CPCTC.

5. $\triangle PAT \cong \triangle IAR$; given. $\overline{PT} \cong \overline{IR}$ and $\angle APT \cong \angle AIR$; CPCTC. $\overline{RT} \cong \overline{RT}$; reflexive property. $\triangle PTR \cong \triangle IRT$; SAS.

6. $\triangle PEQ$ is equilateral; given. $\overline{PE} \cong \overline{QE}$; definition of equilateral triangle. $\overline{RE} \cong \overline{RE}$; reflexive property. $\angle PER \cong \angle QER$; given. $\triangle PER \cong \triangle QER$; SAS. $\angle PRE \cong \angle QRE$; CPCTC. $\overline{RE}$ bisects $\angle PRQ$; an angle bisector divides an angle into two congruent angles.

7. $\angle ERP \cong \angle ERQ$ and $\overline{PR} \cong \overline{QR}$; given. $\overline{RE} \cong \overline{RE}$; reflexive property. $\triangle PER \cong \triangle QER$; SAS. $\overline{PE} \cong \overline{QE}$; CPCTC. $\triangle PEQ$ is isosceles; a triangle with two congruent sides is isosceles.

8. $\triangle PLJ \cong \triangle TLK$; given. $\angle P \cong \angle T$; CPCTC. $\overline{AP} \cong \overline{AT}$; sides opposite congruent angles are congruent. $\overline{AJ} + \overline{JP} \cong \overline{AK} + \overline{KT}$; partition postulate. $\overline{JP} \cong \overline{KT}$; CPCTC. $\overline{AJ} \cong \overline{AK}$; subtraction postulate. $\overline{AL} \cong \overline{AL}$; reflexive property. $\overline{JL} \cong \overline{KL}$; CPCTC. $\triangle JAL \cong \triangle KAL$; SSS.

9. $\overline{AL}$ bisects $\angle PAT$; given. $\angle JAL \cong \angle KAL$; an angle bisector divides an angle into two congruent angles. $\overline{AL} \cong \overline{AL}$; reflexive property. $\angle JLA \cong \angle KLA$; given. $\triangle JLA \cong \triangle KLA$; ASA. $\overline{JL} \cong \overline{KL}$; CPCTC. $\angle JLP \cong \angle KLT$; given. $\angle LJA \cong \angle LKA$; CPCTC. $\angle PJL$ and $\angle LJA$ are a linear pair, and $\angle TKL$ and $\angle LKA$ are a linear pair; definition of a linear pair. $\angle PJL \cong \angle TKL$; angles supplementary to congruent angles are congruent. $\triangle PLJ \cong \triangle TLK$; ASA.

10. $\overline{SP} \cong \overline{PQ}$ and $\overline{SR} \cong \overline{QR}$; given. $\overline{POR} \cong \overline{POR}$; reflexive property. $\triangle PSR \cong \triangle PQR$; SSS. $\angle QPR \cong \angle SPR$; CPCTC. $\overline{POR}$ bisects $\angle SQP$; an angle bisector divides an angle into two congruent angles.

11. $\overline{POR} \perp \overline{SOQ}$; given. $\angle ROS$ and $\angle ROQ$ are right angles; perpendicular lines form right angles. $\triangle ROS$ and $\triangle ROQ$ are right triangles; definition of right triangle. $\overline{SR} \cong \overline{RQ}$; given. $\overline{RO} \cong \overline{RO}$; reflexive property. $\triangle ROS \cong \triangle ROQ$; HL. $\overline{OS} \cong \overline{OQ}$; CPCTC. O is the midpoint of $\overline{QOS}$; definition of a midpoint. $\overline{RO}$ is a median of $\triangle SRQ$; definition of a median.

12. $\overline{LT} \parallel \overline{OV}$; given. $\angle OVT \cong \angle LTE$; corresponding angles formed by parallel lines are congruent. $\overline{LT} \cong \overline{LE}$; given. $\angle LTE \cong \angle E$; angles opposite congruent sides in a triangle are congruent. $\angle OVT \cong \angle E$; transitive property. $\angle E \cong \angle IVO$; given. $\angle OVT \cong \angle IVO$; transitive property. $\overline{OV}$ is an angle bisector; definition of an angle bisector.

13. $\overline{OS}$ bisects $\angle CON$ and $\angle CSN$; given. $\angle COS \cong \angle NOS$ and $\angle OSC \cong \angle OSN$; an angle bisector divides an angle into two congruent angles. $\overline{OS} \cong \overline{OS}$; reflexive property. $\triangle COS \cong \triangle NOS$; ASA. $\overline{OC} \cong \overline{ON}$ and $\angle C \cong \angle N$; CPCTC. $\angle COW \cong \angle NOL$; reflexive property. $\triangle COW \cong \triangle NOL$; ASA. $\angle OLN \cong \angle OWC$; CPCTC.

14. $\overline{DMR}$ bisects $\angle LDN$; given. $\angle NDM \cong \angle MDL$; definition of an angle bisector. $\angle MND \cong \angle MLD$; given. $\overline{MD} \cong \overline{MD}$; reflexive property. $\triangle NDM \cong \triangle LDM$; AAS. $\angle LMD \cong \angle NMD$; CPCTC; $\angle LMD$ and $\angle NMD$ are a linear pair; definition of a linear pair. $\angle LMD$ and $\angle NMD$ are supplementary; linear pairs are supplementary. $\angle LMD$ and $\angle NMD$ are right angles; congruent supplementary angles are right angles. $\overline{DM} \perp \overline{TLM}$; definition of perpendicular lines. $\overline{TLM}$ is an altitude of $\triangle TDR$; definition of altitude.

Section 6.4

A.
1. (3)   2. (2)

B.
3. The perpendicular bisector of a segment is the line of reflection between the two endpoints of the segment. Therefore X is the reflection of F over m, Y is the reflection of G over m, and Z is the reflection of H over m. Since a reflection is a rigid motion, $\triangle XYZ \cong \triangle FHG$.

4. From the definition of an altitude, $\overline{OD} \perp \overline{LG}$. Since $LO = OG$, O is the midpoint of $\overline{LG}$ and $\overline{OD}$ is the perpendicular bisector of $\overline{LG}$. Therefore, G is the reflection of L over $\overline{OD}$. O and D lie on the line of reflection, so they map to themselves under the same reflection. $\triangle OGD$ is the image of $\triangle OLD$ under a reflection over $\overline{OD}$. Since reflections are rigid motions, the triangles are congruent.

5. From the givens, C and D are the images of A and B under the same rotation. O lies on the center of rotation, so it maps to itself. Therefore, $\triangle COD$ is the rotation of $\triangle AOB$. A rotation is a rigid motion, so the triangles are congruent.

6. $\overline{DL} \cong \overline{LV}$ from the definition of a midpoint. Therefore, $\overline{UI}$, $\overline{DL}$, and $\overline{LV}$ are all congruent and parallel. This makes I the image of U, V the image of L, and L the image of D all under the same translation. $\triangle ILV$ is the translation of $\triangle UDL$, making the triangles congruent because translations are rigid motions.

7. I is the midpoint of $\overline{KIH}$ and $\overline{JIG}$, so $KI = IH$ and $JI = IH$. H is the reflection of K through point I, and G is the reflection of J through point I. I is the reflection of I through point I. Since every vertex in $\triangle KIJ$ is reflected through the same point, $\triangle HIG$ is a point reflection through I of $\triangle KIJ$. $\triangle KIJ \cong \triangle HIG$ because a reflection is a rigid motion.

8. C and Q map to themselves after a reflection over $\overline{CQ}$ because they both lie on the line of reflection. A maps to B after the same reflection. Therefore $\triangle AQC \cong \triangle BQC$ because each vertex maps to a corresponding vertex after the same reflection.

9. Since the decagon is regular, every rotation of 36° about P maps a vertex to the next counterclockwise vertex. U is the image of R after a rotation of 216° about P. T is the image of S after a 216° rotation about P, and P maps to itself after the same rotation. Therefore $\triangle PRS \cong \triangle PUT$ because the same rigid motion rotation maps each vertex to the corresponding vertex.

10. *O* is the image of *A* after a rotation about *C* by ∠*ACO*, and *G* is the image of *T* after a rotation about *C* by ∠*TCG*. Since ∠*ACO* ≅ ∠*TCG*, by substitution we have *O* is the image of *T* after a rotation about *C* by ∠*ACO*. *C* maps to itself under the same rotation because the rotation of a point about itself is an identity transformation. △*CAT* ≅ △*COG* because every vertex maps to a corresponding vertex under the same rigid motion rotation.

CHAPTER 7

Section 7.1

A.
  1. (1)   2. (3)   3. (4)   4. (2)   5. (1)   6. (3)
B.
  7. (−4, 5)
  8. (3, 4)
  9. (4, 0)
  10. *M*(−6, 2) and *N*(0, 5)
  11. Find the distances between (5, 6) and each endpoint. If they are the same, (5, 6) lies on the perpendicular bisector.
  12. Yes, each segment has a length equal to √34.

Section 7.2

A.
  1. (3)   2. (1)   3. (2)   4. (3)
B.
  5. Area = 68
  6. Area = 59.5
  7. Area = 48.5

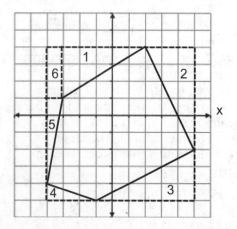

8. Area = 60, using the regions shown. Answers may vary.

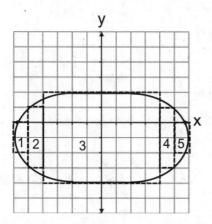

9. Area = 61, using the regions shown. Answers may vary.

10. Area = (0, 5), using the regions shown. Answers may vary.

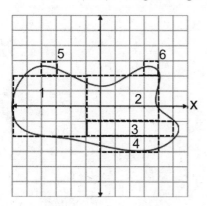

Section 7.3

A.

  1. (4)   **3.** (3)   **5.** (1)   **7.** (3)

  2. (4)   **4.** (1)   **6.** (2)   **8.** (4)

B.

  9. Yes, the slopes between each pair of points is $\dfrac{1}{3}$.

  10. No, the slope between A and B is 1, but the slope between B and C is $\dfrac{4}{5}$.

  11. $y = -\dfrac{1}{2}x + 4$

  12. $y = -7x + 22$

  13. $y - 1 = -\dfrac{3}{4}(x - 3)$ or $y = -\dfrac{3}{4}x + 3\dfrac{1}{4}$

  14. y-intercept = 3

  15. x-intercept = −9

Section 7.4

A.

  1. (3)   **3.** (2)   **5.** (1)   **7.** (3)

  2. (1)   **4.** (4)   **6.** (3)   **8.** (4)

B.

  9. $y = -\dfrac{1}{4}x + 5$

  10. $y = 3x - 1$

  11. $y = -\dfrac{5}{2}x + 5$

  12. $y = -\dfrac{1}{4}x + 2$

  13. $y = -\dfrac{7}{2}x + 4$

  14. $y = -\dfrac{1}{2}x + 4$

  15. Midpoint is (−3, 3), equation is $y = -\dfrac{4}{3}x - 1$

  16. Midpoint is (0, 5), equation is $y = 4x + 5$

Section 7.5

A.
  1. (4)   **2.** (2)   **3.** (3)   **4.** (1)   **5.** (3)

B.
  6. $y = 3x + 2$
  7. $y = -x + 2$

Section 7.6

A.
  1. (3)   **3.** (4)   **5.** (3)   **7.** (3)
  2. (1)   **4.** (4)   **6.** (1)   **8.** (3)

B.
  9. $(x - 5)^2 + (y - 2)^2 = 52$
  10. $(x - 1)^2 + (y - 3)^2 = 25$
  11. $(x - 1)^2 + (y - 4)^2 = 20$
  12. $(x - 3)^2 + (y + 2)^2 = 17$
  13. Diameter $= 2\sqrt{26}$
  14. Center at $(2, -6)$, radius $= 5$
  15. Center at $(-1, -7)$, diameter $= 2\sqrt{30}$
  16. 30

CHAPTER 8

Section 8.1

A.
  1. (2)   **3.** (2)   **5.** (1)   **7.** (4)   **9.** (2)
  2. (4)   **4.** (3)   **6.** (3)   **8.** (1)   **10.** (3)

B.
  11. $AW = 30$
  12. $m\angle U = 93°$
  13. $TO = 12$
  14. 648 in.²
  15. 135 inches
  16. 0.125 gallons
  17. Bobby is correct. Even though one corresponding pair of congruent angles means the other three pairs are congruent as well, the parallelograms are not necessarily similar. A counterexample would be a square whose sides lengths are all 4 and a rectangle with two sides of length 4 and two sides of length 2. They have different shapes even though all pairs of angles measure 90° and are congruent.

Section 8.2

A.
  1. (4)   **3.** (3)
  2. (4)   **4.** (2)

B.
  5. Translate point A to point I, rotate about I by angle $R'IP$, and then dilate with center I and scale factor $\dfrac{5}{3}$.

  6. $\angle DGR$ and $\angle KGL$ are congruent vertical angles $\dfrac{DG}{GL} = \dfrac{RG}{GK} = \dfrac{7}{4}$. Therefore, the triangles are similar by SAS.

  7. $\dfrac{VY}{XT} = \dfrac{VL}{TU} = \dfrac{YL}{XU} = \dfrac{1}{2}$. Therefore, the $\triangle VLY \sim \triangle TUX$ by SSS. $\angle V \cong \angle T$ because corresponding angles in similar triangles are congruent.

  8. $\overline{ACE}$ and $\overline{BC} \parallel \overline{DE}$; given. $\angle EDC \cong \angle BCD$; alternate interior angles formed by parallel lines are congruent. $\angle ABC \cong \angle BCD$; alternate interior angles formed by parallel lines are congruent. $\angle EDC \cong \angle ABC$; transitive property. $\angle CED \cong \angle ACB$; corresponding angles formed by parallel lines are congruent. $\triangle ABC \sim \triangle CDE$; AA.

  9. (a) Square $BFED$ is inscribed in $\triangle ABC$; given. $\overline{BF} \parallel \overline{DE}$; opposite sides of a square are parallel. $\angle A \cong \angle E$, $\overline{BD} \parallel \overline{EF}$; opposite sides of a square are parallel. $\angle DCE \cong \angle FEA$; corresponding angles formed by parallel lines are congruent. $\triangle EFA \sim \triangle CDE$; AA.
    (b) $\sqrt{3}:1$
    (c) Translate vertex E to C, then dilate about C with a scale factor of $\sqrt{3}$.

  10. Parallelogram $ABCD$ and $\overline{FEC}$ intersects $\overline{DEG}$ at E; given. $\overline{AB} \parallel \overline{CD}$; opposite sides of a parallelogram are parallel. $\angle FGE \cong \angle EDC$ and $\angle GFE \cong \angle DCE$; alternate interior angles formed by parallel lines are congruent. $\triangle FEG \sim \triangle CED$; AA.

  11. $11:20$, dilate about E with a $\dfrac{20}{11}$ scale factor and then rotate $180°$ about E.

  12. $\angle RST \cong \angle VSU$ because an angle bisector forms two congruent angles. $\dfrac{RS}{SU} = \dfrac{ST}{SV} = \dfrac{5}{2}$. Therefore, the triangles are similar by SAS.

  13. $\angle RTS \cong \angle YTS$; given. $\overline{TY} \cong \overline{YS}$ and $\overline{TR} \cong \overline{TS}$; given. $\dfrac{TY}{TR} = \dfrac{YS}{TS}$; division property. $\triangle RTS \sim \triangle TRS$; SAS.

14. $\angle O$ and $\angle QPN$ are right angles; given. $\angle O \cong \angle QPN$; all right angles are congruent. $\angle N \cong \angle N$; reflexive property. $\triangle MON \sim \triangle QPN$; AA.
$\dfrac{NP}{NO} = \dfrac{QP}{MO}$; corresponding sides of similar triangles are proportional.
$NP \cdot MO = QP \cdot NO$; cross products of a proportion are equal.

Section 8.3

A.

1. (4)	3. (1)	5. (2)	7. (4)	9. (4)	11. (3)
2. (2)	4. (4)	6. (3)	8. (1)	10. (2)	

B.

12. $LD = 63$

13. Brianna is correct. She can construct segment $\overline{MP}$ and then use the "copy a segment" construction to double its length. Let O be the endpoint of the new segment. The final step is to construct $\overline{NO}$ and bisect it. Let Q be the midpoint of $\overline{NO}$. $\overline{PQ}$ is now a midsegment and is parallel to $\overline{MN}$ from the midsegment theorem.

14. Right triangle RST with hypotenuse $\overline{ST}$ and altitude $\overline{RU}$; given. $\triangle RST \sim \triangle USR \sim \triangle URT$; the altitude to the hypotenuse of a right triangle forms 2 triangles similar to the original. $\dfrac{ST}{RS} = \dfrac{RS}{US}$ and $\dfrac{ST}{RT} = \dfrac{RT}{UT}$; corresponding sides of similar triangles are proportional. $RS^2 = ST \cdot US$ and $RT^2 = ST \cdot UT$; cross products in a proportion are equal. $RS^2 + RT^2 = ST \cdot US + ST \cdot UT$; addition property. $RS^2 + RT^2 = ST(US + UT)$; distributive property. $RS^2 + RT^2 = ST \cdot ST$; partition postulate. $RS^2 + RT^2 = ST^2$; simplify.

15. D is the midpoint of $\overline{AC}$ and E is the midpoint of $\overline{BC}$; given. $\overline{DE} \parallel \overline{GF}$; a segment joining the midpoints of two sides of a triangle is parallel to the third side. $\angle FGD$ and $\angle GFE$ are right angles; given. $\angle FGD$ and $\angle GFE$ are supplementary; two right angles are always supplementary. $\overline{DG} \parallel \overline{EF}$; two lines are parallel if the same side interior angles formed by a transversal are supplementary. $DEFG$ is a rectangle; a quadrilateral with two pairs of parallel opposite sides and a right angle is a rectangle.

16. $\dfrac{TG}{TB} = \dfrac{TK}{TA} = \dfrac{1}{2}$ because a midpoint divides a segment in half. $\angle T$ is a shared angle and is congruent to itself. Therefore, $\triangle TGK \sim \triangle TBA$ by SAS. $GK = \dfrac{1}{2}AB$ because corresponding parts of similar triangles are proportional.

Section 8.4

A.

1. (4)	**3.** (1)	**5.** (4)	**7.** (3)	**9.** (1)	**11.** (3)
2. (4)	**4.** (2)	**6.** (4)	**8.** (3)	**10.** (2)	**12.** (4)

B.

13. 22.7 feet

14. 67.4°

15. 62.35 inches

16. The ramp is 28.7 feet long and must start 28.6 feet from the door.

17. 24.2°

18. 41.4°

19. 923 ft

20. (a) 135 ft (b) 155 ft

21. 42.96 ft

22. Scott is correct. The sine ratio equals $\dfrac{opposite}{hypotenuse}$. The hypotenuse is always the longest side in a right triangle. So the denominator of the fraction is always greater than the numerator, making the ratio less than 1.

23. (a) 188 ft (b) 73 ft

24. 15 ft/sec

CHAPTER 9

Section 9.1

A.

1. (3)	**3.** (4)	**5.** (4)	**7.** (2)
2. (2)	**4.** (1)	**6.** (4)	

B.

8. 69°

Section 9.2

A.

1. (3)	**2.** (2)

B.

3. $ABCD$ is a parallelogram. $\triangle ABD$ is an isosceles triangle with m$\angle BAD$ = m$\angle ADB$ = 54°. m$\angle BDC$ = 126° − 72° = 54°. $\overline{AB} \parallel \overline{CD}$ because alternate interior angles are congruent. One pair of opposite sides is both congruent and parallel.

4. $\overline{FY} \cong \overline{YS}$ because a midpoint forms two congruent segments. $\angle HFY \cong \angle ISY$ and $\angle FHY \cong \angle SIY$ because alternate interior angles formed by 2 parallel lines are congruent. $\triangle FHY \cong \triangle SIY$ by AAS. $\overline{FH} \cong \overline{SI}$ by CPCTC. $FISH$ is a parallelogram because it has a pair of sides that are congruent and parallel.

5. $\angle DFE$ and $\angle BFC$ are congruent vertical angles. $\angle CBF$ and $\angle DEF$ are congruent alternate interior angles. $\overline{DF} \cong \overline{FC}$ because a midpoint forms two congruent segments. $\triangle CFB \cong \triangle DFE$ by AAS. $\overline{BC} \cong \overline{DE}$ by CPCTC. Since $\overline{DA}$ is also congruent to $\overline{BC}$, $\overline{DA} \cong \overline{DE}$, which makes D a midpoint.

6. Parallelogram $RSTU$; given. $\overline{UR} \cong \overline{ST}$; opposite sides of a parallelogram are congruent. W is the midpoint of $\overline{US}$ and $\overline{TR}$; diagonals bisect each other in a parallelogram. $\overline{UW} \cong \overline{WS}$ and $\overline{RW} \cong \overline{WT}$; a midpoint divides a segment into two congruent segments. $\triangle UWR \cong \triangle SWT$; SSS.

7. $\triangle AFD \cong \triangle FCE$; given. $\angle A \cong \angle EFC$ and $\angle AFD \cong \angle FCE$; CPCTC. $\overline{DB} \parallel \overline{FE}$ and $\overline{EB} \parallel \overline{FD}$; 2 lines are parallel if their corresponding angles are congruent. $DBEF$ is a parallelogram; a quadrilateral with 2 pairs of parallel sides is a parallelogram.

8. Regular hexagon $ABCDEF$; given. $\overline{DC} \cong \overline{AF}$, $\angle E \cong \angle B$, $\angle BAF \cong \angle CDG$, and $\angle EFA \cong \angle BCD$; all angles and sides are congruent in a regular hexagon. $\angle HAF \cong \angle GDC$ and $\angle HFA \cong \angle DCG$; angles congruent to supplementary angles are congruent. $\triangle HAF \cong \triangle GDC$; ASA. $\angle H \cong \angle G$; CPCTC. $HBGE$ is a parallelogram; a quadrilateral with opposite angles congruent is a parallelogram.

9. $\overline{ID} \perp \overline{BI}$ and $\overline{ID} \perp \overline{RD}$; given. $\angle BID$ and $\angle RDI$ are right angles; perpendicular lines form right angles. $\angle BID \cong \angle RDI$; right angles are congruent. $\angle RID \cong \angle BDI$; given. $\overline{BD} \parallel \overline{RI}$ and $\overline{BI} \parallel \overline{RD}$; two lines are parallel if alternate interior angles are congruent. $BIRD$ is a parallelogram; a quadrilateral with two pairs of parallel sides is a parallelogram.

10. $\triangle FMI \cong \triangle NHG$; given. $\angle F \cong \angle H$ and $\angle FIM \cong \angle HGN$; CPCTC. $\angle MIN \cong \angle NGM$; given. $\angle FIM + \angle MIN \cong \angle HGN + \angle NGM$; addition postulate. $\angle FIN \cong \angle HGM$; partition postulate. $FGHI$ is a parallelogram; a quadrilateral in which both pairs of opposite angles are congruent.

11. $\triangle WZB \cong \triangle YXA$; given. $\overline{WZ} \cong \overline{XY}$ and $\overline{AX} \cong \overline{ZB}$; CPCTC. $\overline{WA} \cong \overline{BY}$; given. $\overline{WA} + \overline{AX} \cong \overline{ZB} + \overline{BY}$; addition property. $\overline{WX} \cong \overline{ZY}$; partition postulate. *WXYZ* is a parallelogram; a quadrilateral with opposite sides congruent is a parallelogram.

12. *QRST* is a parallelogram; given. $\overline{QR} \cong \overline{ST}$; opposite sides of a parallelogram are congruent. *R* is the midpoint of $\overline{QU}$; given. $\overline{QR} \cong \overline{RU}$; a midpoint forms two congruent segments. $\overline{RU} \cong \overline{ST}$; transitive property. $\overline{QRU} \parallel \overline{ST}$; opposite sides of a parallelogram are parallel. *RUST* is a parallelogram; a quadrilateral with a pair of sides parallel and congruent is a parallelogram.

Section 9.3

A.

1. (3)	**3.** (2)	**5.** (3)	**7.** (1)	**9.** (4)	**11.** (1)
2. (1)	**4.** (4)	**6.** (4)	**8.** (2)	**10.** (2)	

B.

12. $16\sqrt{5}$

13. The diagonals bisect each other and are perpendicular. So the 4 triangles formed are all right triangles with legs of lengths $\frac{1}{2}a$ and $\frac{1}{2}b$. The area of each triangle is $\frac{1}{2}bh = \frac{1}{2}\left(\frac{1}{2}a\right)\left(\frac{1}{2}b\right) = \frac{1}{8}ab$. The four triangles have a total area of $4 \cdot \frac{1}{8}ab = \frac{1}{2}ab$.

14. perimeter $= 16\sqrt{2}$

15. 7

16. $6 + 6\sqrt{5}$

Section 9.4

A.

1. (3)	**2.** (2)	**3.** (1)	**4.** (3)	**5.** (4)	**6.** (2)

B.

7. $m\angle CBD = m\angle ABD = 31°$, $m\angle DCB + m\angle CBA = 180°$, $m\angle DCB = 118°$. $m\angle CDA = m\angle CBA = 62°$. $m\angle EDC = 58° + 62° = 120°$. *EBCD* is *not* an isosceles trapezoid because the base angles are not congruent.

8. $PM = 10$

9. $m\angle KQN = 78°$ and $m\angle LNO = 51°$

Section 9.5

A.
 1. (1) **2.** (4) **3.** (4) **4.** (2) **5.** (3) **6.** (3)

B.
 7. $\overline{AB} \parallel \overline{CD}$, $\overline{AB} \cong \overline{CD}$; given. $ABCD$ is a parallelogram; a quadrilateral with a pair of sides parallel and congruent is a parallelogram. $\overline{AB} \perp \overline{BC}$; given. $\angle ABC$ is a right angle; perpendicular lines form right angles. $ABCD$ is a rectangle; a parallelogram with a right angle is a rectangle. $\angle DEC \cong \angle DEA$; given. $\angle CED$ and $\angle DEA$ are right angles; congruent adjacent angles are right angles. $\overline{AC} \perp \overline{DB}$; lines intersecting at right angles are perpendicular. $ABCD$ is a rhombus; a parallelogram with perpendicular diagonals is a rhombus. $ABCD$ is a square; a parallelogram that is a rectangle and a rhombus is also a square.

 8. Rectangle $ABCD$; given. $\overline{AB} \cong \overline{CD}$ and $\overline{BC} \cong \overline{AD}$; opposite sides of a rectangle are congruent; Points E, F, G, and H are midpoints; given. $\overline{AE} \cong \overline{EB} \cong \overline{GC} \cong \overline{DG}$; halves of congruent segments are congruent. $\overline{HA} \cong \overline{HD} \cong \overline{CF} \cong \overline{FB}$; halves of congruent segments are congruent. $\angle A \cong \angle B \cong \angle C \cong \angle D$; all angles in a rectangle are congruent. $\triangle EAH \cong \triangle EBF \cong \triangle GCF \cong \triangle GDH$; SAS. $\overline{HE} \cong \overline{EF} \cong \overline{GF} \cong \overline{GH}$; CPCTC. $EFGH$ is a rhombus; a quadrilateral with 4 congruent sides is a rhombus.

 9. $\angle BAC \cong \angle DCA \cong \angle ABD$; given. $\angle AEB \cong \angle DEC$; vertical angles are congruent. $\triangle BAE \cong \triangle DCE$; ASA. $\overline{AE} \cong \overline{EC}$; given. $\overline{BE} \cong \overline{DE}$; CPCTC. $\overline{AE} \cong \overline{BE}$; sides opposite congruent angles in a triangle are congruent. $\overline{AE} \cong \overline{EC} \cong \overline{BE} \cong \overline{DE}$; transitive property. $\overline{AE} + \overline{EC} \cong \overline{BE} + \overline{DE}$; addition property. $\overline{AC} \cong \overline{BD}$; partition postulate. E is the midpoint of $\overline{AC}$ and $\overline{BD}$; a midpoint divides a segment into two congruent segments. $\overline{AC}$ and $\overline{BD}$ bisect each other; bisectors intersect at a midpoint. $ABCD$ is a rectangle; a quadrilateral whose diagonals bisect each other and are congruent is a rectangle.

 10. Rhombus $ABED$; given. $\overline{DE} \cong \overline{BE}$; sides of a rhombus are congruent. $\angle AED \cong \angle AEB$; diagonals of a rhombus bisect the angles. $\angle CED \cong \angle CEB$; angles supplementary to congruent angles are congruent. $\overline{EC} \cong \overline{EC}$; reflexive property. $\triangle CED \cong \triangle CEB$; SAS. $\angle ECB \cong \angle ECD$; CPCTC.

11. Q is the midpoint of $\overline{RT}$ and P is the midpoint of $\overline{ST}$; given. $\overline{PQ}$ is a midsegment of $\triangle RST$; definition of midsegment. $\overline{QP} \parallel \overline{RS}$; a midsegment is parallel to the opposite side of a triangle. $PQRS$ is a trapezoid; a quadrilateral with at least one pair of parallel sides is a trapezoid. $\overline{RT} \cong \overline{ST}$; given. $\angle R \cong \angle S$; angles opposite congruent sides of a triangle are congruent. $PQRS$ is an isosceles trapezoid; a trapezoid with congruent base angles is an isosceles trapezoid.

12. Parallelogram $EFGH$ and $\overline{HJ} \cong \overline{IG}$; given. $\overline{HE} \cong \overline{GF}$; opposite sides of a parallelogram are congruent. $\overline{JE} \cong \overline{JF}$; given. $\triangle JHE \cong \triangle IGF$; SSS. $\angle EHJ \cong \angle FGI$; CPCTC. $\angle EHJ$ and $\angle FGI$ are supplementary; consecutive angles in a parallelogram are supplementary. $\angle EHJ$ and $\angle FGI$ are right angles; congruent supplementary angles are right angles. $EFGH$ is a rectangle; a parallelogram with a right angle is a rectangle.

13. Kevin is correct. A counterexample would be an isosceles trapezoid with perpendicular diagonals. The opposite legs are congruent and the diagonals are perpendicular.

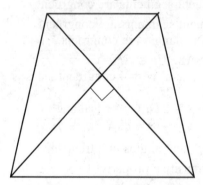

14. In regular pentagon $ABCDE$, all sides are congruent and all angles measure 108°. In isosceles triangle $\triangle AED$, m$\angle EAD = (180° - 108°)/2 = 36°$. m$\angle BAF = 108° - 36° = 72°$. m$\angle AFC = 108°$. Since the sum of the interior angles of a quadrilateral equals 360°, $\angle BCF = 360° - (108° + 108° + 72°) = 72°$. $ABCF$ is a rhombus because opposite angles are congruent and consecutive sides are congruent.

Section 9.6

A.
1. (2)   3. (2)   5. (3)   7. (2)   9. (4)
2. (4)   4. (1)   6. (4)   8. (1)

B.
10. *RSS'R'* is a square. The opposite sides are all congruent, and all angles are right angles.
11. *FGHG'* is a rhombus. *FG* = *GH* because $\triangle AFGH$ is isosceles and a reflection is a rigid motion. So *FG* = *GG'* = *GH* = *HG'*. All four sides are congruent, making it a rhombus.
12. The reflection preserves distance, so *JK* = *KJ'* and *LK* = *KL'*. Therefore *K* is the midpoint of each diagonal of *JLJ'L'* and the diagonals bisect each other.
13. 45°
14. The two translations form congruent and parallel sides $\overline{AM}$ and $\overline{NO}$, making *ANOM* a parallelogram. $\overline{AN}$ is congruent to $\overline{AM}$ because the rotation is a rigid motion, and the 90° rotation forms a right angle at *A*. A parallelogram with a right angle and a pair of consecutive congruent sides is a square.

CHAPTER 10

Section 10.2

A.
1. (3)   2. (1)   3. (2)

B.
4. The midpoints of the diagonals $\overline{AC}$ and $\overline{BD}$ are both (0, 2). The diagonals of *ABCD* bisect each other. So *ABCD* is a parallelogram.

5. The midpoints of the diagonals $\overline{FO}$ and $\overline{RG}$ are both $\left(1, \frac{7}{2}\right)$. The diagonals of *FROG* bisect each other. So *FROG* is a parallelogram.

6. The slope of $\overline{BL} = -6$. The slope of $\overline{UE} = -6$. The slope of $\overline{LU} = \frac{1}{6}$.

The slope of $\overline{EB} = \frac{1}{6}$. The slopes of opposite sides are equal, so opposite sides are parallel. Therefore *BLUE* is a parallelogram. The slopes of consecutive sides are negative reciprocals, so they are perpendicular and the angles are right angles. A parallelogram with right angles is a rectangle.

7. The midpoints of the diagonals $\overline{CA}$ and $\overline{RB}$ are both $\left(1, \dfrac{5}{2}\right)$. The diagonals of $CRAB$ bisect each other, so $CRAB$ is a parallelogram. $CA = RB = \sqrt{85}$, making $\overline{CA} \cong \overline{RB}$. $CRAB$ is a parallelogram with congruent diagonals, so it is also a rectangle.

8. The lengths of the sides are congruent, $AB = BC = CD = DA = \sqrt{37}$. $ABCD$ is a parallelogram because opposite sides are congruent. It is also a rhombus because it is a parallelogram with consecutive congruent sides.

9. The lengths of the sides are congruent, $CL = LU = UE = EC = \sqrt{68}$. $CLUE$ is a parallelogram because opposite sides are congruent. It is also a rhombus because it is a parallelogram with consecutive congruent sides.

10. The lengths of the sides are congruent, $JU = UM = MP = PJ = \sqrt{29}$. $JUMP$ is a parallelogram because opposite sides are congruent. It is a rhombus because it is a parallelogram with consecutive congruent sides. The lengths of the diagonals are congruent, $JM = PU = \sqrt{58}$. So $JUMP$ is a rectangle. Since $JUMP$ is both a rectangle and a rhombus, it is also a square.

11. The lengths of the sides are congruent, $PQ = QR = RS = SP = \sqrt{34}$. $PQRS$ is a parallelogram because opposite sides are congruent. It is a rhombus because it is a parallelogram with consecutive congruent sides. The lengths of the diagonals are congruent, $PR = SQ = \sqrt{68}$. So $PQRS$ is a rectangle. Since $PQRS$ is both a rectangle and a rhombus, it is also a square.

12. Yes. The slope of $5y = 2x + 30$ is 2/5. The slope of $3x + 2y = 20$ is $-3/2$. The slope of $3x + 2y = 1$ is $-3/2$. The slope of $y = \dfrac{2}{5}x + 12$ is 2/5. The vertices form a quadrilateral whose opposite sides have the same slopes, so the opposite sides must be parallel.

13. $D(12, 7)$.

Section 10.3

A.
  1. (1)   **2.** (3)   **3.** (2)
B.
  4. Slope of $\overline{QB} = 1$ and slope of $\overline{QC} = -1$, so $\overline{QB} \perp \overline{QC}$ and $\angle BQC$ is a right angle. A triangle with a right angle is a right triangle. Therefore $\triangle QBC$ is a right triangle.
  5. $TG = TH = 5$, so two sides are congruent. A triangle with two congruent sides is an isosceles triangle.
  6. $GH = FH = 5$, so two sides are congruent. The slope of $\overline{GH} = \dfrac{3}{4}$ and the slope of $\overline{FH} = -\dfrac{4}{3}$. The slopes are negative reciprocals, so $\overline{GH} \perp \overline{FH}$ and $\angle GHF$ is a right angle. A triangle with two congruent sides and a right angle is a right isosceles triangle.
  7. Prove $G(-2, 4)$ is the centroid by finding the midpoints of any two sides and graphing the corresponding medians. The two medians will intersect at $G(-2, 4)$. Demonstrate G divides a median in a $2:1$ ratio by calculating the lengths of the two parts of the median. $M(-5, 4)$ is the midpoint of $\overline{AB}$, and $\overline{CM}$ is a median. $GC = 6$ and $CM = 3$. $\dfrac{GC}{CM} = \dfrac{2}{1}$, so G divides $\overline{CM}$ in a $2:1$ ratio.
  8. Slope of $\overline{BD} = 1$ and slope of $\overline{AC} = -1$. The slopes are negative reciprocals, so $\overline{BD} \perp \overline{AC}$, making $\overline{BD}$ an altitude.
  9. $MO = ON = \sqrt{41}$, so $\overline{MO} \cong \overline{ON}$. The slope of $\overline{MO} = -\dfrac{4}{5}$, and the slope of $\overline{ON} = \dfrac{5}{4}$. The slopes are negative reciprocals, making $\overline{MO} \perp \overline{ON}$ and $\angle O$ a right angle. $\triangle MON$ is therefore an isosceles right triangle with a right angle at $\angle MON$ and $45°$ angles at $\angle M$ and $\angle N$.

10. $DA = DB = DC = \sqrt{13}$. Point D is therefore equidistant from each of the vertices of $\triangle ABC$. A point that is equidistant from each of the vertices of a triangle is the circumcenter.

11. $DF = DG = DH = \sqrt{34}$. P is equidistant from all three points, making it the center of the circumscribed circle of $\triangle FGH$.

12. D will be the orthocenter if it lies on each altitude. The slope of $\overline{DJ} = 7$ and the slope of $\overline{KL}$ is $-\frac{1}{7}$, making $\overline{DJ} \perp \overline{KL}$. The slope of $\overline{DK} = -3$ and the slope of $\overline{JL} = \frac{1}{3}$, making $\overline{DK} \perp \overline{JL}$. The slope of $\overline{DL}$ is $-\frac{1}{2}$ and the slope of $\overline{JK}$ is 2, making $\overline{DL} \perp \overline{JK}$. Since an altitude of a triangle is a segment from a vertex perpendicular to the opposite side, point D lies on each altitude and is the orthocenter.

13. $a = 0$ or $a = 2$

CHAPTER 11

Section 11.1

A.

1. (4)   4. (4)   7. (2)   10. (4)   13. (3)
2. (4)   5. (2)   8. (3)   11. (1)
3. (3)   6. (1)   9. (2)   12. (4)

B.

14. $m\angle R = 94°$, $m\angle S = 99°$, $m\angle M = 86°$

15. $120°$

16. Carly is correct. Each angle of the rectangle measures 90° and is also an inscribed angle of the circle. A 90° inscribed angle intercepts a 180° arc, or a semicircle. So the diagonal must be coincident with a diameter.

17. Rayshawn is correct. Opposite angles in a parallelogram are congruent. The opposite angles are also inscribed angles that intercept two arcs spanning the entire circle. The arc measures sum to 360°, so the inscribed angles must sum to 180°. If the angles are congruent, they must each measure 90°.

18. $\overarc{AR} \cong \overarc{IN}$; given. $\angle AIR \cong \angle NAI$; inscribed angles that intercept congruent arcs are congruent. $\angle INA \cong \angle ARI$; inscribed angles that intercept the same arc are congruent. $\overline{AI} \cong \overline{AI}$; reflexive property. $\triangle RAI \cong \triangle NIA$; AAS.

Section 11.2

A.

  1. (2)   **3.** (2)   **5.** (4)   **7.** (2)   **9.** (2)
  2. (3)   **4.** (1)   **6.** (4)   **8.** (3)

B.

10. 55°

11. 72°

12. $\overline{MN} \cong \overline{PQ}$ and $\overline{MN} \parallel \overline{PQ}$; given. $m\overset{\frown}{NQ} = m\overset{\frown}{MP}$; arcs between parallel chords are congruent. $m\overset{\frown}{PQ} = m\overset{\frown}{MN}$; arcs intercepted by congruent chords are congruent. $m\overset{\frown}{NQ} + m\overset{\frown}{PQ} + m\overset{\frown}{MP} + m\overset{\frown}{MN} = 360°$; the sum of the arc measures around a circle equals 360°. $2m\overset{\frown}{NQ} + 2m\overset{\frown}{MP} = 360°$; substitution property. $m\overset{\frown}{NQ} + m\overset{\frown}{MP} = 180°$, division property.

13. Jason is correct. The congruent chords must intercept congruent arcs $\overset{\frown}{AB}$ and $\overset{\frown}{CD}$. These are the arcs between $\overline{AC}$ and $\overline{BD}$, so $\overline{AC} \parallel \overline{BD}$. The figure has one pair of parallel sides and a pair of congruent legs.

14. $\overline{AC} \cong \overline{BD}$; given. $\overset{\frown}{AC} \cong \overset{\frown}{BD}$; congruent chords intercept congruent arcs. $m\overset{\frown}{AC} = m\overset{\frown}{AB} + m\overset{\frown}{BC}$ and $m\overset{\frown}{BD} = m\overset{\frown}{BC} + m\overset{\frown}{CD}$; partition postulate. $m\overset{\frown}{AB} + m\overset{\frown}{BC} = m\overset{\frown}{BC} + m\overset{\frown}{CD}$, substitution. $m\overset{\frown}{BC} = m\overset{\frown}{BC}$; reflexive property. $m\overset{\frown}{AB} = m\overset{\frown}{CD}$; subtraction property. $\overset{\frown}{AB} \cong \overset{\frown}{CD}$; definition of congruence.

15. $\overline{HO} \cong \overline{HE}$, $\overline{OS} \cong \overline{ES}$; given. $m\overset{\frown}{HO} = m\overset{\frown}{HE}$ and $m\overset{\frown}{OS} = m\overset{\frown}{ES}$; congruent chords intercept congruent arcs. $m\overset{\frown}{HO} + m\overset{\frown}{HE} + m\overset{\frown}{OS} + m\overset{\frown}{ES} = 360°$; the sum of the arcs around a circle equals 360°. $2m\overset{\frown}{HO} + 2m\overset{\frown}{OS} = 360°$; substitution. $m\overset{\frown}{HO} + m\overset{\frown}{OS} = 180°$; division postulate. $m\overset{\frown}{HOS} = 180°$; partition postulate. $\overline{HS}$ is a diameter; a diameter intercepts a semicircle.

Section 11.3

A.

  1. (1)   **3.** (2)   **5.** (4)   **7.** (4)
  2. (4)   **4.** (3)   **6.** (2)

B.

8. Equilateral $\triangle WXY$ and radius $\overline{QP}$ extended to Y; given. $\overline{QPY} \perp \overline{WQX}$, a radius is $\perp$ to a tangent of a circle. $\overline{WSY} \cong \overline{XRY}$; all sides of an equilateral triangle are congruent. $\overline{QY} \cong \overline{QY}$; reflexive property. $\triangle WQY \cong \triangle XQY$; HL. $\angle WYQ \cong \angle XYQ$; CPCTC. $\overline{QPY}$ bisects $\angle WYX$; a segment that divides an angle into two $\cong$ angles is an angle bisector.

Section 11.4

A.

1. (3)	**4.** (1)	**7.** (4)	**10.** (1)	**13.** (4)
2. (2)	**5.** (3)	**8.** (4)	**11.** (2)	**14.** (2)
3. (4)	**6.** (3)	**9.** (3)	**12.** (1)	**15.** (1)

B.

16. 134°
17. 36°
18. 68.5°
19. 47°

20. $\text{m}\angle W = \dfrac{1}{2}\left(\text{m}\widehat{XZY} - \text{m}\widehat{XY}\right)$

$64° = \dfrac{1}{2}(360 - x - x)°$

$x = 116°$

$\widehat{XZY} = 360° - 116° = 244°$

$\text{m}\widehat{XZ} = \widehat{XZY} - \widehat{YZ} = 244° - 64° = 180°$

$\widehat{XAZ}$ is a diameter because it intercepts an arc of 180°.

Section 11.5

A.

1. (1)	**3.** (4)	**5.** (1)	**7.** (4)	**9.** (1)	**11.** (4)
2. (3)	**4.** (1)	**6.** (3)	**8.** (2)	**10.** (2)	

B.

12. $\overline{JK}$ and $\overline{LM}$ intersect at A; given. $\angle J \cong \angle L$ and $\angle K \cong \angle M$; inscribed angles that intercept the same arc are congruent. $\triangle JMA \cong \triangle LKA$; AA. $\dfrac{JA}{MA} = \dfrac{LA}{KA}$; corresponding sides in similar triangles are proportional. $JA \cdot KA = LA \cdot MA$; cross multiply.

13. $CE = 5$ and $EB = 12.5$

14. $EC = 4\sqrt{5}$ and $DE = \dfrac{8\sqrt{5}}{5}$

Section 11.6

A.

1. (3)	**3.** (2)	**5.** (1)	**7.** (3)	**9.** (3)	**11.** (4)
2. (4)	**4.** (1)	**6.** (2)	**8.** (2)	**10.** (4)	**12.** (1)

B.

13. $s = r\theta$, $s_1 = 7(0.28) = 1.96$, $s_2 = 3(0.6) = 1.8$. Arc s_1 is longer.

14. 6,480 square ft

15. 8.8

16. 49.1

CHAPTER 12

Section 12.1

A.

1. (4)	**3.** (2)	**5.** (2)	**7.** (3)	**9.** (2)
2. (3)	**4.** (1)	**6.** (4)	**8.** (3)	**10.** (1)

B.

11. 5 cans of soup

12. 50 ft^2

13. 600,000 gallons

14. 25 hours

Section 12.2

A.

1. (4)	**2.** (1)	**3.** (3)	**4.** (1)	**5.** (2)

B.

6. 0.25 m^3

7. $\dfrac{2}{3}$

8. 230

9. $R_{\text{cone}} : R_{\text{cylinder}} = \sqrt{3} : 1$

10. Diameter = 2.5 cm

11. 144 scoops

12. height = 13.5

Section 12.3

A.

1. (1)	**3.** (2)	**5.** (1)	**7.** (3)	**9.** (2)	**11.** (4)
2. (2)	**4.** (3)	**6.** (2)	**8.** (3)	**10.** (2)	

B.
12. A cone with radius = 4 and height = 6. Volume = 32π.
13. cylinder, volume = 360π
14. $1,152\pi$
15. 2.67 ft^3

Section 12.4

A.
1. (3) 3. (4) 5. (3) 7. (2) 9. (4)
2. (2) 4. (1) 6. (3) 8. (2) 10. (1)

B.
11. Each card has the same cross sectional area and thickness. Cavalieri's principle states that two solids with the same cross-sectional area and thickness perpendicular to that cross section will have the same volume.
12. Two prisms can be contained between the same two parallel planes since they have the same height. Any plane parallel to the bases will intercept the same rectangular cross-sectional area. Cavalieri's principle states that two solids contained between a pair of parallel planes will be equal in volume if any other parallel plane intercepts cross sections of the same area.
13. Start with $A = \dfrac{1}{2}$ perimeter · apothem. As the number of sides of the inscribed polygon increases and approaches infinity, the perimeter approaches the circumference of the circle and the apothem approaches the radius. So $A = \dfrac{1}{2}C \cdot r$. Substitute $C = 2\pi r$ to get

$$A = \frac{1}{2}(2\pi r)r = \pi r^2.$$

Section 12.5

A.
1. (3) 2. (3) 3. (4)

B.
4. $45,000$ ft^2
5. 124
6. Natural gas is cheaper. Heating oil costs $24.34 per million BUT compared to $10 per million BTU for natural gas.
7. $4,590$ ft^3
8. $3,575
9. (a) Beaker B has a greater volume, 63.6 cm^3 vs. 37.7 cm^3
   (b) The liquid has a volume of 56.2 cm^3, so will fit in beaker B.

10. 26,764 blocks weighing 132,422 tons.
11. $0.53/ounce
12. 14 fish
13. $656,925
14. 10,900
15. 20.2 ft
16. 314 boxes
17. $86,400
18. 11,000 apples
19. $829
20. 33 truck loads

Glossary of Geometry Terms

A

Acute angle – An angle whose measure is greater than 0° and less than 90°.

Acute triangle – A triangle whose angles are all acute angles.

Adjacent angles – Two angles that share a common vertex and one side but do not share interior points.

Altitude – A segment from a vertex of a triangle perpendicular to the opposite side.

Angle – A figure formed by two rays with a common endpoint. Symbol is ∠.

Angle bisector – A line, segment, or ray that divides an angle into two congruent angles.

Angle measure – The amount of opening of an angle, measured in degrees or radians.

Angle of depression – The angle formed by the horizontal and the line of sight when looking downward to an object.

Angle of elevation – The angle formed by the horizontal and the line of sight when looking upward to an object.

Angle of rotation – The angle measure by which a figure or point spins around a center point.

Arc – A portion of a circle with two endpoints on the circle. A **major arc** measures more than 180°; a **minor arc** measures less than 180°; a **semicircle** measures 180°. Symbol is ⌒.

Arc length – The distance between the endpoints of an arc. Arc length equals $r\theta$ where θ is the arc measure in radians and r is the radius.

Arc measure – The angle measure of an arc, equal to the measure of the central angle that intercepts the arc.

Axis of symmetry – A line that divides a figure into two congruent parts.

B

Base (of a circular cone) – The circular face of a cone; it is opposite the apex.

Base (of an isosceles triangle) – The non-congruent side of an isosceles triangle.

Base (of a pyramid) – The polygonal face of a pyramid that is opposite the apex.

Bases (of a prism) – A pair of faces of a prism that are parallel, congruent polygons.

Bases (of a circular cylinder) – The pair of congruent parallel circular faces of a cylinder.

Base Angles – The angles formed at each end of the base of an isosceles triangle.

Bisector (Segment bisector) – A line, segment, or ray that passes through the midpoint of a segment.

C

Cavalieri's principle – If two solids are contained between two parallel planes and every parallel plane between these two planes intercepts regions of equal area, then the solids have equal volume. Also any two parallel planes intercept two solids of equal volume.

Center of a regular polygon – The center of the inscribed or circumscribed circle of a regular polygon.

Center of dilation – The fixed point reference point used to determine the expansion or contraction of lengths in a dilation. The center of dilation is the only invariant point in a dilation.

Center–radius equation of a circle – A circle can be represented on the coordinate plane by $(x - h)^2 + (y - k)^2 = r^2$, where the point (h, k) is the center of the circle and r is the radius.

Central angle – An angle in a circle formed by two distinct radii.

Centroid of a triangle – The point of concurrency of the three medians of a triangle.

Chord – A segment whose endpoints lie on a circle.

Circle – The set of points that are equidistant from a fixed center point. Symbol is $\odot$.

Circumcenter – The point that is the center of the circle circumscribed about a polygon, equidistant from the vertices of a polygon, and the point of concurrency of the perpendicular bisectors of a triangle.

Circumference – The distance around a circle. Circumference = $2\pi \cdot$radius.

Coincide (coincident) – Figures that lie entirely on each other.

Collinear – Points that lie on the same line.

Compass – A tool for drawing accurate circles and arcs.

Complementary angles – Angles whose measures sum to 90°.

Completing the square – A method used to rewrite a quadratic expression of the form $ax^2 + bx + c$ as a squared binomial of the form $(dx + e)^2$.

Composition of transformations – A sequence of transformations in a specified order. Notation is *transformation 2 ∘ transformation 1*. The transformation on the right is performed first.

Concave polygon – A polygon with at least one diagonal outside the polygon.

Concentric circles – Circles with the same center.

Concurrent – When three or more lines all intersect at a single point.

Cone (circular) – A solid figure with a single circular base and a curved lateral face that tapers to a point called the apex.

Congruent – Figures with the same size and shape. Symbol is $\cong$.

Construction – A figure drawn with only a compass and straightedge.

Convex polygon – A polygon whose diagonals all lie within the polygon.

Coplanar – Figures that lie in the same plane.

Corresponding parts – A pair of parts (usually points, sides, or angles) of two figures that are paired together through a specified relationship, such as a congruence or similarity statement or a transformation function.

Cosine of an angle – In a right triangle, the ratio of the length of the side adjacent to an acute angle to the length of the hypotenuse.

CPCTC – Corresponding parts of congruent triangles are congruent.

Cross section – The intersection of a plane with a solid.

Cube – A prism whose faces are all squares.

Cubic unit – The amount of space occupied by a cube 1 unit of length in each dimension. (1 ft^3 is the amount of space occupied by a 1 ft × 1 ft × 1 ft cube).

Cylinder (circular) – A solid with two parallel, congruent circular bases and a curved lateral face.

D

Degree – A unit of angle measure equal to $\dfrac{1}{360}$ of a complete rotation.

Diagonal – A segment in a polygon whose endpoints are nonconsecutive vertices of the polygon.

Diameter – A chord that passes through the center of a circle.

Dilation – A similarity transformation about a center point O that maps preimage P to image P' such that O, P, and P' are collinear and $OP' = k \cdot OP$. Notation is D_k.

Direct transformation – A transformation that preserves orientation.

Distance from a point to a line – The length of the segment from the point perpendicular to the line.

E

Edge of a solid – The intersection of two polygonal faces of a solid.

Endpoint – The two points that bound a segment.

Equiangular – A figure whose angles all have the same measure.

Equidistant – The same distance from two or more points.

Equilateral – A figure whose sides all have the same length.

Exterior angle of a polygon – The angle formed by a side of a polygon and the extension of an adjacent side.

F

Face of a solid – Any of the surfaces that bound a solid.

Figure – Any set of points. The points may form a segment, line, ray, plane, polygon, curve, solid, and so on.

G

Geometric mean – See *Mean proportional*.

Great circle – The largest circle that can be drawn on a sphere. The circle formed by the intersection of a sphere and a plane passing through the center of the sphere.

H

Hexagon – A 6-sided polygon.

Hypotenuse – The longest side of a right triangle. It is always opposite the right angle.

I

Identity transformation – A transformation in which the preimage and image are coincident.

Image – The figure that results from applying a transformation to an initial figure called the preimage.

Incenter of a triangle – The point that is the center of the inscribed circle of a triangle, equidistant from the 3 sides of a triangle, and the point of concurrency of the three angle bisectors of a triangle.

Inscribed angle – An angle in a circle formed by two chords with a common endpoint on the circle.

Inscribed circle of a triangle – A circle that is tangent to all three sides of the triangle.

Intersecting – Figures that share at least one common point.

Isometry – See *Rigid motion*.

492

Isosceles triangle – A triangle with two congruent sides.

Isosceles trapezoid – A trapezoid with congruent legs.

K

Kite – A polygon with two distinct pairs of adjacent congruent sides. The opposite sides are not congruent.

L

Lateral edge – The intersection between two lateral faces of a polyhedron.

Lateral face – Any face of a polyhedron that is not a base.

Line – One of the undefined terms in geometry. An infinitely long set of points that has no width or thickness. Symbol is ⟷.

Linear pair of angles – Two adjacent supplementary angles.

Line symmetry – A line over which one half of a figure can be reflected and mapped onto the other half of the figure.

Line reflection – A rigid motion in which every point P is transformed to a point P' such that the line is the perpendicular bisector of PP'.

M

Major arc – An arc whose degree measure is greater that 180°.

Map (Mapping) – A pairing of every point in a preimage with a point in an image. Reflections, rotations, translations, and dilations are one-to-one mappings because every point in the preimage maps to exactly one point in the image and every point in the image maps to exactly one point in the preimage.

Mean proportional (Geometric mean) – The geometric mean of two numbers, a and b, is the square root of their product. If $\dfrac{a}{m} = \dfrac{m}{b}$, then m is the geometric mean.

Median of a triangle – The segment from a vertex of a triangle to the midpoint of the opposite side.

Midpoint – A point that divides a segment into two congruent segments.

Midsegment of a triangle – A segment joining the midpoints of two sides of a triangle.

Midsegment (median) of a trapezoid – A segment joining the midpoints of the two legs of a trapezoid.

Minor arc – An arc whose degree measure is less than 180°.

N

Noncollinear – Points that do not lie on the same line.

Noncoplanar – Points or lines that do not lie on the same plane.

O

Obtuse angle – An angle whose measure is greater than 90° and less than 180°.

Octagon – A polygon with 8 sides.

Opposite rays – Two rays with a common endpoint that together form a straight line.

Opposite transformation – A transformation that changes the orientation of a figure.

Orientation – The order in which the points of a figure are encountered when moving around a figure. The orientation can be clockwise or counterclockwise.

Orthocenter – The point of concurrence of the three altitudes of a triangle.

P

Parallel lines – Coplanar lines that do not intersect. Symbol is ∥.

Parallelogram – A quadrilateral with two pairs of opposite parallel sides.

Parallel planes – Planes that do not intersect.

Pentagon – A polygon with 5 sides.

Perimeter – The sum of the lengths of the sides of a polygon.

Perpendicular – Intersecting at right angles. Symbol is ⊥.

Perpendicular bisector – A line, segment, or ray perpendicular to another segment at its midpoint.

Pi – The ratio of the circumference of a circle to its diameter. Pi is an irrational number whose value is approximately 3.14159. Symbol is π.

Plane – One of the undefined terms in geometry. A set of points with no thickness that extends infinitely in two directions. It is usually visualized as a flat surface.

Point – An undefined term in geometry. A location in space with no length, width, or thickness. Symbol is •.

Point of tangency – The point where a tangent intersects a curve.

Point reflection – A transformation in which a specified center point is the midpoint of each of the segments connecting a point in the preimage with a corresponding point in the image. It is equivalent to a rotation of 180°.

Point-slope equation of a line – A form of the equation of a line,

$y - y_1 = m(x - x_1)$, where m is the slope and (x_1, y_1) are the coordinates of a point on the line.

Point symmetry – A figure has point symmetry if every point in the figure is a 180° rotation of a corresponding point in the same figure.

Polygon – A closed planar figure whose sides are segments that intersect only at their endpoints (do not overlap).

Polyhedron – Plural: polyhedra. A solid figure in which each face is a polygon. Prisms and pyramids are examples of polyhedra.

Postulate – A statement that is accepted to be true without proof.

Preimage – The original figure that is acted on by a transformation.

Preserves – Remains unchanged.

Prism – A polyhedron with two congruent, parallel polygons for bases and whose lateral faces are parallelograms.

Proportion – An equation that states two ratios are equal. For example, $\frac{a}{b} = \frac{c}{d}$. The cross products of a proportion are equal, $a \cdot d = b \cdot c$.

Pyramid – A polyhedron with one polygonal base and triangles for lateral faces.

Pythagorean theorem – In a right triangle, the sum of the squares of the two legs equals the square of the hypotenuse, or $a^2 + b^2 = c^2$, where a and b are the lengths of the two legs and c is the length of the hypotenuse.

Q

Quadratic equation – An equation in the form $ax^2 + bx + c = 0$, where $a \neq 0$.

Quadratic formula – A formula for finding the two solutions to a quadratic equation of the form $ax^2 + bx + c = 0$,

$$x = \frac{-b \pm \sqrt{b^2 - 4ac}}{2a}.$$

Quadrilateral – A polygon with four sides.

R

Radian – An angle measure in which one full rotation is 2π radians. Also 1 radian is the measure of an arc such that the arc's length is equal to the radius of that circle.

Radius – A segment from the center of a circle to a point on the circle.

Ray – A portion of a line starting at an endpoint and including all points on one side of the endpoint. Symbol is —.

Rectangle – A parallelogram with right angles.

Reflection – See *Line reflection* and *Point reflection*.

Reflexive property of equality – Any quantity is equal to itself. For figures, any figure is congruent to itself.

Regular polygon – A polygon in which all sides are congruent and all angles are congruent.

Rhombus – A parallelogram with four congruent sides.

Right angle – An angle that measures 90°.

Right circular cone – A cone with a circular base and whose altitude passes through the center of the base.

Right circular cylinder – A cylinder with circular bases and whose altitude passes through the center of the bases.

Right pyramid – A pyramid whose faces are isosceles triangles.

Right triangle – A triangle that contains a right angle.

Rigid motion (Isometry) – A transformation that preserves distance. The image and preimage are congruent under a rigid motion. Translations, reflections, and rotations are isometries.

Rotation – A rigid motion in which every point, P, in the preimage spins by a fixed angle around a center point, C, to a point P'. The distance to the center point is preserved.

Rotational symmetry – A figure has rotational symmetry if it maps onto itself after a rotation of less than 360°.

S

Scale drawing – A drawing of a figure or an object that represents a dilation of the actual figure or object. A drawing of an object in which every length is enlarged or reduced by the same scale factor.

Scale factor – The ratio by which a figure is enlarged or reduced by a dilation.

Scalene triangle – A triangle in which no sides have the same length.

Secant of a circle – A line that intersects a circle in exactly two points.

Sector of a circle – A region bounded by a central angle of a circle and the arc it intersects.

Segment – A portion of a line bounded by two endpoints.

Segment bisector – See *Bisector*.

Semicircle – An arc that measures 180°.

Similar polygons – Two polygons with the same shape but not necessarily the same size.

Similarity transformation – A transformation in which the preimage and the image are similar. A transformation that includes a dilation.

Sine of an angle – In a right triangle, the ratio of the length of the side opposite an acute angle to the length of the hypotenuse.

Skew lines – Two lines that are not coplanar.

Slant height – The distance along a lateral face of a solid from the apex perpendicular to the opposite edge.

Slope – A numerical measure of the steepness of a line. In the coordinate plane, the slope of a line equals the change in the y-coordinates divided by the change in the x-coordinates between any two points. The slope of a vertical line is undefined.

Slope-intercept equation of a line – A form of the equation of a line, $y = mx + b$, where m is the slope and b is the y-intercept.

Solid figure – 3-dimensional figure fully enclosed by surfaces.

Sphere – A solid comprised of the set of points in space that are a fixed distance from a center point.

Square – A parallelogram with four right angles and four congruent sides.

Straightedge – A ruler with no length marking, used for constructing straight lines.

Substitution property of equality – The property that a quantity can be replaced by an equal quantity in an equation.

Subtraction property of equality – If the same or equal quantities are subtracted from the same or equal quantities, the differences are equal.

Supplementary angles – Two angles whose measures add to 180°.

Surface area – The sum of the areas of all the faces or curved surfaces of a solid figure.

T

Tangent of an angle – In a right triangle, the ratio of the length of the side opposite an acute angle to the length of the side adjacent to the angle.

Tangent to a circle – A line, coplanar with a circle, that intersects the circle at only one point.

Theorem – A general statement that can be proved.

Transformation – A one-to-one function that maps a set of points, called the preimage, to a new set of points, called the image.

Transitive property of congruence – The property that states if figure $A \cong$ figure B and figure $B \cong$ figure C, then figure $A \cong$ figure C.

Transitive property of equality – The property that states if $a = b$ and $b = c$, then $a = c$.

Translation – A rigid motion that slides every point in the preimage in the same direction and by the same distance.

Transversal – A line that intersects two or more other lines at different points.

Trapezoid – A quadrilateral that has one and only one pair of parallel sides.

Truth value – The value of true or false that is assigned to a statement.

U

Undefined terms – Terms that cannot be formally defined using previously defined terms. In geometry, these traditionally include point, line, and plane.

V

Vector – A quantity that has both magnitude and direction; represented geometrically by a directed line segment. Symbol is —.

Vertex – Plural: vertices. The point of intersection of two consecutive sides of a polygon.

Vertex angle – The angles formed by the two congruent sides in an isosceles triangle.

Vertical angles – The pairs of opposite angles formed by two intersecting lines.

Volume – The amount of space occupied by a solid figure measured in cubic units (in.^{3}, cm^3, and so on). The number of nonoverlapping unit cubes that can fit into the interior of a solid.

Z

Zero product property – The property that states if $a \cdot b = 0$, then $a = 0$, or $b = 0$, or a and $b = 0$.

Summary of Geometric Relationships and Formulas

ANGLES AND LINES

- The sum of the measures of adjacent angles around a point equals 360°.

$$m\angle 1 + m\angle 2 + m\angle 3 + m\angle 4 = 360°$$

- Vertical angles are congruent.

$$\angle 1 \cong \angle 3 \text{ and } \angle 2 \cong \angle 4$$

- A linear pair adds to 180°.

$$m\angle 1 + m\angle 2 = 180°$$

PARALLEL LINES

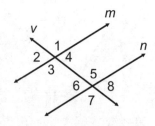

If line $m \parallel$ line n:

- Alternate interior angles are congruent: $\angle 4 \cong \angle 6$ and $\angle 3 \cong \angle 5$.
- Corresponding angles are congruent: $\angle 2 \cong \angle 6$, $\angle 3 \cong \angle 7$, $\angle 1 \cong \angle 5$, and $\angle 4 \cong \angle 8$.
- Same side interior angles are supplementary: $m\angle 3 + m\angle 6 = 180°$ and $m\angle 4 + m\angle 5 = 180°$.

POLYGONS

- The sum of the interior angles of a polygon with n sides equals $180(n - 2)$.
- The sum of the exterior angles of a polygon with n sides equals $360°$.

SPECIAL SEGMENTS IN TRIANGLES

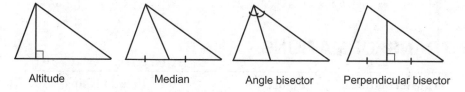

Altitude   Median   Angle bisector   Perpendicular bisector

ANGLE AND SEGMENT RELATIONSHIPS IN TRIANGLES

- Angle sum theorem

$$m\angle A + m\angle B + m\angle C = 180°$$

- Exterior angle theorem

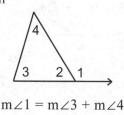

$$m\angle 1 = m\angle 3 + m\angle 4$$

499

- Isosceles triangle

$\angle A \cong \angle B, \ \overline{AC} \cong \overline{BC}$

- Pythagorean theorem

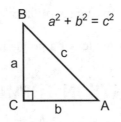

$a^2 + b^2 = c^2$

TRANSFORMATIONS

Transformation	Figure	Relationships
Reflection $r_{\overleftrightarrow{FE}}\left(\overline{AB}\right) \rightarrow \overline{A'B'}$		$\overleftrightarrow{FE}$ is the $\perp$ bisector of $\overline{BB'}$ and $\overline{AA'}$ $\overline{AA'} \parallel \overline{BB'}$
Rotation $R_{C,100°}\left(\overline{AB}\right) \rightarrow \overline{A'B'}$		$\overline{AC} \cong \overline{A'C}$ $\overline{BC} \cong \overline{B'C}$ $\angle ACA' \cong \angle BCB'$
Translation $T_{\overrightarrow{UV}}\left(\overline{AB}\right) \rightarrow \overline{A'B'}$		$\overline{AB} \parallel \overline{A'B'}$ $\overline{AA'} \parallel \overline{BB'}$

Transformation	Figure	Relationships
Dilation $D_k\left(\overline{AB}\right) \rightarrow \overline{A'B'}$		$\overline{AB} \parallel \overline{A'B'}$

Transformations on the Coordinate Plane

Reflection	Rotation About the Origin	Translation	Dilation
$r_{x\text{-axis}}(x, y) \rightarrow (x, -y)$	$R_{90}(x, y) \rightarrow (-y, x)$	$T_{h,k}(x, y) \rightarrow (x + h, y + k)$	$D_{k,\text{origin}}(x, y) \rightarrow (kx, ky)$
$r_{y\text{-axis}}(x, y) \rightarrow (-x, y)$	$R_{180}(x, y) \rightarrow (-x, -y)$		
$r_{y=x}(x, y) \rightarrow (y, x)$	$R_{270}(x, y) \rightarrow (y, -x)$		
$r_{y=-x}(x, y) = (-y, -x)$			

TRIANGLE CONGRUENCE POSTULATES

- SSS
- SAS
- ASA
- AAS
- HL

CPCTC

- Corresponding parts of congruent triangles are congruent.

Coordinate Geometry Formulas

Given two points (x_1, y_1) and (x_2, y_2):

- The distance between the points is $\sqrt{(x_1 - x_2)^2 + (y_1 - y_2)^2}$.

- The midpoint of the segment joining the points is $\dfrac{x_1 + x_2}{2}, \dfrac{y_1 + y_2}{2}$.

- The slope of the segment joining the points is $\dfrac{y_2 - y_1}{x_2 - x_1}$ or $\dfrac{\text{rise}}{\text{run}}$.

- Find the points that divide a segment proportionally using ratio $= \dfrac{x - x_1}{x_2 - x}$ and ratio $= \dfrac{y - y_1}{y_2 - y}$.

EQUATIONS OF CURVES

- Line: $y = mx + b$ m = slope, b = y-intercept
- Circle: $(x - h)^2 + (y - k)^2 = r^2$ (h, k) = coordinates of center, r = radius

TRIANGLE SIMILARITY POSTULATES

- AA
- SSS
- SAS

SIMILARITY RELATIONSHIPS IN TRIANGLES

- Segment parallel to a side forms two similar triangles

$$\overline{PQ} \parallel \overline{AB}$$
$$\triangle ABC \sim \triangle PQC$$

- Side splitter theorem

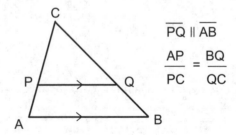

$$\overline{PQ} \parallel \overline{AB}$$
$$\frac{AP}{PC} = \frac{BQ}{QC}$$

- Centroid divides a median in a $1:2$ ratio

$$\frac{EG}{AG} = \frac{1}{2} \text{ and } \frac{DG}{CG} = \frac{1}{2}$$

- Midsegment theorem—A segment joining the midpoints of two sides of a triangle is parallel to the opposite side and its length is $\frac{1}{2}$ the opposite side.

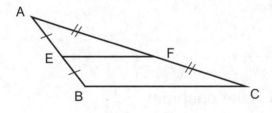

$$\overline{EF} \parallel \overline{BC} \text{ and } EF = \frac{1}{2} BC$$

- Altitude to the hypotenuse of a right triangle theorem—An altitude to the hypotenuse of a right triangle divides the triangle into two similar triangles, each of which is also similar to the original right triangle.

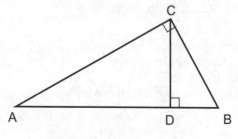

$$\triangle BDC \sim \triangle CDA \sim \triangle BCA$$

SCALING LENGTH, AREA, AND VOLUME

- Length is proportional to the scale factor.
- Area is proportional to the (scale factor)^2.
- Volume is proportional to the (scale factor)^3.

TRIGONOMETRIC RELATIONSHIPS

- $\sin = \dfrac{\text{opposite}}{\text{hypotenuse}}$

- $\cos = \dfrac{\text{adjacent}}{\text{hypotenuse}}$

- $\tan = \dfrac{\text{opposite}}{\text{adjacent}}$

Cofunction Relationships:

- $\sin(A) = \cos(90 - A)$
- $\cos(A) = \sin(90 - A)$

PARALLELOGRAMS

A parallelogram is a quadrilateral whose opposite sides are parallel. All parallelograms have the following properties:

- Opposite sides are congruent.
- Opposite angles are congruent.
- Adjacent angles are supplementary.
- The diagonals bisect each other.
- The two diagonals each divide the parallelogram into two congruent triangles.

Special Parallelograms

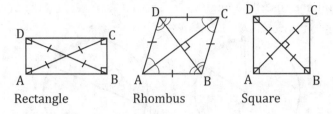

Rectangle   Rhombus   Square

Additional Properties of Special Parallelograms

	Rectangle	Rhombus	Square
4 right angles	✓		✓
Congruent diagonals	✓		✓
4 congruent sides		✓	✓
Perpendicular diagonals		✓	✓
Diagonals bisect the angles		✓	✓

Trapezoids

$\overline{AB} \parallel \overline{CD}$

$m\angle BAD + m\angle ADC = 180°$

$m\angle ABC + m\angle BCD = 180°$

Isosceles Trapezoids

$\overline{AB} \parallel \overline{CD} \quad \overline{AD} \cong \overline{BC}$

$m\angle BAD + m\angle ADC = 180°$

$m\angle ABC + m\angle DCB = 180°$

$\overline{AC} \cong \overline{BD} \quad \angle BAD \cong \angle ABC$

505

Circle Relationships

- Central and inscribed angles

$$m \overset{\frown}{AB} = m\angle APB \qquad m \overset{\frown}{CD} = 2m\angle CMD$$

- Congruent chords

$$\overline{AB} \cong \overline{CD}, \ \overset{\frown}{AB} \cong \overset{\frown}{CD}$$

- Parallel chords

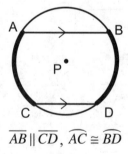

$$\overline{AB} \parallel \overline{CD}, \ \overset{\frown}{AC} \cong \overset{\frown}{BD}$$

- The perpendicular bisector of any chord passes through the center of the circle.
- A diameter or radius that is perpendicular to a chord bisects the chord.

- A diameter or radius that bisects a chord is perpendicular to the chord.

- A tangent is perpendicular to a radius.

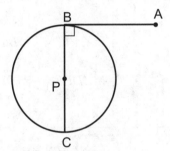

- Tangents from the same point are congruent.

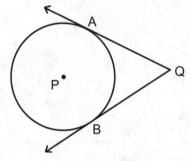

• Angles formed by chords, secants, and tangents:

Vertex inside the circle	Angle measure = $\frac{1}{2}$ the sum of the intercepted arcs
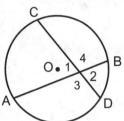	$m\angle 1 = m\angle 2 = \frac{1}{2}\left(m\widehat{AC} + m\widehat{DB}\right)$ $m\angle 3 = m\angle 4 = \frac{1}{2}\left(m\widehat{BC} + m\widehat{AD}\right)$
Vertex on the circle	Angle measure = $\frac{1}{2}$ of the intercepted arc
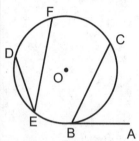	$m\angle ABC = \frac{1}{2}m\widehat{BC}$ $m\angle DEF = \frac{1}{2}m\widehat{DF}$
Vertex outside the circle	Angle measure = $\frac{1}{2}$ the difference of the intercepted arcs
  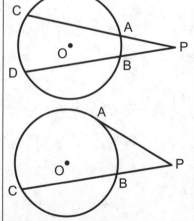	$m\angle P = \frac{1}{2}\left(m\widehat{ACB} - m\widehat{AB}\right)$ $m\angle P = \frac{1}{2}\left(m\widehat{CD} - m\widehat{AB}\right)$ $m\angle P = \frac{1}{2}\left(m\widehat{AC} - m\widehat{AB}\right)$

• Segments formed by chords, secants, and tangents:

Intersecting chords	Products of the parts are equal $a \cdot b = c \cdot d$
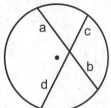	
Intersecting secants	outside · whole = outside · whole $PA \cdot PC = PB \cdot PD$
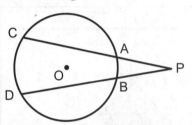	
Intersecting tangent and secant	outside · whole = outside · whole $PA^2 = PB \cdot PC$
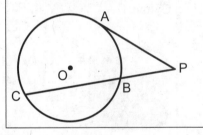	

Radian Measure

• $\text{radians} = \dfrac{\pi}{180°} \cdot \text{degrees}$

• $\text{degrees} = \dfrac{180°}{\pi} \cdot \text{radians}$

Arc Length and Sector Area

Arc Length	Sector Area
$r \cdot \theta$	$\dfrac{1}{2} r^2 \cdot \theta$

where θ is the measure of the central angle in radians and r is the radius.

Volume

- Prism: $V = Bh$
- Cylinder: $V = \pi r^2 h$
- Cone and pyramids: $V = \dfrac{1}{3} Bh$
- Sphere: $V = \dfrac{4}{3} \pi r^3$

where B is the area of the base, h is the height, r is the radius.

- Solids of revolution

Figure Rotated	Solid
Right triangle, rotate 360° about leg $\overline{AB}$	Cone
Rectangle $ABCD$, rotate 360° around side $\overline{CD}$	Cylinder
Circle P, rotate 180° about diameter $\overline{AB}$	Sphere

Cavalieri's Principle

If two solids are contained between two parallel planes and every parallel plane between these two planes intercepts regions of equal area, then the solids have equal volume. Also any two parallel planes intercept two solids of equal volume.

Modeling with Density

Mass = volume × density

About the Exam

TEST LAYOUT AND STANDARDS

The Geometry Regents is a 3-hour exam and consists of 4 parts:

Part	Number of Questions	Type of Question	Points per Question	Total Number of Points
I	24	Multiple choice	2	48
II	7	Constructed response	2	14
III	3	Constructed response	4	12
IV	2	Constructed response	6	12
Total	36	—	—	86

Part I consists of 24 multiple-choice questions worth 2 points each. Part II consists of 7 constructed-response questions worth 2 points each. Part III consists of 3 constructed-response questions worth 4 points each. Part IV consists of 2 constructed-response questions worth 6 points each. One question in Part IV will either involve a multiple-step proof or ask you to develop an extended logical argument. The second question in Part IV will require you to use modeling to solve a real-world problem.

The complete set of Geometry Learning Standards are listed in Appendix I. They are grouped in domains. Each domain accounts for a specified percentage of the total points on the exam, shown in the accompanying table. Note that some domains account for a far greater percentage of points than others. The domains are further divided into clusters with different levels of emphasis on the exam—major, supporting, and additional.

Domain	Cluster	Emphasis	Standard[1]
Congruence (27%–34%)	Experiment with transformations in the plane	Supporting	G-CO.1 G-CO.2 G-CO.3 G-CO.4 G-CO.5
	Understand congruence in terms of rigid motions Prove geometric theorems	Major	G-CO.6 G-CO.7 G-CO.8 G-CO.9 G-CO.10 G-CO.11
	Make geometric constructions	Supporting	G-CO.12 G-CO.13
Similarity and Right Triangle Trigonometry (29%–37%)	Understand similarity in terms of similarity transformations	Major	G-SRT.1 G-SRT.2 G-SRT.3
	Prove theorems involving similarity	Major	G-SRT.4 G-SRT.5
	Define trigonometric ratios and solve problems involving right triangles	Major	G-SRT.6 G-SRT.7 G-SRT.8
Circles (2%–8%)	Understand and apply theorems about circles	Additional	G-C.1 G-C.2 G-C.3
	Find arc lengths and areas of sectors of circles	Additional	G-C.5
Expressing Geometric Properties with Equations (12%–18%)	Translate between the geometric description and the equation for a conic section	Additional	G-GPE.1
	Use coordinates to prove simple geometric theorems algebraically	Major	G-PE.4 G-PE.5 G-PE.6 G-PE.7

Domain	Cluster	Emphasis	Standard[1]
Geometric Measurements and Dimensions (2%–8%)	Explain volume formulas and use them to solve problems	Additional	G-GMD.1 G-GMD.3
	Visualize relationships between 2-D and 3-D objects	Additional	G-GMD.4
Modeling with Geometry (8%–15%)	Apply geometric concepts in modeling situations	Major	G-MG.1 G-MG.2 G-MG.3

[1]See Appendix I for a complete list of the Geometry Learning Standards.

How the Test Is Scored

All multiple-choice questions are worth 2 points each and no partial credit is awarded. The constructed-response questions are worth 2, 4, or 6 points each, depending on the part in which they appear. These are graded according to the scoring rubric issued by the New York State Department of Education. The rubric allows for partial credit if an answer is partially correct.

When determining partial credit, the rubrics often distinguish between *computational* errors and *conceptual* errors. A computational error might be an error in your algebra, graphing, or rounding. Computational errors will generally cost you 1 point, regardless if the question is worth 2 points or 6 points. Examples of conceptual errors are using the incorrect formula (volume of a cone instead of a prism) or applying an incorrect relationship (congruent instead of supplementary alternate interior angles). Half the credit of the problem is generally deducted for conceptual errors.

Because partial credit is awarded on the constructed-response questions, it is extremely important to show all your work. A correct answer with no work shown will usually cost you most of the points available for that problem.

The total number of points you earn for all four parts will be added to determine your raw score. A conversion table specific for each exam is then used to convert your raw score to a final scaled score, which is reported to your school. The actual conversion charts are included at the end of each of the Regents exams in this book so you can convert your raw score to a final scaled score. The percentage of points needed to earn a final score of 65% is usually less than 65%, but the effect of the curve diminishes as your raw score increases.

Calculator, Compass, Straightedge, Pen, and Pencil

Graphing calculators are required for the Geometry Regents Exam, and schools must provide one to any student who doesn't have his/her own. You may bring your own calculator if you own one. Many students feel more comfortable using their own familiar calculator. Be aware, though, that test administrators may clear the memory and any stored programs you might have saved on your calculator before the exam begins.

Any calculator provided to you will likely have had its memory cleared as well. This process will restore the calculator to its default settings, which may not be the familiar ones you are used to. The most common setting that may affect your work is the degree versus radian mode. Some calculators have radian mode as the default. It is a good idea to find out what calculator will be provided to you and how to switch between degree and radian mode.

Having a good working knowledge of the graphing calculator will let you apply more than one method to solve a problem. Some calculator techniques worth knowing are the following:

- The graph and intersect feature to solve linear and quadratic equations
- Tables to find patterns or quickly apply trial and error to solve a problem
- Finding the square root and cube root of a number
- Using the trigonometric and inverse trigonometric functions

You will also be provided with a compass and straightedge if you do not have your own. Bringing your own compass is highly recommended since the compasses provided by your school are of unknown quality. It will also be less stressful on test day to work with a compass that you are familiar with.

The straightedge provided to you should be used only for making straight lines when graphing or doing constructions. The straightedge may have length markings in inches or centimeters. However, these should not be used for determining the length of a segment or checking if two segments are congruent.

You are responsible for bringing your own pen and pencil to the exam. All work must be done in blue or black pen with the exception of graphs and diagrams, which may be done in pencil.

Scrap Paper and Graph Paper

You are not allowed to bring your own scrap paper or graph paper into the exam. The exam booklet contains one page of scrap paper and one page of graph paper. These are perforated and can be removed from the booklet to make working with them easier. Any work you put on these sheets is *not* graded. If you want a grader to consider any work there, you must copy it into the appropriate space in the booklet.

Reference Sheet

The following reference sheet is provided to you during the Regents exam. The same reference sheet is used for the Algebra I, Geometry, and Algebra II exams. The three sequence and series formulas and the exponential growth and decay formulas are not part of the Geometry curriculum, so do not be concerned with them. You should be familiar with all the other formulas. Remember: anytime you are asked to calculate an area or volume, check the reference sheet. Many of those formulas are provided.

HIGH SCHOOL MATH REFERENCE SHEET

(Algebra I, Geometry, Algebra II)
Conversions

1 inch = 2.54 centimeters	1 kilometer = 0.62 mile	1 cup = 8 fluid ounces
1 meter = 39.37 inches	1 pound = 16 ounces	1 pint = 2 cups
1 mile = 5,280 feet	1 pound = 0.454 kilograms	1 quart = 2 pints
1 mile = 1,760 yards	1 kilogram = 2.2 pounds	1 gallon = 4 quarts
1 mile = 1.609 kilometers	1 ton = 2,000 pounds	1 gallon = 3.785 liters
		1 liter = 0.264 gallon
		1 liter = 1,000 cubic centimeters

Formulas

Formulas

Triangle	$A = \dfrac{1}{2}bh$	Pythagorean Theorem	$a^2 + b^2 = c^2$
Parallelogram	$A = bh$	Quadratic Formula	$x = \dfrac{-b \pm \sqrt{b^2 - 4ac}}{2a}$
Circle	$A = \pi r^2$	Arithmetic Sequence	$a_n = a_1 + (n-1)d$
Circle	$C = \pi d$ or $C = 2\pi r$	Geometric Sequence	$a_n = a_1 r^{n-1}$
General Prism	$V = Bh$	Geometric Series	$S_n = \dfrac{a_1 - a_1 r^n}{1-r}$ where $r \neq 1$
Cylinder	$V = \pi r^2 h$	Radians	$1 \text{ radian} = \dfrac{180}{\pi} \text{ degrees}$
Sphere	$V = \dfrac{4}{3}\pi r^3$	Degrees	$1 \text{ degree} = \dfrac{\pi}{180} \text{ radians}$
Cone	$V = \dfrac{1}{3}\pi r^2 h$	Exponential Growth/Decay	$A = A_o\, e^{k(t-t_o)} + B_o$
Pyramid	$V = \dfrac{1}{3}Bh$		

The Geometry Learning Standards

The New York State Geometry Regents is aligned to the Geometry Learning Standards. All topics listed below may be found on the Regents exam, with the exception of topics indicated with a (+). Those items are optional extension topics.

CONGRUENCE G-CO

Experiment with Transformations in the Plane

1. Know precise definitions of angle, circle, perpendicular line, parallel line, and line segment, based on the undefined notions of point, line, distance along a line, and distance around a circular arc.

2. Represent transformations in the plane using, e.g., transparencies and geometry software; describe transformations as functions that take points in the plane as inputs and give other points as outputs. Compare transformations that preserve distance and angle to those that do not (e.g., translation versus horizontal stretch).

3. Given a rectangle, parallelogram, trapezoid, or regular polygon, describe the rotations and reflections that carry it onto itself.

4. Develop definitions of rotations, reflections, and translations in terms of angles, circles, perpendicular lines, parallel lines, and line segments.

5. Given a geometric figure and a rotation, reflection, or translation, draw the transformed figure using, e.g., graph paper, tracing paper, or geometry software. Specify a sequence of transformations that will carry a given figure onto another.

Understand Congruence in Terms of Rigid Motions

6. Use geometric descriptions of rigid motions to transform figures and to predict the effect of a given rigid motion on a given figure; given two figures, use the definition of congruence in terms of rigid motions to decide if they are congruent.

7. Use the definition of congruence in terms of rigid motions to show that two triangles are congruent if and only if corresponding pairs of sides and corresponding pairs of angles are congruent.

8. Explain how the criteria for triangle congruence (ASA, SAS, and SSS) follow from the definition of congruence in terms of rigid motions.

Prove Geometric Theorems

9. Prove theorems about lines and angles. *Theorems include: vertical angles are congruent; when a transversal crosses parallel lines, alternate interior angles are congruent and corresponding angles are congruent; points on a perpendicular bisector of a line segment are exactly those equidistant from the segment's endpoints.*

10. Prove theorems about triangles. *Theorems include: measures of interior angles of a triangle sum to 180°; base angles of isosceles triangles are congruent; the segment joining midpoints of two sides of a triangle is parallel to the third side and half the length; the medians of a triangle meet at a point.*

11. Prove theorems about parallelograms. *Theorems include: opposite sides are congruent, opposite angles are congruent, the diagonals of a parallelogram bisect each other, and conversely, rectangles are parallelograms with congruent diagonals.*

Make Geometric Constructions

12. Make formal geometric constructions with a variety of tools and methods (compass and straightedge, string, reflective devices, paper folding, dynamic geometric software, etc.). *Copying a segment; copying an angle; bisecting a segment; bisecting an angle; constructing perpendicular lines, including the perpendicular bisector of a line segment; and constructing a line parallel to a given line through a point not on the line.*

13. Construct an equilateral triangle, a square, and a regular hexagon inscribed in a circle.

SIMILARITY, RIGHT TRIANGLES, & TRIGONOMETRY G-SRT

Understand Similarity in Terms of Similarity Transformations

1. Verify experimentally the properties of dilations given by a center and a scale factor:
   a. A dilation takes a line not passing through the center of the dilation to a parallel line, and leaves a line passing through the center unchanged.
   b. The dilation of a line segment is longer or shorter in the ratio given by the scale factor.

2. Given two figures, use the definition of similarity in terms of similarity transformations to decide if they are similar; explain using similarity transformations the meaning of similarity for triangles as the equality of all corresponding pairs of angles and the proportionality of all corresponding pairs of sides.

3. Use the properties of similarity transformations to establish the AA criterion for two triangles to be similar.

Prove Theorems Involving Similarity

4. Prove theorems about triangles. *Theorems include: a line parallel to one side of a triangle divides the other two proportionally, and conversely; the Pythagorean theorem proved using triangle similarity.*

5. Use congruence and similarity criteria for triangles to solve problems and to prove relationships in geometric figures.

Define Trigonometric Ratios and Solve Problems Involving Right Triangles

6. Understand that by similarity, side ratios in right triangles are properties of the angles in the triangle, leading to definitions of trigonometric ratios for acute angles.

7. Explain and use the relationship between the sine and cosine of complementary angles.

8. Use trigonometric ratios and the Pythagorean theorem to solve right triangles in applied problems.

Apply Trigonometry to General Triangles

9. (+) Derive the formula $A = \frac{1}{2}ab\sin(C)$ for the area of a triangle by drawing an auxiliary line from a vertex perpendicular to the opposite side.

10. (+) Prove the Laws of Sines and Cosines and use them to solve problems.

11. (+) Understand and apply the Law of Sines and the Law of Cosines to find unknown measurements in right and non-right triangles (e.g., surveying problems, resultant forces).

CIRCLES G-C

Understand and Apply Theorems About Circles

1. Prove that all circles are similar.

2. Identify and describe relationships among inscribed angles, radii, and chords. *Include the relationship between central, inscribed, and circumscribed angles; inscribed angles on a diameter are right angles; the radius of a circle is perpendicular to the tangent where the radius intersects the circle.*

3. Construct the inscribed and circumscribed circles of a triangle, and prove properties of angles for a quadrilateral inscribed in a circle.

4. (+) Construct a tangent line from a point outside a given circle to the circle.

Find Arc Lengths and Areas of Sectors of Circles

5. Derive using similarity the fact that the length of the arc intercepted by an angle is proportional to the radius, and define the radian measure of the angle as the constant of proportionality; derive the formula for the area of a sector.

EXPRESSING GEOMETRIC PROPERTIES WITH EQUATIONS G-GPE

Translate Between the Geometric Description and the Equation for a Conic Section

1. Derive the equation of a circle of given center and radius using the Pythagorean theorem; complete the square to find the center and radius of a circle given by an equation.

2. Derive the equation of a parabola given a focus and directrix.

3. (+) Derive the equations of ellipses and hyperbolas given the foci, using the fact that the sum or difference of distances from the foci is constant.

USE COORDINATES TO PROVE SIMPLE GEOMETRIC THEOREMS ALGEBRAICALLY

4. Use coordinates to prove simple geometric theorems algebraically. *For example, prove or disprove that a figure defined by four given points in the coordinate plane is a rectangle; prove or disprove that the point*

(1, √3) lies on the circle centered at the origin and containing the point (0, 2).

5. Prove the slope criteria for parallel and perpendicular lines and use them to solve geometric problems (e.g., find the equation of a line parallel or perpendicular to a given line that passes through a given point).

6. Find the point on a directed line segment between two given points that partitions the segment in a given ratio.

7. Use coordinates to compute perimeters of polygons and areas of triangles and rectangles, e.g., using the distance formula.

GEOMETRIC MEASUREMENT & DIMENSION G-GMD

Explain Volume Formulas and Use Them to Solve Problems

1. Give an informal argument for the formulas for the circumference of a circle, area of a circle, volume of a cylinder, pyramid, and cone. *Use dissection arguments, Cavalieri's principle, and informal limit arguments.*

2. (+) Give an informal argument using Cavalieri's principle for the formulas for the volume of a sphere and other solid figures.

3. Use volume formulas for cylinders, pyramids, cones, and spheres to solve problems.

Visualize Relationships Between Two-dimensional and Three-dimensional Objects

4. Identify the shapes of two-dimensional cross-sections of three-dimensional objects, and identify three-dimensional objects generated by rotations of two-dimensional objects.

MODELING WITH GEOMETRY G-MG

Apply Geometric Concepts in Modeling Situations

1. Use geometric shapes, their measures, and their properties to describe objects (e.g., modeling a tree trunk or a human torso as a cylinder).

2. Apply concepts of density based on area and volume in modeling situations (e.g., persons per square mile, BTUs per cubic foot).

3. Apply geometric methods to solve design problems (e.g., designing an object or structure to satisfy physical constraints or minimize cost; working with typographic grid systems based on ratios).

Practice and Test-Taking Tips

PREPARING FOR THE EXAM

- Preparing to earn a high grade on the Geometry Regents begins well before exam day. Don't try to cram for the Geometry Regents the day before the exam or even the week before. Successful students begin preparing months ahead of time!
- Make the most of your class time. Take good notes, attempt all homework and classwork, and—most importantly—ask questions when you encounter something you don't understand. Check and correct your work, especially exams and quizzes. Save your exams and quizzes in a folder or binder. They are an excellent resource for review and can help you document your progress.
- You should be working with a good review book throughout the school year. The recommended review book is *Barron's Let's Review: Geometry*. You will be able to start reviewing with that book topic by topic before you reach the end of the curriculum in your geometry class.
- Use the self-analysis charts at the end of each Regents exam in this book to record the number of points you earn within each topic area. Use those results to help identify areas for additional practice.
- Work through at least one Regents exam under exam conditions—no distractions, no breaks, and a 3-hour time limit. Measure how much time you spend on each part. Use that as a guide to help establish a time management plan for the actual exam.
- Review any vocabulary words you are unfamiliar with. Geometry is a vocabulary-intensive subject. Understanding the full meaning and implication of a word encountered in a problem is critical to being able to solve the problem. The glossary at the end of this book is a good resource. One good strategy is to make flash cards with a word on one side and a figure on the other.
- Work through as many practice problems as you can. The greater the variety of problems that you encounter beforehand means the greater the probability of seeing familiar problems on the Regents.
- Become familiar with the formula sheet. You should know what is on the formula sheet and what's not. You should also know what each of the symbols and variables represent.

- Read through the step-by-step solutions to the Regents questions for any question you do not get right while practicing. Try to identify why you were not able to answer the question correctly. You will learn from your errors only if you correct them as soon as possible. Be aware of why the errors occur. Common errors include:
  - Not remembering a formula or relationship
  - Difficulty applying multiple steps or concepts in a single problem
  - Weak algebra skills
  - Weak vocabulary
  - Not reading the question carefully
  - Not checking your work

TEST-TAKING TIPS

1. *Be mentally prepared for the exam.*
   i.   Don't try to cram the night before the exam—prepare ahead of time.
   ii. Walk into the exam room with confidence.
   iii. You can retake the exam if the grade you earn does not meet the expectation you set for yourself; so don't panic.

2. *Be physically prepared for the exam.*
   i.   Get plenty of sleep the night before, and set an alarm if necessary so you wake on time.
   ii. Eat a good breakfast the day of the test.
   iii. Dress in comfortable clothes.
   iv. Give yourself plenty of time to get to the test. Late students may be admitted up to a certain time but will not be given any additional time at the end of the exam. Check with your teachers for the current policies.

3. *Come prepared with all materials.*
   i.   Be sure to have:
      - Any identification required by your school
      - The time and room number of your exam
      - Two pens, two pencils, an eraser, a compass, a ruler, and a graphing calculator (if you have one)
      - Do **not** bring any restricted items such as cell phones, music or recording devices, food, or drinks. Ask your teacher ahead of time what items are prohibited. If you do need to bring a cell phone or other electronic device to school, plan to leave it in your locker or other secure place. Using a prohibited device during the exam will invalidate your score.

4. *Have a consistent plan for working through each problem.*
   i.   Start by reading the question carefully. Common errors to watch out for include:
        - Not writing your answer in the correct form, such as a simplified radical or a rounded decimal
        - Not using the correct method, such as graphical versus algebraic
        - Solving for a variable but not substituting to find a requested angle or segment measure
        - Missing key words that provide necessary information (parallel, isosceles, regular, not necessarily true, and so on)
   ii. Sketch a figure if none is provided. Mark up the figure with relevant dimensions, congruence markings, parallel markings, and so on.
   iii. Try to break down complex multistep problems into smaller pieces, looking for relationships that you can apply.
   iv. For proofs, decide if you are going to write a two-column or a paragraph proof.
   v.   If you don't see how to get to a final answer at first, try applying any formulas or relationships that you recognize. Sometimes finding a certain length or an angle may help you see the path to the final answer. Don't panic if you cannot immediately answer a question!

5. *Budget your time.*
   i.   Don't spend too much time on any one problem. If you are having difficulty with a problem, circle it and come back to it at the end. There are going to be a mix of easy, moderate, and difficult questions. You don't want to run out of time before doing all the easy problems.
   ii. Before the test day, record the time you take to complete each part of one of the practice Regents exams in this book. Your goal should be to finish all parts in $2\frac{1}{2}$ hours, which will give you 30 minutes to check your work.

6. *Express you answer in the form requested.*
   i.   When required to round to a specified place value, do not round until you get your final answer.
   ii. When using formulas that involve π, check if the answer is to be "expressed in terms of π" or "rounded" to a certain place value. "Expressed in terms of π" means leave π out of the calculation and then put in the π when writing your final answer, for example "32π." If the answer is to be rounded, then you should type π into your calculator along with the appropriate formula. Remember,

always use the π button, not 3.14 or any other approximation of π.
iii. Simplify radicals when necessary.
iv. If your answer to a multiple-choice question is not one of the choices, check if your answer can be rewritten as one of the choices by simplifying a fraction or radical or by rearranging an equation. For example, the equation of a line may be written in multiple forms or an answer may be in terms of π instead of a rounded equivalent.

7. *Maximize your partial credit.*
   i. Show all work for parts II, III, and IV. Most of the credit is lost if you write a correct answer with no supporting work.
   ii. Write any formulas used in parts II, III, and IV. Then show each step of the evaluation.
   iii. Work through the entire problem even if you are not sure you are correct! If you make an error along the way but the work that follows is consistent with the error, you will likely receive partial credit for using the correct method.
   iv. Do not leave any questions blank. There is no penalty for guessing.

8. *Consider alternative methods to solving a problem.*
   i. Trial and error is a valid method for solving an equation or a problem. However, you must clearly write the equation you are using along with at least *three* guesses with an appropriate check. Even if your first guess is correct, demonstrate that you understand the method by showing work for two incorrect guesses and the checks.
   ii. Some problems can be solved graphically even though a coordinate plane grid is not provided with the problem. You can use the provided scrap graph paper in these cases. The scrap graph paper is not graded. However, this is not a concern for multiple-choice problems since work does not need to be shown for them. For some constructed-response questions, sketching the problem and solution on the graph paper may be a good way to check your result.

9. *Eliminate obviously incorrect choices on multiple-choice questions.*
   i. You can sometimes rule out choices if you do the following:
      - Expect an obtuse or an acute angle for an answer
      - Expect a length to be longer or shorter than another known length
   ii. Use the fact that unless otherwise specified, figures are approximately to scale.

10. *Make your paper easy to grade.*
    i.   Work neatly. Your work should be legible and flow from top to bottom.
    ii. Cross out any incorrect work you do not want the graders to consider on constructed-response questions. You will lose credit if you show both a correct and an incorrect answer.
    iii. Follow directions if you need to change an answer on a multiple-choice bubble sheet. Ask a proctor if you are unsure.

11. *Check your work.*
    i.   If you finish the exam with time left, go back and check as much of your work as you can. After taking the time to prepare for the exam, it would be a waste of that effort not to make use of any extra time you might have.

Examination
June 2018
Geometry

GEOMETRY REFERENCE SHEET

1 inch = 2.54 centimeters	1 ton = 2000 pounds
1 meter = 39.37 inches	1 cup = 8 fluid ounces
1 mile = 5280 feet	1 pint = 2 cups
1 mile = 1760 yards	1 quart = 2 pints
1 mile = 1.609 kilometers	1 gallon = 4 quarts
1 kilometer = 0.62 mile	1 gallon = 3.785 liters
1 pound = 16 ounces	1 liter = 0.264 gallon
1 pound = 0.454 kilogram	1 liter = 1000 cubic centimeters
1 kilogram = 2.2 pounds	

Triangle	$A = \frac{1}{2}bh$
Parallelogram	$A = bh$
Circle	$A = \pi r^2$
Circle	$C = \pi d$ or $C = 2\pi r$
General Prisms	$V = Bh$
Cylinder	$V = \pi r^2 h$
Sphere	$V = \frac{4}{3}\pi r^3$

Cone	$V = \frac{1}{3}\pi r^2 h$
Pyramid	$V = \frac{1}{3}Bh$
Pythagorean Theorem	$a^2 + b^2 = c^2$
Quadratic Formula	$x = \dfrac{-b \pm \sqrt{b^2 - 4ac}}{2a}$
Arithmetic Sequence	$a_n = a_1 + (n-1)d$
Geometric Sequence	$a_n = a_1 r^{n-1}$
Geometric Series	$S_n = \dfrac{a_1 - a_1 r^n}{1-r}$ where $r \neq 1$
Radians	$1 \text{ radian} = \dfrac{180}{\pi} \text{ degrees}$
Degrees	$1 \text{ degree} = \dfrac{\pi}{180} \text{ radians}$
Exponential Growth/Decay	$A = A_0 e^{k(t - t_0)} + B_0$

PART I

Answer all 24 questions in this part. Each correct answer will receive 2 credits. No partial credit will be allowed. For each statement or question, write in the space provided the numeral preceding the word or expression that best completes the statement or answers the question. [48 credits]

1 After a counterclockwise rotation about point *X*, scalene triangle *ABC* maps onto △*RST*, as shown in the diagram below.

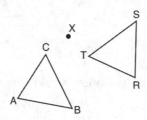

Which statement must be true?

(1) ∠*A* ≅ ∠*R*   (3) $\overline{CB} \cong \overline{TR}$

(2) ∠*A* ≅ ∠*S*   (4) $\overline{CA} \cong \overline{TS}$ 1 _____

2 In the diagram below, $\overline{AB} \parallel \overleftrightarrow{DEF}$, $\overline{AE}$ and $\overline{BD}$ intersect at *C*, m∠*B* = 43°, and m∠*CEF* = 152°.

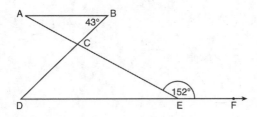

Which statement is true?

(1) m∠*D* = 28°   (3) m∠*ACD* = 71°

(2) m∠*A* = 43°   (4) m∠*BCE* = 109°   2 _____

531

3 In the diagram below, line *m* is parallel to line *n*. Figure 2 is the image of Figure 1 after a reflection over line *m*. Figure 3 is the image of Figure 2 after a reflection over line *n*.

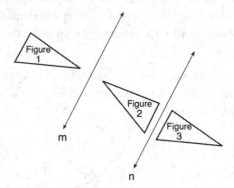

Which single transformation would carry Figure 1 onto Figure 3?

(1) a dilation   (3) a reflection
(2) a rotation   (4) a translation   3 _____

4 In the diagram below, $\overline{AF}$ and $\overline{DB}$ intersect at C, and $\overline{AD}$ and $\overline{FBE}$ are drawn such that m∠D = 65°, m∠CBE = 115°, DC = 7.2, AC = 9.6, and FC = 21.6.

What is the length of $\overline{CB}$?

(1) 3.2   (3) 16.2

(2) 4.8   (4) 19.2   4 ____

5 Given square $RSTV$, where RS = 9 cm. If square $RSTV$ is dilated by a scale factor of 3 about a given center, what is the perimeter, in centimeters, of the image of $RSTV$ after the dilation?

(1) 12   (3) 36

(2) 27   (4) 108   5 ____

6 In right triangle ABC, hypotenuse $\overline{AB}$ has a length of 26 cm, and side $\overline{BC}$ has a length of 17.6 cm. What is the measure of angle B, to the *nearest degree*?

(1) 48°   (3) 43°

(2) 47°   (4) 34°   6 ____

7 The greenhouse pictured below can be modeled as a rectangular prism with a half-cylinder on top. The rectangular prism is 20 feet wide, 12 feet high, and 45 feet long. The half-cylinder has a diameter of 20 feet.

To the *nearest cubic foot*, what is the volume of the greenhouse?

(1) 17,869   (3) 39,074

(2) 24,937   (4) 67,349   7 _____

8 In a right triangle, the acute angles have the relationship $\sin(2x + 4) = \cos(46)$.

What is the value of x?

(1) 20   (3) 24

(2) 21   (4) 25   8 _____

9 In the diagram below, $\overline{AB} \parallel \overline{DFC}$, $\overline{EDA} \parallel \overline{CBG}$, and $\overline{EFB}$ and $\overline{AG}$ are drawn.

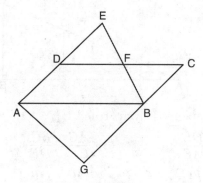

Which statement is always true?

(1) $\triangle DEF \cong \triangle CBF$ (3) $\triangle BAG \sim \triangle AEB$

(2) $\triangle BAG \cong \triangle BAE$ (4) $\triangle DEF \sim \triangle AEB$ 9 _____

10 The base of a pyramid is a rectangle with a width of 4.6 cm and a length of 9 cm. What is the height, in centimeters, of the pyramid if its volume is 82.8 cm³?

(1) 6   (3) 9

(2) 2   (4) 18   10 _____

11 In the diagram below of right triangle AED, $\overline{BC} \parallel \overline{DE}$.

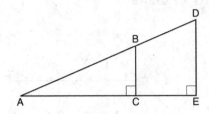

Which statement is always true?

(1) $\dfrac{AC}{BC} = \dfrac{DE}{AE}$

(3) $\dfrac{AC}{CE} = \dfrac{BC}{DE}$

(2) $\dfrac{AB}{AD} = \dfrac{BC}{DE}$

(4) $\dfrac{DE}{BC} = \dfrac{DB}{AB}$

11 _____

12 What is an equation of the line that passes through the point $(6, 8)$ and is perpendicular to a line with equation $y = \dfrac{3}{2}x + 5$?

(1) $y - 8 = \dfrac{3}{2}(x - 6)$

(3) $y + 8 = \dfrac{3}{2}(x + 6)$

(2) $y - 8 = -\dfrac{2}{3}(x - 6)$

(4) $y + 8 = -\dfrac{2}{3}(x + 6)$

12 _____

13 The diagram below shows parallelogram *ABCD* with diagonals $\overline{AC}$ and $\overline{BD}$ intersecting at *E*.

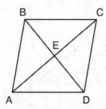

What additional information is sufficient to prove that parallelogram *ABCD* is also a rhombus?

(1) $\overline{BD}$ bisects $\overline{AC}$.

(2) $\overline{AB}$ is parallel to $\overline{CD}$.

(3) $\overline{AC}$ is congruent to $\overline{BD}$.

(4) $\overline{AC}$ is perpendicular to $\overline{BD}$.　　　　　13 ____

14 Directed line segment *DE* has endpoints *D*(–4, –2) and *E*(1, 8). Point *F* divides $\overline{DE}$ such that *DF* : *FE* is 2 : 3. What are the coordinates of *F*?

(1) (–3, 0)　　　　　(3) (–1, 4)

(2) (–2, 2)　　　　　(4) (2, 4)　　　　　14 ____

15 Triangle *DAN* is graphed on the set of axes below. The vertices of △*DAN* have coordinates $D(-6, -1)$, $A(6, 3)$, and $N(-3, 10)$.

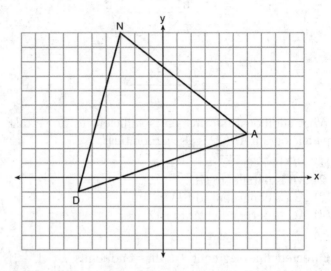

What is the area of △*DAN*?

(1) 60

(3) $20\sqrt{13}$

(2) 120

(4) $40\sqrt{13}$

15 _____

16 Triangle *ABC*, with vertices at *A*(0, 0), *B*(3, 5), and *C*(0, 5), is graphed on the set of axes shown below.

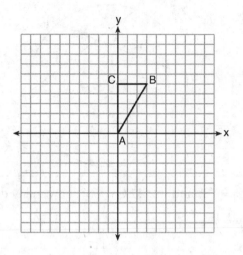

Which figure is formed when △*ABC* is rotated continuously about $\overline{BC}$?

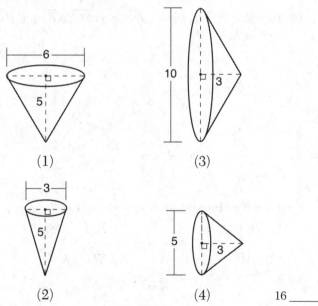

(1)

(2)

(3)

(4)

16 _____

17 In the diagram below of circle O, chords $\overline{AB}$ and $\overline{CD}$ intersect at E.

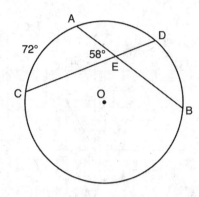

If $m\widehat{AC} = 72°$ and $m\angle AEC = 58°$, how many degrees are in $m\widehat{DB}$?

(1) 108°   (3) 44°
(2) 65°   (4) 14°   17 _____

18 In triangle SRK below, medians $\overline{SC}$, $\overline{KE}$, and $\overline{RL}$ intersect at M.

Which statement must always be true?

(1) $3(MC) = SC$ (3) $RM = 2MC$

(2) $MC = \dfrac{1}{3}(SM)$ (4) $SM = KM$ 18 _____

19 The regular polygon below is rotated about its center.

Which angle of rotation will carry the figure onto itself?

(1) 60°   (3) 216°

(2) 108°   (4) 540°   19 _____

20 What is an equation of circle O shown in the graph below?

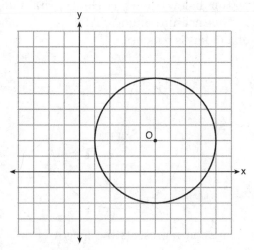

(1) $x^2 + 10x + y^2 + 4y = -13$

(2) $x^2 - 10x + y^2 - 4y = -13$

(3) $x^2 + 10x + y^2 + 4y = -25$

(4) $x^2 - 10x + y^2 - 4y = -25$ 20 _____

21 In the diagram below of $\triangle PQR$, $\overline{ST}$ is drawn parallel to $\overline{PR}$, $PS = 2$, $SQ = 5$, and $TR = 5$.

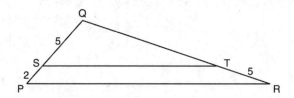

What is the length of $\overline{QR}$?

(1) 7

(3) $12\frac{1}{2}$

(2) 2

(4) $17\frac{1}{2}$

21 _____

22 The diagram below shows circle O with radii $\overline{OA}$ and $\overline{OB}$. The measure of angle AOB is 120°, and the length of a radius is 6 inches.

Which expression represents the length of arc AB, in inches?

(1) $\frac{120}{360}(6\pi)$

(3) $\frac{1}{3}(36\pi)$

(2) $120(6)$

(4) $\frac{1}{3}(12\pi)$

22 _____

23 Line segment *CD* is the altitude drawn to hypotenuse $\overline{EF}$ in right triangle *ECF*. If *EC* = 10 and *EF* = 24, then, to the *nearest tenth*, *ED* is

(1) 4.2   (3) 15.5

(2) 5.4   (4) 21.8   23 _____

24 Line *MN* is dilated by a scale factor of 2 centered at the point (0, 6). If $\overleftrightarrow{MN}$ is represented by $y = -3x + 6$, which equation can represent $\overleftrightarrow{M'N'}$, the image of $\overleftrightarrow{MN}$?

(1) $y = -3x + 12$ (3) $y = -6x + 12$

(2) $y = -3x + 6$ (4) $y = -6x + 6$ 24 _____

PART II

Answer all 7 questions in this part. Each correct answer will receive 2 credits. Clearly indicate the necessary steps, including appropriate formula substitutions, diagrams, graphs, charts, etc. For all questions in this part, a correct numerical answer with no work shown will receive only 1 credit. [14 credits]

25 Triangle $A'B'C'$ is the image of triangle ABC after a translation of 2 units to the right and 3 units up. Is triangle ABC congruent to triangle $A'B'C'$? Explain why.

26 Triangle ABC and point $D(1, 2)$ are graphed on the set of axes below.

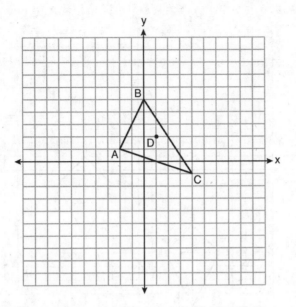

Graph and label $\triangle A'B'C'$, the image of $\triangle ABC$, after a dilation of scale factor 2 centered at point D.

27 Quadrilaterals *BIKE* and *GOLF* are graphed on the set of axes below.

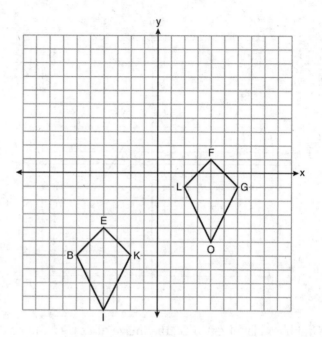

Describe a sequence of transformations that maps quadrilateral *BIKE* onto quadrilateral *GOLF*.

28 In the diagram below, secants $\overline{RST}$ and $\overline{RQP}$, drawn from point
R, intersect circle O at S, T, Q, and P.

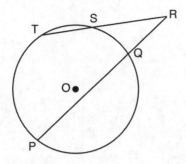

If $RS = 6$, $ST = 4$, and $RP = 15$, what is the length of $\overline{RQ}$?

29 Using a compass and straightedge, construct the median to side $\overline{AC}$ in $\triangle ABC$ below. [Leave all construction marks.]

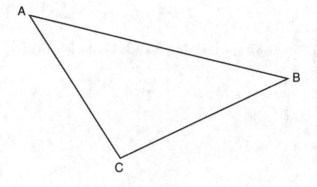

30 Skye says that the two triangles below are congruent. Margaret says that the two triangles are similar.

Are Skye and Margaret both correct? Explain why.

31 Randy's basketball is in the shape of a sphere with a maximum circumference of 29.5 inches. Determine and state the volume of the basketball, to the *nearest cubic inch*.

PART III

Answer all 3 questions in this part. Each correct answer will receive 4 credits. Clearly indicate the necessary steps, including appropriate formula substitutions, diagrams, graphs, charts, etc. For all questions in this part, a correct numerical answer with no work shown will receive only 1 credit. [12 credits]

32 Triangle ABC has vertices with coordinates $A(-1, -1)$, $B(4, 0)$, and $C(0, 4)$. Prove that $\triangle ABC$ is an isosceles triangle but *not* an equilateral triangle. [The use of the set of axes below is optional.]

33 The map of a campground is shown below. Campsite C, first aid station F, and supply station S lie along a straight path. The path from the supply station to the tower, T, is perpendicular to the path from the supply station to the campsite. The length of path $\overline{FS}$ is 400 feet. The angle formed by path $\overline{TF}$ and path $\overline{FS}$ is 72°. The angle formed by path $\overline{TC}$ and path $\overline{CS}$ is 55°.

Campground Map

Determine and state, to the *nearest foot*, the distance from the campsite to the tower.

34 Shae has recently begun kickboxing and purchased training equipment as modeled in the diagram below. The total weight of the bag, pole, and unfilled base is 270 pounds. The cylindrical base is 18 inches tall with a diameter of 20 inches. The dry sand used to fill the base weighs 95.46 lbs per cubic foot.

To the *nearest pound*, determine and state the total weight of the training equipment if the base is filled to 85% of its capacity.

PART IV

Answer the question in this part. A correct answer will receive 6 credits. Clearly indicate the necessary steps, including appropriate formula substitutions, diagrams, graphs, charts, etc. A correct numerical answer with no work shown will receive only 1 credit. [6 credits]

35 Given: Parallelogram $ABCD$, $\overline{BF} \perp \overline{AFD}$, and $\overline{DE} \perp \overline{BEC}$

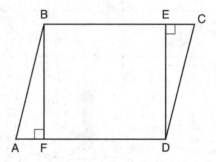

Prove: $BEDF$ is a rectangle

Answers
June 2018
Geometry

Answer Key

PART I

1. (1)	**5.** (4)	**9.** (4)	**13.** (4)	**17.** (3)	**21.** (4)
2. (3)	**6.** (2)	**10.** (1)	**14.** (2)	**18.** (1)	**22.** (4)
3. (4)	**7.** (1)	**11.** (2)	**15.** (1)	**19.** (3)	**23.** (1)
4. (3)	**8.** (1)	**12.** (2)	**16.** (3)	**20.** (2)	**24.** (2)

PART II

25. $\overline{AB} \cong \overline{A'B'}$, $\overline{BC} \cong \overline{B'C'}$, and $\overline{CD} \cong \overline{C'D'}$ so the triangles are congruent by SSS.

26. $A'(-5, 0)$ $B'(-1, 8)$ $C'(7, -4)$

27. A reflection over the y-axis followed by a translation 5 units up

28. $RQ = 4$

29. Construct the midpoint of side $\overline{AC}$. Connect this midpoint to point B to form the median.

30. Use the Pythagorean theorem to show the triangles both have side lengths of 5, 12, and 13. The triangles are congruent by SSS and similar with a scale factor of 1. Skye and Meghan are both correct.

31. Volume = 434 in^3

PART III

32. Use the distance formula to show $AB = \sqrt{26}$, $BC = \sqrt{32}$, and $CA = \sqrt{26}$. Only two sides are congruent so the triangle is isosceles but not equilateral.

33. The distance from the campsite to the tower is 1503 ft.

34. The total weight is 536 lbs.

PART IV

35. Prove *BEDF* is a parallelogram with a right angle. See the complete solutions for the proof.

In **PARTS II–IV** you are required to show how you arrived at your answers. For sample methods of solutions, see Barron's *Regents Exams and Answers* book for Geometry.

Examination
June 2019
Geometry

HIGH SCHOOL MATH REFERENCE SHEET

Conversions

1 inch = 2.54 centimeters	1 cup = 8 fluid ounces
1 meter = 39.37 inches	1 pint = 2 cups
1 mile = 5280 feet	1 quart = 2 pints
1 mile = 1760 yards	1 gallon = 4 quarts
1 mile = 1.609 kilometers	1 gallon = 3.785 liters
	1 liter = 0.264 gallon
1 kilometer = 0.62 mile	1 liter = 1000 cubic centimeters
1 pound = 16 ounces	
1 pound = 0.454 kilogram	
1 kilogram = 2.2 pounds	
1 ton = 2000 pounds	

Triangle	$A = \frac{1}{2}bh$
Parallelogram	$A = bh$
Circle	$A = \pi r^2$
Circle	$C = \pi d$ or $C = 2\pi r$

General Prisms	$V = Bh$
Cylinder	$V = \pi r^2 h$
Sphere	$V = \dfrac{4}{3}\pi r^3$
Cone	$V = \dfrac{1}{3}\pi r^2 h$
Pyramid	$V = \dfrac{1}{3}Bh$
Pythagorean Theorem	$a^2 + b^2 = c^2$
Quadratic Formula	$x = \dfrac{-b \pm \sqrt{b^2 - 4ac}}{2a}$
Arithmetic Sequence	$a_n = a_1 + (n-1)d$
Geometric Sequence	$a_n = a_1 r^{n-1}$
Geometric Series	$S_n = \dfrac{a_1 - a_1 r^n}{1 - r}$ where $r \neq 1$
Radians	1 radian $= \dfrac{180}{\pi}$ degrees
Degrees	1 degree $= \dfrac{\pi}{180}$ radians
Exponential Growth/Decay	$A = A_0 e^{k(t-t_0)} + B_0$

PART I

Answer all 24 questions in this part. Each correct answer will receive 2 credits. No partial credit will be allowed. For each statement or question, write in the space provided the numeral preceding the word or expression that best completes the statement or answers the question. [48 credits]

1 On the set of axes below, triangle *ABC* is graphed. Triangles *A′B′C′* and *A″B″C″*, the images of triangle *ABC*, are graphed after a sequence of rigid motions.

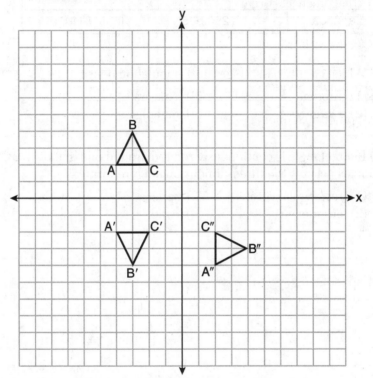

Identify which sequence of rigid motions maps △*ABC* onto △*A′B′C′* and then maps △*A′B′C′* onto △*A″B″C″*.

(1) a rotation followed by another rotation

(2) a translation followed by a reflection

(3) a reflection followed by a translation

(4) a reflection followed by a rotation   1 _____

2 The table below shows the population and land area, in square miles, of four counties in New York State at the turn of the century.

County	2000 Census Population	2000 Land Area (mi²)
Broome	200,536	706.82
Dutchess	280,150	801.59
Niagara	219,846	522.95
Saratoga	200,635	811.84

Which county had the greatest population density?

(1) Broome   (3) Niagara

(2) Dutchess   (4) Saratoga   2 _____

3 If a rectangle is continuously rotated around one of its sides, what is the three-dimensional figure formed?

(1) rectangular prism   (3) sphere

(2) cylinder   (4) cone   3 _____

4 Which transformation carries the parallelogram below onto itself?

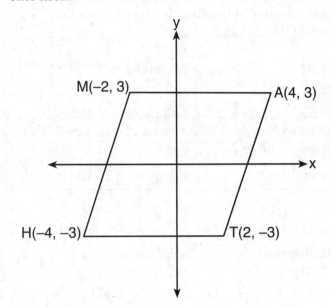

(1) a reflection over $y = x$

(2) a reflection over $y = -x$

(3) a rotation of 90° counterclockwise about the origin

(4) a rotation of 180° counterclockwise about the origin

4 _____

5 After a dilation centered at the origin, the image of $\overline{CD}$ is $\overline{C'D'}$. If the coordinates of the endpoints of these segments are $C(6, -4)$, $D(2, -8)$, $C'(9, -6)$, and $D'(3, -12)$, the scale factor of the dilation is

(1) $\dfrac{3}{2}$ (3) 3

(2) $\dfrac{2}{3}$ (4) $\dfrac{1}{3}$

5 _____

6 A tent is in the shape of a right pyramid with a square floor. The square floor has side lengths of 8 feet. If the height of the tent at its center is 6 feet, what is the volume of the tent, in cubic feet?

(1) 48   (3) 192

(2) 128   (4) 384   6 ____

7 The line $-3x + 4y = 8$ is transformed by a dilation centered at the origin. Which linear equation could represent its image?

(1) $y = \frac{4}{3}x + 8$ (3) $y = -\frac{3}{4}x - 8$

(2) $y = \frac{3}{4}x + 8$ (4) $y = -\frac{4}{3}x - 8$ 7 ____

8 In the diagram below, $\overline{AC}$ and $\overline{BD}$ intersect at E.

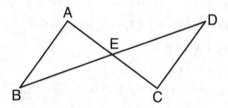

Which information is always sufficient to prove $\triangle ABE \cong \triangle CDE$?

(1) $\overline{AB} \parallel \overline{CD}$

(2) $\overline{AB} \cong \overline{CD}$ and $\overline{BE} \cong \overline{DE}$

(3) E is the midpoint of $\overline{AC}$.

(4) $\overline{BD}$ and $\overline{AC}$ bisect each other.   8 ____

9 The expression $\sin 57°$ is equal to

(1) $\tan 33°$ (3) $\tan 57°$

(2) $\cos 33°$ (4) $\cos 57°$ 9 ____

10 What is the volume of a hemisphere that has a diameter of 12.6 cm, to the *nearest tenth of a cubic centimeter?*

(1) 523.7   (3) 4189.6

(2) 1047.4   (4) 8379.2   10 _____

11 In the diagram below of $\triangle ABC$, D is a point on $\overline{BA}$, E is a point on $\overline{BC}$, and $\overline{DE}$ is drawn.

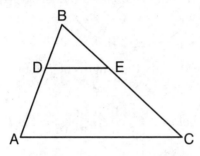

If $BD = 5$, $DA = 12$, and $BE = 7$, what is the length of $\overline{BC}$ so that $\overline{AC} \parallel \overline{DE}$?

(1) 23.8   (3) 15.6

(2) 16.8   (4) 8.6   11 _____

12 A quadrilateral must be a parallelogram if

(1) one pair of sides is parallel and one pair of angles is congruent

(2) one pair of sides is congruent and one pair of angles is congruent

(3) one pair of sides is both parallel and congruent

(4) the diagonals are congruent   12 _____

13 In the diagram below of circle O, chords $\overline{JT}$ and $\overline{ER}$ intersect at M.

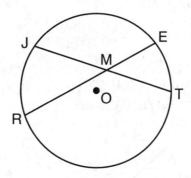

If $EM = 8$ and $RM = 15$, the lengths of $\overline{JM}$ and $\overline{TM}$ could be

(1) 12 and 9.5   (3) 16 and 7.5

(2) 14 and 8.5   (4) 18 and 6.5   13 _____

14 Triangles JOE and SAM are drawn such that $\angle E \cong \angle M$ and $\overline{EJ} \cong \overline{MS}$. Which mapping would *not* always lead to $\triangle JOE \cong \triangle SAM$?

(1) $\angle J$ maps onto $\angle S$ (3) $\overline{EO}$ maps onto $\overline{MA}$

(2) $\angle O$ maps onto $\angle A$ (4) $\overline{JO}$ maps onto $\overline{SA}$ 14 _____

15 In $\triangle ABC$ shown below, $\angle ACB$ is a right angle, E is a point on $\overline{AC}$, and $\overline{ED}$ is drawn perpendicular to hypotenuse $\overline{AB}$.

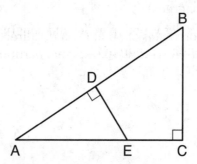

If $AB = 9$, $BC = 6$, and $DE = 4$, what is the length of $\overline{AE}$?

(1) 5   (3) 7

(2) 6   (4) 8   15 _____

16 Which equation represents a line parallel to the line whose equation is $-2x + 3y = -4$ and passes through the point $(1, 3)$?

(1) $y - 3 = -\frac{3}{2}(x - 1)$ (3) $y + 3 = -\frac{3}{2}(x + 1)$

(2) $y - 3 = \frac{2}{3}(x - 1)$ (4) $y + 3 = \frac{2}{3}(x + 1)$ 16 ____

17 In rhombus *TIGE*, diagonals $\overline{TG}$ and $\overline{IE}$ intersect at *R*. The perimeter of *TIGE* is 68, and $TG = 16$.

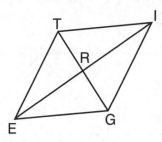

What is the length of diagonal $\overline{IE}$?

(1) 15   (3) 34
(2) 30   (4) 52   17 ____

18 In circle *O* two secants, $\overline{ABP}$ and $\overline{CDP}$, are drawn to external point *P*. If $m\widehat{AC} = 72°$, and $m\widehat{BD} = 34°$, what is the measure of $\angle P$?

(1) 19°   (3) 53°
(2) 38°   (4) 106°   18 ____

19 What are the coordinates of point *C* on the directed segment from $A(-8, 4)$ to $B(10, -2)$ that partitions the segment such that *AC:CB* is 2:1?

(1) $(1, 1)$ (3) $(2, -2)$
(2) $(-2, 2)$ (4) $(4, 0)$ 19 ____

20 The equation of a circle is $x^2 + 8x + y^2 - 12y = 144$.
What are the coordinates of the center and the length
of the radius of the circle?

(1) center (4, −6) and radius 12

(2) center (−4, 6) and radius 12

(3) center (4, −6) and radius 14

(4) center (−4, 6) and radius 14   20 ____

21 In parallelogram $PQRS$, $\overline{QP}$ is extended to point T and
$\overline{ST}$ is drawn.

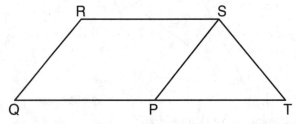

If $\overline{ST} \cong \overline{SP}$ and m$\angle R = 130°$, what is m$\angle PST$?

(1) 130°   (3) 65°

(2) 80°   (4) 50°   21 ____

22 A 12-foot ladder leans against a building and reaches a
window 10 feet above ground. What is the measure of
the angle, to the *nearest degree*, that the ladder forms
with the ground?

(1) 34   (3) 50

(2) 40   (4) 56   22 ____

23 In the diagram of equilateral triangle ABC shown below, E and F are the midpoints of $\overline{AC}$ and $\overline{BC}$, respectively.

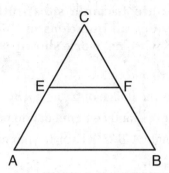

If $EF = 2x + 8$ and $AB = 7x - 2$, what is the perimeter of trapezoid $ABFE$?

(1) 36   (3) 100

(2) 60   (4) 120   23 _____

24 Which information is *not* sufficient to prove that a parallelogram is a square?

(1) The diagonals are both congruent and perpendicular.

(2) The diagonals are congruent and one pair of adjacent sides are congruent.

(3) The diagonals are perpendicular and one pair of adjacent sides are congruent.

(4) The diagonals are perpendicular and one pair of adjacent sides are perpendicular.   24 _____

PART II

Answer all 7 questions in this part. Each correct answer will receive 2 credits. Clearly indicate the necessary steps, including appropriate formula substitutions, diagrams, graphs, charts, etc. For all questions in this part, a correct numerical answer with no work shown will receive only 1 credit. [14 credits]

25 Triangle $A'B'C'$ is the image of triangle ABC after a dilation with a scale factor of $\frac{1}{2}$ and centered at point A. Is triangle ABC congruent to triangle $A'B'C'$? Explain your answer.

26 Determine and state the area of triangle PQR, whose vertices
have coordinates $P(-2, -5)$, $Q(3, 5)$, and $R(6, 1)$.

[The use of the set of axes below is optional.]

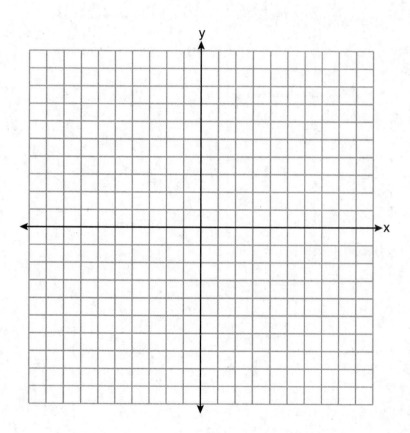

27 A support wire reaches from the top of a pole to a clamp on the ground. The pole is perpendicular to the level ground and the clamp is 10 feet from the base of the pole. The support wire makes a 68° angle with the ground. Find the length of the support wire to the *nearest foot*.

28 In the diagram below, circle O has a radius of 10.

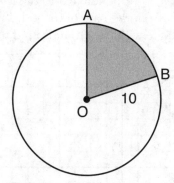

If $m\widehat{AB} = 72°$, find the area of shaded sector AOB, in terms of π.

29 On the set of axes below, $\triangle ABC \cong \triangle STU$.

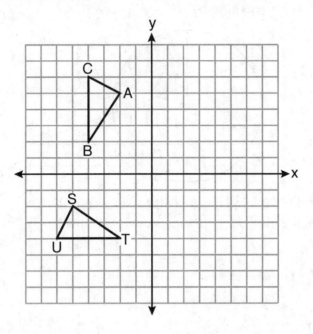

Describe a sequence of rigid motions that maps $\triangle ABC$ onto $\triangle STU$.

30 In right triangle *PRT*, m∠*P* = 90°, altitude $\overline{PQ}$ is
    drawn to hypotenuse $\overline{RT}$, *RT* = 17, and *PR* = 15.

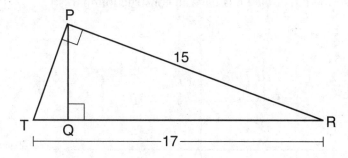

Determine and state, to the *nearest tenth*, the length
of $\overline{RQ}$.

31 Given circle O with radius $\overline{OA}$, use a compass and
    straightedge to construct an equilateral triangle
    inscribed in circle O. [Leave all construction marks.]

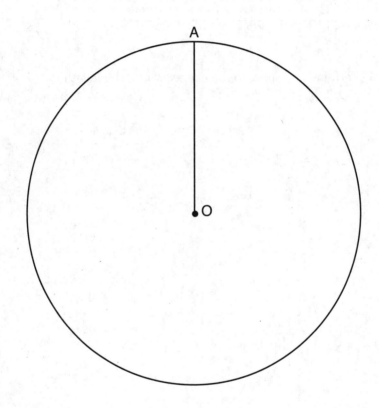

PART III

Answer all 3 questions in this part. Each correct answer will receive 4 credits. Clearly indicate the necessary steps, including appropriate formula substitutions, diagrams, graphs, charts, etc. For all questions in this part, a correct numerical answer with no work shown will receive only 1 credit. [12 credits]

32 Riley plotted $A(-1, 6)$, $B(3, 8)$, $C(6, -1)$, and $D(1, 0)$ to form a quadrilateral. Prove that Riley's quadrilateral $ABCD$ is a trapezoid. [The use of the set of axes on the next page is optional.]

Question 32 is continued on the next page.

Question 32 continued

Riley defines an isosceles trapezoid as a trapezoid with congruent diagonals. Use Riley's definition to prove that *ABCD* is *not* an isosceles trapezoid.

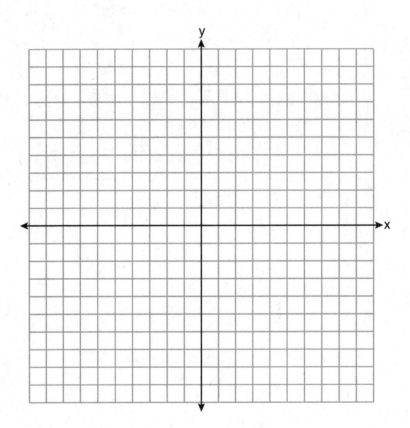

33 A child-sized swimming pool can be modeled by a cylinder. The pool has a diameter of $6\frac{1}{2}$ feet and a height of 12 inches. The pool is filled with water to $\frac{2}{3}$ of its height. Determine and state the volume of the water in the pool, to the *nearest cubic foot*.

One cubic foot equals 7.48 gallons of water. Determine and state, to the *nearest gallon*, the number of gallons of water in the pool.

34 Nick wanted to determine the length of one blade of the windmill pictured below. He stood at a point on the ground 440 feet from the windmill's base. Using surveyor's tools, Nick measured the angle between the ground and the highest point reached by the top blade and found it was 38.8°. He also measured the angle between the ground and the lowest point of the top blade, and found it was 30°.

Determine and state a blade's length, x, to the *nearest foot*.

PART IV

Answer the question in this part. A correct answer will receive 6 credits. Clearly indicate the necessary steps, including appropriate formula substitutions, diagrams, graphs, charts, etc. For the question in this part, a correct numerical answer with no work shown will receive only 1 credit.   [6 credits]

35 Given: Quadrilateral *MATH*, $\overline{HM} \cong \overline{AT}$, $\overline{HT} \cong \overline{AM}$, $\overline{HE} \perp \overline{MEA}$, and $\overline{HA} \perp \overline{AT}$

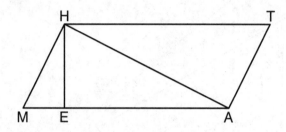

Prove: $TA \bullet HA = HE \bullet TH$

Work space for question 35 is continued on the next page.

Question 35 continued

Answers
June 2019
Geometry

Answer Key

PART I

1. (4)	**5.** (1)	**9.** (2)	**13.** (3)	**17.** (2)	**21.** (2)
2. (3)	**6.** (2)	**10.** (1)	**14.** (4)	**18.** (1)	**22.** (4)
3. (2)	**7.** (2)	**11.** (1)	**15.** (2)	**19.** (4)	**23.** (3)
4. (4)	**8.** (4)	**12.** (3)	**16.** (2)	**20.** (4)	**24.** (3)

PART II

25. The triangles are not congruent because corresponding sides are not congruent.

26. Area $= 25$

27. Length $= 27$ ft

28. Area $= 20\pi$

29. Rotate 90° counterclockwise about the origin.

30. $RQ = 13.2$

31. See the detailed solution for the construction.

PART III

32. $ABCD$ is a trapezoid because $AD \parallel BC$.

ABCD is not isosceles because AC is not congruent to BD.

33. Volume $= 22$ ft^3, number of gallons $= 165$

34. Length $= 100$ ft

PART IV

35. See the detailed solution for the proof.

In **PARTS II–IV** you are required to show how you arrived at your answers. For sample methods of solutions, see Barron's *Regents Exams and Answers* book for Geometry.

Index

A

AAS criteria
  for triangle
    congruence, 159,
    163–165
  for triangle similarity,
    245–248
Acute angle, 4–5
Acute triangle, 7–8
Addition, of radicals,
  19
Addition property
  of equality, 94,
  97–98
Adjacent, 271–272,
  504
Alternate exterior angle
  theorem, 49
Alternate interior angle
  theorem, 49
Altitude, 8, 83, 91, 182,
  340, 499
Altitude to a hypotenuse
  theorem, 264–267,
  503
Angle(s)
  acute, 4–5
  central, 44, 350–351,
    506
  classification of,
    4–5
  compass used to copy,
    68–69
  complementary,
    31–32, 105, 173,
    280
  congruence of, 12
  definition of, 4
  degrees of, 4
  inscribed, 350–351,
    354, 506
  naming of, 4
  obtuse, 4–5

overlapping, 177
in polygons, 41–47
proving relationships
  with, 182
relationships among,
  30–35
right, 4–5
straight, 4–5
sum of the angles
  about a point,
  30–31
summary of, 498
supplementary,
  31–32, 103–104,
  311
in triangles, 56–66
with vertex on the
  circle, 380–382
with vertex outside the
  circle, 382–386
vertical, 32–33,
  103–105, 498
Angle addition postulate,
  36
Angle bisector, 8, 37,
  70, 79, 82, 182,
  499
Angle of depression,
  281–285
Angle of elevation,
  281–285
Angle of rotation,
  130–131
Angle pairs, 47–48
Angle subtraction
  postulate, 37,
  43
Angle sum theorem,
  56–57, 115–116,
  499
Angle-Angle Similarity
  criterion. See AAS
  criteria

Angle-arc theorem
  for chords and
    tangents, 380–382
  for intersecting chords,
    377–378
Arc
  addition of, 353–356
  definition of,
    350–351
Arc length, 402–406,
  408–409, 509
Arc sum theorem, 353
Area, 240–242, 449–450,
  504
  of circle, 400–402,
    437
  of polygons,
    203–205
  of sector, 406,
    408–409
  of sphere, 421–422
  proving, 431–440
ASA criteria, for triangle
  congruence, 159,
  163–165
Auxiliary line, 51

B

Bases, 311, 424
Bisectors, 88–92
  angle, 8, 37, 70, 79,
    82, 182, 499
  perpendicular, 39,
    70–71, 82, 220–221,
    340, 499
  segment, 38–39

C

Cavalieri's principle,
  431–432, 434–436,
  510
Center of rotation,
  130

Center-radius form of the equation of a circle, 227–231

Central angles, 44, 350–352, 506

Centroid, 81, 83, 260–263, 341, 503

Chord(s)
angle-arc theorem for, 380
congruent, 360–362, 506
definition of, 346
intersecting. *See* Intersecting chords
parallel, 363–365, 506
perpendicular, 365–366, 507

Chord–perpendicular bisector theorem, 365

Circle, 75–76
arc of, 350–351
area of, 400–402, 436–437
center-radius form of the equation of, 227–231
central angle, 350–352
chord. *See* Chord
circumference of, 400–402, 436–437
circumscribed, 79
concentric, 348
congruence, 346–350
definition of, 346
equation of, 227–232
inscribed, 79
radius, 346–350
sector, 406–408
segments in, 346
similar, 348, 408–409

tangents, 370–376
vertex inside, 376–377, 508

Circular cone, 418–420

Circular cylinders, 415–416

Circumcenter, 80, 82, 341

Circumference of circle, 400–402, 436–437

Circumscribed circle, 79

Cofunctions, 280–281, 504

Coincide, 13

Collinear, 13

Collinearity, 210–211

Compass, 67–71

Complementary angles, 31–32, 105, 173, 280

Complements/supplements of congruent angles theorem, 103

Completing the square, 24–25

Compositions, 154–156

Concentric circles, 348

Concurrent lines, 78

Cone
circular, 418–420
volume of, 419, 439, 510

Congruent chords, 360–362, 506

Congruent inscribed angles theorem, 352

Congruent tangent theorem, 372

Congruent/congruence circles, 346–350
definition of, 12
transformations used to prove, 188–192
triangle. *See* Triangle congruence

Conic section, 227

Constructions, 67–85

Coordinate geometry formulas, 501
proofs for, 330–331

Coordinate plane
circle in, 228
dilations in, 142–144
reflections in, 137–138
review of, 193–194
rotations in, 140–141
transformations in, 136–148, 501
translations in, 139–140

Coplanar, 13–14

Corresponding angle postulate, 49

Corresponding parts of congruent polygons are congruent. *See* CPCTC

Corresponding point, 121

Cosine, 272, 276, 280, 504

CPCTC, 181–188, 501

Cross sections, 424–426, 428

Cyclic quadrilaterals, 356

Cylinders, 415–416, 424, 437–438, 448, 510

D

Decagon, 6

Definition, 11

Degree, 4, 400

Denominator, rationalizing the, 20

Density, 449–450, 511

Diameter, 346

Difference of perfect squares, 24

Dilations
　description of, 121,
　　132
　equation of a line
　　after, 225
　summary of, 501
Direct rigid motion,
　124
Directed segment, 3,
　198–200
Dissection, 431–432
Distance formula,
　194–196
Division, of radicals,
　19
Division property of
　equality, 94,
　96–97
Double congruence
　proofs, 184–186
Double distributive
　property, 22

E
Edge, 412
Equality
　addition property of,
　　94, 97–98
　congruent versus,
　　12
　division property of,
　　94, 96–97
　multiplication property
　　of, 94, 96–97
　partition postulate of,
　　95, 97–99
　properties of,
　　94–102
　reflexive property of,
　　94, 99–100
　substitution property
　　of, 95–96
　subtraction property
　　of, 94, 97–98
　transitive property of,
　　95–96

Equation
　of a line, 214
　of circle, 227–232
　of parallel lines,
　　216–219
Equilateral triangle, 7–8,
　61, 69, 75
Exterior angle theorem,
　114–115, 499
Exterior angles of
　polygons, 43–44

F
Face, 412
Factoring
　methods for, 23–26
　of quadratic equations,
　　23
FOIL method, 22–23
Formula(s)
　central angles, 44
　coordinate geometry,
　　501
　distance, 194–196
　midpoint, 196
　quadratic, 25
　slope, 208
Fractions, in radicand,
　20

G
General form of the
　equation of a circle,
　　231–232
"Great circle," 428
Greatest common factor,
　23
Grouping with a = 1, 24

H
Hexagon, 6, 75
Horizontal stretch, 144
Hypotenuse
　altitude to a
　　hypotenuse theorem,
　　264–267

definition of, 271–272,
　504
hypotenuse-leg
　criterion, for
　triangle congruence,
　159, 166

I
Image, 121
Image of a point after a
　reflection, 224
Incenter, 79–80, 82
Indirect rigid motion,
　124
Inscribed angle, 350–
　351, 354, 506
Inscribed circle, 79
Interior angles, 42–43
Intersecting, 13–14
Intersecting chords, 509
　angle-arc theorem for,
　377–378
　segments formed by,
　391–394
Intersecting secants, 509
Invariance, 122–123
Inverse trigonometric
　functions, 275–277
Isometry, 122
Isosceles trapezoid,
　312–313, 316, 505
Isosceles triangle, 7–8,
　59–61, 500
Isosceles triangle
　theorem, 116–118

L
Lateral face, 412
Legs, 311
Length, 193, 330, 504
Limit, 432–434
Line(s)
　auxiliary, 51
　in circles, 346
　concurrent, 78
　definition of, 1

equations of, 211–214
intersecting, 13–14
of reflection, 223–224
parallel, 47–56, 74–75,
    108–113, 216–219,
    498–499
perpendicular, 14,
    73–74, 88–92,
    216–219
point-slope form of,
    211
skew, 15–16
slope of, 208–210
slope-intercept form
    of, 211
Line of reflection, 127,
    223
Line segment, 3
Line symmetry, 149–150
Linear equations
    definition of, 21
    solving of, 21–22
Linear function, 211
Linear pairs, 31, 103–
    104, 173, 498

M
Major arc, 350
Mapping of
    quadrilaterals,
    324–329
Median, 8, 83, 91, 182,
    260, 340, 499
Midpoints, 38–39,
    88–92, 193, 196–
    198, 330
Midsegment theorem,
    503
Midsegments, 263–264,
    340
Minor arc, 350
Modeling, 443–448, 511
Monomials, 22
Multiplication
    of polynomials, 22–23
    of radicals, 19

Multiplication property
    of equality, 94,
    96–97

N
Named angle pairs,
    47–48
Noncollinear, 13
Noncoplanar, 13

O
Oblique cylinder, 415
Oblique prism, 413
Obtuse angle, 4–5
Obtuse triangle, 7–8
Octagon, 6
Opposite, 271–272,
    504
Opposite rays, 3
Orientation of polygons,
    124
Origin, 193
Orthocenter, 82–83
Overlapping triangles,
    175–178

P
Parabola, 502
Paragraph proof, 87, 91
Parallel chords, 363–365,
    506
Parallel lines
    description of, 47–56,
    74–75, 108–113
    equations of, 216–219
    slope of, 217–219
    summary of, 498–499
Parallelograms
    definition of, 289,
    504
    proofs with, 295–303
    properties of,
    289–295
    proving, 331–337
    quadrilateral as,
    296–297

rectangle, 304–305,
    316, 333–334, 427,
    505, 510
rhombus, 305–307,
    317, 333–334, 505
square, 24–25,
    307–309, 317, 334,
    505
    theorems associated
    with, 289–291
    transformations and,
    324–329
Partition postulate of
    equality, 95, 97–99
Pentagon, 6, 43
Perimeter, 202–203,
    240–242
Perpendicular bisector
    description of, 8, 39,
    70–71, 82, 340, 499
    equation of, 220–221
Perpendicular chords,
    365–366, 507
Perpendicular lines
    description of, 14,
    73–74, 88–92
    equations of,
    216–219
    slope of, 217–219
Planar figures, 427
Plane(s)
    definition of, 1
    intersecting, 14
Point, 1, 14
Point of tangency, 79,
    346
Point-slope form, 211
Polygons
    angles in, 41–47
    area of, 203–205
    central angles of, 44
    definition of, 6
    exterior angles of,
    43–44
    interior angles of,
    42–43

orientation of, 124
perimeter of, 202–203
regular, 42
summary of, 499
Polyhedron, 412
Polynomials, 22–23
Population density, 449
Postulates
  angle addition, 36
  angle subtraction, 37
  corresponding angle,
    49
  definition of, 11, 86
  segment addition,
    35–36
  segment subtraction,
    37
  triangle congruence,
    169–174, 501
Power density, 449
Preimage, 121, 132
Prism, 413–415, 424,
  434, 510
Proof(s)
  coordinate geometry,
    330–331
  for CPCTC, 182
  definition of, 86
  double congruence,
    184–186
  of area, 431–440
  of volume, 431–440
  paragraph, 87, 91
  parallelogram,
    331–337
  with parallelograms,
    295–303
  for perpendicular
    lines, 90
  triangles, 339–343
  two-column, 87
Proportional parts,
  235
Proportions
  definition of, 26
  solving, 26–27

Pyramid, 418, 420–421,
  438–439, 510
Pythagorean theorem,
  62–63, 194, 202,
  227, 265, 267, 273,
  500

Q
Quadratic equations,
  23
Quadratic formula, 25
Quadrilaterals
  as parallelogram,
    296–297
  classification of,
    315–321
  cyclic, 356
  definition of, 6
  mapping of,
    324–329
  trapezoids, 311–316

R
Radian, 275, 400,
  402–406, 509
Radicals
  addition of, 19
  division of, 19
  multiplication of, 19
  operations with,
    19–21
  simplifying of,
    19–20
  subtraction of, 19
Radicand, 19–20
Radius, 346–350,
  370–371
"Rationalizing the
  denominator," 20
Ray, 3
Reciprocals, 218
Rectangle, 304–305, 316,
  333–334, 427, 505,
  510
Rectangular prism,
  434

Reflections
  definition of, 127–128
  image of a point after,
    224
  line of, 127, 223
  over a line in the
    coordinate plane,
    137–138
  summary of, 500
  through a point in the
    coordinate plane,
    138
Reflexive property
  of equality, 94,
    99–100
Rhombus, 305–307, 317,
  333–334, 505
Right angle, 4–5
Right circular cylinder,
  415–416
Right prism, 413
Right triangle, 427, 510
  description of, 7–8,
    91
  similarity in,
    264–267
Right triangle theorem,
  503
Rigid motions
  compositions of,
    154–156
  definition of, 121–123
Rise, 208
Rotational symmetry,
  150–152
Rotations
  in coordinate plane,
    140–141
  description of,
    130–132
  summary of, 500
Run, 208

S
Same side interior angle
  theorem, 49

SAS criteria
  for triangle
    congruence,
    159–162
  for triangle similarity,
    248–249
Scale factor, 235–237,
    504
Scaled drawing, 239–240
Scalene triangle, 7–8
Secant, 346, 382,
    395–396, 509
Sector, 406–408
Sector area, 408–409,
    509
Segment(s)
  in circles, 346
  compass used to copy,
    68
  congruent, 12
  definition of, 3
  directed, 3, 198–200
  division of, 35
  intersecting chords in
    formation of,
    391–394
  proving relationships
    with, 182
  secants in formation
    of, 395–396
  tangents in formation
    of, 395–396
  in triangles, 8, 91–92,
    499
Segment addition
    postulate, 35–36
Segment bisector, 38–39
Segment parallel to a
    side theorem,
    257–260
Segment subtraction
    postulate, 37
Semicircle, 350,
    353–354
Side splitter theorem,
    258–260, 502

Side-Angle-Side
    criterion. See SAS
    criteria
Side-Side-Side criterion.
    See SSS criteria
Similar circles, 348,
    408–409
Similar figures, 235–244
Similarity transformation,
    121, 124, 250–253
Sine, 272, 276, 280, 504
Skew lines, 15–16
Slope, 208–210, 217–
    219, 330
Slope-intercept form,
    211
Solid(s)
  cross section of, 424
  definition of, 412
  of revolution, 426–428
Sphere, 418, 421–422,
    424, 428
Square, 76, 505
  completing the, 24–25
  definition of, 307
  properties of, 307–
    309, 317
  proving, 334
Square root, 19–20
SSS criteria
  for triangle
    congruence, 159,
    163–165
  for triangle similarity,
    249–250
Straight angle, 4–5
Straightedge, 67
Substitution property of
    equality, 95–96
Subtraction, of radicals,
    19
Subtraction property
    of equality, 94,
    97–98
Sum of the angles about
    a point, 30–31

Supplementary angles,
    31–32, 103–104,
    311
Symmetry
  line, 149–150
  rotational, 150–152

T
Tangent, 78, 272, 276,
    346, 370–376, 380,
    382, 395–396, 504,
    507, 509
Theorems
  alternate exterior
    angle, 49
  alternate interior angle,
    49
  altitude to a
    hypotenuse, 264–
    267, 503
  arc length, for similar
    circles, 409
  arc sum, 353
  centroid, 261
  chord–perpendicular
    bisector, 365
  complements/
    supplements of
    congruent angles,
    103
  congruent chord,
    360–361
  congruent inscribed
    angles, 352
  congruent tangent,
    372
  converse of the
    isosceles triangle,
    59
  definition of, 11,
    86
  exterior angle,
    114–115, 499
  isosceles triangle, 59,
    116–118
  midsegment, 263

parallel chord, 363

parallelograms, 289–291

Pythagorean, 62–63, 193, 202, 227, 265, 267, 273, 500

radius to a point of tangency, 79

same side interior angle, 49

sector area, 409

segment parallel to a side, 257–260

side splitter, 258–260, 502

slope of parallel lines, 217

slope of perpendicular lines, 218

tangent from a common point, 372

tangent-radius, 370

triangle angle sum, 56–57, 115–116, 499

triangle exterior angle, 57–59, 114–115, 499

Transformations

compositions of, 154–156

congruence proven using, 188–192

in coordinate plane, 136–148, 501

definitions, 121–122

dilations, 121, 132, 501

notation, 121–122

parallelograms and, 324–329

properties of, 127–136

reflections, 127–128, 137–138, 500

rigid motion, 121–123

rotations, 130–132, 140–141, 500

similarity, 121, 124, 250–253

summary of, 500–501

translations, 129–130, 139–140, 500

triangle congruence proven using, 189

Transitive property of equality, 95–96

Translations, 129–130, 139–140, 500

Transversal, 47, 109

Trapezoids, 311–316, 505

Triangle(s)

angles in, 56–66

classification of, 7–8

definition of, 6

equilateral, 7–8, 61, 69, 75

isosceles, 7–8, 59–61, 500

median of, 260

obtuse, 7–8

overlapping, 175–178

proving, 339–343

right, 7–8, 91, 264–267, 427, 510

scalene, 7–8

segment parallel to a side of a, 257–260

segments in, 8, 91–92, 499

sides in, 56–66

Triangle angle sum theorem, 56–57, 115–116, 499

Triangle congruence

AAS criteria, 159, 163–165

ASA criteria, 159, 163–165

CPCTC, 181–188, 501

HL criteria, 159, 166

justification for criteria used in, 165–166

postulates for, 169–174, 501

proving of, 169–181

SAS criteria, 159–162

SSS criteria, 159, 163–165

transformations used to prove, 189

Triangle exterior angle theorem, 57–59, 114–115, 499

Triangle similarity, 502–503

Angle-Angle criterion for proving, 245–248

Side-Angle-Side criterion for, 248–249

Side-Side-Side criterion for, 249–250

Triangular prism, 434, 446

Trigonometry

cofunctions, 280–281, 504

inverse functions, 275–277

relationships, 504

Trigonometry ratios

definition of, 271–277

unknown side found using, 277–279

Two-column proof, 87

V
Vector, 129
Venn diagram, 315–316
Vertex, 4, 6, 326,
    376–377, 508
Vertical angles, 32–33,
    103–105, 498
Vertical stretch,
    144
Volume, 240–242, 504,
    510
  definition of, 412

modeling with,
    444–448
of circular cone, 419,
    439
of circular cylinders,
    415–416
of cone, 439, 510
of cylinder, 437–438,
    510
of prism, 413, 510
of pyramid, 420,
    438–439, 510

of sphere, 421–422,
    510
proving, 431–440

W
Word problems,
    27–28

Z
Zero product rule, 23

BARRON'S

Regents Exams and Answers

Geometry

ANDRE CASTAGNA, PH.D.
Mathematics Teacher
Albany High School
Albany, New York

Published by Kaplan, Inc., d/b/a Barron's Educational Series
750 Third Avenue
New York, NY 10017
www.barronseduc.com

ISBN: 978-1-5062-5403-6

PRINTED IN CANADA
10 9 8 7 6 5 4 3 2 1

Kaplan, Inc., d/b/a Barron's Educational Series print books are available at special quantity discounts to use for sales promotions, employee premiums, or educational purposes. For more information or to purchase books, please call the Simon & Schuster special sales department at 866-506-1949.

Dedication

To my loving wife Loretta, who helped make this endeavor possible with her unwavering support; my geometry buddy, Eva; and my future geometry buddies, Rose and Henry, and my professional colleagues for their patience and understanding.

Contents

Preface   vii

About the Exam   1

Test-Taking Tips   15

A Brief Review of Key Geometry Facts and Skills   21

  1. Angle, Line, and Plane Relationships 21
  2. Triangle Relationships 38
  3. Constructions 52
  4. Transformations 67
  5. Triangle Congruence 86
  6. Coordinate Geometry 98
  7. Similar Figures 116
  8. Trigonometry 137
  9. Parallelograms 153
 10. Coordinate Geometry Proofs 168
 11. Circles 184
 12. Solids 205

Glossary of Geometry Terms   231

Regents Examinations, Answers, and
Self-Analysis Charts   241

August 2017 Exam 243

June 2018 Exam 293

August 2018 Exam 349

June 2019 Exam 395

August 2019 Exam 441

The Geometry Learning Standards   485

Preface

This book is designed to prepare you for the New York State Geometry Common Core Regents and is aligned to the Common Core Geometry Learning Standards. The key features you will find in this book are:

- **Test-Taking Tips** Advice and test-day strategies are provided to help you earn as many points as possible on the exam.

- **Brief Review of Geometry Concepts** A concise summary of the theorems, formulas, and math skills needed for success on the Regents exam is provided. Much of the information is presented in bulleted or table form to make it easy for you to use as a refresher and study aid. This summary is intended to supplement the more thorough presentation of high school geometry found in the Barron's *Lets Review: Geometry* book.

- **Glossary** A full glossary is found at the end of the Brief Review section. Be sure to look up any word that you are unfamiliar with as you work through the practice problems. Geometry is a vocabulary-intensive subject.

- **Actual Regents Exams and Step-by-Step Solutions with Explanations** Each of the problems in this book has a detailed solution. All steps in the solution are shown, along with an explanation of the relevant geometry theorems and concepts applied. Solutions to many of the multi-step problems begin with a summary of the big-picture strategy to be used in that problem. These solutions are an excellent way to strengthen your geometry knowledge. Understanding why particular strategies and concepts were chosen will make you a much more effective problem solver when you come across similar problems. Note that solving linear equations is an algebraic tool that is used frequently in Regents problems. It is expected that you have mastered that skill, and not all steps in solving linear equations are explicitly explained. For example, if you have the equation $2x + 4 = 12$, then the next line of the solution may read $2x = 8$. It is assumed that you understand that we subtracted 4 from each side in order to separate the variable from the constants in the equation.

- **Self-Analysis Charts** There is a self-analysis chart at the end of each Regents exam. The chart lists the main topic areas and shows which questions fall into which category. Use the chart to record how many points were earned on that practice exam, broken down by each topic area. You can then target areas of weakness for follow-up study and practice. You can also use the charts to find problems covering similar topics on other Regents exams in this book.

About the Exam

TEST LAYOUT AND STANDARDS

The common core geometry Regents is a 3-hour exam and consists of four parts:

Part	Number of Questions	Type of Question	Points per Question	Total Number of Points
I	24	Multiple-choice	2	48
II	7	Constructed-response	2	14
III	3	Constructed-response	4	12
IV	2	Constructed-response	6	12
Total	36	—	—	86

Part I consists of 24 multiple-choice questions worth 2 points each. Part II consists of 7 constructed-response questions worth 2 points each, and part III consists of 3 constructed-response questions worth 4 points each. Part IV consists of 2 constructed-response questions worth 6 points each. One part IV question will either involve a multiple-step proof or ask you to develop an extended logical argument. The second question will require you to use modeling to solve a real-world problem.

The complete set of Geometry Common Core Learning Standards are listed in the Appendix. They are grouped in domains, and each domain accounts for a specified percentage of the total points on the exam, shown in the accompanying table. Note that some domains account for a far greater percentage of points than others. The domains are further divided into clusters with different levels of emphasis on the exam—major, supporting, and additional.

Domain	Cluster	Emphasis	Standard[1]
Congruence (27–34%)	Experiment with transformations in the plane	Supporting	C-CO.1 through G-CO.5
	Understand congruence in terms of rigid motions Prove geometric theorems	Major	G-CO.6 through G-CO.11
	Make geometric constructions	Supporting	G-CO.12 G-CO.13
Similarity and right triangle trigonometry (29–37%)	Understand similarity in terms of similarity transformations	Major	G-SRT.1 G-SRT.2 G-SRT.3
	Prove theorems involving similarity	Major	G-SRT.4 G-SRT.5
	Define trigonometric ratios and solve problems involving right triangles	Major	G-SRT.6 G-SRT.7 G-SRT.8
Circles (2–8%)	Understand and apply theorems about circles	Additional	G-C.1 G-C.2 G-C.3
	Find arc lengths and areas of sectors of circles	Additional	G-C.5
Expressing geometric properties with equations (12–18%)	Translate between the geometric description and the equation for a conic section	Additional	G.GPE-1
	Use coordinates to prove simple geometric theorems algebraically	Major	G-PE.4 through G-PE.7
Geometric measurements and dimensions (2–8%)	Explain volume formulas and use them to solve problems	Additional	G-GMD.1 G-GMD.3
	Visualize relationships between 2D and 3D objects	Additional	G-GMD.4
Modeling with geometry (8–15%)	Apply geometric concepts in modeling situations	Major	G-MG.1 G-MG.2 G-MG.3

Frequency of Topics

The actual distribution of questions in the last few Regents exams is shown in the following table.

Topic	June 2015	August 2015	June 2016	August 2016	June 2017	August 2017
1. *Angle and Segment Relationships* (define geometric terms and figures, vertical angles, complementary angles, supplementary angles, bisectors and midpoints, parallel and perpendicular lines, proofs of angle and segment relationships)	17, 32			1, 11		
2. *Angle and Segment Relationships in Triangles and Polygons* (angle sum theorem, exterior angle theorem, isosceles triangle theorem, equilateral triangles, interior and exterior angles of polygons, midsegments, centroids, central angles, proofs of triangle and polygon theorems)	34	8, 11	9, 13	4, 8	10, 17	11, 16

Topic	June 2015	August 2015	June 2016	August 2016	June 2017	August 2017
3. *Constructions* (copy a segment, copy an angle; bisect a segment, bisect an angle; perpendicular bisector, perpendicular lines, parallel lines, equilateral triangle, square, regular hexagon, circle inscribed in a triangle, circle circumscribed about a triangle)	25	26	31	28, 32	2	28
4. *Transformations* (definitions and properties of the rigid motions and dilations, preserved properties, compositions of transformations, mapping a figure onto itself with rotations)	2, 4, 10, 16, 33	6, 7, 13, 20, 34	4, 8, 16, 25, 34	2, 9, 27, 29	1, 6, 7, 14, 22, 30, 31, 32	2, 6, 22, 27
5. *Transformations on the Coordinate Plane* (transformations of lines, transformation of points and figures)	18, 22	5, 23, 24	3, 6, 12, 14, 22, 23, 26	5, 26, 33a		

Topic	June 2015	August 2015	June 2016	August 2016	June 2017	August 2017
6. *Triangle Congruence* (SAS, SSS, ASA, AAS, HL criteria; proofs involving triangle congruence; explain the congruence criteria in terms of rigid motions; prove congruence using rigid motions)	24, 30	2, 30, 34	7, 35		9, 33	30, 35
7. *Lines and Segments on the Coordinate Plane* (slope, distance, midpoint, dividing a segment in a specified ratio, equations of lines, determine if a point lies on curve given its equation, area, and perimeter)	3, 9, 27	10, 31	3, 6, 12, 14, 22, 23, 26	14, 15, 16, 18, 21, 30	2, 12, 15, 19	3, 10, 17, 24, 29, 31
8. *Circles on the Coordinate Plane* (equations and graphs of circles, use completing the square to identify the center and radius of a circle)	14	9	3, 6, 12, 14, 22, 23, 26		2, 12, 15, 19	3, 10, 17, 24, 29, 31

Topic	June 2015	August 2015	June 2016	August 2016	June 2017	August 2017
9. *Similarity* (proportional corresponding parts, similar triangle proofs, scaled drawings, perimeter and area of similar figures, proportions in right triangles, prove the Pythagorean theorem)	11, 15, 21, 31, 34	12, 14, 17, 19, 27, 29	2, 5, 17, 21, 27	10, 12, 22	5, 24, 29	5, 7, 9, 18
10. *Trigonometry* (finding a missing side of a right triangle, applications of the sine, cosine, and tangent in word problems, angle of elevation and depression, finding an angle using an inverse trigonometric function, cofunction relationships)	5, 12, 28, 31	4, 32	11, 15, 28, 30	3, 6, 34	3, 13, 21, 36	15, 19, 21

Topic	June 2015	August 2015	June 2016	August 2016	June 2017	August 2017
11. *Parallelograms and Trapezoids* (classifying figures, properties of parallelograms and trapezoids, proofs involving parallelograms and trapezoids, transformations of parallelograms and trapezoids)	13, 26, 33	1, 8, 28, 35	9, 33	7, 24	11, 20	8, 14, 26
12. *Coordinate Geometry Proofs* (proofs on the coordinate plane)	36	22, 33		33b	35	32
13. *Circles* (similarity of circles, central and inscribed angles, angles and segments formed by chords, tangents, and secants, quadrilaterals inscribed in a circle, radian measure, arc length, area of a sector)	8, 20, 29	11, 12, 15, 18	10, 24, 29	19, 23, 25, 35	4, 8, 26	4, 12, 23, 33

Topic	June 2015	August 2015	June 2016	August 2016	June 2017	August 2017
14. *Volume* (prisms, cylinders, cones, pyramids, spheres, properties of solids, calculating volume of solids, justify the formulas for circumference and area of a circle, justify the volume formulas, Cavalieri's principle, identify cross sections and solids of revolution)	1, 6, 23	3, 21	1, 20, 36	3, 13, 20	16, 18, 27, 28	1, 13, 25
15. *Modeling* (modeling real-world situations with solids and planar figures, applications of density, design problems involving area and volume)	7, 19, 35	16, 25, 36	18, 32	17, 36	23, 34	20, 34, 36

HOW THE TEST IS SCORED

All multiple-choice questions are worth 2 points each, and no partial credit is awarded. The constructed-response questions are worth 2, 4, or 6 points each. These are graded according to the scoring rubric issued by the New York State Department of Education. The rubric allows for partial credit if an answer is partially correct.

When determining partial credit, the rubrics often distinguish between *computational* errors and *conceptual* errors. A computational error might be an error in algebra, graphing, or rounding. Computational errors generally will cost you 1 point, regardless of whether the question is a 2-point or 6-point question.

Conceptual errors include using the incorrect formula (volume of a cone instead of a prism) or applying an incorrect relationship (congruent instead of supplementary same side interior angles). Half the credit of the problem is generally deducted for conceptual errors.

Because partial credit is awarded on the constructed-response questions, it is extremely important to show all your work. A correct answer with no work shown will usually cost you most of the points available for that problem.

Your total number of points earned for all four parts will be added to determine your raw score. A conversion table specific for each exam will then be used to convert your raw score to a final scaled score, which is reported to your school. The actual conversion charts are included at the end of each of the Regents exams in this book so you can convert your raw score to a final scaled score. The percentage of points needed to earn a final score of 65% is usually less than 65%, but the effect of the "curve" diminishes as your raw score increases.

CALCULATOR, COMPASS, STRAIGHTEDGE, PEN, AND PENCIL

Graphing calculators are required for the Geometry Common Core exam, and schools must provide one to any students who don't have their own. You may bring your own calculator if you own one, and many students feel more comfortable using their own familiar calculator. Be aware, though, that test administrators may clear the memory and any stored programs you might have saved on your calculator before the exam begins.

Any calculator provided to you will likely have had its memory cleared as well. This process will restore the calculator to its default settings, which may not be the familiar ones you are accustomed to using. The most common setting that may affect your work is the degree versus radian mode. Some calculators have radian mode as the default. It is a good idea to find out what calculator will be provided to you and how to switch between degree and radian mode.

A good working knowledge of the graphing calculator will let you apply more than one method to solving a problem. Some calculator techniques worth knowing are

- Using the graph and intersect feature to solve linear and quadratic equations
- Using tables to find patterns or to quickly apply trial and error to solve a problem
- Finding the square root and cube root of a number
- Using the trigonometric and inverse trigonometric functions

You will also be provided with a compass and straightedge if you do not have your own. Bringing your own compass is highly recommended, since the compasses provided by your school are of unknown quality. Also, it will be less stressful on test day to work with a compass that you are familiar with.

The straightedge provided to you should be used only for making straight lines when graphing or doing constructions. The straightedge may have length markings in inches or centimeters, but these should not be used for determining the length of a segment or checking if two segments are congruent.

You are responsible for bringing your own pen and pencil to the exam. All work must be done in blue or black pen, with the exception of graphs and diagrams, which may be done in pencil.

SCRAP PAPER AND GRAPH PAPER

You are not allowed to bring your own scrap paper or graph paper into the exam. The exam booklet contains one page of scrap paper and one page of graph paper. These are perforated and can be removed from the booklet to make working with them easier. Any work you put on these sheets is *not* graded. If you want a grader to consider any work there, you must copy it into the appropriate space in the booklet.

REFERENCE SHEET

The following reference sheet is provided to you during the Regents exam. The same reference sheet is used for the Algebra I, Geometry, and Algebra II exams. The three sequence and series formulas and the exponential growth and decay formulas are not part of the Common Core Geometry curriculum, so do not be concerned with them. You should be familiar with all the other formulas. Remember, anytime you are asked to calculate an area or volume, check the reference sheet. Many of those formulas are provided.

COMMON CORE HIGH SCHOOL MATH REFERENCE SHEET

(Algebra I, Geometry, Algebra II)

Conversions

1 inch = 2.54 centimeters	1 ton = 2000 pounds
1 meter = 39.37 inches	1 cup = 8 fluid ounces
1 mile = 5280 feet	1 pint = 2 cups
1 mile = 1760 yards	1 quart = 2 pints
1 mile = 1.609 kilometers	1 gallon = 4 quarts
1 kilometer = 0.62 mile	1 gallon = 3.785 liters
1 pound = 16 ounces	1 liter = 0.264 gallon
1 pound = 0.454 kilogram	1 liter = 1000 cubic centimeters
1 kilogram = 2.2 pounds	

Formulas

Triangle	$A = \frac{1}{2}bh$
Parallelogram	$A = bh$
Circle	$A = \pi r^2$
Circle	$C = \pi d$ or $C = 2\pi r$
General Prisms	$V = Bh$
Cylinder	$V = \pi r^2 h$
Sphere	$V = \frac{4}{3}\pi r^3$
Cone	$V = \frac{1}{3}\pi r^2 h$

Pyramid	$V = \frac{1}{3}Bh$
Pythagorean Theorem	$a^2 + b^2 = c^2$
Quadratic Formula	$x = \dfrac{-b \pm \sqrt{b^2 - 4ac}}{2a}$
Arithmetic Sequence	$a_n = a_1 + (n-1)d$
Geometric Sequence	$a_n = a_1 r^{n-1}$
Geometric Series	$S_n = \dfrac{a_1 - a_1 r^n}{1 - r}$ where $r \neq 1$
Radians	$1 \text{ radian} = \dfrac{180}{\pi} \text{ degrees}$
Degrees	$1 \text{ degrees} = \dfrac{\pi}{180} \text{ radians}$
Exponential Growth/Decay	$A = A_0 e^{k(t - t_0)} + B_0$

PREPARING FOR THE EXAM

- Preparing to earn a high grade on the Geometry Regents begins well before exam day. Don't try to cram for the Geometry Regents the day before the exam, or even the week before. Successful students begin preparing months ahead of time!

- Make the most of your class time. Take good notes, attempt all homework and classwork, and—most importantly—ask questions when you encounter something you don't understand. Check and correct your work, especially exams and quizzes. Save your exams and quizzes in a folder or binder—they are an excellent resource for review and can help you document your progress.

- You should be working with a good review book throughout the school year. The recommended review book is Barron's *Let's Review: Geometry*. You will be able to start reviewing with that book topic by topic before you reach the end of the curriculum in your geometry class.

- Use the self-analysis charts at the end of each Regents exam in this book to record the number of points you earn within each topic area, and use those results to help identify areas for additional practice.

- Work through at least one Regents exam under "exam conditions"—no distractions, no breaks, and a 3-hour time limit. Measure how much time you spend on each part, and use that as a guide to help establish a time-management plan for the actual exam.

- Review any vocabulary words you are unfamiliar with. Geometry is a vocabulary-intensive subject, and understanding the full meaning and implication of a word encountered in a problem is critical to being able to solve the problem. The glossary at the end of this book is a good resource. One good strategy is to make flash cards with a word on one side and figure on the other.

- Work through as many practice problems as you can. A greater variety of problems that you encounter beforehand means a greater probability of seeing familiar problems on the Regents.

- Become familiar with the formula sheet. You should know what is on the formula sheet and what's not. Also, you should know what each of the symbols and variables represents.

- Read through the step-by-step solutions to the Regents questions for any question you do not get right while practicing. Try to identify why you were not able to correctly answer the question. You will learn from your errors only if you correct them as soon as possible. Be aware of why the errors occur. Common errors are

  - Not remembering a formula or relationship
  - Having difficulty applying multiple steps or concepts in a single problem
  - Having weak algebra skills
  - Having a weak vocabulary
  - Not reading the question carefully
  - Not checking your work

Test-Taking Tips

TIP 1

Be mentally prepared for the exam

SUGGESTIONS

1. Don't try to cram the night before the exam—prepare ahead of time.
2. Walk into the exam room with confidence.
3. You can retake the exam if the grade you earn does not meet the expectation you set for yourself, so don't panic.

TIP 2

Be physically prepared for the exam

SUGGESTIONS

1. Get plenty of sleep the night before, and set an alarm if necessary so you wake up on time.
2. Eat a good breakfast the day of the test.
3. Dress in comfortable clothes.
4. Give yourself plenty of time to get to the test. Late students may be admitted up to a certain time, but they will not be given any additional time at the end of the exam. Check with your teachers for the current policies.

TIP 3

Come prepared with all materials

SUGGESTIONS

1. Be sure to have
   - Any identification required by your school
   - Time and room number of your exam
   - Two pens, two pencils, eraser, compass, ruler, and graphing calculator (if you have one)
   - Do **not** bring any restricted items such as cell phones, music or recording devices, food, and drinks. Ask your teacher ahead of time what items are prohibited. If you do need to bring a cell phone or other electronic device to school, plan to leave it in your locker or other secure place. Using a prohibited device during the exam will invalidate your score.

TIP 4

Have a consistent plan for working through each problem

SUGGESTIONS

1. Start by reading the question carefully. Common errors to watch out for are
   - Not writing your answer in the correct form, such as a simplified radical or rounded decimal
   - Using the wrong method, such as graphical versus algebraic
   - Solving for a variable but not substituting to find a requested angle or segment measure
   - Missing key words that provide necessary information (parallel, isosceles, regular, "not necessarily true," etc.)

2. Sketch a figure if none is provided. Mark up the figures with relevant dimensions, congruence markings, parallel markings, and so on.

3. Try to break down complex multi-step problems into smaller pieces, looking for relationships that you can apply.

4. For proofs, decide if you are going to write a two-column or paragraph proof.

5. If you don't see how to get to a final answer at first, try applying any formulas or relationships that you recognize. Sometimes finding a certain length or angle may help you see the path to the final answer. Don't panic if you cannot immediately answer a question!

TIP 5

Budget your time

SUGGESTIONS

1. Don't spend too much time on any one problem. If you are having difficulty with a problem, circle it and come back to it at the end. There are going to be a mix of easy, moderate, and difficult questions. You don't want to run out of time before doing all the easy problems.

2. Before the test day, record the time it takes you to do each part of one of the practice Regents exams in this book. Your goal should be to finish all parts in $2\frac{1}{2}$ hours, which will give you 30 minutes to check your work.

TIP 6

Express your answer in the form requested

SUGGESTIONS

1. When required to round to a specified place value, do not round until you get your final answer.

2. When using formulas that involve π, check if the answer is to be "expressed in terms of π" or "rounded" to a certain place value. Expressed in terms of π means leave π out of the calculation, and then put in the π when writing your final answer, for example "32π." If the answer is to be rounded, then you should type π into your calculator along with the appropriate formula. Remember, always use the π button, not 3.14 or any other approximation of π.

3. Simplify radicals when necessary.

4. If your answer to a multiple-choice question is not one of the choices, check to see if it can be rewritten as one of the choices by simplifying a fraction or radical or rearranging an equation. For example, the equation of a line may be written in multiple forms, or an answer may be expressed in terms of π instead of a rounded equivalent.

TIP 7

Maximize your partial credit

SUGGESTIONS

1. Show all work for parts II, III, and IV. Most of the credit is lost if you write a correct answer with no supporting work.

2. Write any formulas used in parts II, III, and IV, and then show each step of the evaluation.

3. Work through the entire problem even if you are not sure if you are correct! If you make an error along the way, but the following work is consistent with the error, you will likely receive partial credit for the correct method.

4. Do not leave any questions blank. There is no penalty for guessing.

TIP 8

Consider alternative methods to solving a problem

SUGGESTIONS

1. Trial and error is a valid method for solving an equation or problem. However, you must clearly write the equation you are using along with at least *three* guesses with an appropriate check. Even if your first guess is correct, demonstrate that you understand the method by showing work for two incorrect guesses and the checks.

2. Some problems can be solved graphically even though a coordinate plane grid is not provided with the problem. You can use the provided scrap graph paper in these cases. The scrap graph paper is not graded, but this is not a concern for multiple-choice problems since work does not need to be shown. For some constructed-response questions, sketching the problem and solution on the graph paper may be a good way to check your result.

TIP 9

Eliminate obviously incorrect choices on multiple-choice questions

SUGGESTIONS

1. You can sometimes rule out choices if you
   - Expect an obtuse or acute angle for an answer
   - Expect a length to be longer or shorter than another known length
2. Use the fact that unless otherwise specified, figures are approximately to scale

---

TIP 10

Make your paper easy to grade

SUGGESTIONS

1. Work neatly. Your work should be legible and flow from top to bottom.
2. Cross out any incorrect work you do not want the graders to consider on constructed-response questions. You will lose credit if there is both a correct and incorrect answer.
3. Follow directions if you need to change an answer on a multiple-choice bubble sheet. Ask a proctor if you are unsure.

TIP 11

Check your work

SUGGESTIONS

1. If you finish the exam with time left, go back and check as much of your work as you can. After taking the time to prepare for the exam, it would be a waste of that effort to not make use of any extra time you might have.

A Brief Review of Key Geometry Facts and Skills

3.1 ANGLE, LINE, AND PLANE RELATIONSHIPS

NOTATION

Points, segments, lines, and planes are represented using the following notation

Name	Definition	Figure	Notation
Point	A location in space with no length, width, or thickness	$\bullet$ A	Point A
Line	An infinitely long set of points that has no width or thickness	$\overleftrightarrow{A\ B}$ $\overleftrightarrow{M\ N\ O}$ $\overleftrightarrow{\quad}$ m	$\overleftrightarrow{AB}$ $\overleftrightarrow{MN}, \overleftrightarrow{NO},$ or $\overleftrightarrow{MO}$ Line m
Ray	A portion of a line with one endpoint and including all points on one side of the endpoint	$\overrightarrow{A\ B}$	$\overrightarrow{AB}$

Segment	A portion of a line bounded by two endpoints		$\overline{FG}$
Length of a segment	The distance between the two endpoints of a segment		FG
Plane	A flat set of points with no thick-ness that extends infinitely in all directions		Plane ABC Plane R
Angle	A figure formed by two rays with a common end-point		$\angle RST$ or $\angle S$
Angle measure	The amount of opening of an angle, measured in degrees or radians		$\angle 1$ $m\angle 1$

An angle can be classified as a

- **Right angle**—measures 90°
- **Acute angle**—measures less than 90°
- **Obtuse angle**—measures greater than 90° and less than 180°
- **Straight angle**—measures 180°

INTERSECTING, PARALLEL, PERPENDICULAR, AND SKEW

- The intersection of two lines is one point (lines r and s intersecting at point M).
- The intersection of two planes is one line (planes ABC and ABD intersect along $\overleftrightarrow{AB}$).

- *Coplanar lines* lie in the same plane, they are either parallel ($\overleftrightarrow{AD}$ and $\overleftrightarrow{BC}$) or intersect ($\overleftrightarrow{AD}$ and $\overleftrightarrow{AB}$).
- *Parallel lines* are coplanar and never intersect. The symbol for parallel is $//$ ($\overleftrightarrow{AD}$ $//$ $\overleftrightarrow{BC}$).
- *Skew lines* are lines that are not coplanar. $\overleftrightarrow{AE}$ and $\overleftrightarrow{GF}$ are skew.
- *Perpendicular lines* intersect at right angles, which measure 90°. The symbol for perpendicular is $\perp$ ($r \perp s$). Line $r \perp$ line s in the accompanying figure. The small square at the point of intersection is a symbol for a right angle. $m\angle ABC + m\angle DEF = 180°$.

BASIC ANGLE RELATIONSHIPS

- *Supplementary* angles have measures that sum to 180°. ∠ABC and ∠DEF are supplementary.

- *Complementary* angles have measures that sum to 90°. ∠1 and ∠2 are complementary. m∠1 + m∠2 = 90°.

- The sum of the measures of adjacent angles around a point equals 360°. m∠1 + m∠2 + m∠3 + m∠4 = 360°.

- The sum of the measures of adjacent angles around a line equals 180°. Two adjacent angles that form a straight line are called a *linear pair*, and are supplementary. m∠1 + m∠2 = 180°. ∠1 and ∠2 are a linear pair.

m∠1 + m∠2 = 180°

- The measure of a whole equals the sum of the measures of its parts. m∠ABC = m∠1 + m∠2 + m∠3.

- *Vertical angles* are the congruent opposite angles formed by intersecting lines. $\angle 1$ and $\angle 3$ are vertical angles; therefore, $\angle 1 \cong \angle 3$. $\angle 2$ and $\angle 4$ are vertical angles; therefore, $\angle 2 \cong \angle 4$.

- *Angle bisectors* divide angles into two congruent angles. $\overrightarrow{CD}$ bisects $\angle ACB$, $\angle ACD \cong \angle BCD$.

ANGLES FORMED BY PARALLEL LINES

When two parallel lines (lines m and n) are intersected by a transversal (line v), eight angles are formed. Any pair of these angles will be either supplementary or congruent. Certain pairs are given special names. The named pairs have the following properties:

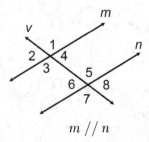

$$m \; // \; n$$

Alternate interior angles are congruent. Alternate interior angles lie between parallel lines m and n on opposite sides of transversal v.

Same side interior angles are supplementary. Same side interior angles lie between parallel lines m and n on the same side of transversal v. $\angle 4$ and $\angle 5$ are supplementary, $\angle 3$ and $\angle 6$ are supplementary. $m\angle 4 + m\angle 5 = 180°$, $m\angle 3 + m\angle 6 = 180°$.

Corresponding angles are congruent. Corresponding angles are the interior/exterior pair of congruent angles on the same side of the transversal. $\angle 4 \cong \angle 8$, $\angle 1 \cong \angle 5$, $\angle 2 \cong \angle 6$, $\angle 3 \cong \angle 7$.

These relationships can also be used to prove two lines are parallel. For example, if the alternate interior angles formed by two lines are congruent, then the lines are parallel.

If a problem involves two parallel lines, but no transversal that intersects both lines, it may be helpful to add an additional line called an auxiliary line. In order to find $m\angle BED$ in the accompanying figure, the auxiliary line $\overleftrightarrow{FEG}$ is drawn parallel to $\overline{CD}$ and $\overline{AB}$. Now the two alternate interior angle relationships can be used:

$$m\angle FED = m\angle CDE = 38° \qquad \text{alternate interior angles}$$
$$m\angle FEB = m\angle ABE = 32° \qquad \text{alternate interior angles}$$
$$m\angle DEB = m\angle FED + m\angle FEB \quad \text{angle addition}$$
$$= 38° + 32°$$
$$= 70°$$

BASIC LINE/SEGMENT/RAY RELATIONSHIPS

In segments, the point that divides a segment into two congruent segments is called a *midpoint*. In the accompanying figure, S is the midpoint of $\overline{RT}$, and $\overline{RS} \cong \overline{ST}$. Any line, segment, or ray that intersects a segment at its midpoint is called a *segment bisector*, or simply *bisector*. Line m bisects $\overline{RT}$ at its midpoint S. Lines can bisect segments, but lines cannot be bisected themselves because they do not have a finite length.

- A line, segment, or ray that is perpendicular to and bisects a segment is called a perpendicular bisector. $\overleftrightarrow{CD}$ is the perpendicular bisector of $\overline{AB}$. The four angles formed by their intersection are right angles, and $\overline{AE} \cong \overline{BE}$.

As with angles, the length of a divided segment is equal to the sum of its parts. In the accompanying figure, $MN + NO = MO$. If point N is also a midpoint, then the additional relationship $MN = NO$ would be true.

ANGLES IN POLYGONS

Definitions:

- **Polygon**—a closed planar figure with straight sides
- **Regular polygon**—a polygon whose sides and angles are all congruent
- **Interior angle**—the angle inside the polygon formed by two adjacent sides
- **Exterior angle**—the angle formed by a side and the extension of an adjacent side in a polygon
- **Triangle**—a polygon with 3 sides
- **Quadrilateral**—a polygon with 4 sides
- **Pentagon, hexagon, octagon, decagon**—polygons with 5, 6, 8, and 10 sides

In the accompanying pentagon, angles 1 through 5 are interior angles and angles 6 through 10 are exterior angles.

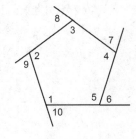

The following relationships can be used to find interior and exterior angles of a polygon:

- The sum of the measures of interior angles of a polygon with n sides equals $180° \times (n - 2)$.

- The measure of one interior angle of a regular polygon with n sides equals $\dfrac{180°(n-2)}{n}$.
- The sum of the measures of exterior angles of a polygon with n sides equals $360°$.
- The measure of one exterior angle of a regular polygon with n sides equals $\dfrac{360°}{n}$.

Using the accompanying figure of the pentagon on page 27:

$$m\angle 1 + m\angle 2 + m\angle 3 + m\angle 4 + m\angle 5 = 180 \times (n-2).$$
$$= 180(5-2)$$
$$= 540°$$
$$m\angle 6 + m\angle 7 + m\angle 8 + m\angle 9 + m\angle 10 = 360°$$

If the pentagon is regular,

$$m\angle 1 = \frac{180°(n-2)}{n}$$
$$= \frac{180°(5-2)}{5}$$
$$= 108°$$
$$m\angle 6 = \frac{360°}{n}$$
$$= \frac{360°}{5}$$
$$= 72°$$

- A central angle is the angle formed by connecting the center of the polygon to two adjacent vertices, as shown. The measure of a central angle in a regular polygon with n sides equals $\dfrac{360°}{n}$. For pentagon $ABCDE$, $m\angle APB = \dfrac{360°}{5} = 72°$.

Practice Exercises

1 In the figure of the rectangular prism, which of the following is true?

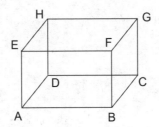

   (1) Points E, H, D, and A are coplanar and collinear.

   (2) $\overline{HD}$ is skew to $\overline{CD}$, and $\overline{CD} \perp \overline{CG}$.

   (3) $\overline{EA}$ // $\overline{CG}$, and $\overline{EH}$ is skew to $\overline{FB}$.

   (4) $\overline{EA} \perp \overline{BC}$, and $\overline{AB}$ // $\overline{CD}$.

2 Which of the following would be the best definition of an angle?

   (1) the union of two rays with a common endpoint

   (2) a geometric figure measured in degrees

   (3) one-third of a triangle

   (4) a line that bends

3 $\overrightarrow{BFA} \perp \overrightarrow{CF}$ and m$\angle CFE = 42°$. Find m$\angle BFD$.

   (1) 42°   (3) 48°

   (2) 45°   (4) 52°

4 $\overrightarrow{BC}$ bisects $\angle ABD$. If m$\angle ABD = (8x - 12)°$ and m$\angle ABC = (3x + 4)°$, find m$\angle ABD$.

   (1) 10°   (3) 34°

   (2) 20°   (4) 68°

5 Line p intersects lines m and n. For what values of x could you conclude that m is parallel to n?

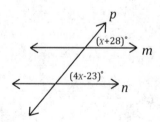

   (1) 12°   (3) 35°

   (2) 17°   (4) 62°

6 Points F, G, H, and I are collinear. G is the midpoint of $\overline{FH}$, H is the midpoint of $\overline{GI}$, $FG = (x^2 + 5x)$, and $\overline{HI} = 3x + 8$.

What is the length of $\overline{FG}$?

(1) 14   (3) 24

(2) 20   (4) 32

7 In the accompanying figure, lines m and n are parallel. If $m\angle 1 = 32°$ and $m\angle 2 = 78°$, what is the measure of $\angle 3$?

(1) 102°   (3) 134°

(2) 110°   (4) 148°

8 A regular polygon has interior angles that each measure 156°. How many sides does the polygon have?

9 Regular octagon $ABCDEFGH$ is shown in the figure below.

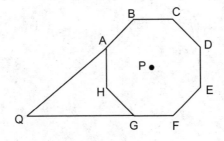

Sides $\overline{AB}$ and $\overline{FG}$ are both extended to Q.

What is the measure of $\angle Q$?

(1) 22.5°   (3) 45°

(2) 30°   (4) 60°

10 In regular hexagon $ABCDEF$, $\overline{AD}$ and $\overline{FC}$ intersect at O. What is the measure of $\angle AOF$?

11 In the accompanying figure, $\overline{WPX}$ intersects $\overline{YPZ}$ at P.

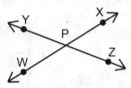

Prove the theorem that the opposite angles formed by intersecting lines are congruent, $\angle WPY \cong \angle ZPX$.

Solutions

1 Choice (1) is not correct since E, H, and D do not lie on a single straight line and are not collinear.

Choice (2) is not correct since $\overline{HD}$ and $\overline{CD}$ are perpendicular, so they cannot be skew.

Choice (4) is not correct since $\overline{EA}$ and $\overline{BC}$ do not intersect, so they cannot be perpendicular.

The correct choice is **(3)**.

2 Considering each choice:

(1) This definition is precise and uses previously defined geometric terms.

(2) This definition is not precise. A counterexample would be an arc, which can also have a degree measure.

(3) Not all angles are part of triangles.

(4) This definition is not precise, since "bend" is not a defined geometric term.

The correct choice is **(1)**.

3 $m\angle CFE + m\angle EFA = 90°$ Perpendicular lines form 90° angles.

$\qquad 42° + m\angle EFA = 90°$

$\qquad\qquad m\angle EFA = 48°$

$\qquad\qquad m\angle BFD = m\angle EFA$ Vertical angles are congruent

$\qquad\qquad m\angle BFD = 48°$

The correct choice is **(3)**.

4 $m\angle ABD = m\angle ABC + m\angle CBD$ angle addition

   $m\angle ABC = m\angle CBD$ bisector forms

                                              2 $\cong$ segments

   $m\angle ABC + m\angle ABC = m\angle ABD$ substitute $m\angle ABC$ for

                                              $m\angle CBD$

$$3x + 4 + 3x + 4 = 8x - 12$$
$$6x + 8 = 8x - 12$$
$$x = 10$$

$$m\angle ABD = (8 \cdot 10 - 12)^\circ \quad \text{evaluate } m\angle ABD \text{ for}$$
$$x = 10$$
$$= 68^\circ$$

The correct choice is **(4)**.

5 The two angles indicated in the figure are corresponding angles, so lines m
   and n are parallel if the corresponding angles are congruent, or have the
   same angle measure.

$$4x - 23 = x + 28$$
$$3x = 51$$
$$x = 17$$

The correct choice is **(2)**.

6 A bisector divides a segment into two congruent halves, so $FG = GH$ and $GH = HI$. Combine these relationships to obtain $FG = HI$.

$$x^2 + 5x = 3x + 8$$
$$x^2 + 2x - 8 = 0$$

$(x - 2)(x + 4) = 0$ factor

$(x - 2) = 0$ $(x + 4) = 0$ zero product property

       $x = 2$ $x = -4$

Eliminate the solution $x = -4$, because that would lead to a negative or zero length; therefore, $x = 2$.

$$FG = x^2 + 5x$$
$$= 2^2 + 5 \cdot 2$$
$$= 14$$

The correct choice is (**1**).

7 Sketch the auxiliary line parallel to m and n as shown.

The 78° angle is divided into a lower and upper part. The lower part is an alternate interior angle to $\angle 1$, so it must measure 32°. The measure of the upper part is equal to $78° - 32° = 46°$.

$\angle 3$ and the 46° angle are same side interior angles, so they must be supplementary.

$$m\angle 3 + 46° = 180°$$
$$m\angle 3 = 134°$$

The correct choice is (**3**).

8 Each interior angle of a polygon measures $\dfrac{180°(n-2)}{n}$ degrees.

$$\frac{180°(n-2)}{n} = 156$$
$$180(n-2) = 156n$$
$$180n - 360 = 156n$$
$$24n = 360$$
$$n = 15$$

The polygon has 15 sides.

9 The sum of the exterior angles of a polygon equals 360°, and in a regular polygon each exterior angle measures $\dfrac{360°}{n}$.

The strategy is to find each interior angle of quadrilateral $QAHG$.

$$m\angle QAH = m\angle QGH = \frac{360°}{8}$$
$$= 45°$$

Each interior angle of a regular polygon is the supplement of an exterior angle.

$$m\angle AHG = 180° - 45° = 135°$$

The exterior convex angle $\angle AHG = 360° - 135° = 225°$.

The sum of the measures of the interior angles of

$$QAHG = (n-2) \cdot 180°$$
$$= (4-2) \cdot 180°$$
$$= 360°$$

$$m\angle Q + m\angle QAH + m\angle AHG + m\angle QGH = 360°$$
$$m\angle Q + 45° + 225° + 45° = 360°$$
$$m\angle Q + 315° = 360°$$
$$m\angle Q = 45°$$

The correct choice is **(3)**.

10 The intersection of the two diagonals $\overline{FC}$ and $\overline{AD}$ is the center of the regular polygon, so $\angle AOF$ is a central angle.

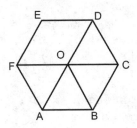

$$m\angle AOF = \frac{360°}{n} \quad \text{central angle formula for a regular polygon}$$

$$= \frac{360°}{6}$$

$$= 60°$$

11 The strategy is to show $\angle WPY$ and $\angle ZPX$ are supplementary to the same angle.

Statement	Reason
1. $\overleftrightarrow{WPX}$ intersects $\overleftrightarrow{YPZ}$ at P.	1. Given
2. $\angle WPY$ and $\angle YPX$ are a linear pair. $\angle ZPX$ and $\angle YPX$ are a linear pair.	2 Definition of a linear pair
3. $\angle WPY$ and $\angle YPX$ are supplementary. $\angle ZPX$ and $\angle YPX$ are supplementary.	3. Linear pairs are supplementary
4. $\angle WPY \cong \angle ZPX$	4. Angles supplementary to the same angle are congruent

3.2 TRIANGLE RELATIONSHIPS

CLASSIFYING TRIANGLES

A triangle is a polygon with three sides.
Triangles can be classified by their sides or by their angles.

- By Sides
  Scalene—no sides are congruent.
  Isosceles—two sides are congruent.
  Equilateral—three sides are congruent.

- By Angle
  Acute—all angles are acute.
  Right—one angle is a right angle.
  Obtuse—one angle is obtuse.

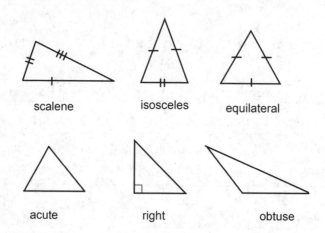

scalene   isosceles   equilateral

acute   right   obtuse

SPECIAL SEGMENTS AND POINTS OF CONCURRENCY IN TRIANGLES

There are four special segments that can be drawn in a triangle, and every triangle has three of each. These are the altitude, median, angle bisector, and perpendicular bisector, shown in the accompanying figure.

Altitude—a segment from a vertex perpendicular to the opposite side
Median—a segment from a vertex to the midpoint of the opposite side
Angle bisector—a line, segment, or ray passing through the vertex of a triangle and bisecting that angle

Perpendicular bisector—a line, segment, or ray that is perpendicular to and passes through the midpoint of a side

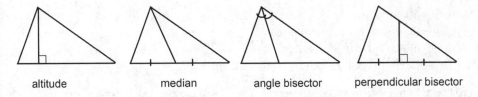

| altitude | median | angle bisector | perpendicular bisector |

Special segments in triangles

The three medians of a triangle will all intersect at a single point called the centroid. We say the medians are *concurrent* at the centroid, or the centroid is the *point of concurrency*. The other special segments will be concurrent as well. The following table illustrates each of the points of concurrency.

Segment	Concurrent at	Feature
Perpendicular bisectors	Circumcenter	Center of the circumscribed circle, equidistant from each of the three vertices
Angle bisectors	Incenter	Center of the inscribed circle, equidistant from the three sides of the triangle
Medians	Centroid	Divides each median in 2:1 ratio and is the center of gravity of the triangle
Altitudes	Orthocenter	Located inside acute triangles, on a vertex of right triangles, and outside obtuse triangles

The following phrase can help you remember the points of concurrency:

"**All of my c**hildren **are b**ringing **in p**eanut **b**utter **c**ookies"

AO—altitudes/orthocenter
MC—medians/centroid
ABI—angle bisectors/incenter
PBC—perpendicular bisectors/circumcenter

ANGLE AND SEGMENT RELATIONSHIPS IN TRIANGLES

Angle sum theorem	The sum of the measures of the interior angles of a triangle equals 180°.	$m\angle A + m\angle B + m\angle C = 180°$
Exterior angle theorem	The measure of any exterior angle of a triangle equals the sum of the measures of the nonadjacent interior angles.	$m\angle 1 = m\angle 3 + m\angle 4$
Isosceles triangle theorem and its converse	If two sides of a triangle are congruent, then the angles opposite them are congruent. If two angles in a triangle are congruent, then the sides opposite them are congruent.	$\angle A \cong \angle B, \overline{AC} \cong \overline{BC}$

Equilateral triangle theorem	All interior angles of an equilateral triangle measure 60°.	 $\overline{AB} \cong \overline{BC} \cong \overline{AC}$, $m\angle A = m\angle B = m\angle C = 60°$
Pythagorean theorem	In a right triangle, the sum of the squares of the legs equals the square of the hypotenuse.	 $a^2 + b^2 = c^2$

Practice Exercises

1 In $\triangle FGH$, K is the midpoint of $\overline{GH}$. What type of segment is $\overline{FK}$?
   (1) median   (3) angle bisector
   (2) altitude   (4) perpendicular bisector

2 The side lengths of a triangle are 6, 8, and 10.

   The triangle can be classified as
   (1) equilateral   (3) obtuse
   (2) acute   (4) right

3 The angle measures of a triangle are $(6x + 6)°$, $(8x - 8)°$, and $(12x)°$. The triangle can be classified as
   (1) obtuse   (3) isosceles
   (2) equilateral   (4) scalene

4 Line m // line n, and $\triangle ABC$ is isosceles with $AC = BC$. What is the measure of $\angle 3$?

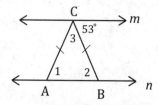

   (1) 53°   (3) 63.5°
   (2) 60°   (4) 74°

5 In △ABC, $\overline{CY}$ is an angle bisector, m∠AYC = 71° and m∠B = 23°. What is
   the measure of angle A?

(1) 61°   (3) 76°

(2) 71°   (4) 86°

6 △RST is a right triangle with right angle ∠RST. △PSR is equilateral. Find the
   measure of angle T.

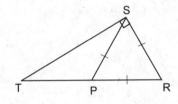

(1) 15°   (3) 30°

(2) 25°   (4) 45°

7 In the figure below, triangle ABC is isosceles with AC = BC, m∠A = 70°, and
   $\overline{AC}$ // $\overrightarrow{BE}$. Find the measure of ∠CBE.

(1) 20°   (3) 35°

(2) 25°   (4) 40°

8 $\overline{JHFK}$ intersects $\overline{IHG}$ at H. If m$\angle KFG = 156°$, m$\angle G = 117°$, m$\angle JHI = (2x + 15)°$, and m$\angle J = (4x + 6)°$, what is the measure of angle J?

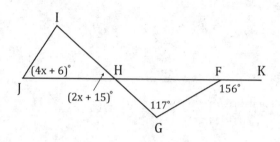

(1) 42°   (3) 62°
(2) 54°   (4) 70°

9 $\angle A$ of $\triangle ABC$ is a right angle, and $\overline{CD}$ is a median. Find the length CD if $AC = 9$ and $BC = 13$.

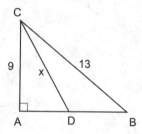

10 In $\triangle ABC$ shown below, m$\angle B = 36°$, $AB = BC$, and $\overline{AD}$ bisects $\angle BAC$.

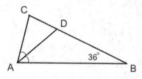

Is $\triangle CAD$ isosceles? Justify your answer.

11 $\overline{RTV}$ intersects $\overline{STU}$ at T.

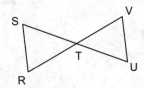

What is the relationship between the sum ($m\angle R$ + $m\angle S$) and the sum ($m\angle V$ + $m\angle U$)? Explain your reasoning.

Solutions

1 A median is a segment that joins a vertex of a triangle to the midpoint of the opposite side.

The correct choice is **(1)**.

2 The three side lengths satisfy the Pythagorean theorem:

$$a^2 + b^2 = c^2$$
$$6^2 + 8^2 = 10^2$$
$$36 + 64 = 100$$
$$100 = 100$$

Therefore, the triangle is a right triangle.

The correct choice is **(4)**.

3 $6x + 6 + 8x - 8 + 12x = 180$ angle sum theorem
$$26x - 2 = 180$$
$$26x = 182$$
$$x = 7$$

Substitute $x = 7$ to find the three angle measures.

$$(6 \cdot 7 + 6)^\circ = 48^\circ$$
$$(8 \cdot 7 - 8)^\circ = 48^\circ$$
$$(12 \cdot 7)^\circ = 84^\circ$$

Since the triangle has two congruent angles, it must have two congruent sides, so it is isosceles.

The correct choice is **(3)**.

4    $m\angle 2 = 53°$ alternate interior angles
formed by parallel lines
are congruent

$m\angle 1 = m\angle 2$ angles opposite congruent
sides in a triangle are
congruent

$m\angle 1 = 53°$

$$m\angle 1 + m\angle 2 + m\angle 3 = 180° \quad \text{angle sum theorem}$$
$$53° + 53° + m\angle 3 = 180°$$
$$106° + m\angle 3 = 180°$$
$$m\angle 3 = 74°$$

The correct choice is (**4**).

5              $m\angle AYC = m\angle B + m\angle BCY$ exterior angle
theorem in $\triangle YBC$

$$71° = 23° + m\angle BCY$$
$$m\angle BCY = 48°$$
$$m\angle ACY = m\angle BCY$$ angle bisector forms
2 congruent angles

$$m\angle ACY = 48°$$
$$m\angle A + m\angle AYC + m\angle ACY = 180°$$ angle sum theorem
in $\triangle AYC$

$$m\angle A + 71° + 48° = 180°$$
$$m\angle A + 119° = 180°$$
$$m\angle A = 61°$$

The correct choice is (**1**).

6   $m\angle R = m\angle RPS = m\angle PSR = 60°$ equilateral triangle theorem

$m\angle RST = 90°$ right angles measure 90°

$m\angle T + m\angle RST + m\angle R = 180°$ angle sum theorem in $\triangle RST$

$$m\angle T + 90° + 60° = 180°$$
$$m\angle T + 150° = 180°$$
$$m\angle T = 30°$$

The correct choice is **(3)**.

7 The strategy is to first find $m\angle C$, which must be congruent to $\angle CBE$.

$$m\angle ABC = m\angle A = 70°$$ isosceles triangle theorem

$$m\angle C + m\angle A + m\angle ABC = 180°$$ triangle angle sum theorem

$$m\angle C + 70° + 70° = 180°$$
$$m\angle C + 140° = 180°$$
$$m\angle C = 40°$$
$$m\angle CBE = m\angle C$$ congruent alternate interior angles

$$m\angle CBE = 40°$$

The correct choice is **(4)**.

8 Start by finding m∠GHK by applying the exterior angle theorem in △FGH.

$$m\angle GHF + m\angle FGH = m\angle KFG \qquad \text{exterior angle theorem}$$

$$m\angle GHF + 117° = 156°$$

$$m\angle GHF = 39°$$

$$m\angle JHI = m\angle GHF \qquad \text{vertical angles are}$$
$$\text{congruent}$$

$$(2x + 15)° = 39°$$

$$2x = 24$$

$$x = 12$$

$$m\angle J = (4 \cdot 12 + 6)° \qquad \text{substitute } x = 12$$

$$= 54°$$

The correct choice is **(2)**.

9 Start by using the Pythagorean theorem in △ABC to find AB. From AB, calculate AD, and then use the Pythagorean theorem again on △ADC to find CD.

$$a^2 + b^2 = c^2 \qquad \text{Pythagorean theorem}$$

$$AB^2 + AC^2 = BC^2$$

$$AB^2 + 9^2 = 13^2$$

$$AB^2 + 81 = 169$$

$$AB^2 = 88$$

$$AB = \sqrt{88}$$

$$= \sqrt{4 \cdot 22}$$

$$= 2\sqrt{22}$$

Median $\overline{CD}$ intersects midpoint D, so

$$AD = \frac{1}{2}AB$$
$$AD = \frac{1}{2}\left(2\sqrt{22}\right)$$
$$= \sqrt{22}$$

Now apply the Pythagorean theorem in $\triangle ADC$:

$$a^2 + b^2 = c^2$$
$$AD^2 + AC^2 = CD^2$$
$$\left(\sqrt{22}\right)^2 + 9^2 = CD^2$$
$$22 + 81 = CD^2$$
$$103 = CD^2$$
$$CD = \sqrt{103}$$

10 $\triangle BAC$ has $2 \cong$ sides $\overline{AB}$ and $\overline{BC}$, so it is isosceles. The strategy is to find the measures of $\angle C$, $\angle ADC$, and $\angle CAD$.

$$\text{m}\angle BAC = \text{m}\angle C \quad \text{isosceles triangle theorem}$$
$$\text{in } \triangle ABC$$
$$\text{m}\angle BAC + \text{m}\angle C + \text{m}\angle B = 180° \quad \text{angle sum theorem in}$$
$$\triangle ABC$$
$$\text{m}\angle BAC + \angle BAC + \text{m}\angle B = 180° \quad \text{substitute m}\angle BAC \text{ for m}\angle C$$
$$2\text{m}\angle BAC + 36° = 180°$$
$$2\text{m}\angle BAC = 144°$$
$$\text{m}\angle BAC = 72°$$
$$\text{m}\angle C = 72°$$

From the angle bisector, we know

$$m\angle CAD = \frac{1}{2}m\angle BAC$$
$$= \frac{1}{2}(72°)$$
$$= 36°$$

Find $m\angle CDA$ by applying the angle sum theorem to $\triangle CAD$.

$$m\angle CDA + m\angle C + m\angle CAD = 180°$$
$$m\angle CDA + 72° + 36° = 180°$$
$$m\angle CDA + 108° = 180°$$
$$m\angle CDA = 72°$$

Both $\angle C$ and $\angle CDA$ measure $72°$; therefore, $\overline{AC} \cong \overline{AD}$, and $\triangle CAD$ is isosceles.

11   $m\angle RTS + m\angle R + m\angle S = 180°$ from the angle sum theorem. Also, $m\angle VTU + m\angle V + m\angle U = 180°$ for the same reason; therefore,

$$m\angle RTS + m\angle R + m\angle S = m\angle VTU + m\angle V + m\angle U$$

$\angle RTS$ and $\angle VTU$ are congruent vertical angles, so we can subtract their measure from each side of the equation, resulting in

$$m\angle R + m\angle S = m\angle V + m\angle U$$

Therefore, the two sums are equal.

3.3 CONSTRUCTIONS

COPY AN ANGLE

- Given angle $\angle ABC$ and $\overrightarrow{DE}$, construct $\angle EDF$ congruent to $\angle ABC$.
    1. Place point on B and make an arc intersecting the angle at R and S.
    2. With the same compass opening, place point on E and make an arc intersecting at T.
    3. Place point on R and pencil on S and make a small arc.
    4. With the same compass opening, place point on T and make an arc intersecting the previous one at F.
    5. $\angle DEF$ is congruent to $\angle ABC$.

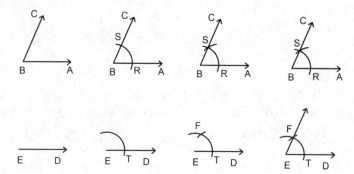

EQUILATERAL TRIANGLE

- Given segment $\overline{AB}$, construct an equilateral triangle with side length AB.
    1. Place point on A and pencil on B and make a quarter circle.
    2. Place point on B and pencil on A and make a quarter circle that intersects the first circle at C.
    3. $\triangle ABC$ is an equilateral triangle.

ANGLE BISECTOR

- Given angle $\angle ABC$, construct angle bisector $\overrightarrow{BD}$.
  1. Place point on B and make an arc intersecting $\overrightarrow{BA}$ and $\overrightarrow{BC}$ at R and S.
  2. Place point on R and make an arc in interior of the angle.
  3. With the same compass opening, place point on S and make an arc that intersects the previous arc at point D.
  4. $\overrightarrow{BD}$ is the angle bisector.

PERPENDICULAR BISECTOR

- Given segment $\overline{AB}$, construct the perpendicular bisector of $\overline{AB}$.
  1. With compass open more than half the length of $\overline{AB}$, place the point at A and make a semicircle running above and below $\overline{AB}$.

  With the same compass opening, place the point at B and make a semicircle running above and below $\overline{AB}$ so that it intersects the first semicircle at R and S.
  2. $\overleftrightarrow{RS}$ is the perpendicular bisector of $\overline{AB}$.

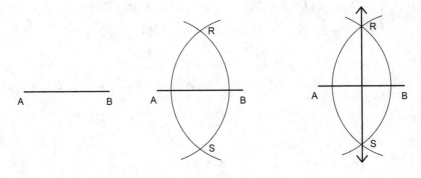

PERPENDICULAR TO LINE FROM A POINT NOT ON THE LINE

- Given line $\overleftrightarrow{AB}$ and point P not on $\overleftrightarrow{AB}$, construct a line perpendicular to $\overleftrightarrow{AB}$ passing through P.

  1. Place point at P and make an arc intersecting $\overleftrightarrow{AB}$ at R and S (extend $\overleftrightarrow{AB}$ if necessary).
  2. Place point at R and make an arc on opposite side of line as P.
  3. Place point at S and make an arc intersecting the previous arc at Q.
  4. $\overleftrightarrow{PQ}$ is perpendicular to $\overleftrightarrow{AB}$ and passes through P.

PERPENDICULAR TO LINE FROM A POINT ON THE LINE

- Given line $\overleftrightarrow{AB}$ and point P on $\overleftrightarrow{AB}$ between A and B, construct a line perpendicular to $\overleftrightarrow{AB}$ and passing through P.

  1. Place point at P and make an arc intersecting $\overleftrightarrow{AB}$ at R and S.
  2. Place point at R and make an arc.
  3. Place point at S and make an arc intersecting the previous arc at Q.
  4. $\overleftrightarrow{PQ}$ is perpendicular to $\overleftrightarrow{AB}$ and passes through P.

PARALLEL TO LINE THROUGH POINT OFF THE LINE

- Given $\overleftrightarrow{AB}$ and point P not on $\overleftrightarrow{AB}$, construct a line parallel to $\overleftrightarrow{AB}$ and passing through P.

  1. Construct a line passing through P and intersecting $\overleftrightarrow{AB}$ at R.

  2. With point at R, make an arc intersecting $\overleftrightarrow{PR}$ and $\overleftrightarrow{AB}$ at S and T.

  3. With point at P and the same compass opening, make an arc intersecting $\overleftrightarrow{PR}$ at U.

  4. With point at T, make an arc intersecting at S.

  5. With point at U and the same compass opening, make an arc intersecting at V.

  6. $\overleftrightarrow{PV}$ is parallel to $\overleftrightarrow{AB}$ and passes through point P.

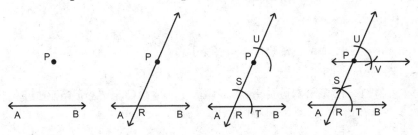

INSCRIBE A REGULAR HEXAGON OR EQUILATERAL TRIANGLE IN A CIRCLE

  1. Construct circle P of radius PA.

  2. Using the same radius as the circle, place point on A and make an arc intersecting circle at B.

  3. Place point at B and make an arc intersecting circle at C. Continue making arcs intersecting at D, E, and F.

  4. Connect each point to form regular hexagon $ABCDEF$ or every other point to form equilateral triangle ACE.

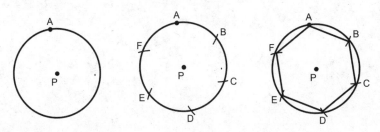

INSCRIBE A SQUARE IN A CIRCLE

- Given circle *T*
    1. Construct a diameter through the center point *T*.
    2. Construct the perpendicular bisector of the first diameter, which will also be a diameter.
    3. The intersections of the diameters with the circle are the vertices of the square.

INSCRIBED CIRCLE

- Given △*ABC*, construct the inscribed circle.
    1. Construct the three angle bisectors of the triangle.
    2. The point of concurrency of the angle bisectors, or incenter *I*, is the center of the inscribed circle.
    3. Construct a line perpendicular to one of the sides of the triangle that passes through the incenter *I*. Label the point of intersection with the side of the triangle *P*.
    4. Construct the inscribed circle with center *I* and radius *IP*.

CIRCUMSCRIBED CIRCLE

- Given $\triangle ABC$, construct the circumscribed circle.
    1. Construct the three perpendicular bisectors of the triangle.
    2. The point of concurrency of the perpendicular bisectors, or circumcenter P, is the center of the circumscribed circle.
    3. Construct the circumscribed circle with center P and radius PA. (The circle should pass through points B and C as well.)

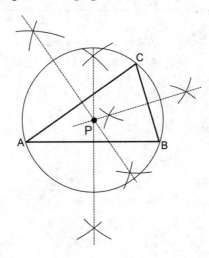

Practice Exercises

Directions: Each of the constructions described in this chapter is a required construction and should be practiced. The constructions in this set of practice problems use the required constructions in various combinations. The first step in each problem is to identify which of the required constructions is called for.

1 Construct a triangle whose angles measure 30°, 60°, and 90°.

2 Given △*JOT*, construct a median from vertex *O*.

3 Construct the centroid of △*HOT* using a compass and straightedge. Use the compass to confirm your construction by demonstrating the 2:1 ratio. Leave all marks of construction.

4 Given △*XYZ*, construct the altitude from vertex *X*.

5 Given △*XYZ*, construct a midsegment that intersects sides $\overline{XZ}$ and $\overline{YZ}$.

6 Given $\overline{AB}$ and $\overline{BC}$, construct parallelogram *ABCD*.

7 Construct a triangle whose angles measure 45°, 45°, and 90° and has $\overline{AB}$ as one of its legs.

A   B

8 Construct a 15° angle.

9 $\overline{JK}$ and $\overline{LM}$ are chords in circle P. Construct the location of point P.

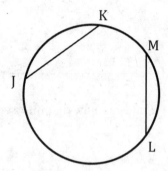

10 Given $\overline{AB}$ and point P, construct the dilation of $\overline{AB}$ with a scale factor of 2 and center at P.

11 Construct the translation of point P by vector $\overrightarrow{AB}$.

12 P' is the reflection of P over line m. Construct line m.

P'

P •

13 Construct point Q', the rotation of point Q about center P by $\angle ABC$.

Solutions

1 Construct an equilateral triangle, then bisect one of the angles.

2 Construct A, the midpoint of $\overline{JT}$, using the "perpendicular bisector" construction. Connect O to A. $\overline{OA}$ is a median of $\triangle JOT$.

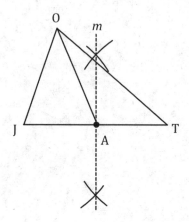

3 The centroid is the point of concurrency of the medians.

- First construct midpoints of two sides of the triangle.
- Then use those midpoints to construct the medians.
- The centroid, G, is the point where the medians intersect.

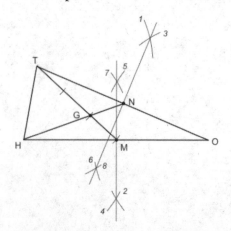

To confirm the centroid divides a median in a 2:1 ratio, measure the distance *GM* and show the distance *TG* is twice *GM*:

- Place the compass point at *M* and make an arc at *G*.
- With the same compass opening, place the point at *G* and make a 2nd arc in the direction of *T*.
- With the same compass opening, you should be able to make a 3rd arc starting at the 2nd and ending at *T*.

4 Using the "perpendicular to line from point not on the line" construction, construct a line perpendicular to $\overline{YZ}$ through point *X*. Let *A* be the intersection of the perpendicular line with $\overline{YZ}$. $\overline{AX}$ is the altitude of △*XYZ*.

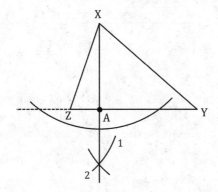

5 Using the "perpendicular bisector" construction, construct the midpoints of $\overline{XZ}$ and $\overline{YZ}$. Label these *A* and *B*. $\overline{AB}$ is a midsegment of △*XYZ*.

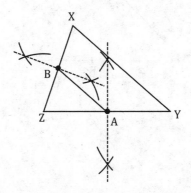

6 The steps are

- Extend $\overline{BA}$.
- Use the "parallel line" construction to construct a line parallel to $\overline{BC}$ and through point A.
- Measure the length BC by making arc 1 with point at B and pencil at C.
- Locate point D by measuring a segment of the same length from point A. D is the fourth vertex of parallelogram $ABCD$.

7 Extend $\overline{AB}$, and locate point C so that $AB = BC$ using the "copy a segment" construction. Construct line m, the perpendicular bisector of $\overline{AC}$. Locate point D so that $AB = BD$ using the "copy a segment" construction. $\triangle ABD$ is an isosceles right triangle with angles measuring $45°$, $45°$, and $90°$.

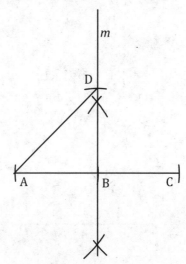

8 Construct an equilateral triangle. Bisect one of its angles to get a pair of 30° angles, and then bisect one of those angles to create two 15° angles.

9 Construct the perpendicular bisectors of $\overline{JK}$ and $\overline{LM}$. The perpendicular bisector of a chord always passes through the center of the circle, so the intersection of the two perpendicular bisectors is point P, the center of the circle.

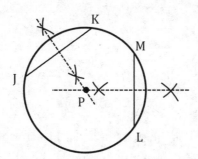

10 The distance from a point to the center is multiplied by the scale factor of the dilation, so $PA' = 2PA$ and $PB' = 2PB$. Construct rays $\overrightarrow{PA}$ and $\overrightarrow{PB}$. With the point at P, measure the distance PA. Move the point to A, and measure to same distance to locate A'. Repeat for point B and B'.

11 Use the "parallel to a line through point off the line" construction to construct a ray through P and parallel to $\overleftrightarrow{AB}$. With the point at A, measure length
AB. Move the point to P, and measure the same distance along the new ray
to locate P'. P' is the translation of P by vector $\overrightarrow{AB}$.

12 The line of reflection is the perpendicular bisector of the segment through a
point and its image. Construct the perpendicular bisector of $\overline{PP'}$.

13 Construct a ray through P and Q, and apply the "copy an angle" construction
to copy $\angle ABC$ to $\angle PQD$. Then, with the point at P, measure the distance PQ.
Use the same distance to locate point Q' along $\overrightarrow{PD}$. Q' is the rotation of Q
about point P by $\angle ABC$.

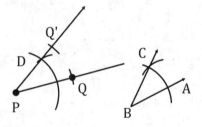

3.4 TRANSFORMATIONS

Transformations can be thought of as functions that operate on geometric figures. The original figure is called the pre-image and the new figure is called the image. There are four basic transformations to be familiar with—translations, reflections, rotations, and dilations. The first three are rigid motions that do not change the size of the figure and result in a congruent figure. The last one, dilations, can change the size of a figure and results in a similar figure.

TRANSLATIONS

A *translation* slides a figure from one position to another. Translations can be specified by a vector, with the notation $T_{\overrightarrow{FG}}$. A vector consists of a magnitude and a direction. The magnitude is represented by the length of the vector and the direction is indicated by following the endpoint, point F, toward the second point. A translation can also be specified by stating one point and the point it maps to. For example, a translation such that maps E to G. The accompanying figure illustrates $T_{\overrightarrow{FG}}(ABCDE) \rightarrow A'B'C'D'E'$. $ABCDE$ is translated a distance FG parallel to the direction of $\overrightarrow{FG}$. The translation is a rigid motion so $ABCDE \cong A'B'C'D'E'$. Also, segments joining corresponding points are congruent and parallel; $\overline{AA'}$ is congruent and parallel to $\overline{BB'}$, $\overline{CC'}$, $\overline{DD'}$, and $\overline{EE'}$.

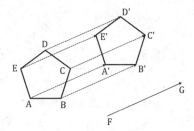

LINE REFLECTIONS

A *line reflection* flips a figure over a line, so that it appears as a mirror image. The line is called the *line of reflection*. The notation r_ℓ indicates a reflection over the line ℓ. The accompanying figure illustrates the reflection of $ABCD$ over line ℓ. Reflections are a rigid motion, so $ABCD \cong A'B'C'D'$. The line of reflection is also the perpendicular bisector of each segment joining corresponding points in the pre-image and image. Therefore, every point on the pre-image and its corresponding point on the image are equidistant from the line of reflection.

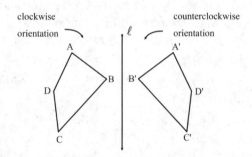

Line reflections do not preserve orientation, which is the direction one must travel to move from one vertex to the next. In the figure, the orientation changes from clockwise to counterclockwise. Noting a change in orientation is one way to identify a reflection.

POINT REFLECTION

A *point reflection* maps each point in the pre-image to a point such that the center of the reflection is the midpoint of the segment through any corresponding pair of points. $\overline{PQ}$ is the image of $\overline{ST}$ after a reflection through point R. Therefore R is the midpoint of $\overline{SP}$ and $\overline{TQ}$. A point reflection is equivalent to a 180° rotation.

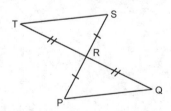

ROTATIONS

A *rotation* is the spinning of a figure about a pivot point called the *center of rotation*. The angle of rotation is measured counterclockwise unless otherwise specified. The accompanying figure shows $ABCD$ rotated about point O to $A'B'C'D'$. The notation $R_{C,a}$ is used to specify a rotation, where a is the angle of rotation and C is the center of rotation. Rotations are rigid motions, so $ABCD \cong A'B'C'D'$. The angle formed by any two corresponding points with the center of rotation as the vertex is equal to the angle of rotation. $m\angle AOA' = m\angle BOB' = m\angle COC' = m\angle DOD' = $ angle of rotation.

DILATIONS

A *dilation* is a similarity transformation that enlarges or reduces the size of a figure without changing its shape. Angle measures are preserved, and the lengths of segments in the image are proportional to lengths in the pre-image. The ratio of lengths is called the scale factor. Dilations are specified about a center point. The distance from a point to the center is also multiplied by the scale factor after a dilation. $D_{C,k}$ is a dilation with a center point C and scale factor k. The accompanying figure shows the dilation $D_{C,2}\overline{(PQ)} \rightarrow \overline{P'Q'}$.

CP' = 2CP, CQ' = 2CQ

P'Q' = 2PQ

HORIZONTAL AND VERTICAL STRETCH

Two examples of transformations that are neither rigid motions nor similarity transformations are the horizontal stretch and vertical stretch. A *horizontal stretch* will elongate a figure only in the horizontal direction. On the coordinate plane, the x-coordinate of every point will be multiplied by a scale factor. A *vertical stretch* will do the same thing to the y-coordinate.

pre-image   vertical stretch   horizontal stretch

Summary of the Properties of Transformations

Transformation	What's Preserved	Segments Between Corresponding Points Congruent?	Other Relationships
Reflection $r_{\overline{FE}}(\overline{AB}) \rightarrow \overline{A'B'}$	Corresponding lengths and angles, parallelism	Not necessarily $AA' \neq BB'$	$\overleftrightarrow{FE}$ is the $\perp$ bisector of $\overline{BB'}$ and $\overline{AA'}$ $\overline{AA'} \mathbin{//} \overline{BB'}$
Rotation $R_{C,100°}(\overline{AB}) \rightarrow \overline{A'B'}$	Corresponding lengths and angles, parallelism, orientation	Not necessarily $AA' \neq BB'$	$\overline{AC} \cong \overline{A'C}$ $\overline{BC} \cong \overline{B'C}$ $\angle ACA' \cong \angle BCB'$

Transformation	What's Preserved	Segments Between Corresponding Points Congruent?	Other Relationships
Translation $T_{\overrightarrow{UV}}(\overline{AB}) \to \overline{A'B'}$	Corresponding lengths and angles, slope, parallelism, orientation	Yes $AA' \neq BB'$	$\overline{AB} \ // \ \overline{A'B'}$ $\overline{AA'} \ // \ \overline{BB'}$
Dilation $D_{C,3/2}(\overline{AB}) \to \overline{A'B'}$	Corresponding angles, slope, parallelism, orientation	Not necessarily $AA' \neq BB'$	$\overline{AB} \ // \ \overline{A'B'}$

Transformations on the Coordinate Plane

The following set of rules can be used to apply a transformation to a point on the coordinate plane.

Reflection	Rotation About the Origin
$r_{x\text{-axis}}(x, y) \to (x, -y)$	$R_{90}(x, y) \to (-y, x)$
$r_{y\text{-axis}}(x, y) \to (-x, y)$	$R_{180}(x, y) \to (-x, -y)$
$r_{y=x}(x, y) \to (y, x)$	$R_{270}(x, y) \to (y, -x)$
$r_{y=-x}(x, y) = (-y, -x)$	

Translation	Dilation
$T_{h,k}(x, y) \rightarrow (x + h, y + k)$	$D_{k,\text{origin}}(x, y) \rightarrow (kx, ky)$

- Rotations of 180° and 270° can also be found by repeating the rule for a 90° rotation.

COMPOSITIONS

A **composition of transformation** is a sequence of transformations where the image found after the first transformation becomes the pre-image for the next transformation. The symbol ° is used to indicate a composition, as in $T_{3,4} \, ° \, D_{3,\text{origin}}$. The transformation on the right is performed first. The same composition can also be expressed as $T_{3,4}(D_{3,\text{origin}})$.

SYMMETRY

A figure has *line symmetry* if a line of reflection can be drawn which divides the figure into two congruent mirror images. For example:

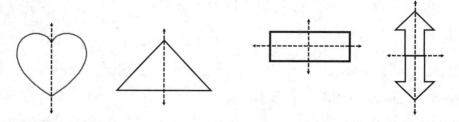

vertical line symmetry   horizontal and vertical line symmetry

A figure has *rotational symmetry* if it can be rotated by some angle $0 < \theta < 360°$ about its center and have every point map to another point on the image. In other words, the image will look identical to the pre-image after the rotation.

The accompanying figure shows rectangle $ABCD$ after rotations of 0°, 90°, 180°, 270°, and 360°. Rotations of 0°, 180°, and 360° result in figures that look identical to the original. We say the figure has 180° rotational symmetry—the 0° and 360° rotations are not considered rotational symmetries. Note that the location of individual points has changed. After a 180° rotation, point A is mapped to point C, B is mapped to D, C is mapped to A, and D is mapped to B.

Some figures have more than one rotation that results in an identical figure. An equilateral triangle has 120° and 240° rotational symmetry when rotated about its center.

A regular polygon with n sides always has rotational symmetry, with rotations in increments equal to its central angle of $\dfrac{360°}{n}$.

Example

By how many degrees must a regular octagon be rotated so that it maps onto itself? List all the rotations less than 360° that will map an octagon onto itself.

Solution:

$n = 8$ for an octagon.

$$\frac{360°}{n} = \frac{360°}{8} = 45°$$

Any rotation between 0° and 360° that is a multiple of 45° will map the octagon onto itself—45°, 90°, 135°, 180°, 225°, 270°, and 315°.

Example

By how many degrees must pentagon *ABCDE* be rotated about its center to map point *A* to point *C*?

Solution:

First calculate the rotation to map one vertex to the adjacent vertex.

$$\frac{360°}{n} = \frac{360°}{5} = 72°$$

Since rotations are counterclockwise, the pentagon must be rotated three increments of 72°, or 216°, to map point *A* to point *C*.

Practice Exercises

1 △*ABC* undergoes a transformation such that its image is △*A′B′C′*. If the side lengths and angles are preserved, but the slopes of the sides are not, the transformation could be a

(1) reflection or rotation.
(3) translation or rotation.

(2) reflection or translation.
(4) dilation or translation.

2 Which of the following transformations is not a rigid motion?

(1) reflection
(3) rotation

(2) dilation
(4) translation

3 Which of the following letters has more than one line of symmetry?

(1) **A**
(3) **H**

(2) **C**
(4) **L**

4 Triangles *AMN* and *APQ* share vertex *A*. A circle can be constructed that has a center at *A* and passes through points *M*, *N*, *P*, and *Q*. If m∠*MAN* = m∠*PAQ*, which of the following is *not* necessarily true?

(1) ∠*NAQ* is a right angle.

(2) ∠*M* ≅ ∠*Q*.

(3) △*APQ* is the image of △*AMN* after a rotation about point *A*.

(4) △*AMN* and △*AQP* are isosceles.

5 Which transformation is represented in the figure below?

(1) $R_{\text{origin, }90°}$ (3) $r_{x\text{-axis}}$

(2) $R_{\text{origin, }180°}$ (4) $r_{y\text{-axis}}$

6 Which of the following will always map a rectangle onto itself?

   (1) reflection over one of its sides

   (2) reflection over one of its diagonals

   (3) rotation of 180° about its center

   (4) rotation of 90° about one of its vertices

7 The area of a rectangle is 4 cm². The rectangle then undergoes a rotation of
   90° about one of its vertices, followed by a dilation with a scale factor of 3,
   centered at the same vertex as the rotation. What is the area of the image?

   (1) 12 cm²   (3) 32 cm²

   (2) 16 cm²   (4) 36 cm²

8 Regular hexagon *ABCDEF* is shown in the figure below. Which of the following transformations will map the hexagon onto itself such that point *D* maps to point *E*?

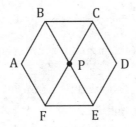

(1) a 60° rotation about point *P* followed by a reflection of line $\overleftrightarrow{BE}$
(2) reflection over line $\overleftrightarrow{BE}$ followed by 60° rotation about point *P*
(3) reflection over line $\overleftrightarrow{CF}$ followed by a translation by vector $\overrightarrow{FE}$
(4) reflection over line $\overleftrightarrow{CF}$ followed by a 120° rotation about point *P*

9 After a certain transformation the image of △*BAD* with coordinates *B*(7, 1) *A*(0, 3) *D*(12, 4) is *B*′(13, 3), *A*′(6, 5), *D*′(18, 1). Is the transformation a translation? Explain why or why not.

10 The vertices of △*MNO* have coordinates *M*(−5, 2), *N*(1, 2), *P*(0, 5). △*M*′*N*′*P*′ is the image of △*MNP* after a translation. Find the coordinates of *N*′ and *P*′ if vertex *M* is mapped to *M*′(3, 6).

11 In the figure below, △*DAB* ≅ △*BED* ≅ △*EBC* ≅ △*CFE*. Which of the following sequences of rigid motions will always map △*DAB* to △*CFE*?

(1) a translation by vector $\overrightarrow{BE}$ followed by a translation by vector $\overrightarrow{EC}$
(2) a rotation of 180° about point *B*, followed by a translation by vector $\overrightarrow{CF}$
(3) a reflection over $\overline{BE}$ followed by a translation by vector $\overrightarrow{BC}$
(4) a translation by vector $\overrightarrow{AC}$ followed by a reflection over $\overline{FC}$

12 Which of the following will map parallelogram $ABCD$ onto itself?

(1) a rotation of 180° about the point of intersection of the two diagonals of $ABCD$

(2) a reflection over $\overline{CD}$, followed by a translation along vector $\overrightarrow{DA}$

(3) a reflection over diagonal $\overline{AC}$

(4) a reflection over $\overline{BC}$, followed by a translation by vector $\overrightarrow{BA}$

13 In the figure below, $\triangle SDR \sim \triangle RCP$. Which of the following compositions of transformations will map $\triangle RCP$ to $\triangle SDR$?

(1) a translation by vector $\overrightarrow{PR}$, followed by a 180° rotation about point R, followed by a dilation centered at R with scale factor $\dfrac{RS}{RP}$

(2) a dilation centered at R with scale factor $\dfrac{RS}{RP}$, followed by a reflection over $\overline{RS}$

(3) a translation by vector $\overrightarrow{CR}$, followed by a reflection over $\overline{PC}$, followed by a dilation centered at R with scale factor $\dfrac{RS}{RP}$

(4) a dilation centered at R with scale factor $\dfrac{RP}{DS}$, followed by a translation along vector $\overrightarrow{PD}$, followed by a reflection over $\overline{DR}$

14 Graph $\triangle A(-2, 1)$, $B(0, 2)$, $C(-1, 4)$. Graph $\triangle A''B''C''$, the image of $\triangle ABC$ after the transformation $T_{1,3} \cdot r_{x=y}$. State the coordinates of $A''B''C''$.

15 Parallelogram *STAR* has coordinates $S(-8, -4)$, $T(-5, -2)$, $A(1, -5)$, and $R(-2, -7)$.

   (a) Graph and state the coordinates of $S'T'A'R'$, the image of *STAR* after $r_{y=-1}$.

   (b) Graph and state the coordinates of $S''T''A''R''$, the image of $S'T'A'R'$ after $T_{7,-2}$.

   (c) Name a single reflection that would map T to T''.

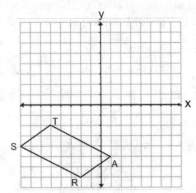

Solutions

1 A translation always preserves slope, so choices (2), (3), and (4) cannot be correct.

The correct choice is (**1**).

2 Reflections, rotations, and translations are always rigid motions because the image and pre-image are congruent. A dilation may result in an image with a different size, so it is not a rigid motion.

The correct choice is (**2**).

3 **H** has both horizontal and vertical lines of symmetry.

The correct choice is (**3**).

4 If points M, N, P, and Q all lie on the same circle with center A, then $\overline{AM}$, $\overline{AN}$, $\overline{AP}$, and $\overline{AQ}$ are all radii of circle A, and $AM = AN = AP = AQ$. The triangles are congruent by SAS and are isosceles, so $\angle M \cong \angle N \cong \angle P \cong \angle Q$ and choices (2) and (4) are true.

Since $\angle MAN \cong \angle PAQ$ and $\angle NAP \cong \angle NAP$ by the reflexive property, then $m\angle MAN + m\angle NAP = m\angle NAP + m\angle PAQ$. This is equivalent to $\angle MAP \cong \angle NAQ$. Therefore, P is the rotation of M and Q is the rotation of point N about point A, so one triangle is the rotation of the other. Choice (3) is true. The only choice that cannot be proven is choice (1); we do not know the measures of any of the angles and cannot tell if $\angle NAQ$ measures 90°.

The correct choice is (**1**).

5 Approach this problem by simply applying each transformation and checking if $A(1, -1)$ maps to $A'(1, 1)$ and $B(3, -4)$ maps to $B'(4, 3)$. If the correct image results for both, then the choice is correct.

$R_{\text{origin, }90°}(1, -1) \to (1, 1)$ correct	$R_{\text{origin, }90°}(3, -4) \to (4, 3)$
$R_{\text{origin, }180°}(1, -1) \to (-1, 1)$ A' and B' are incorrect	$R_{\text{origin, }180°}(3, -4) \to (-3, 4)$
$R_{x\text{-axis}}(1, -1) \to (1, 1)$ B' is incorrect	$R_{x\text{-axis}}(3, -4) \to (3, 4)$
$r_{y\text{-axis}}(1, -1) \to (-1, -1)$ A' and B' are incorrect	$r_{y\text{-axis}}(3, -4) \to (-3, -4)$

The correct choice is **(1)**.

6 Since a rectangle has twofold symmetry, a rotation of 180° about its center will map it to itself.

The correct choice is **(3)**.

7 Under a dilation, the area of a polygon is proportional to the scale factor squared, so the area becomes 3^2 times larger, or 9 times larger. The new area is $9 \cdot 4 \text{ cm}^2 = 36 \text{ cm}^2$.

The correct choice is **(4)**.

8 A reflection over $\overleftrightarrow{BE}$ will map point D to point F. Each central angle measures $\dfrac{360°}{60°} = 60°$. Therefore, a rotation of 60° will map any vertex to the adjacent vertex in the counterclockwise direction. The 60° rotation will map a point at vertex F to vertex E.

The correct choice is **(2)**.

9 If the transformation is a translation, then every point in the pre-image would undergo the same translation $(x, y) \rightarrow (x + h, y + k)$. Calculate each h and k by working backwards from the pre-image and image:

$B'(13, 3) = B(7 + h, 1 + k)$	$13 = 7 + h, h = 6$	$3 = 1 + k, k = 2$
$A'(6, 5) = A(0 + h, 3 + k)$	$6 = 0 + h, h = 6$	$5 = 3 + k, k = 2$
$D'(18, 1) = D(12 + h, 4 + k)$	$18 = 12 + h, h = 6$	$1 = 4 + k, k = -3$

The transformation is not a translation because A and B undergo the translation $T_{6,2}$ while D undergoes the translation $T_{6,-3}$.

10 First, find the translation rule.

$T_{h,k} \, M(-5, 2) \rightarrow M'(-5 + h, 2 + k)$

Using the coordinates $(3, 6)$ for M', solve for h and k.

$-5 + h = 3$	$2 + k = 6$
$h = 8$	$k = 4$

The translation is $T_{8,4}$. Now apply that translation to N and P.

$T_{8,4} \, N(1, 2) \rightarrow N'(1 + 8, 2 + 4) \rightarrow N'(9, 6)$
$T_{8,4} \, P(0, 5) \rightarrow P'(0 + 8, 5 + 4) \rightarrow P'(8, 9)$

The correct choice is **(3)**.

11 The figure shows the $\triangle DAB$ after a 180° rotation about B. A translation by $\overrightarrow{CF}$ will then slide the image up and map it to $\triangle CFE$.

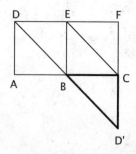

The correct choice is **(2)**.

12 Rotating 180° about the point of intersection of the diagonals will map:
$A \rightarrow C, B \rightarrow D, C \rightarrow A$, and $D \rightarrow B$.

The other choices will not map $ABCD$ onto itself. Choices (2), (3), and (4) result in the following:

choice (2)

choice (3)

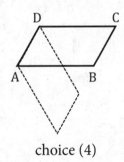

choice (4)

The correct choice is (**1**).

13 The figures show the figure after the translation by $\overrightarrow{PR}$, and then the rotation of 180° about R. The final dilation will map C'' to D and R'' to S.

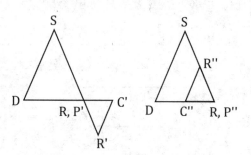

The correct choice is (**1**).

14 In a composition always do the rightmost transformation first. For $T_{1,3} \cdot r_{x=y}$, start with $r_{y=x}$. For a reflection over the line $y = x$, switch the order of the coordinates.

$r_{y=x}$: $A(-2, 1) \rightarrow A'(1, -2)$ $B(0, 2) \rightarrow B'(2, 0)$ $C(-1, 4) \rightarrow C'(4, -1)$

Now use the image from the reflection as the pre-image of the translation.

For a translation $T_{1,3}$, add 1 to each x-coordinate and 3 to each y-coordinate.

$T_{1,3}$: $A'(1, -2) \rightarrow A''(2, 1)$ $B'(2, 0) \rightarrow B''(3, 3)$ $C'(4, -1) \rightarrow C''(5, 2)$

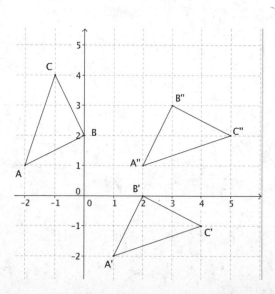

15 a) The x-coordinates of each point in *STAR* are unchanged after a reflection over the line $y = -1$. To find the y-coordinates, count the vertical distance from the point to $y = -1$, then continue by the same distance on the other side of $y = -1$. $S(-8, -4)$ is 3 units below $y = -1$.

The y-coordinate of S' is $-1 + 3 = 2$. $T(-5, -2)$ is 1 unit below $y = -1$.	$S'(-8, 2)$
The y-coordinate of T' is $-1 + 1 = 0$. $A(1, -5)$ is 4 units below $y = -1$.	$T'(-5, 0)$
The y-coordinate of A' is $-1 + 4 = 3$. $R(-2, -7)$ is 6 units below $y = -1$.	$A'(1, 3)$
The y-coordinate of R' is $-1 + 6 = 5$. $S'(-8, 2)$ $T'(-5, 0)$ $A'(1, 3)$ $R'(-2, 5)$	$R'(-2, 5)$

b) Apply the translation $T_{7,-2}$ by adding 7 to each x-coordinate and subtracting 2 from each y-coordinate.

$$S''(-1,0) T''(2,-2) A''(8,1) R''(5,3)$$

c) To reflect $T(-5, -2)$ to $T''(2, -2)$, note that the y-coordinates are the same, so we need a reflection over a vertical line that is halfway between $x = -5$ and $x = 2$.

$$\frac{1}{2}(-5 + 2) = -1\frac{1}{2}$$

A reflection over the line $x = -1\frac{1}{2}$ will map $T(-5, -2)$ to $T''(2, -2)$.

3.5 TRIANGLE CONGRUENCE

CONGRUENCE

Two figures are congruent if one figure can be mapped onto another by a sequence of rigid motions. The congruent figures will have exactly the same size and shape, which means for polygons that all *corresponding* pairs of sides and angles are congruent.

A congruence statement will match up the corresponding pairs of congruent parts.

For example, from the statement $\triangle BUS \cong \triangle CAR$ we can conclude:

$$\angle B \cong \angle C \quad \overline{BU} \cong \overline{CA}$$
$$\angle U \cong \angle A \quad \overline{US} \cong \overline{AR}$$
$$\angle S \cong \angle R \quad \overline{SB} \cong \overline{RC}$$

Once two figures have been proven congruent, we know all pairs of parts are congruent. In triangles, this theorem is abbreviated by CPCTC—corresponding parts of congruent triangles are congruent.

PROVING POLYGONS CONGRUENT BY TRANSFORMATIONS

Polygons can be proven congruent by identifying a sequence of rigid motions that map one onto the other. One method is to show each vertex of the pre-image maps to a corresponding vertex of the image using the same transformation, or sequence of transformations. The angle and segment relationships between pre-image, image, lines of reflection, center of rotation, and translation vectors can help establish specific transformations that map one point onto another:

- If the segments joining corresponding vertices of the two triangles are congruent and parallel, then one triangle is a translation of the other.
- If a single line is the perpendicular bisector of segments formed by corresponding vertices, then one triangle is the reflection of the other.
- If angles formed by corresponding vertices and a center point are all congruent, and corresponding distances to the center point are equal, then one triangle is a rotation of the other.

For example, given $\overline{ADBE}$, $\overline{AD} \cong \overline{BE}$, $\overline{AD} \cong \overline{CF}$, $\overline{AD} \,/\!/\, \overline{CF}$, we can show that $\triangle ABC \cong \triangle DEF$ because one is a translation of the other. Point A translates to point D along segment $\overline{ADBE}$. B translates to E by the same distance along the same segment, and C translates by the same distance along a parallel segment. Each point in $\triangle ABC$ is mapped to a corresponding point in $\triangle DEF$ by the same translation (distance AD along a vector parallel to $\overline{ADBE}$). Therefore, $\triangle DEF$ is a translation of $\triangle ABC$, and the two triangles are congruent because translations are a rigid motion.

TRIANGLE CONGRUENCE POSTULATES

We know two polygons are congruent if all pairs of corresponding sides and angles are congruent. When proving two triangles congruent we do not need to prove all three pairs of sides and all three pairs of angles congruent. There are five shortcuts commonly used—the SAS, SSS, ASA, AAS, and the HL criteria. Each criteria requires only a total of three pairs of parts to be congruent for us to conclude the triangles are congruent.

Criteria	What It Looks Like
SAS—two pairs of sides and the included angles are congruent.	
SSS—three pairs of sides are congruent.	
ASA—two pairs of angles and the included sides are congruent.	
AAS—two pairs of angles and the nonincluded sides are congruent.	
HL—the hypotenuses and one pair of legs are congruent in a pair of right triangles.	

Once a pair of triangles has been proven congruent using one of these postulates, then CPCTC can be used to justify why any of the remaining pairs of corresponding parts are congruent.

You do not need to limit yourself strictly to a rigid motion or a congruence postulate. Sometimes the given information will lead to a combination of these approaches.

STRATEGIES FOR WRITING CONGRUENCE PROOFS

1. Mark congruent parts on the figure using given information and anything else you can conclude from the figure.
2. Look for the following common relationships:
   a. Vertical angles
   b. Shared sides and angles
   c. Linear pairs
   d. Supplementary and complementary angles
   e. Parallel lines and perpendicular lines
   f. Midpoints and bisectors of segments
   g. Angle bisectors
   h. Altitudes, medians, and perpendicular bisectors in triangles
   i. Isosceles triangle theorem, exterior angle theorem, and triangle angle sum theorem
3. Statements should always refer to specific named parts of the figure (such as $\triangle ABC$, $\overline{AB}$, $\angle C$). Reasons may only involve definitions, postulates, and theorems. Never name a specific part of the figure in the reasons column.
4. There must be a line in your proof for each part of the postulate you use. You can label each line that corresponds to a part of the postulate with (A) or (S), or use checkboxes:

S	A	S
✓	✓	✓

OVERLAPPING TRIANGLES

Angle addition or segment addition can be used to show parts are congruent when triangles overlap. For example, if given $\overline{ABCD}$ and $\overline{AB} \cong \overline{CD}$, we can prove $\overline{AC} \cong \overline{BD}$ by stating that $\overline{BC}$ is congruent to itself by the reflexive property, then adding the two congruence statements:

$$\overline{BC} \cong \overline{BC}$$ reflexive property

$$\overline{AC} \cong \overline{BD}$$ given

$$\overline{AB} + \overline{BC} \cong \overline{BC} + \overline{CD}$$ addition property

$$\overline{AC} \cong \overline{BD}$$ partition property

Another strategy is to sketch the two triangles separately. By pulling them apart, you may more easily see the parts required to complete the proof.

Practice Exercises

1 $A'B'C'D'$ is the image of $ABCD$ after a rotation about the origin, and $A''B''C''D''$ is the image of $A'B'C'D'$ after a translation. Which of the following must be true?

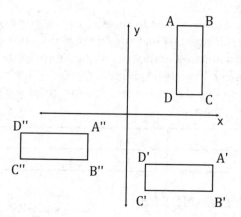

(1) All three figures are rectangles.

(2) $ABCD$ is congruent to $A'B'C'D'$, but not congruent to $A''B''C''D''$.

(3) $ABCD$ is congruent to $A''B''C''D''$, but not congruent to $A'B'C'D'$.

(4) $ABCD$ is congruent to both $A'B'C'D'$ and $A''B''C''D''$.

2 Which triangle congruence criteria can be used to prove the two triangles are congruent?

(1) SAS   (3) SSS

(2) ASA   (4) HL

3 Which of the following pieces of information would *not* allow you to conclude that $\triangle ACD \cong \triangle ACB$?

(1) $\overline{AC}$ bisects $\angle BCD$ and $\overline{CD} \cong \overline{CB}$.

(2) $\overline{AC}$ bisects $\angle BCD$ and $\overline{AD} \cong \overline{AB}$.

(3) $\overline{AC}$ bisects $\angle BCD$ and $\angle BAD$.

(4) $\overline{AC}$ bisects $\angle BAD$ and $\overline{AD} \cong \overline{AB}$.

4 Which of the following is sufficient to prove △*SMA* ≅ △*S'M'A'*?

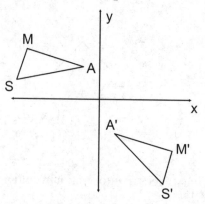

(1) There is a sequence of rigid motions that maps
     ∠*M* to ∠*M'* and ∠*S* to ∠*S'*.

(2) There is a sequence of rigid motions that maps
     ∠*M* to ∠*M'* and $\overline{MS}$ to $\overline{M'S'}$.

(3) There is a sequence of rigid motions that maps
     ∠*S* to ∠*S'* and $\overline{MS}$ to $\overline{M'S'}$.

(4) There is a sequence of rigid motions that maps
     ∠*A* to ∠*A'* and point *A* to point *A'*.

5 Given: $\overline{AB}$ // $\overline{CD}$, $\overline{AB} \cong \overline{CD}$

Prove: △*ABE* ≅ △*CDE*

6 Given: $\overline{BC}$ bisects $\overline{AD}$ at *E*, ∠*A* ≅ ∠*D*

Prove: △*ABE* ≅ △*DCE*

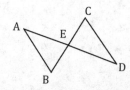

7 *ABCDEF* is a regular hexagon. State a sequence of rigid motions that could
   be used to justify why △*ABC* is congruent to △*AEF*.

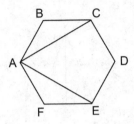

8 In the accompanying figure of △*FGH* and △*GIJ*, there is a sequence of rigid
   motions that maps *F* to *G*, *G* to *I*, and *H* to *J*. Explain why ∠*H* must be congruent to ∠*J*.

9 Given $\overline{RPB}$ and line *m* is the perpendicular bisector of $\overline{TJ}$ and $\overline{RB}$, explain
   why △*TRP* ≅ △*JBP* and why $\overline{TR}$ ≅ $\overline{JB}$.

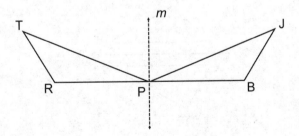

10 Given: △WXY and perpendicular bisector $\overline{YZ}$

Explain why △WZY ≅ △XZY in terms of a rigid motion.

11 Given: $\overline{AB} \cong \overline{AD}$, $\overline{AC} \cong \overline{AE}$, m∠BAD = 60°, and m∠CAE = 60°

Prove: △BAC ≅ △DAE in terms of a rigid motion

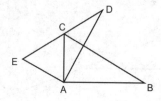

12 Given: line m is the perpendicular bisector of $\overline{FX}$, $\overline{GY}$, and $\overline{HZ}$

Prove: △XYZ ≅ △FHG

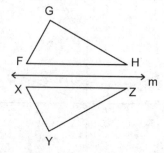

Solutions

1 Both rotations and translations are rigid motions, and any sequence of rigid motions will result in a congruent image. Therefore, $ABCD$ is congruent to both $A'B'C'D'$ and $A''B''C''D''$.

The correct answer choice is (**4**).

2 Both triangles are right triangles, a corresponding pair of legs are congruent, and the two hypotenuses are congruent. Therefore HL applies. SAS is not correct here because the angle does not lie between the two congruent sides.

The correct choice is (**4**).

3 The shared side $\overline{AC}$ is a pair of congruent sides. Check each choice to find one that does *not* provide enough information to complete one of the congruent postulates.

Choice 1: The angle bisector gives $\angle DCA \cong \angle BCA$, and SAS applies.

Choice 2: The angle bisector gives $\angle DCA \cong \angle BCA$, and we have two sides and an angle, but the angle is not the included angle. None of the congruence postulates apply.

Choice 3: The angle bisector gives $\angle DCA \cong \angle BCA$ and $\angle DAC \cong \angle BAC$. ASA applies.

Choice 4: The angle bisector gives $\angle DAC \cong \angle BAC$, and SAS applies.

The correct choice is (**2**).

4 Choice 1 maps two consecutive angles onto corresponding angles, as well as M to M' and S to S'. This means $\overline{SM} \cong \overline{S'M'}$ because a sequence of rigid motions maps the endpoints of $\overline{SM}$ onto corresponding endpoints. $\overline{SM}$ is the included side between the congruent angles, therefore, $\triangle SMA \cong \triangle S'M'A'$ by the SAS postulate.

The correct choice is (**1**).

5

Statement	Reason
1. $\overline{AB} \cong \overline{CD}$, $\overline{AB} // \overline{CD}$	1. Given
2. $\angle A \cong \angle C$ $\angle B \cong \angle D$	2. Alternate interior angles formed by the parallel lines are congruent
3. $\triangle ABE \cong \triangle CDE$	3. ASA

6

Statement	Reason
1. $\overline{BC}$ bisects $\overline{AD}$ at E	1. Given
2. E is the midpoint of $\overline{AD}$	2. A bisector intersects a segment at its midpoint
3. $\overline{AE} \cong \overline{ED}$	3. A midpoint divides a segment into two congruent segments
4. $\angle AEB \cong \angle DEC$	4. Vertical angles are congruent
5. $\angle A \cong \angle D$	5. Given
6. $\triangle ABE \cong \triangle DCE$	6. ASA

7   $\triangle ABC$ is mapped to $\triangle AFE$ by a rotation about point A followed by a reflection over $\overline{AE}$. The angle of rotation is $\angle CAE$. From the figure each of the diagonals $\overline{AC}$, $\overline{AE}$, and $\overline{CE}$ are congruent, so $\triangle ACE$ is equilateral, and its angles measure 60°. The angle of rotation is 60°. The sequence of rigid motions is a 60° rotation about point A followed by a reflection over $\overline{AE}$.

8 The two triangles are congruent because there exists a rigid motion that maps each vertex of $\triangle JGH$ to a corresponding vertex $\triangle GIJ$. Since the triangles are congruent, all pairs of corresponding parts are congruent (CPCTC). $\angle G$ and $\angle H$ are congruent corresponding angles.

9 Since line m is the perpendicular bisector of both $\overline{TJ}$ and $\overline{RB}$, it is a line of reflection that maps T to J and R to B. The same reflection will map P to itself because it lies on the line of reflection. A single reflection maps $\triangle TRP$ onto $\triangle JBP$, and reflections are rigid motions, so the two triangles must be congruent. $\overline{TR} \cong \overline{JB}$ because corresponding parts of congruent triangles are congruent.

10 Point Y is the image of itself after a reflection over $\overline{YZ}$ because Y lies on the line of reflection. Point Z is the image of itself for the same reason. X is the image of point W after a reflection over $\overline{YZ}$ because a line of reflection is the perpendicular bisector of the segment joining the pre-image and the image after the reflection. Therefore, the triangles are congruent because one is mapped to the other by a reflection over $\overline{YZ}$, which is a rigid motion.

11 $\overline{AB} \cong \overline{AD}$; therefore, D is the image of B after a 60° rotation about A. $\overline{AC} \cong \overline{AE}$; therefore, C is the image of E after a 60° rotation about A. Point A will map to itself after the same rotation because it is the center of rotation. Therefore, $\triangle BAC$ is the image of $\triangle DAE$ after a 60° rotation about A. $\triangle BAC \cong \triangle DAE$ because rotations are rigid motions.

12 A perpendicular bisector of a segment is also the line of reflection that maps one endpoint to the other. Therefore, after a reflection over line m, X is the image of F, Z is the image of H, and Y is the image of G. $\triangle FHG$ is mapped to $\triangle XZY$ by a reflection, which is a rigid motion; therefore, $\triangle XZY \cong \triangle FHG$.

3.6 COORDINATE GEOMETRY

COORDINATE GEOMETRY FORMULAS

Given two points (x_1, y_1) and (x_2, y_2):

- The distance between the points is $\sqrt{(x_2 - x_1)^2 + (y_2 - y_1)^2}$

- The midpoint of the segment joining the points is $\left(\dfrac{x_1 + x_2}{2}, \dfrac{y_1 + y_2}{2}\right)$

- The slope of the segment joining the points is $\dfrac{y_2 - y_1}{x_2 - x_1}$, or $\dfrac{rise}{run}$

DIVIDING A SEGMENT PROPORTIONALLY

A directed segment is one that has a specified starting point and ending point. We can divide a directed segment into parts that are in any given ratio using the two proportions ratio $= \dfrac{x - x_1}{x_2 - x}$ and ratio $= \dfrac{y - y_1}{y_2 - y}$.

For example, given $J(1, -2)$ and $K(11, 3)$, find the coordinate (x, y) of point L that divides $\overline{JK}$ in a 2:3 ratio.

J(1, -2)   L(x,y)   K(11, 3)

- Use the coordinates of J for point 1 and the coordinates of K for point 2.

$$x_1 = 1 \qquad x_2 = 11 \qquad y_1 = -2 \text{ and } y_2 = 3$$

- Apply the formula for the x-coordinate.

$$\text{ratio} = \frac{x - x_1}{x_2 - x}$$

$$\frac{2}{3} = \frac{x - 1}{11 - x}$$
$$2(11 - x) = 3(x - 1)$$
$$22 - 2x = 3x - 3$$
$$25 = 5x$$
$$x = 5$$

- Repeat the process for the y-coordinate.

$$\text{ratio} = \frac{y - y_1}{y_2 - y}$$

$$\frac{2}{3} = \frac{y - (-2)}{3 - y}$$
$$3y + 6 = 6 - 2y$$
$$5y = 0$$
$$y = 0$$

Point $L(5, 0)$ divides $\overline{JK}$ in a 2:3 ratio.

AREA AND PERIMETER

The lengths found with the distance formula can be used to calculate the perimeter and area of figures. If the figure is irregular, three strategies can be used:

- Divide the figure into shapes whose areas can be calculated easily (squares, rectangles, triangles, trapezoids, and circles).
- Sketch a bounding rectangle around the figure. Calculate the area of the rectangle, and then subtract the area of all the triangles that fall outside the figure but with the rectangle.
- If the figure has curves, estimate the area by modeling the curved sides with straight segments. The more segments that are used to model a curve, the more accurate the result will be.

Example

Find the area of polygon $ABCDE$, with vertices $A(-3, -5)$, $B(2, -5)$, $C(5, 3)$, $D(-2, 5)$, and $E(-5, -1)$.

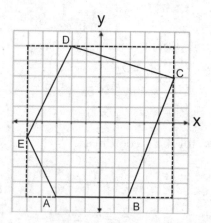

Solution:

Sketch the bounding rectangle in around $ABCDE$.

The length is 10 and the width is 10, giving an area of 100. The triangles have areas:

$$\text{upper left triangle} = \frac{1}{2}(3 \cdot 6) = 9$$

$$\text{upper right triangle} = \frac{1}{2}(7 \cdot 2) = 7$$

$$\text{lower left triangle} = \frac{1}{2}(4 \cdot 2) = 4$$

$$\text{lower right triangle} = \frac{1}{2}(3 \cdot 8) = 12$$

The area of $ABCDE = 100 - 9 - 7 - 4 - 12 = 68$

COLLINEARITY

Three points are **collinear** if the slopes between any two pairs are equal. For example, points A, B, and C are collinear if the slope of $\overline{AB}$ equals the slope of $\overline{BC}$.

EQUATIONS OF LINES

- The slopes of parallel lines are equal.
- The slopes of perpendicular lines are negative reciprocals. (If the slope of line m is $\frac{2}{3}$, then the slope of any line perpendicular to m is $-\frac{3}{2}$.)
- The equation of a line in slope-intercept form is $y = mx + b$ where m is the slope and b is the y-intercept. To graph the line, plot a point on the y-axis at the y-intercept. From that point, plot additional points using the rise and run from the slope.
- The equation of a line in point-slope form is $y - y_1 = m(x - x_1)$ where m is the slope and (x_1, y_1) are the coordinates of any point on the line. To graph the line, plot the first point at (x_1, y_1). From that point, plot additional points using the rise and run from the slope.

Strategy for writing the equation of a line in slope-intercept form:

1. Find the slope

Given two points: use $\frac{y_2 - y_1}{x_2 - x_1}$

Given a parallel line: use the slope of the parallel line

Given a perpendicular line: use the negative reciprocal of the slope of the perpendicular line

2. Substitute the slope and (x,y) coordinate pair into y = mx + b. Solve for b

3. Write the equation with the correct slope and y-intercept

TRANSFORMATIONS AND LINES

Translations and Dilations

Translations and dilations preserve slope, so the slope of the image will be the same as the slope of the pre-image.

To translate or dilate a line given its equation,

1. Choose any point on the line (the y-intercept is often an easy choice).

2. Apply the translation or dilation to that point.

3. Find the equation of the line that has the same slope as the original line and passes through the transformed point.

Rotations

Rotations of 90° will result in a line perpendicular to the original, so the slope will be the negative reciprocal. To write the equation of a line after a 90° rotation, use the same procedure for translations and dilations, except use the negative reciprocal of the slope.

EQUATION OF THE CIRCLE

Center Radius Form of the Equation of a Circle

$(x - h)^2 + (y - k)^2 = r^2$ where the center has coordinates (h, k) and radius has length r.

- To graph a circle, first identify the center and radius from the equation. Plot a point at the center. Then plot points up, down, left, and right a distance r from the center.

Example

Graph the equation $(x - 2)^2 + (y + 1)^2 = 9$.

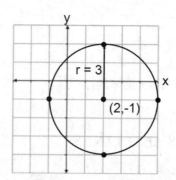

The center is located at $(2, -1)$, and $r^2 = 9$, so $r = 3$. We plot the center point at $(2, -1)$; then plot points up, down, right, and left 3 units from the center. Use these four points as a guide to complete the circle.

General Form of the Equation of a Circle

$$x^2 + y^2 + Cx + Dy + E = 0$$

To find the coordinates of the center and the radius from the general form of the equation, you will need to convert it to the center-radius form using the following procedure:

1. Group the x-terms and y-terms on one side of the equation, and the constant on the other side of the equation.

2. Complete the square with the x-terms, and then complete the square with the y-terms.

Example

Find the coordinates of the center and the length of the radius of a circle whose equation is $x^2 + 4x + y^2 - 6y + 7 = 0$.

Solution:

Bring the constant term to the right.

$$x^2 + 4x + y^2 - 6y = -7$$

The coefficient of x is 4, so a constant term of $\left(\frac{4}{2}\right)^2$, or 4, is needed to complete the square with the x-terms. The coefficient of y is -6, so a constant term of $\left(\frac{-6}{2}\right)^2$, or 9, is needed to complete the square with the y-terms.

$$x^2 + 4x + 4 + y^2 - 6y + 9 = -7 + 4 + 9$$
$$(x + 2)^2 + (y - 3)^2 = 6$$

The center has coordinates $(-2, 3)$ and the radius has a length of $\sqrt{6}$.

Practice Exercises

1 Points $A(2, -1)$ and $B(8, -3)$ lie on line m. After a rotation of $90°$ about the origin, the images of A and B are A' and B'. If A' and B' lie on line n, what is the equation of line n?

(1) $y = -3x + 2$

(3) $y = 3x - 1$

(2) $y = -\frac{1}{3}x - \frac{1}{3}$

(4) $y = \frac{1}{3}x + 6$

2 What is the equation of the line $6x + 2y = 12$ after a dilation by a scale factor of 5?

(1) $y = -3x + 30$

(3) $y = -15x + 30$

(2) $y = -3x + 6$

(4) $y = -15x + 6$

3 Which of the following lines is perpendicular to the line $x + 4y = 8$?

(1) $y = -\frac{1}{4}x + 2$

(3) $y = \frac{1}{4}x + 2$

(2) $y = 4x + 3$

(4) $y = -4x + 3$

4 Which of the following is the equation of a line parallel to $2x + 3y + 6 = 0$ and passes through the point $(6, 1)$?

(1) $y = -\frac{2}{3}x + 5$

(3) $y = -2x + 13$

(2) $y = 2x - 11$

(4) $y = -\frac{2}{3}x + 3$

5 What are the coordinates of the midpoint of a segment whose endpoints have coordinates $(3, 1)$ and $(15, -7)$?

(1) $(27, -15)$ (3) $(6, -4)$

(2) $(-6, 4)$ (4) $(9, -3)$

6 The diameter of a circle has endpoints with coordinates $(4, -1)$ and $(8, 3)$. Which of the following is an equation of the circle?

(1) $(x - 2)^2 + (y - 1)^2 = 8$ (3) $(x - 6)^2 + (y - 1)^2 = 8$

(2) $(x - 2)^2 + (y - 1)^2 = 32$ (4) $(x - 6)^2 + (y - 1)^2 = 32$

7 Are the segments $\overline{AB}$ and $\overline{TU}$ congruent, given coordinates $A(1, 4)$, $B(-3, 6)$, $T(2, 5)$, and $U(4, 1)$? Justify your answer.

8 Find the coordinates of the point W that divides directed segment $\overline{UV}$ in a 1:5 ratio, given coordinates $U(-3, 7)$ and $V(9, 1)$.

9 Point A has coordinates $(-2, 7)$ and point B has coordinates $(6, 3)$. Line m has the property that every point on the line is equidistant from points A and B. Find the equation of line m.

10 A circle is described by the equation $x^2 + 6x + y^2 - 12y + 25 = 0$. Find the radius of the circle and the coordinates of its center.

11 Circle P has center $P(4, -5)$ and a radius with length $\sqrt{65}$. Does the point $A(8, 2)$ lie on circle P? Justify your answer.

12 Parallelogram $ABCD$ has coordinates $A(2, -1)$, $B(5, 1)$, $C(a, b)$, and $D(3, 4)$. Write the equation of the line that contains side $\overline{CD}$.

13 Estimate the area of the ellipse shown in the accompanying figure by modeling it as the sum of 5 rectangles.

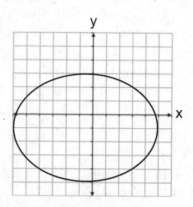

14 Ben wants to estimate the area of the curved region shown. He would like to
do so by calculating the average of the areas of two different circles.

a) Graph the two circles that Ben could use.

b) Using the two circles from part (a), estimate the area of the curved region.
Each grid unit represents 1 centimeter. Round to the nearest 1 cm^2.

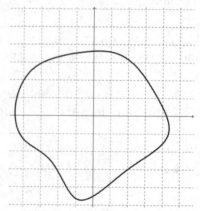

15 $\triangle SKY$ has coordinates $S(0, 2)$, $K(6, 0)$, $Y(4, 4)$.

a) Find the equation of the perpendicular bisector to side $\overline{SK}$.

b) Show that the perpendicular bisector is also an altitude.

Solutions

1 Applying a 90° rotation, A maps to $A'(1, 2)$ and B maps to $B'(3, 8)$.

$$\text{Slope } \overleftrightarrow{A'B'} = \frac{y_2 - y_1}{x_2 - x_1}$$

$$= \frac{8 - 2}{3 - 1}$$

$$= 3$$

Write the equation of the line using the point-slope form and the coordinates of A' for x_1 and y_1.

$$y - y_1 = m(x - x_1)$$
$$y - 2 = 3(x - 1)$$
$$y - 2 = 3x - 3$$
$$y = 3x - 1$$

The correct choice is **(3)**.

2 A dilation with a positive scale factor will preserve slope, but the distance of each point from the center will be multiplied by the scale factor. The strategy is to rewrite the equation in $y = mx + b$ form. The dilated line will have the same slope, but the y-intercept will be multiplied by the scale factor.

$$6x + 2y = 12$$
$$2y = -6x + 12$$
$$y = 3x + 6$$

The y-intercept of the original line is 6, and the y-intercept of the dilated line is $6 \cdot 5$, or 30.

$$y = -3x + 30$$

The correct choice is **(1)**.

3 Perpendicular lines have negative reciprocal slopes. Rewrite the equation in slope-intercept form to identify the slope.

$$x + 4y = 8$$
$$4y = -x + 8$$
$$y = -\frac{1}{4}x + 2$$

The negative reciprocal of $-\frac{1}{4}$ is 4, so the perpendicular line must have a slope of 4. The y-intercept can be any value since a particular perpendicular line was not specified. The only line with a slope of 4 is $y = 4x + 3$.

The correct choice is (2).

4 Parallel lines have equal slopes, so the first step is to rewrite the equation in slope-intercept form to help identify the slope.

$$2x + 3y + 6 = 0$$
$$3y = -2x - 6$$
$$y = -\frac{2}{3}x - 2$$

The slope is $-\frac{2}{3}$.

Now substitute $x = 6$, $y = 1$, and the slope into $y = mx + b$ and solve for the new y-intercept.

$$y = mx + b$$
$$y = -\frac{2}{3}x + b$$
$$1 = -\frac{2}{3}(6) + b$$
$$1 = -4 + b$$
$$b = 5$$

The equation is $y = -\frac{2}{3}x + 5$.

The correct choice is (1).

5 Apply the midpoint formula:

$$x_{MP} = \frac{1}{2}(x_1 + x_2) \quad y_{MP} = \frac{1}{2}(y_1 + y_2)$$

$$x_{MP} = \frac{1}{2}(3 + 15) \quad y_{MP} = \frac{1}{2}(1 + (-7))$$

$$x_{MP} = \frac{1}{2}(18) \quad\quad y_{MP} = \frac{1}{2}(-6)$$

$$x_{MP} = 9 \quad\quad\quad\quad y_{MP} = -3$$

The correct choice is **(4)**.

6 The center and radius of the circle are needed to write its formula. The midpoint of the diameter gives the center:

$$x_{MP} = \frac{1}{2}(x_1 + x_2) \quad y_{MP} = \frac{1}{2}(y_1 + y_2)$$

$$x_{MP} = \frac{1}{2}(4 + 8) \quad y_{MP} = \frac{1}{2}(-1 + 3)$$

$$x_{MP} = \frac{1}{2}(12) \quad\quad y_{MP} = \frac{1}{2}(2)$$

$$x_{MP} = 6 \quad\quad\quad\quad y_{MP} = 1$$

The radius is the distance from the center point to either endpoint of the diameter. Apply the distance formula with points $(6, 1)$ and $(8, 3)$.

$$\text{distance} = \sqrt{(x_1 - x_2)^2 + (y_1 - y_2)^2}$$

$$= \sqrt{(8 - 6)^2 + (3 - 1)^2}$$

$$= \sqrt{(2)^2 + (2)^2}$$

$$= \sqrt{8}$$

The radius of the circle is $\sqrt{8}$, and its center has coordinates $(6, 1)$. Substitute these values for r, h, and k in the equation of a circle:

$$(x - h)^2 + (y - k)^2 = R^2$$
$$(x - 6)^2 + (y - 1)^2 = \sqrt{8}^2$$
$$(x - 6)^2 + (y - 1)^2 = 8$$

The correct choice is **(3)**.

7 Two segments are congruent if their lengths are equal, so apply the distance formula to determine the length of each segment.

$$d = \sqrt{(x_1 - x_2)^2 + (y_1 - y_2)^2}$$

$$AB = \sqrt{(-3 - 1)^2 + (6 - 4)^2} \qquad TU = \sqrt{(4 - 2)^2 + (1 - 5)^2}$$
$$= \sqrt{(-4)^2 + (2)^2} \qquad\qquad = \sqrt{(2)^2 + (-4)^2}$$
$$= \sqrt{16 + 4} \qquad\qquad\qquad = \sqrt{4 + 16}$$
$$= \sqrt{20} \qquad\qquad\qquad\qquad = \sqrt{20}$$

$AB = TU$; therefore, the 2 segments are congruent.

8

$$\frac{UW}{WV} = \frac{1}{5} = \frac{x - (-3)}{9 - x}$$
$$9 - x = 5(x + 3)$$
$$9 - x = 5x + 15$$
$$-6 = 6x$$
$$x = -1$$

Repeating for the y-coordinate:

$$\frac{UW}{WV} = \frac{1}{5} = \frac{y - 7}{1 - y}$$
$$1 - y = 5(y - 7)$$

The coordinates of W are $(-1, 6)$.

9 Line m is the line of reflection that maps A to B, so it must be the perpendicular bisector of $\overline{AB}$. To find the perpendicular bisector, calculate the midpoint and slope of $\overline{AB}$. Then write the equation of the line with the negative
reciprocal slope that passes through the midpoint.

$$x_{MP} = \frac{1}{2}(x_1 + x_2), \quad y_{MP} = \frac{1}{2}(y_1 + y_2)$$

$$x_{MP} = \frac{1}{2}(-2 + 6), \quad y_{MP} = \frac{1}{2}(7 + 3)$$

$$x_{MP} = 2, \quad\quad\quad\quad y_{MP} = 5$$

$$\text{slope} = \frac{y_2 - y_1}{x_2 - x_1}$$

$$\text{slope}_{AB} = \frac{3 - 7}{6 - (-2)}$$

$$= -\frac{1}{2}$$

$\text{slope}_{\text{line } m} = 2$ $\quad\quad\quad\quad\quad\quad$ $\perp$ lines have negative
reciprocal slopes

$$y - y_1 = m(x - x_1)$$ point-slope equation of
a line

substitute the coordinates

$$y - 5 = 2(x - 2) \text{ or } y = 2x + 1$$ of the midpoint for x_1 and y_1,
and 2 for m

10 Apply the completing the square procedure to rewrite the circle in $(x - h)^2 + (y - k)^2 = R^2$ form. Rewrite the equation with the variables on the left and
constant on the right.

$$x^2 + 6x + y^2 - 12y + 25 = 0$$

$$x^2 + 6x + y^2 - 12y = -25$$

The constant needed to complete the square is $\left(\frac{1}{2}b\right)^2$, where b is the coefficient of the linear x- and y-terms. For the x-terms, the necessary constant is $\left(\frac{1}{2}(6)\right)^2$, or 9. For the y-terms $\left(\frac{1}{2}(-12)\right)^2$, or 36, is needed. Add the required constants to each side of the equation.

$$x^2 + 6x + 9 + y^2 - 12y + 36 = -25 + 9 + 36$$

$(x + 3)^2 + (y - 6)^2 = 20$ factor the x-terms and the y-terms

The center has coordinates $(-3, 6)$ and the radius is $\sqrt{20}$.

11 A point that lies on a curve will satisfy the equation of the curve, so the strategy is to write the equation of the circle. Then substitute the coordinates of point A for x and y in the equation and check if the equation is balanced.

$(x - h)^2 + (y - k)^2 = R^2$ equation of a circle

$(x - 4)^2 + (y + 5)^2 = \sqrt{65}^2$ substitute $h = 4$, $k = -5$, and $R = \sqrt{65}$

$(8 - 4)^2 + (2 + 5)^2 = \sqrt{65}^2$ substitute coordinates of A for x and y

$$4^2 + 7^2 = \sqrt{65}^2$$
$$16 + 49 = 65$$
$$65 = 65$$

The equation is balanced, so the point $A(8, 2)$ does lie on circle P.

12 Opposite sides of a parallelogram are parallel, so side $\overline{CD}$ is parallel to side $\overline{AB}$ and must have the same slope.

Start by finding the slope of $\overline{AB}$.

$$\text{slope } \overline{AB} = \frac{y_2 - y_1}{x_2 - x_1}$$
$$= \frac{1 - (-1)}{5 - 2}$$
$$= \frac{2}{3}$$

The desired line also passes through point $D(3, 4)$; $x = 3$ and $y = 4$ must satisfy the equation of $\overline{CD}$. Apply the point-slope equation of a line with slope $= \frac{2}{3}$, which represents the slope of $\overline{AB}$ using $x_1 = 3$ and $y_1 = 4$:

$$y - y_1 = m(x - x_1)$$
$$y - 4 = \frac{2}{3}(x - 3)$$

or, in slope-intercept form $y = \frac{2}{3}x + 2$.

13 We can estimate the area by filling in the ellipse with 5 rectangles. Triangles could have been used to get a more accurate estimate of the area.

The areas of the rectangles are

    rectangle 1 area $= 7 \cdot 6 = 42$

    rectangle 2 area $= 4 \cdot 2 = 8$

    rectangle 3 area $= 5 \cdot 1 = 5$

    rectangle 4 area $= 4 \cdot 2 = 8$

    rectangle 5 area $= 5 \cdot 1 = 5$

The approximate area of the ellipse is $42 + 8 + 5 + 8 + 5 = 68$ square units.

14 The two circles are the largest that will fit inside the region, and the smallest that will fit outside the region.

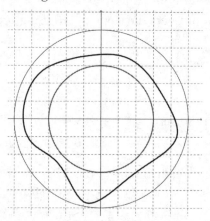

The inner circle has a radius of 3 cm and the outer circle has a radius of 5 cm. The estimate is the average of the areas of the two circles.

$$A_{estimate} = \frac{1}{2}\left(\pi R^2_{inner} + \pi R^2_{outer}\right)$$

$$= \frac{1}{2}\left(\pi \cdot 3^2 + \pi \cdot 5^2\right)$$

$$= \frac{1}{2}(106.8141)$$

$$\approx 53.4 \text{ cm}^2$$

15 a) To find the equation of a perpendicular bisector, the slope and midpoint of $\overline{SK}$ are needed.

$$\text{slope } \overline{SK} = \frac{y_2 - y_1}{x_2 - x_1}$$

$$= \frac{0 - 2}{6 - 0}$$

$$= -\frac{1}{3}$$

midpoint $\overline{SK}$ is $\dfrac{x_1 + x_2}{2}, \dfrac{y_1 + y_2}{2}$

$$\left(\frac{0 + 6}{2}, \frac{2 + 0}{2}\right)$$

$$(3, 1)$$

Perpendicular lines have negative reciprocal slopes, so the slope of the perpendicular bisector is 3, and the line must pass through the midpoint (3, 1). Substitute these values in the point-slope equation of a line:

$$y - y_1 = m(x - x_1)$$
$$y - 1 = 3(x - 3)$$

or $y = 3x - 8$ in slope-intercept form.

b) An altitude will pass through the opposite vertex, so check if the coordinates of point $Y(4, 4)$ satisfy the equation of the perpendicular bisector.

$$y - 1 = 3(x - 3),$$
$$4 - 1 = 3(4 - 3)$$
$$3 = 3(1)$$
$$3 = 3$$

The equation is balanced, so the perpendicular bisector passes through the opposite vertex and is also an altitude.

3.7 SIMILAR FIGURES

PROPERTIES AND DEFINITION OF SIMILAR FIGURES

Two figures are similar if there exists a sequence of similarity transformations that maps one figure onto another. Remember, the sequence of similarity transformations often includes a dilation.

Similar figures have the following properties:

- Same shape but possibly different sizes
- All corresponding lengths proportional
- All corresponding angles congruent

The ratio of proportionality is called the *scale factor*. A similarity statement is written using the symbol ∼. In the accompanying figure, $\triangle ABC \sim \triangle DEF$, from the similarity statement, we have the following relationships:

$$\angle A \cong \angle D \quad \angle B \cong \angle E \quad \angle C \cong \angle F$$

$$\frac{AB}{DE} = \frac{BC}{EF} = \frac{AC}{DF} = \frac{1}{3}$$

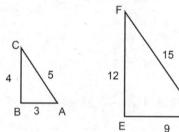

PROVING TRIANGLES SIMILAR USING THE SIMILARITY POSTULATES

- Two triangles can be proven similar using the following postulates:
  AA (angle–angle)
  SAS (side–angle–side)
  SSS (side–side–side)

Unlike congruence, pairs of corresponding sides must be in the same ratio.

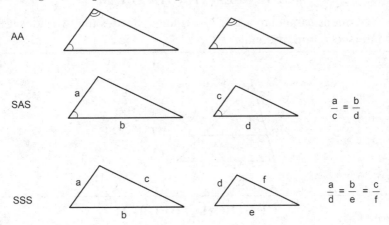

PERIMETER, AREA, AND VOLUME IN SIMILAR FIGURES

Perimeter, area, and volume in similar figures and solids are related to the ratio of corresponding sides.

$\dfrac{\text{Perimeter}_A}{\text{Perimeter}_B}$	$\dfrac{\text{Area}_A}{\text{Area}_B}$	$\dfrac{\text{Volume}_A}{\text{Volume}_B}$
scale factor	(scale factor)^2	(scale factor)^3

SIMILARITY RELATIONSHIPS IN TRIANGLES

> ### SEGMENT PARALLEL TO A SIDE THEOREM
> - A segment parallel to a side of a triangle forms a triangle similar to the original triangle.
> - If a segment intersects two sides of a triangle such that a triangle similar to the original is formed, the segment is parallel to the third side of the original triangle.

$\overline{PQ} \,/\!/\, \overline{AB}$

$\triangle ABC \sim \triangle PQC$

SIDE SPLITTER THEOREM

A segment parallel to a side in a triangle divides the two sides it intersects proportionally.

$$\overline{PQ} \,/\!/\, \overline{AB}$$

$$\frac{AP}{PC} = \frac{BQ}{QC}$$

CENTROID THEOREM

The centroid of a triangle divides each median in a 1 : 2 ratio, with the longer segment having a vertex as one of its endpoints.

In the accompanying figure, D is the midpoint of $\overline{AB}$, and E is the midpoint of $\overline{BC}$, making $\overline{AE}$ and $\overline{CD}$ medians. Their point of intersection, G, is the centroid and divides the medians in a 1 : 2 ratio.

$$\frac{EG}{AG} = \frac{1}{2} \text{ and } \frac{DG}{CG} = \frac{1}{2}$$

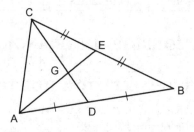

MIDSEGMENT THEOREM

A segment joining the midpoints of two sides of a triangle (a midsegment) is parallel to the opposite side, and its length is equal to $\frac{1}{2}$ the length of the opposite side.

In the accompanying figure, E is the midpoint of $\overline{AB}$, and F is the midpoint of $\overline{AC}$. $\overline{EF}$ // $\overline{BC}$ and $EF = \frac{1}{2}BC$.

ALTITUDE TO THE HYPOTENUSE OF A RIGHT TRIANGLE THEOREM

The altitude to the hypotenuse of a right triangle forms two triangles that are similar to the original triangle.

In the accompanying figure, $\overline{CD}$ is an altitude to hypotenuse $\overline{AB}$, $\triangle BDC \sim \triangle CDA \sim \triangle BCA$.

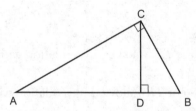

Two useful strategies for dealing with these overlapping similar triangles is to sketch the triangles separately or to make a table with sides classified as the "leg 1," "leg 2," and "hypotenuse." The altitude will be leg 1 of one of the interior triangles, and leg 2 of the other.

	Leg 1	Leg 2	Hypotenuse
Small △	$\overline{BD}$	$\overline{CD}$	$\overline{BC}$
Medium △	$\overline{CD}$	$\overline{AD}$	$\overline{AC}$
Large △	$\overline{BC}$	$\overline{AC}$	$\overline{AB}$

Practice Exercises

1 In the accompanying figure, $\triangle PQR \sim \triangle TSR$ and points P, R, and T are collinear. Which of the following transformations will map $\triangle TSR$ to $\triangle PQR$?

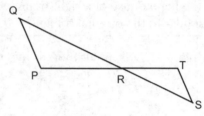

(1) reflection over line $\overline{TP}$ followed by a dilation with a scale factor of $\dfrac{QR}{TR}$ centered at Q

(2) translation by vector $\overline{TR}$ followed by a dilation with a scale factor of $\dfrac{PQ}{ST}$ centered at R

(3) rotation of $180°$ about point R followed by a dilation with a scale factor of $\dfrac{PQ}{ST}$ centered at R

(4) rotation of $180°$ about point R followed by a dilation with a scale factor of $\dfrac{QR}{TR}$ centered at R

2 $ABCDE$ and $VWXYZ$ are similar pentagons. If $BC = 4$ and $WX = 6$, what is the ratio of the area of $ABCDE$ to the area of $VWXYZ$?

(1) $\dfrac{4}{9}$ (3) $\dfrac{2}{5}$

(2) $\dfrac{2}{3}$ (4) $\sqrt{\dfrac{2}{3}}$

3 If $\triangle XYZ \sim \triangle XVW$, which of the following transformations will map $\triangle XVW$ to $\triangle XYZ$?

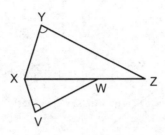

(1) reflect $\triangle XVW$ over $\overline{XW}$ and then dilate it with a center at X and scale factor of $\dfrac{XW}{WZ}$

(2) reflect $\triangle XVW$ over $\overline{XW}$ and then dilate it with a center at X and scale factor of $\dfrac{XZ}{XW}$

(3) rotate $\triangle XVW$ 180° about point X and then dilate it with a center at X and scale factor of $\dfrac{XW}{WZ}$

(4) rotate $\triangle XVW$ 180° about point X and then dilate it with a center at X and scale factor of $\dfrac{XZ}{XW}$

4 Which of the following triangles can be proven to be similar?

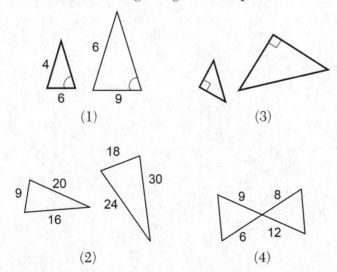

5 Given triangles FOG and BAT with $\angle G \cong \angle T$, $\angle O \cong \angle B$, $FG = 5x$, $FO = 42$, $AB = 14$, and $AT = x + 12$, find the length FG.

   (1) 18   (3) 45

   (2) 30   (4) 90

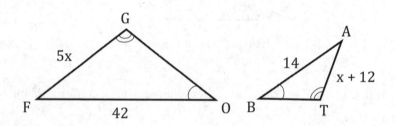

6 $\overline{UV}$ is parallel to side $\overline{ST}$ of $\triangle RST$. If $RU = x - 6$, $US = x$, $RV = x$, and $VT = 2x - 9$, what is the value of x?

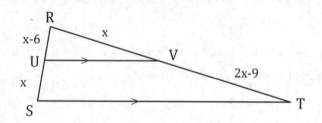

   (1) 3   (3) 18

   (2) 11   (4) 39

7 In the figure below, D lies on $\overline{AB}$ and E lies on $\overline{AC}$. If $AD = 3$, $BD = x + 3$, $DE = 5$, and $BC = 2x + 2$, find the value of x that would let you conclude that $\overline{DE} \parallel \overline{BC}$.

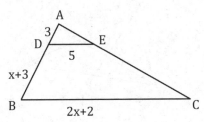

(1) 2   (3) 18

(2) 9   (4) 24

8 Mollie wants to rearrange the furniture in her living room. She measures the rectangular room to be 16 ft by 12 ft and makes a scaled drawing of the room that measures 10 in by 7.5 in. She wants to make scaled paper cutouts to represent her furniture so she can determine her favorite arrangement without having to move the furniture. If her sofa is rectangular and measures 6 ft by 2 ft, what should be the dimensions of the cutout?

(1) $1\frac{1}{3}$ in $\times$ 1 in   (3) $3\frac{3}{4}$ in $\times$ $1\frac{1}{4}$ in

(2) $2\frac{3}{4}$ in $\times$ 1 in   (4) 9.6 in $\times$ 3.2 in

9 Given: $\overline{LSI}$ and $\overline{FSP}$ intersecting at S, $\overline{FL} \parallel \overline{IP}$, $IP = 4$ and $FL = 6$

(1) Explain why the two triangles formed are similar.

(2) Find the ratio $\dfrac{PS}{FS}$.

(3) State a specific similarity transformation that will map the smaller triangle onto the larger.

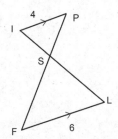

10 In $\triangle VTP$, F is the midpoint of $\overline{VP}$, G is the midpoint of $\overline{PT}$, and $\overline{VG}$ and $\overline{FT}$ intersect at K. If $VK = 5x + 5$ and $KG = 2x + 8$, find the length VG.

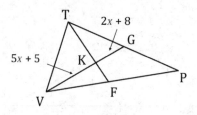

11 The midpoints of $\overline{PG}$, $\overline{PI}$, and $\overline{GI}$ are C, O, and W, respectively. If $OW = 7$, $CW = 5$, and $OC = 6$, find the perimeter of $\triangle PIG$.

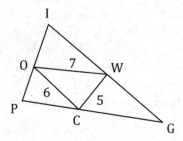

12 In the figure below, $\angle ABC$ and $\angle BDC$ are right angles, $CB = 6$, and $AB = 12$. Find the length of $\overline{BD}$. Express your answer in simplest radical form.

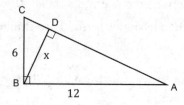

13 Rectangle $AEFG$ is the image of rectangle $ABCD$ after the transformation clockwise $R_{90°,A} \circ D_{\frac{2}{3},A}$. If $GA = 8$ and $DE = 12$, what is the perimeter of $ABCD$?

14 $ABCDEF$ and $ARSTUV$ are both regular hexagons and R is the midpoint of $\overline{AB}$. Name a sequence of transformations that will map $ARSTUV$ to $ABCDEF$.

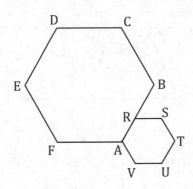

15 Given: $\overline{RUT}$, $\overline{SVT}$, and $\overline{UV} \parallel \overline{RS}$

Prove: $\dfrac{UR}{TU} = \dfrac{VS}{TV}$

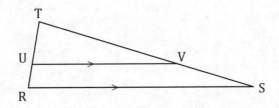

16 Given: $\overline{ABD}$, $\angle C \cong \angle E$, $\angle DEB \cong \angle EBC$

Prove: $\triangle ABC \sim \triangle BDE$

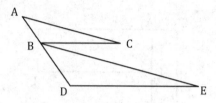

17 Prove the Pythagorean theorem.

Given: right triangle $\triangle BCA$ with a right angle at C, altitude $\overline{CD}$ drawn to hypotenuse $\overline{AB}$

Prove: $(AC)^2 + (BC)^2 = (AB)^2$

Solutions

1 We want to map the smaller triangle to the larger, so the scale factor will be greater than 1. The correct scale factor among the choices is $\frac{PQ}{ST}$. A rotation of 180° will rotate $\triangle TSR$ to $\triangle PQR$. The correct sequence is: a rotation of 180° centered at R, followed by a dilation by a scale factor of $\frac{PQ}{ST}$ centered at R. Note that the order of the sequence does not matter in this situation.

The correct choice is (**3**).

2 The ratio of the areas of similar figures is equal to the ratio of lengths of corresponding sides squared. $\overline{BC}$ and $\overline{WX}$ are corresponding sides, and are in a ratio of $\frac{4}{6}$, or $\frac{2}{3}$.

$$\text{ratio of areas} = \text{ratio of sides}^2$$
$$= \left(\frac{2}{3}\right)^2$$
$$= \frac{4}{9}$$

The correct choice is (**1**).

3 A reflection over $\overline{XW}$ will map $\overline{XV}$ onto $\overline{XZ}$. Then a dilation centered about X with a scale factor of $\frac{XZ}{XW}$ will match the sizes.

The correct choice is (**2**).

4 The triangle similarity postulates are **AA**, **SAS**, and **SSS**, where each S represents the *ratio* of a pair of corresponding sides. Examine each of the choices.

Choice 1: The two pairs of corresponding sides are in a 2:3 ratio, but the congruent angle is not the included angle.

Choice 2: Corresponding sides are not in the same ratio, $\frac{9}{18} \neq \frac{16}{24}$.

Choice 3: Lengths of the sides are not given.

Choice 4: 2 pairs of corresponding sides are in the same ratio, $\dfrac{6}{8} = \dfrac{9}{12}$, and the congruent vertical angles are the included angles. The triangles are similar by SAS.

The correct choice is (4).

5 The two triangles are similar by AA, and corresponding sides of similar triangles are in the same ratio. The similarity statement can be written as $\triangle FOG \sim \triangle ABT$, so the following proportion can be written:

$$\frac{FG}{AT} = \frac{FO}{AB}$$

$$\frac{5x}{x+12} = \frac{42}{14}$$

$$14(5x) = 42(x+12) \quad \text{cross products are equal}$$

$$70x = 42x + 504$$

$$28x = 504$$

$$x = 18 \qquad \text{substitute } x = 18$$

$$FG = 5(18)$$

$$= 90$$

The correct choice is (4).

6 When a segment is drawn parallel to a side of a triangle, the two triangles are similar, and the side splitter theorem states that the sides are divided proportionally.

$$\frac{RU}{US} = \frac{RV}{VT} \qquad \text{side splitter theorem}$$

$$\frac{x-6}{x} = \frac{x}{2x-9}$$

$$(2x - 9)(x - 6) = x^2 \quad \text{cross products are equal}$$
$$2x^2 - 21x + 54 = x^2$$
$$x^2 - 21x + 54 = 0 \quad \text{solve by moving all terms to one side and factoring}$$
$$(x - 18)(x - 3) = 0 \quad \text{factor the trinomial}$$
$$x - 18 = 0 \quad x - 3 = 0 \quad \text{zero product theorem}$$
$$x = 18 \qquad x = 3$$

The two solutions of the quadratic are *possible* solutions to the problem. Eliminate any results that lead to negative lengths or any other impossible situations. The solution $x = 3$ would result in a negative length for RU and VT, so we eliminate this result, and the value of x is 18.

The correct choice is **(3)**.

7 From the original figure, $\overline{DE} \mathbin{/\!/} \overline{BC}$ when $\angle CBA \cong \angle EDA$ (corresponding angles are congruent). The triangles $\triangle ADE$ and $\triangle ABC$ are similar by AA, and corresponding sides will be proportional. Sketch the two triangles separately to help identify the corresponding sides. Note that $AB = AD + DB$.

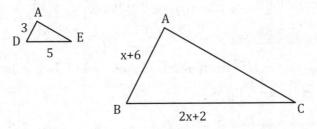

$$\frac{AD}{AB} = \frac{DE}{BC} \quad \text{corresponding sides are proportional}$$
$$\frac{3}{x + 6} = \frac{5}{2x + 2}$$
$$3(2x + 2) = 5(x + 6) \quad \text{cross products are equal}$$
$$6x + 6 = 5x + 30$$
$$x = 24$$

The correct choice is **(4)**.

8 First convert feet to inches so all units are consistent.

$$\text{Length of room} \quad 16 \text{ ft} \times \frac{12 \text{ in}}{\text{ft}} = 192 \text{ in}$$

$$\text{Length of sofa} \quad 6 \text{ ft} \times \frac{12 \text{ in}}{\text{ft}} = 72 \text{ in}$$

$$\text{Width of sofa} \quad 2 \text{ ft} \times \frac{12 \text{ in}}{\text{ft}} = 24 \text{ in}$$

The scaled drawing is similar to the actual room, so the ratio of corresponding sides is

$$\frac{\text{length of drawing}}{\text{length of room}} = \frac{10}{192}$$

Apply this scale factor to the length and width of the sofa to determine the dimensions of the cutout of the sofa.

$$\frac{\text{length}_{\text{cutout}}}{\text{length}_{\text{actual}}} = \frac{10}{192} \qquad \frac{\text{width}_{\text{cutout}}}{\text{width}_{\text{actual}}} = \frac{10}{192}$$

$$\frac{\text{length}_{\text{cutout}}}{72 \text{ in}} = \frac{10}{192} \qquad \frac{\text{width}_{\text{cutout}}}{24 \text{ in}} = \frac{10}{192}$$

$$\text{length}_{\text{cutout}} = \frac{72 \cdot 10}{192} \qquad \text{width}_{\text{cutout}} = \frac{24 \cdot 10}{192}$$

$$\text{length}_{\text{cutout}} = 3.75 \text{ in} \qquad \text{width}_{\text{cutout}} = 1.25 \text{ in}$$

Converting decimals to fractions, the dimensions of the cutout are $3\frac{3}{4} \times 1\frac{1}{4}$.
The correct choice is **(3)**.

9 a) Given $\overline{FL} \parallel \overline{IP}$, we know $\angle F \cong \angle P$ and $\angle L \cong \angle I$ because they are the alternate interior angles formed by parallel lines and a transversal. Therefore, $\triangle SIP \sim \triangle SLF$ by AA. (Note the correct order of vertices in the similarity statement.)

b) Corresponding parts are in the same ratio, so

$$\frac{PS}{FS} = \frac{IP}{LF}$$

$$= \frac{4}{6}$$

$$= \frac{2}{3}$$

c) We have to use the reciprocal of the ratio in part (b) because we want to enlarge the smaller triangle and not reduce it. A rotation of $180°$ about point S will align $\overline{SP}$ with $\overline{FS}$ and $\overline{IS}$ with $\overline{SL}$. A dilation with center at S and a scale factor of $\frac{3}{2}$ will then map $\triangle SIP$ to $\triangle SLF$.

10 $\overline{VG}$ and $\overline{TF}$ are medians that intersect at centroid K. The centroid divides $\overline{VG}$ in a $1:2$ ratio, so

$$VK = 2KG$$
$$5x + 5 = 2(2x + 8)$$
$$5x + 5 = 4x + 16$$
$$5x = 4x + 11$$
$$x = 11$$

VG is equal to the sum of VK and KG, which can be evaluated at $x = 11$.

$$VG = VK + KG$$
$$= 5x + 5 + 2x + 8$$
$$= 7x + 13$$
$$= 7(11) + 13$$
$$= 90$$

11 $\overline{CO}$, $\overline{OW}$, and $\overline{WC}$ are all midsegments, and their lengths are $\frac{1}{2}$ the length of the opposite side of $\triangle PIG$. Therefore,

$$PI = 2 \cdot WC$$
$$= 2 \cdot 5$$
$$= 10$$

$$IG = 2 \cdot OC$$
$$= 2 \cdot 6$$
$$= 12$$

$$GP = 2 \cdot WO$$
$$= 2 \cdot 7$$
$$= 14$$

The perimeter of $\triangle PIG = PI + IG + GP$
$$= 10 + 12 + 14$$
$$= 36$$

12 An altitude to the hypotenuse of a right triangle forms 3 similar triangles. The three similar triangles are $\triangle CDB$, $\triangle BDA$, and $\triangle CBA$. Fill in the table of side lengths, and look for two pairs of corresponding sides that can be used to write a proportion involving the unknown length x.

	Leg 1	Leg 2	Hypotenuse
Small $\triangle CDB$		x	6
Medium $\triangle BDA$	x		12
Large $\triangle CBA$	6	12	

We don't have a full set of four corresponding sides, so we need to use the Pythagorean theorem on the large triangle to find the hypotenuse AC.

$$BC^2 + AB^2 = AC^2$$
$$6^2 + 12^2 = AC^2$$
$$AC^2 = 180$$
$$AC = \sqrt{180}$$
$$= 6\sqrt{5}$$

Now the table indicates two pairs of corresponding parts can be used to write a proportion.

	Leg 1	Leg 2	Hypotenuse
Small $\triangle CDB$		x	6
Medium $\triangle BDA$	x		12
Large $\triangle CBA$	6	12	$6\sqrt{5}$

$$\frac{x}{12} = \frac{6}{6\sqrt{5}}$$

$6x\sqrt{5} = 72$ \qquad cross products are equal

$$x = \frac{72}{6\sqrt{5}}$$

$$= \frac{12}{\sqrt{5}}$$

$$= \frac{12}{\sqrt{5}} \cdot \frac{\sqrt{5}}{\sqrt{5}} \qquad \text{rationalize the denominator}$$

$$= \frac{12\sqrt{5}}{5}$$

13 The dilation tells us that the ratio of corresponding parts between rectangles *AEFG* and *ABCD* is 2:3. *GA* and *AD* are corresponding parts, so the following proportion can be written:

$$\frac{GA}{AD} = \frac{2}{3}$$

$$\frac{8}{AD} = \frac{2}{3}$$

$2AD = 24$ \qquad cross products are equal

$AD = 12$

$AE = AD + DE$ \qquad segment addition

$AE = 12 + 12$

$\quad = 24$

Now we can find AB by writing a proportion involving corresponding sides AE and AB.

$$\frac{AE}{AB} = \frac{2}{3}$$

$$3AE = 2AB \quad \text{cross products are equal}$$

$$3(24) = 2AB$$

$$72 = 2AB$$

$$AB = 36$$

Since $ABCD$ is a rectangle, opposite sides are congruent, and the perimeter is equal to

$$\text{perimeter}_{ABCD} = 2AB + 2BC$$

$$= 2(36) + 2(12)$$

$$= 72 + 24$$

$$= 96$$

14 A dilation and a reflection are required to map $ARSTUV$ to $ABCDEF$. Since R is a midpoint, we know $AR = \frac{1}{2}AB$, and a dilation with a scale factor of 2 centered at A is needed to map $\overline{AR}$ to $\overline{AB}$. The second transformation needed is a reflection over line $\overleftrightarrow{AB}$.

The transformation is a dilation centered at A with a scale factor of 2, followed by a reflection over $\overline{AB}$. Alternatively, the order could be reversed.

15 The strategy is to prove $\triangle TUV \sim \triangle TRS$, which leads to the proportion of corresponding sides $\dfrac{TR}{TU} = \dfrac{TS}{TV}$. Algebraic rearranging gives the final result.

Statement	Reason
1. $\overline{RUT}$, $\overline{SVT}$, and $\overline{UV} \parallel \overline{RS}$	1. Given
2. $\angle TUV \cong \angle R$ and $\angle TVU \cong \angle S$	2. Corresponding angles formed by parallel lines are congruent
3. $\triangle TUV \sim \triangle TRS$	3. AA
4. $\dfrac{TR}{TU} = \dfrac{TS}{TV}$	4. Corresponding sides in similar triangles are proportional
5. $TR = TU + UR$ $TS = TV + VS$	5. Partition
6. $\dfrac{TU + UR}{TU} = \dfrac{TV + VS}{TV}$	6. Substitution
7. $1 + \dfrac{UR}{TV} = 1 + \dfrac{VS}{TV}$	7. Division
8. $\dfrac{UR}{TU} = \dfrac{VS}{TV}$	8. Subtraction

16 The strategy is to show that $\overline{BC} \parallel \overline{DE}$ and then to use the corresponding angle relationship to show $\angle ABC \cong \angle BDE$.

Statement	Reason
1. $\overline{ABD}$, $\angle C \cong \angle E$, $\angle DEB \cong \angle EBC$	1. Given
2. $\overline{BC} \parallel \overline{DE}$	2. Two lines are parallel if the alternate interior angles formed by a transversal are congruent
3. $\angle ABC \cong \angle BDE$	3. The corresponding angles formed by two parallel lines are congruent
4. $\triangle ABC \sim \triangle BDE$	4. AA

17 The strategy is to use the theorem that an altitude to the hypotenuse of a right triangle forms three similar triangles. Two proportions from these triangles can be written and combine to yield the Pythagorean theorem.

Statement	Reason
1. Right triangle $\triangle ABC$ with a right angle at C, altitude $\overline{CD}$ drawn to hypotenuse $\overline{AC}$	1. Given
2. $\triangle BDC \sim \triangle CDA \sim \triangle BCA$	2. The altitude to the hypotenuse of a right triangle forms 3 similar right triangles
3. $\dfrac{AB}{AC} = \dfrac{AC}{AD}, \dfrac{AB}{BC} = \dfrac{BC}{BD}$	3. Corresponding sides in similar triangles are proportional
4. $(AC)^2 = AB \cdot AD,$ $(BC)^2 = AB \cdot BD$	4. The cross products of a proportion are equal
5. $(AC)^2 + (BC)^2 = AB \cdot AD + AB \cdot BD$	5. Addition property
6. $(AC)^2 + (BC)^2 = AB(AD + BD)$	6. Factor
7. $(AC)^2 + (BC)^2 = AB(AB)$	7. Partition property
8. $(AC)^2 + (BC)^2 = (AB)^2$	8. Simplify

3.8 TRIGONOMETRY

DEFINITION OF SIDES IN A RIGHT TRIANGLE RELATIVE TO AN ANGLE

- Hypotenuse—the side across from the right angle
- Opposite—the side across from the specified acute angle
- Adjacent—the side included between the right angle and the specified angle

a) Opposite, adjacent, and hypotenuse relative to $\angle A$

b) Opposite, adjacent, and hypotenuse relative to $\angle B$

TRIGONOMETRIC RATIOS

Ratio	Abbreviation	Definition
Sine	Sin	$\dfrac{\text{opposite}}{\text{hypotenuse}}$
Cosine	Cos	$\dfrac{\text{adjacent}}{\text{hypotenuse}}$
Tangent	Tan	$\dfrac{\text{opposite}}{\text{adjacent}}$

The ratios sine, cosine, and tangent will have fixed values for a given acute angle in a right triangle. This is because any two right triangles with a pair of congruent acute angles are similar, and corresponding parts of similar triangles are in the same ratio. These three trigonometric ratios can be used to find a missing side of a triangle given one angle and side by setting up the correct proportion. Be sure your calculator is in degree mode if the angle is given in degrees.

Example

$\triangle SOX$ has a right angle at O. The measure of angle $S = 72°$ and $XO = 12$. Find the length OS to the nearest tenth.

Solution:

Relative to $\angle S$, the given side and the unknown side are the opposite and adjacent, which suggest using the tangent ratio.

$$\tan(S) = \frac{\text{opposite}}{\text{adjacent}} = \frac{OX}{OS}$$

$$\tan(72°) = \frac{12}{OS}$$

$$OS = \frac{12}{\tan(72°)}$$

$$OS = \frac{12}{3.07768} = 3.89903$$

$$= 3.9$$

INVERSE TRIGONOMETRIC RATIOS

The inverse of a function "undoes" the original function. For example, $f(x) = \sqrt{x}$ and $g(x) = x^2$ are inverse functions. If $x = 5$, then $f(g(5)) = \sqrt{5^2} = 5$. The trigonometric functions have the following inverse functions:

$\sin(x)$	$\arcsin(x)$
$\cos(x)$	$\arccos(x)$
$\tan(x)$	$\arctan(x)$

- These three inverse functions are often abbreviated $\sin^{-1}(x)$, $\cos^{-1}(x)$, and $\tan^{-1}(x)$. They can be used to find the measure of an angle given a ratio of any two sides in the triangle.

Example

Using the accompanying figure, find the measure of angles M and N.

Solution:

$$\cos(M) = \frac{3}{7}$$

$$m\angle M = \cos^{-1}\left(\frac{3}{7}\right)$$

$$= 64.6°$$

$$\sin(N) = \frac{3}{7}$$

$$m\angle N = \sin^{-1}\left(\frac{3}{7}\right)$$

$$= 25.4°$$

COFUNCTIONS

Cofunction Relationship

The sine of an angle is equal to the cosine of its complement: $\sin(A) = \cos(90 - A)$.

The cosine of an angle is equal to the sine of its complement: $\cos(A) = \sin(90 - A)$.

$$\sin(A) = \cos(B) = \cos(90 - A) = \frac{BC}{AB}$$

$$\sin(B) = \cos(A) = \sin(90 - A) = \frac{AC}{AB}$$

In other words, the cofunction relationship states that if a sine is equal to a cosine, then the two angles must sum to 90°.

Example

Given $\sin(2x + 4) = \cos(3x + 21)$. Solve for x.

Solution:

The cofunctions are equal, so the angles must be complementary.

$$
\begin{aligned}
2x + 4 + 3x + 21 &= 90 \\
5x + 25 &= 90 \\
5x &= 65 \\
x &= 13
\end{aligned}
$$

MODELING WITH TRIGONOMETRY

- *Angle of elevation*—The angle formed by the horizontal and the line directed upward to an object
- *Angle of depression*—The angle formed by the horizontal and the line directed downward to an object

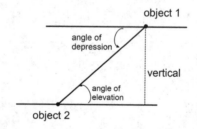

Strategy for trigonometry modeling problems:

- Start with a detailed sketch.
- Identify the triangles formed and the relevant trigonometric ratios.
- Consider parallel lines and other relationships if you are still missing parts. Notice that the angle of elevation and angle of depression are congruent alternate interior angles.
- Consider working with more than one triangle, especially if the problem involves an object that moves from one position to another.

Practice Exercises

1 In $\triangle ABC$, m$\angle B = 90°$, m$\angle C = 51°$, and $BC = 13$. What is the length AB?

(1) 16.1   (3) 18.2

(2) 16.7   (4) 20.7

2 $\triangle JPL$ is a right triangle with a right angle at L. If m$\angle J = 19°$ and $\overline{LJ} = 21$, what is the length of $\overline{PJ}$?

(1) 21.5   (3) 61.0

(2) 22.2   (4) 64.5

3 A 24 ft ladder is leaning against a wall. The base of the ladder is 5 ft from the house. What angle does the ladder make with the ground?

(1) 12°   (3) 67°

(2) 65°   (4) 78°

4 In right triangle PQR, m$\angle Q = 90°$. If $\sin(P) = 4x^2 + 5x + 2$ and $\cos(R) = 5x + 3$, what is the measure of $\angle P$?

(1) 5°   (3) 25°

(2) 8°   (4) 30°

5 John constructs three right triangles of different sizes, each having angles that measure 30°, 60°, and 90°. He then measures the side lengths of each triangle and calculates the value of $\sin(60°)$ from the side lengths using each of the triangles. Which of the following theorems would best explain why the value he calculates is the same for all three triangles?

(1) The sum of the measures of the interior angles of a triangle equals 180°.

(2) Corresponding parts of congruent triangles are congruent.

(3) Corresponding sides of similar triangles are proportional.

(4) The longest side of a triangle is always opposite the largest angle.

6 The side of Rayquan's house is 30 ft long. Rayquan is planning to install a gutter along the roof as shown. He wants the angle of depression to be 3° so the water will drain out at the end of the gutter. How long should the gutter be?

(1) 31 ft 7 in   (3) 30 ft 4 in

(2) 31 ft 2 in   (4) 30 ft 0.5 in

7 In $\triangle ABC$, $m\angle B = 90°$, $\tan(A) = m$, and $\sin(C) = \dfrac{1}{p}$. What is the value of $\cos(C)$?

(1) $\dfrac{m}{p}$

(3) $\dfrac{p}{m}$

(2) $m \cdot p$

(4) $m + p$

8 Jack is watching the launch of a rocket from a viewing area 4,500 feet from the launch pad. At 15 seconds after launch, he measures a 67° angle of elevation from the ground to the rocket. What is the average speed of the rocket during the first 15 seconds of its flight? Assume the rocket travels upward, perpendicular to the ground, and give your answer to the nearest foot per second.

9 Rick is flying a kite in the park. He holds the spool of string 3 ft above the ground, and lets out 200 ft of string. The kite is initially flying with a 72° angle of elevation.

(1) What is the altitude of the kite?

(2) The wind shifts and the angle of elevation altitude decreases to 52°. How many feet does the kite drop?

10 Frank is standing at the top of a water tower and looks down at the town below. Frank measures an angle of depression of 52° to his house, and he knows the water tower is 120 ft tall. How far along the ground is the tower from his house? Round to the nearest foot.

11 A lighthouse sits on a 100 ft cliff. An observer in a boat in the harbor notes an angle of elevation to the base of the cliff of 28°, and an angle of elevation to the top of the lighthouse of 44°. What is the height of the lighthouse, rounded to the nearest foot?

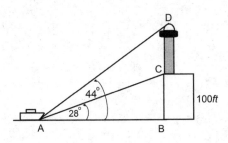

Solutions

1 Relative to $\angle C$, $\overline{AB}$ is the opposite and $\overline{BC}$ is the adjacent, so the tangent ratio applies.

$$\text{tangent} = \frac{\text{opposite}}{\text{adjacent}}$$

$$\tan(C) = \frac{AB}{BC}$$

$$\tan(51°) = \frac{x}{13}$$

$$x = 13\tan(51°)$$

$$= 16.0536$$

$$= 16.1$$

The correct choice is (**1**).

2 Relative to $\angle J$, $\overline{LJ}$ is the adjacent and $\overline{JP}$ is the hypotenuse. The cosine ratio applies here.

$$\text{cosine} = \frac{\text{adjacent}}{\text{hypotenuse}}$$

$$\cos(J) = \frac{LJ}{JP}$$

$$\cos(19°) = \frac{21}{x}$$

$$x = \frac{21}{\cos(19°)}$$

$$= 22.2100$$

$$= 22.2$$

The correct choice is (**2**).

3 The ladder, wall, and ground form a right triangle as shown.

Relative to the desired angle, the 5 ft distance is the adjacent and the 24 ft length is the hypotenuse. Adjacent and hypotenuse suggest using the cosine. Since we are looking for the angle, apply the inverse cosine function.

$$\cos(x) = \frac{\text{adjacent}}{\text{hypotenuse}}$$

$$\cos(x) = \frac{5}{24}$$

$$x = \cos^{-1}\left(\frac{5}{24}\right)$$

$$= 77.9753°$$

$$= 78°$$

The correct choice is **(4)**.

4 The sine of an angle and the cosine of its complement are always equal. Since $\triangle PQR$ is a right triangle with the right angle at Q, angles P and R must be complementary. Setting $\sin(P)$ and $\cos(R)$ equal, we get:

$$\sin(P) = \cos(R)$$
$$4x^2 + 5x + 2 = 5x + 3$$
$$4x^2 = 1$$
$$x^2 = \frac{1}{4}$$
$$x = \pm\frac{1}{2}$$

Substituting $x = -\frac{1}{2}$ into the expression for $\sin(P)$, we get

$$\sin(P) = 4x^2 + 5x + 2$$
$$= 4\left(-\frac{1}{2}\right)^2 + 5\left(-\frac{1}{2}\right) + 2$$
$$= 1 - 2\frac{1}{2} + 2$$
$$= \frac{1}{2}$$

Now substitute $x = \dfrac{1}{2}$.

$$\sin(P) = 4\left(\frac{1}{2}\right)^2 + 5\left(\frac{1}{2}\right) + 2$$

$$= 1 + 2\frac{1}{2} + 2$$

$$= 5\frac{1}{2}$$

The sine function cannot be greater than 1, so $x = \dfrac{1}{2}$ is not a solution. Using the solution $x = -\dfrac{1}{2}$ and $\sin(P) = \dfrac{1}{2}$, find $m\angle P$ using the inverse sine function.

$$\sin(P) = \frac{1}{2}$$

$$m\angle P = \sin^{-1}\left(\frac{1}{2}\right)$$

$$m\angle P = 30°$$

The correct choice is **(4)**.

5 The three triangles are all similar by the angle–angle theorem. The sine ratio is equal to the $\dfrac{\text{opposite}}{\text{hypotenuse}}$. Since all pairs of corresponding sides in similar triangles are proportional, the ratio $\dfrac{\text{opposite}}{\text{hypotenuse}}$ will be equal to the same value in all three triangles.

The correct choice is **(3)**.

6 Relative to the 3° angle, the horizontal distance is the adjacent and the gutter is the hypotenuse, so we can apply the cosine ratio.

$$\cos(3°) = \frac{\text{adjacent}}{\text{hypotenuse}}$$

$$\cos(3°) = \frac{30\,\text{ft}}{x}$$

$$x = \frac{30\,\text{ft}}{\cos(3°)}$$

$$= 30.04117\,\text{ft}$$

To convert to feet and inches, multiply the decimal part of the number by 12.

$0.4117\,(12) = 0.49405\,\text{in}$

The gutter length should be 30 ft 0.49405 in. Rounded to the nearest 0.1 inch, the length is 30 ft 0.5 in.

The correct choice is **(4)**.

7 Making a sketch of $\triangle ABC$, we can fill in the lengths of AB, BC, and AC as shown in the figure.

The tangent is equal to $\dfrac{\text{opposite}}{\text{adjacent}}$, so

$$\tan(A) = m$$
$$\frac{BC}{AB} = \frac{m}{1}$$

Since we are only concerned with ratios, we can assume $BC = m$ and $AB = 1$.

The second ratio states

$$\sin(C) = \frac{1}{p}$$
$$\frac{AB}{AC} = \frac{1}{p}$$

Since we assumed $AB = 1$, we can assume $AC = p$. $\text{Cos}(C)$ can now be calculated.

$$\cos(C) = \frac{\text{adjacent}}{\text{hypotenuse}}$$
$$= \frac{BC}{AC}$$
$$= \frac{m}{p}$$

The correct choice is **(1)**.

8 The path of the rocket traveling upward forms a right triangle with the ground. The distance to the rocket is the adjacent, and the height of the rocket is the opposite, so the tangent ratio can be applied.

4500 ft

$$\tan(67°) = \frac{h}{4,500}$$

$$h = 4,500\tan(67°)$$

$$= 4,500(2.3558)$$

$$= 10,601.33 \text{ ft}$$

To find the speed, apply the relationship rate $= \dfrac{\text{distance}}{\text{time}}$.

$$\text{speed} = \frac{10,601.33\,\text{ft}}{15\,\text{sec}}$$

$$= 706.73\,\text{ft/sec}$$

$$= 707\,\text{ft/sec}$$

9 First make a sketch. A right triangle is formed by the kite string, the altitude measured from the spool (x), and the horizontal distance to the kite. The total altitude will be $(x + 3)$ ft. To find how many feet the kite drops, calculate the altitude at the two angles of elevation, and then find the difference.

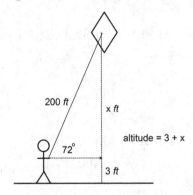

a) Relative to the angle of elevation, the 200 ft kite string is the hypotenuse, and the altitude measured from the spool is opposite. The opposite and hypotenuse suggest the sine ratio. Find x using the following:

$$\sin(72°) = \frac{\text{opposite}}{\text{hypotenuse}}$$

$$\sin(72°) = \frac{x}{200}$$

$$x = 200\sin(72°)$$

$$x = 200(0.951056)$$

$$= 190.2112\,\text{ft}$$

Next, calculate the altitude from x.

$$\text{Altitude} = 3\,\text{ft} + x\,\text{ft}$$

$$= 3\,\text{ft} + 190.21123\,\text{ft}$$

$$= 193.21123\,\text{ft}$$

$$= 193\,\text{ft}$$

b) We can use the same equations to calculate the new altitude if the angle of elevation decreases to 52°.

$$\sin(52°) = \frac{x}{200}$$

$$x = 200\sin(52°)$$

$$x = 200(0.788010)$$

$$= 157.60215\,\text{ft}$$

$$\text{Altitude} = 3\,\text{ft} + x\,\text{ft}$$

$$= 3 + 157.60215\,\text{ft}$$

$$= 160.60215\,\text{ft}$$

The change in altitude is

$$193.21123 \text{ ft} - 160.60215 \text{ ft}$$
$$= 32.6090 \text{ ft}$$
$$= 33 \text{ ft}$$

The kite drops 33 ft.

10 Start with a sketch, and fill in the appropriate dimensions, horizontals, and verticals.

The angle of depression is 52°. It is easier in the problem to find the angle complementary to the angle of depression and work with that triangle. The complementary angle is 38°. The height of the tower is adjacent to the 38° angle, and the distance along the ground is opposite from the 38° angle, so we use the tangent ratio.

$$\tan(38°) = \frac{d}{120\,\text{ft}}$$
$$d = 120\,\text{ft} \cdot \tan(38°)$$
$$d = 93.75\,\text{ft}$$

The house is 94 ft from the water tower.

11 First find length AB using $\triangle ABC$; then find length BD using $\triangle ABD$.

In $\triangle ABC$, BC is opposite $\angle A$, and AB is the adjacent, so we can use the tangent.

$$\tan = \frac{\text{opposite}}{\text{adjacent}}$$

$$\tan(\angle BAC) = \frac{BC}{AB}$$

$$\tan(28°) = \frac{100}{AB}$$

$$AB = \frac{100}{\tan(28°)}$$

$$AB = 188.0726$$

In $\triangle ABD$, AB is the adjacent and BD is the opposite, so we use the tangent again.

$$\tan(\angle BAD) = \frac{BD}{AD}$$

$$\tan(44°) = \frac{BD}{188.0726}$$

$$BD = 188.0726 \tan(44°)$$

$$BD = 181.6196$$

The height of the lighthouse, DC, is found by subtracting the height of the cliff from BD.

$$DC = BD - BC$$

$$DC = 181.6196 - 100$$

$$DC = 81.6196$$

The height of the lighthouse is 82 ft.

3.9 PARALLELOGRAMS

PROPERTIES OF PARALLELOGRAMS

A **parallelogram** is a quadrilateral whose opposite sides are parallel. All parallelograms have the following properties:

1. Opposite sides are congruent.

2. Opposite angles are congruent.

3. Adjacent angles are supplementary.

4. The diagonals bisect each other.

5. The two diagonals each divide the parallelogram into two congruent triangles.

Congruent Sides:
$AB = CD, BC = AD$

Diagonals Bisect Each Other:
$AE = EC, BE = ED$

Congruent angles:
$\angle A \cong \angle C, \angle B \cong \angle D$

Diagonals Form Congruent Triangles:
$\triangle BAD \cong \triangle DCB, \triangle ADC \cong \triangle CBA$

Supplementary angles:
$\angle A$ and $\angle B$, $\angle B$ and $\angle C$
$\angle C$ and $\angle D$, $\angle D$ and $\angle A$

SPECIAL PARALLELOGRAMS

Rectangle—a parallelogram with right angles
Rhombus—a parallelogram with consecutive congruent sides
Square—a parallelogram with right angles and consecutive congruent sides

Rectangles, rhombuses, and squares have all the properties of parallelograms plus the following properties:

	Rectangle	Rhombus	Square
4 right angles	✓		✓
Congruent diagonals	✓		✓
4 congruent sides		✓	✓
Perpendicular diagonals		✓	✓
Diagonals bisect the angles		✓	✓

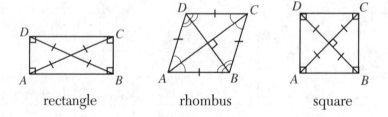

rectangle   rhombus   square

TRAPEZOIDS

A *trapezoid* is a quadrilateral with exactly one pair of parallel sides. A trapezoid with one pair of parallel sides is shown in the accompanying figure. The parallel sides are called the bases and the nonparallel sides are called the legs. The same side interior angles formed with the parallel bases are supplementary.

$\overline{AB} \parallel \overline{CD}$
$m\angle BAD + m\angle ADC = 180°$
$m\angle ABC + m\angle BCD = 180°$

CLASSIFYING PARALLELOGRAMS AND TRAPEZOIDS

The following table lists the properties sufficient to prove a quadrilateral is a parallelogram, rectangle, rhombus, square, or trapezoid.

Classification	What's Needed to Classify the Figure
Trapezoid	One pair of parallel sides
Parallelogram	Any *one* of the following: • Two pairs of opposite sides parallel • Two pairs of opposite sides congruent • Two pairs of opposite angles congruent • Consecutive angles supplementary • Diagonals that bisect each other • One pair of sides congruent *and* parallel
Rectangle	Any one property from the parallelogram list, plus any one of the following: • One right angle • Diagonals that are congruent
Rhombus	Any one property from the parallelogram list, plus any one of the following: • Diagonals are perpendicular • One pair of consecutive sides is congruent • A diagonal bisects one of the angles
Square	Any one property from the parallelogram list *plus* Any one property from the rectangle list *plus* Any one property from the rhombus list

The parallelogram properties needed for a proof are illustrated in the following sketches.

Property of Parallelogram	What It Looks Like
Two pairs of opposite sides are parallel.	
Two pairs of opposite sides are congruent.	
One pair of parallel sides is congruent and parallel.	
Two pairs of opposite angles are congruent.	
Two pairs of consecutive angles are supplementary.	∠1 and ∠2, ∠2 and ∠3 are supplementary
The diagonals bisect each other.	

Practice Exercises

1 Parallelogram *PARK* has congruent and perpendicular diagonals. Which of the following describes *PARK*?

   (1) square

   (2) rectangle, but not rhombus

   (3) rhombus, but not rectangle

   (4) trapezoid, but not rectangle or rhombus

2 Quadrilateral *JUMP* has the following properties: $JU = MP$, $UM = PJ$, and $JU = UM$. How can *JUMP* be classified?

   (1) parallelogram only

   (2) parallelogram and rectangle

   (3) parallelogram and rhombus

   (4) not enough information to make any classification

3 In parallelogram *RSTU*, $m\angle R = 8x + 12$ and $m\angle T = 4x + 24$. What is $m\angle R$?

   (1) 144°   (3) 36°

   (2) 72°   (4) 18°

4 In parallelogram *ABCD*, the measures of $\angle B$ and $\angle C$ are in a 3:2 ratio. What is the measure of $\angle B$?

   (1) 120°   (3) 72°

   (2) 108°   (4) 60°

5 What are the measures of ∠1 and ∠2 in the given trapezoid?

(1) m∠1 = 24° and m∠2 = 24°

(2) m∠1 = 66° and m∠2 = 24°

(3) m∠1 = 66° and m∠2 = 66°

(4) m∠1 = 24° and m∠2 = 66°

6 Given rhombus $ABCD$ with diagonals $\overline{AC}$ and $\overline{BD}$ intersecting at E. If m∠DAE = 41°, what is the measure of ∠CBE?

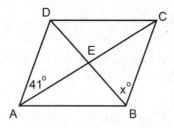

(1) 20.5°   (3) 45°

(2) 41°   (4) 49°

7 $ABCD$ is a rhombus with diagonals $\overline{AC}$ and $\overline{BD}$ intersecting at E. If $AC = 16$ and $BD = 12$, what is the length of $\overline{AB}$?

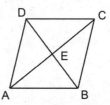

(1) 14   (3) 10

(2) $4\sqrt{14}$ (4) 28

8 Find the measure of ∠C in the parallelogram *ABCD*.

(1) 42°   (3) 69°
(2) 58°   (4) 71°

9 Diagonal $\overline{QS}$ is drawn in rectangle *QRST*. If m∠*RQS* = $(4x + 2)°$ and m∠*QSR* = $(3x + 11)°$, what is m∠*RQS*?

(1) 11°   (3) 46°
(2) 44°   (4) 90°

10 In rectangle *CARS*, diagonals $\overline{SA}$ and $\overline{CR}$ intersect at *X*. If m∠*RSX* = 28°, what is the measure of ∠*SXC*?

(1) 56°   (3) 66°
(2) 62°   (4) 68°

11 Square *STAR* has diagonals that intersect at *P*. If the perimeter of *STAR* is 24, what is the length of $\overline{RP}$ in simplest radical form?

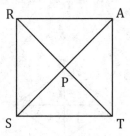

12 Given: $\angle E \cong \angle G$ and $\angle EDF \cong \angle GFD$

Prove: $DEFG$ is a parallelogram

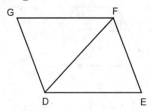

13 Given: $\overline{DKB}, \overline{DLC}, PLDK$ is a rhombus, and $\overline{BK} \cong \overline{CL}$

Prove: $\angle B \cong \angle C$

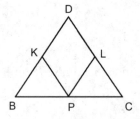

14 Given: Rectangle $ABCD$ with diagonals $\overline{AC}$ and $\overline{BD}$

Prove: The diagonals of a rectangle are congruent $\left(\overline{AC} \cong \overline{BD} \right)$

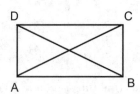

15 Given: $\angle BCA \cong \angle DAC$, $\angle BAC \cong \angle BCA$, $\overline{BC} \cong \overline{AD}$

Prove: $ABCD$ is a rhombus

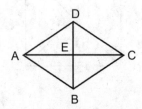

Solutions

1 A parallelogram with perpendicular diagonals is a rhombus, and a parallelogram with congruent diagonals is a rectangle. A figure with the properties of a rectangle and a rhombus is a square.

The correct choice is (**1**).

2 Opposites sides $\overline{JU}$ and $\overline{MP}$, and $\overline{UM}$ and $\overline{PJ}$ are congruent, making $JUMP$ a parallelogram. Consecutive sides $\overline{JU}$ and $\overline{UM}$ are congruent, making $JUMP$ a rhombus.

The correct choice is (**3**).

3    $m\angle R = m\angle T$ Opposite angles of a parallelogram are congruent

$$8x + 12 = 4x + 24$$
$$4x + 12 = 24$$
$$4x = 12$$
$$x = 3$$
$$m\angle R = 8(3) + 12 \quad \text{Substitute } x = 3$$
$$= 24 + 12$$
$$= 36°$$

The correct choice is (**3**).

4 Let $m\angle B = 3x$ and $m\angle C = 2x$, and apply the parallelogram property that consecutive angles are supplementary.

$$m\angle B + m\angle C = 180°$$
$$3x + 2x = 180$$
$$5x = 180$$
$$x = 36$$
$$m\angle B = 3(36) \quad \text{Substitute } x = 36$$
$$= 108°$$

The correct choice is (**2**).

5 The diagonal is a transversal forming a pair of congruent alternate interior angles, so m∠1 = 24°. The triangle angle sum theorem can be used on the bottom triangle to find ∠2.

$$m\angle 1 + m\angle 2 + 90° = 180°$$
$$24° + m\angle 2 + 90° = 180°$$
$$m\angle 2 + 114° = 180°$$
$$m\angle 2 = 66°$$

The correct choice is (**4**).

6   m∠BCE = 41°   ∠DAE and ∠BEC are congruent alternate interior angles

    m∠BEC = 90°   Diagonals of a rhombus are perpendicular

    m∠CBE + m∠BEC + m∠BCE = 180°   Triangle angle sum theorem

$$m\angle CBE + 90° + 41° = 180°$$
$$m\angle CBE + 131° = 180°$$
$$m\angle CBE = 49°$$

The correct choice is (**4**).

7 Use the rhombus property that the diagonals bisect each other and are perpendicular.

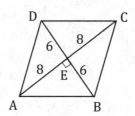

$$AE = \frac{1}{2}AC \quad \text{Diagonals of a rhombus bisect}$$
$$\text{each other}$$
$$= \frac{1}{2}(16)$$
$$= 8$$
$$BE = \frac{1}{2}BD$$
$$= \frac{1}{2}(12)$$
$$= 6$$

$\triangle AEB$ is a right triangle   Diagonals of a rhombus are perpendicular

$$(AE)^2 + (BE)^2 = (AB)^2 \quad \text{Pythagorean theorem}$$
$$6^2 + 8^2 = (AB)^2$$
$$100 = (AB)^2$$
$$AB = 10$$

The correct choice is (**3**).

8 $m\angle A = \dfrac{180° - 42°}{2}$ Isosceles triangle theorem
$$= 69°$$
$$m\angle C = m\angle A \qquad \text{Opposite angles are congruent}$$
$$= 69°$$

The correct choice is (**3**).

9 Start by sketching the figure.

$\triangle QSR$ is a right triangle with a right angle at $\angle R$, so $\angle RQS$ and $\angle QSR$ must be complementary.

$$\text{m}\angle RQS + \text{m}\angle QSR = 90°$$
$$4x + 2 + 3x + 11 = 90$$
$$7x + 13 = 90$$
$$7x = 77$$
$$x = 11$$
$$\text{m}\angle RQS = 4(11) + 2 \quad \text{Substitute } x = 11$$
$$\text{m}\angle RQS = 46°$$

The correct choice is **(3)**.

10 $\text{m}\angle CSX + \text{m}\angle RSX = 90°$ Angles in a rectangle are right
angles
$\text{m}\angle CSX + 28° = 90°$
$\text{m}\angle CSX = 62°$

The diagonals of a rectangle are congruent and bisect each other; therefore, $\overline{SX} \cong \overline{CX}$, making $\triangle SXC$ isosceles.

$\text{m}\angle SCX = \text{m}\angle CSX$ Base angles of an isosceles triangle
are congruent

$= 62°$

$\text{m}\angle CSX + \text{m}\angle SCX + \text{m}\angle SXC = 180°$ Triangle angle sum
theorem

$62° + 62° + \text{m}\angle SXC = 180°$
$124° + \text{m}\angle SXC = 180°$
$\text{m}\angle SXC = 56°$

The correct choice is **(1)**.

11 $RS = \dfrac{1}{4}$ perimeter   All sides of a square are congruent

$\qquad = \dfrac{1}{4}(24)$

$\qquad = 6$

$\qquad\qquad RP = PS$ — Diagonals of a square bisect each other and are congruent

$(RP)^2 + (PS)^2 = (RS)^2$ — Pythagorean theorem

$(RP)^2 + (RP)^2 = 6^2$

$\qquad\quad 2(RP)^2 = 36$

$\qquad\qquad (RP)^2 = 18$

$\qquad\qquad\quad RP = \sqrt{18}$

$\qquad\qquad\qquad = 3\sqrt{2}$

12

Statement	Reason
1. $\angle EDF \cong \angle GFD$	1. Given
2. $\overline{ED} \,/\,/\, \overline{FG}$	2. Two lines are parallel if the alternate interior angles formed are congruent
3. $\angle E \cong \angle G$	3. Given
4. $\overline{FD} \cong \overline{FD}$	4. Reflexive property
5. $\triangle EDF \cong \triangle GFD$	5. AAS
6. $\angle EFD \cong \angle GDF$	6. CPCTC
7. $\overline{GD} \,/\,/\, \overline{FE}$	7. Two lines are parallel if the alternate interior angles formed are congruent
8. $DEFG$ is a parallelogram	8. A quadrilateral with two pairs of opposite parallel sides is a parallelogram

13

Statement	Reason
1. $\overline{DKB}$, $\overline{DLC}$, and $PLDK$ is a rhombus	1. Given
2. $\overline{KD} \cong \overline{LD}$	2. Consecutive sides of a rhombus are congruent
3. $\overline{BK} \cong \overline{CL}$	3. Given
4. $\overline{BK} + \overline{KD} \cong \overline{CL} + \overline{LD}$	4. Addition postulate
5. $\overline{BD} \cong \overline{CD}$	5. Partition property
6. $\angle B \cong \angle C$	6. Angles opposite congruent sides in triangles are congruent

14 The strategy is to prove $\triangle ABC \cong \triangle BAD$ and then apply CPCTC.

Statement	Reason
1. Rectangle $ABCD$ with diagonals $\overline{AC}$ and $\overline{BD}$	1. Given
2. $\angle ABC$ and $\angle BAD$ are right angles	2. All angles in a rectangle are right angles
3. $\angle ABC \cong \angle BAD$	3. Right angles are congruent
4. $\overline{AD} \cong \overline{BC}$	4. Opposite sides of a rectangle are congruent
5. $\overline{AB} \cong \overline{AB}$	5. Reflexive property
6. $\triangle ABC \cong \triangle BAD$	6. SAS
7. $\overline{AC} \cong \overline{BD}$	7. CPCTC

15 The strategy is to first prove $ABCD$ is a parallelogram and then prove $\triangle ABC$ is an isosceles triangle, with legs $\overline{AB} \cong \overline{BC}$.

Statement	Reason
1. $\angle BCA \cong \angle DAC$	1. Given
2. $\overline{BC} \ // \ \overline{AD}$	2. Two lines are parallel if the alternate interior angles formed by a transversal are congruent
3. $\overline{BC} \cong \overline{AD}$	3. Given
4. $ABCD$ is a parallelogram	4. A quadrilateral is a parallelogram if one pair of sides is parallel and congruent
5. $\angle BAC \cong \angle BCA$	5. Given
6. $\triangle ABC$ is isosceles	6. A triangle with congruent base angles is isosceles
7. $\overline{AB} \cong \overline{BC}$	7. Sides opposite congruent angles in a triangle are congruent
8. $ABCD$ is a rhombus	8. A parallelogram with a pair of consecutive congruent sides is a rhombus

3.10 COORDINATE GEOMETRY PROOFS

Distance, midpoint, and slope are the tools used on the coordinate plane to prove properties of a figure. They are used as follows:

Quantity	Formula	Use
Slope	$m = \dfrac{y_2 - y_1}{x_2 - x_1}$ or $\dfrac{\text{rise}}{\text{run}}$	Prove segments or lines are parallel (slopes are equal) or perpendicular (slopes are negative reciprocals)
Midpoint	$x_{\text{MP}} = \dfrac{1}{2}(x_1 + x_2)$ $y_{\text{MP}} = \dfrac{1}{2}(y_1 + y_2)$	Prove segments bisect each other (midpoints are concurrent)
Length	$d = \sqrt{(x_2 - x_1)^2 + (y_2 - y_1)^2}$	Prove segments are congruent (distances between endpoints are equal)

The steps to writing a good coordinate geometry proof are

- *Graph the points*—The graph will help you plan a strategy, check your work, and help with some calculations.
- *Plan a strategy*—Determine what property you will demonstrate and what calculations are needed.
- *Perform the calculations*—Write down the general equations you are using and clearly label which segments correspond to which calculations.
- *Write a summary statement*—You must state in words a justification that explains why your calculations justify the proof.

PROVING PARALLELOGRAMS ON THE COORDINATE PLANE

A given quadrilateral can be proven to be a parallelogram by demonstrating one of the parallelogram properties. The three properties that can be easily proven are parallel sides, congruent sides, and diagonals bisecting each other. The calculations required and suggested summary statements are shown in the following table.

Property	Calculations	Summary Statement
2 pairs of opposite sides are parallel	Slope of each side (4 slope calculations)	The slopes of opposite sides are equal; therefore, both pairs of opposite sides are parallel. A quadrilateral with two pairs of opposite sides parallel is a parallelogram.
2 pairs of opposite sides are congruent	Length of each side (4 distance calculations)	The lengths of opposite sides are equal; therefore, both pairs of opposite sides are congruent. A quadrilateral with two pairs of opposite congruent sides is a parallelogram.
The diagonals bisect each other	Midpoints of the diagonals (2 midpoint calculations)	The diagonals have the same midpoint; therefore, they bisect each other. A quadrilateral whose diagonals bisect each other is a parallelogram.

PROVING SPECIAL PARALLELOGRAMS ON THE COORDINATE PLANE

To prove a quadrilateral is a special parallelogram, a good strategy is to first prove the figure is a parallelogram and then to demonstrate one of the special properties shown in the table. For squares, you need to show that the figure is both a rectangle and a rhombus.

First show the figure is a parallelogram, then...

Figure	Property	Calculation	Example of Summary Statement
Rhombus	Perpendicular diagonals	Slope of each diagonal	The slopes of the diagonals are negative reciprocals; therefore, they are ⊥. A parallelogram with ⊥ diagonals is a rhombus.
	Two consecutive congruent sides	Length of two consecutive sides	The lengths of 2 consecutive sides are equal, so they are ≅. A parallelogram with consecutive ≅ sides is a rhombus.
Rectangle	A right angle	Slope of two consecutive sides	The slopes of two consecutive sides are negative reciprocals, so the sides are ⊥ and form a right angle. A parallelogram with a right angle is a rectangle.
	Congruent diagonals	Length of each diagonal	The lengths of the diagonals are ≅, so they are congruent. A parallelogram with congruent diagonals is a rectangle.
Square	Show the figure is both a rectangle and a rhombus.		A figure that is both a rectangle and a rhombus is a square.

One strategy to aid in remembering which properties to prove is to work only with the diagonals:

- Midpoints for a parallelogram
- Midpoints and slope for a rhombus
- Midpoints and length for a rectangle

OTHER COORDINATE GEOMETRY PROOFS

A quadrilateral can be shown to be a trapezoid using slope calculations to show two sides are parallel.

Triangles can be classified using coordinate geometry calculations:

Isosceles triangle—distance calculation to show two sides are congruent
Equilateral—distance calculation to show three sides are congruent
Right triangle—slope calculation to show two sides are perpendicular

Coordinate geometry can be used to prove special segments in triangles. The following table summarizes some of the segments and points of concurrency, and the calculations needed.

Segment	Calculation Needed	Summary Statement
Median	Midpoint	The segment has endpoints at a vertex and midpoint, so it is a median.
Altitude	Two slopes	The slopes are negative reciprocals, so the segment is ⊥ to the opposite side, making it an altitude.
Perpendicular bisector	Slope and midpoint	The slopes are negative reciprocals, making the segment ⊥, and the segment passes through the midpoint, making it a bisector. Therefore, it is a perpendicular bisector.
Midsegment	Two midpoints	The segment's endpoints are the midpoints of sides of the triangle; therefore, it is a midsegment.
Circumcenter	Three lengths	The distance from the point to the three vertices are equal, so it is equidistant from each vertex. Therefore, the point is the circumcenter.
Centroid	Two lengths	The distances from the point to the vertices and the distance from the point to the opposite midpoint are in a 1:2 ratio; therefore, the point is a centroid.

Practice Exercises

1 The coordinates of quadrilateral *BGRM* are $B(0, 2)$, $G(6, 0)$, $R(9, 9)$, and $M(3, 11)$.

Prove *BGRM* is a rectangle.

2 △*GYP* has coordinates $G(1, 1)$ $Y(7, 5)$ and $P(3, 11)$.

Prove △*GYP* is an isosceles right triangle.

3 Given parallelogram *FLIP* with vertices $F(2, 9)$, $L(h, k)$, $I(11, 6)$, and $P(1, 2)$, find the values of h and k.

4 Given quadrilateral *CUTE* with coordinates $C(-5, -4)$, $U(-3, 4)$, $T(5, 6)$, and $E(3, -2)$, prove *CUTE* is a rhombus.

5 Quadrilateral *PLOW* has coordinates $P(1, -4)$, $L(-1, 4)$, $O(7, 6)$, and $W(9, -2)$.

Prove *PLOW* is a square.

6 Quadrilateral *ABCD* has coordinates $A(1, 2)$, $B(9, 4)$, $C(5, 8)$, and $D(1, 7)$.

Prove *ABCD* is a trapezoid.

7 Segment $\overline{MG}$ has coordinates $M(0, 2)$ and $G(4, 0)$. Point T is the midpoint of $\overline{MG}$. $M''G''$ is the image of $\overline{MG}$ after a 90° rotation about T followed by a dilation centered about T with a scale factor of 2.

a) State the coordinates of M'' and G'' and graph $\overline{M''G''}$.

b) Prove $MG''GM''$ is a rhombus.

8 $\triangle ABC$ has coordinates $A(1, 5)$, $B(7, -1)$, and $C(13, 11)$. If point G has coordinates $(7, 5)$, prove G is the centroid of $\triangle ABC$.

9 $\triangle DOG$ has coordinates $D(4, 3)$, $O(9, 8)$, and $G(1, 12)$. If point A has coordinates $(3, 6)$, prove $\overline{OA}$ is an altitude of $\triangle DOG$.

10 Rectangle $ABCD$ has coordinates $A(4, 12)$, $B(13, 3)$, $C(9, -1)$, and $D(0, 8)$. Points F and G have coordinates $F(2, 10)$ and $G(7, 5)$, and $AHGF$ is also a rectangle. Is rectangle $ABCD$ similar to rectangle $AHGF$? Justify your answer.

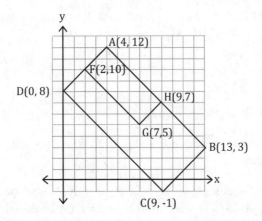

Solutions

1 There are several different approaches to proving a quadrilateral is a rectangle. Here, we will calculate the slopes of the 4 sides. This will let us show opposite sides have the same slope and are parallel, and consecutive sides have negative reciprocal slopes and are perpendicular. *BGRM* is graphed in the accompanying diagram.

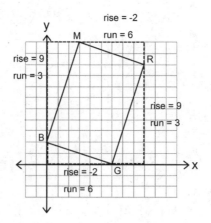

Using the grid, we can calculate slope from slope $= \dfrac{\text{rise}}{\text{run}}$. The rise and run for each side of *BGRM* are shown in the figure.

$$\text{Slope } \overline{BM} = \frac{\text{rise}}{\text{run}} \qquad \text{Slope } \overline{GR} = \frac{\text{rise}}{\text{run}}$$

$$= \frac{9}{3} \qquad\qquad\qquad = \frac{9}{3}$$

$$= 3 \qquad\qquad\qquad\quad = 3$$

$$\text{Slope } \overline{BG} = \frac{\text{rise}}{\text{run}} \qquad \text{Slope } \overline{MR} = \frac{\text{rise}}{\text{run}}$$

$$= \frac{-2}{6} \qquad\qquad\qquad = \frac{-2}{6}$$

$$= -\frac{1}{3} \qquad\qquad\qquad = -\frac{1}{3}$$

The slopes of opposite sides are equal, therefore they are parallel and *BRGM* is a parallelogram.

The slopes of consecutive sides are negative reciprocals, making them perpendicular. *BGRM* is a rectangle because it is a parallelogram with right angles.

2 The strategy is to show that $\overline{GY} \cong \overline{YP}$ using distance calculations and $\overline{GY} \perp \overline{YP}$ using slope calculations.

$$\text{slope} = \frac{y_2 - y_1}{x_2 - x_1}$$

$$\text{slope } \overline{GY} = \frac{5-1}{7-1} \qquad\qquad \text{slope } \overline{YP} = \frac{11-5}{3-7}$$

$$= \frac{4}{6} = \frac{2}{3} \qquad\qquad\qquad\qquad = -\frac{6}{4} = -\frac{3}{2}$$

$$\text{length} = \sqrt{(x_1 - x_2)^2 + (y_1 - y_2)^2}$$

$$\text{length } \overline{GY} = \sqrt{(7-1)^2 + (5-1)^2}$$

$$= \sqrt{36 + 16}$$

$$= \sqrt{52}$$

$$\text{length } \overline{YP} = \sqrt{(7-3)^2 + (5-11)^2}$$

$$= \sqrt{16 + 36}$$

$$= \sqrt{52}$$

$\overline{GY}$ is perpendicular to $\overline{YP}$ because their slopes are negative reciprocals, and $\overline{GY} \cong \overline{YP}$ because the lengths GY and YP are equal. Therefore, $\triangle GYP$ is a right isosceles triangle.

3 The opposite sides of a parallelogram are parallel and congruent. The strategy is to locate point L so that $FL = PI$ and the slopes of $\overline{FL}$ and $\overline{PI}$ are equal.

$$\text{slope} = \frac{y_2 - y_1}{x_2 - x_1}$$

$$\text{slope } \overline{PI} = \frac{6 - 2}{11 - 1}$$

$$= \frac{4}{10}$$

$$\text{slope } \overline{FL} = \frac{4}{10} \qquad \text{Parallel lines have equal slopes.}$$

Since slope is also equal to $\dfrac{\text{rise}}{\text{run}}$, locate point L by adding the rise to the y-coordinate of F, and add the run to the x-coordinate of x.

$$x\text{-corrdinate of } L = 2 + 10$$
$$= 12$$
$$y\text{-coordinate of } L = 9 + 4$$
$$= 13$$

The coordinates of L are (12, 13).

4 One strategy is to first prove the figure is a parallelogram by showing the diagonals bisect each other using midpoint calculations. Then prove it is a rhombus by showing the diagonals are perpendicular using slope calculations.

$$\text{midpoint} = \frac{x_1 + x_2}{2}, \frac{y_1 + y_2}{2}$$

midpoint $\overline{CT}$: $\dfrac{-5+5}{2}, \dfrac{-4+6}{2}$ midpoint $\overline{EU}$: $\dfrac{3+(-3)}{2}, \dfrac{-2+4}{2}$

$(0, 1)$ $(0, 1)$

$$\text{slope} = \frac{y_2 - y_1}{x_2 - x_1}$$

$$\text{slope } \overline{CT} = \frac{6 - (-4)}{5 - (-5)} \qquad \text{slope } \overline{EU} = \frac{4 - (-2)}{-3 - 3}$$

$$= \frac{10}{10} \qquad\qquad\qquad = -\frac{6}{6}$$

$$= 1 \qquad\qquad\qquad = -1$$

The diagonals of *CUTE* bisect each because they have the same midpoint, making it a parallelogram. The diagonals are also perpendicular because their slopes are negative reciprocals. Therefore, *CUTE* is a rhombus.

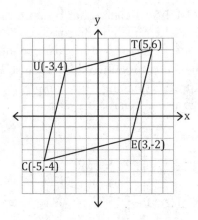

5 First prove the figure is a parallelogram by showing the diagonals bisect each other using midpoint calculations; then show it is a rhombus with perpendicular diagonals using slope. Finally, show it is a rectangle with a pair of perpendicular sides using slope.

$$\text{midpoint} = \frac{x_1 + x_2}{2}, \frac{y_1 + y_2}{2}$$

midpoint $\overline{PO}$: $\frac{1+7}{2}, \frac{-4+6}{2}$ midpoint $\overline{WL}$: $\frac{9+(-1)}{2}, \frac{-2+4}{2}$

$(4,1)$ $(4,1)$

$$\text{slope} = \frac{y_2 - y_1}{x_2 - x_1}$$

slope $\overline{PO} = \frac{6-(-4)}{7-1}$ slope $\overline{WL} = \frac{4-(-2)}{-1-9}$

$= \frac{10}{6}$ $= -\frac{6}{10}$

slope $\overline{PW} = \frac{-2-(-4)}{9-1}$ slope $\overline{LP} = \frac{-4-4}{1-(-1)}$

$= \frac{2}{8}$ $= -\frac{8}{2}$

The diagonals of *PLOW* bisect each other because they have the same midpoint, making it a parallelogram. The diagonals of *PLOW* are perpendicular because their slopes are negative reciprocals, making it a rhombus. *PLOW* has a pair of perpendicular consecutive sides because $\overline{PW}$ and $\overline{LP}$ have slopes that are negative reciprocals, making it a rectangle. *PLOW* is both a rectangle and a rhombus, so it is also a square.

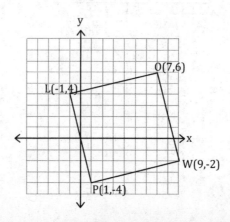

6 A good time-saving strategy is to graph the figure first to determine the pair of sides that are likely to be parallel. The parallel sides appear to be $\overline{AB}$ and $\overline{CD}$, and the legs are $\overline{AD}$ and $\overline{BC}$.

$$\text{slope} = \frac{y_2 - y_1}{x_2 - x_1}$$

$$\text{slope } \overline{AB} = \frac{4 - 2}{9 - 1} \qquad \text{slope } \overline{CD} = \frac{8 - 7}{5 - 1}$$

$$= \frac{2}{8} \qquad\qquad\qquad = \frac{1}{4}$$

$$= \frac{1}{4}$$

$\overline{AB}$ is parallel to $\overline{CD}$ because their slopes are equal, making $ABCD$ a trapezoid.

7 a) First find the coordinates of the midpoint T using the midpoint formula; then use T to graphically find the coordinates of M' and G'.

$$x_{\text{MP}} = \frac{1}{2}(x_1 + x_2)$$

$$= \frac{1}{2}(0 + 4)$$

$$= 2$$

$$y_{\text{MP}} = \frac{1}{2}(y_1 + y_2)$$

$$= \frac{1}{2}(2 + 0)$$

$$= 1$$

The coordinates of T are $(2, 1)$.

Rotating a segment $90°$ about its midpoint results in a segment congruent and perpendicular to the pre-image.

$$\text{slope of } \overline{MG} = \frac{y_2 - y_1}{x_2 - x_1}$$

$$= \frac{0 - 2}{4 - 0}$$

$$= -\frac{1}{2}$$

Let $\overline{M'G'}$ be the image of $\overline{MG}$ after the 90° rotation. The slope of $\overline{M'G'}$ is the negative reciprocal, or 2. The coordinates of M' and G' can be found graphically by starting at T and moving with a slope of 2 in either direction for a length equal to TM and TG. From the graph, the coordinates are $M'(1, -1)$ and $G'(3, 3)$.

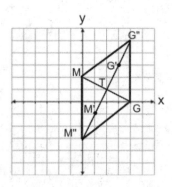

A dilation through midpoint T will extend the segment the same distance on either side of the midpoint because T remains the midpoint. Graph $\overline{M''G''}$ collinear with $\overline{M'G'}$, but twice as long.

b) $MG''GM''$ must be a rhombus. T is the midpoint of both of its diagonals. The diagonals bisect each other making it a parallelogram. The 90° rotation resulted in $\overline{MG} \perp \overline{M''G''}$. A parallelogram with perpendicular diagonals is a rhombus.

8 The centroid is the point of concurrency of the medians of a triangle, so one strategy is to graph two medians and look for the point of intersection. A median is the segment from a vertex to the opposite midpoint, so the first step is to find the midpoints of two sides.

$$\text{midpoint} = \frac{x_1 + x_2}{2}, \frac{y_1 + y_2}{2}$$

let D be the of $\overline{AB}$: $\frac{1+7}{2}, \frac{5+(-1)}{2}$

$D(4,2)$

let E be the of $\overline{AC}$: $\frac{1+13}{2}, \frac{5+11}{2}$

$E(7,8)$

Graphing $\overline{CD}$ and $\overline{BE}$, the point of intersection is $G(7,5)$. Therefore, G is the centroid of $\triangle ABC$.

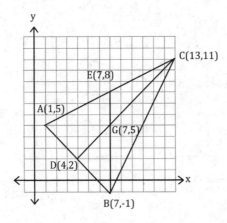

Algebraically, you could have also used the distance formula to calculate DG and GC and then show $DG = \frac{1}{2}GC$.

9 An altitude is a segment from a vertex perpendicular to the opposite side of a triangle. $\overline{AO}$ is an altitude if it is perpendicular to $\overline{GD}$.

$$\text{slope} = \frac{y_2 - y_1}{x_2 - x_1}$$

$$\text{slope } \overline{AO} = \frac{8 - 6}{9 - 3} \qquad \text{slope } \overline{GD} = \frac{12 - 3}{1 - 4}$$

$$= \frac{2}{6} \qquad\qquad = -\frac{9}{3}$$

$$= \frac{1}{3} \qquad\qquad = -\frac{3}{1}$$

$\overline{AO}$ is perpendicular to $\overline{GD}$ because their slopes are negative reciprocals, so $\overline{AO}$ is an altitude of $\triangle DOG$.

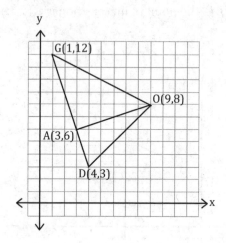

10 Since both figures are rectangles, all pairs of corresponding angles must be congruent. The lengths of two pairs of corresponding sides must be calculated to see if they are in the same ratio. The other two pairs must be in the same ratio since opposite sides of a rectangle are congruent.

The corresponding lengths to be checked are AB and AH, and AD and AF.

$$\text{length} = \sqrt{(x_1 - x_2)^2 + (y_1 - y_2)^2}$$

$$\text{length } AB = \sqrt{(4 - 13)^2 + (12 - 3)^2}$$
$$= \sqrt{81 + 81}$$
$$= \sqrt{162}$$

$$\text{length } AH = \sqrt{(4 - 9)^2 + (12 - 7)^2}$$
$$= \sqrt{25 + 25}$$
$$= \sqrt{50}$$

$$\text{length } AD = \sqrt{(4 - 0)^2 + (12 - 8)^2}$$
$$= \sqrt{16 + 16}$$
$$= \sqrt{32}$$

$$\text{length } AF = \sqrt{(4 - 2)^2 + (12 - 10)^2}$$
$$= \sqrt{4 + 4}$$
$$= \sqrt{8}$$

Calculate the ratios of the two pairs of corresponding sides.

$$\frac{AB}{AH} = \frac{\sqrt{162}}{\sqrt{50}} \qquad \frac{AD}{AF} = \frac{\sqrt{32}}{\sqrt{8}}$$
$$= \sqrt{\frac{81}{25}} \qquad\qquad = \sqrt{4}$$
$$= \frac{9}{5} \qquad\qquad = 2$$

The rectangles are not similar because the corresponding sides are not in the same ratio.

3.11 CIRCLES

DEFINITIONS AND BASIC THEOREMS

Segment and Line Definitions in Circles

Segment or Line	Definition
Radius	A segment with one endpoint at the center of the circle and one endpoint on the circle
Chord	A segment with both endpoints on the circle
Diameter	A chord that passes through the center of the circle
Secant	A line that intersects a circle at exactly two points
Tangent	A line that intersects a circle at exactly one point
Point of tangency	The point at which a tangent intersects a circle

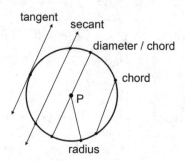

Some basic theorems of circles:

- All radii of a given circle are congruent.
- All circles are similar.
- Two circles are congruent if and only if their radii are congruent.

Since all circles are similar, any circle can be mapped to another using a similarity transformation comprised of a translation to move one center onto the other followed by a dilation to make the radii congruent.

An arc is a section of a circle. An arc may be major, minor, or a semicircle:

- *Minor arc*—an arc spanning less than 180°
- *Semicircle*—an arc spanning exactly 180°
- *Major arc*—an arc spanning more than 180°

When naming a semicircle or major arc, we add an additional point to show which arc is being specified. The accompanying figure shows a minor arc $\overset{\frown}{AB}$, semicircle $\overset{\frown}{ABC}$, and major arc $\overset{\frown}{ABD}$.

minor arc $\overset{\frown}{AB}$ semicircle $\overset{\frown}{ABC}$ major arc $\overset{\frown}{ABD}$

Some useful arc theorems:

- A diameter intercepts a semicircle.
- The sum of the arc measures around a circle equals 360°.
- The sum of the arc measures around a semicircle equals 180°.

CENTRAL AND INSCRIBED ANGLES

Central angle—an angle whose vertex is the center of a circle and whose rays intersect the circle

---

CENTRAL ANGLE THEOREM

The angle measure of an arc equals the measure of the central angle that intercepts the arc.

Inscribed angle—an angle whose vertex is on a circle and whose rays intersect the circle

---

INSCRIBED ANGLE THEOREM

The angle measure of an arc equals twice the measure of the inscribed angle that intercepts the arc.

The accompanying figure shows central angle $\angle APB$ and inscribed angle $\angle CMD$. Each angle cuts off, or intercepts, an arc on the circle. Central $\angle APB$ intercepts $\overset{\frown}{AB}$ and inscribed angle $\angle CMD$ intercepts $\overset{\frown}{CD}$.

$\text{m } \overset{\frown}{AB} = \text{m} \angle APB$ $\text{m } \overset{\frown}{CD} = 2 \text{ m} \angle CMD$

- Since the inscribed angle measures half the intercepted arc, any inscribed angle that intercepts a semicircle is a right angle.

CONGRUENT, PARALLEL, AND PERPENDICULAR CHORDS

CONGRUENT CHORD THEOREM

Congruent chords intercept congruent arcs on a circle.
   Congruent arcs on a circle are intercepted by congruent chords.

$$\overline{AB} \cong \overline{CD}, \quad \overset{\frown}{AB} \cong \overset{\frown}{CD}$$

PARALLEL CHORD THEOREM

The two arcs formed *between* a pair of parallel chords are congruent.

  If the two arcs formed *between* a pair of chords are congruent, then the chords are parallel.

$$\overline{AB} \parallel \overline{CD}, \quad \overarc{AC} \cong \overarc{BD}$$

CHORD–PERPENDICULAR BISECTOR THEOREM

The perpendicular bisector of any chord passes through the center of the circle.

  A diameter or radius that is perpendicular to a chord bisects the chord.

  A diameter or radius that bisects a chord is perpendicular to the chord.

In the accompanying figure, $\overline{AC}$ is perpendicular to diameter $\overline{BD}$ at E; therefore, E is the midpoint of $\overline{AC}$.

You can think of this theorem linking three properties of a diameter:

- Bisects another chord
- Is perpendicular to another chord
- Is a diameter

If any two are true, then the third must also be true.

TANGENT RADIUS THEOREM

- A diameter or radius to a point of tangency is perpendicular to the tangent.
- A line perpendicular to a tangent at the point of tangency passes through the center of the circle.

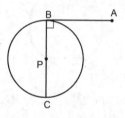

CONGRUENT TANGENT THEOREM

Given a circle and external point Q, segments between the external point and the two points of tangency are congruent.

In the accompanying figure, tangents $\overrightarrow{QA}$ and $\overrightarrow{QB}$ are both constructed from point Q, and $\overline{QA} \cong \overline{QB}$.

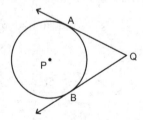

CYCLIC QUADRILATERALS

Quadrilaterals that are inscribed in a circle are called cyclic quadrilaterals.

CYCLIC QUADRILATERAL THEOREM

Opposite angles of cyclic quadrilaterals are supplementary.

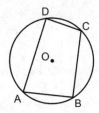

$\angle A$ and $\angle C$ are supplementary.
$\angle B$ and $\angle D$ are supplementary.

This property is proven by considering the intercepted arcs. Together, arcs $\overset{\frown}{ADC}$ and $\overset{\frown}{ABC}$ comprise the entire circle, so $m\overset{\frown}{ADC} + m\overset{\frown}{ABC} = 360°$. Since the intercepted angles measure $\frac{1}{2}$ the arc measures, $m\angle ABC + m\angle ADC = 180°$, and the angles are supplementary. The same justification can be used to show $\angle BAD$ and $\angle BCD$ are supplementary.

ANGLES FORMED BY INTERSECTING CHORDS, SECANTS, AND TANGENTS

Angles formed by intersecting chords, secants, and tangents follow three different relationships, depending on whether the vertex of the angle lies within, on, or outside the circle.

Vertex inside the circle 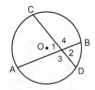	angle measure $= \frac{1}{2}$ the *sum* of the intercepted arcs $$m\angle 1 = m\angle 2 = \frac{1}{2}\left(m\overarc{AC} + m\overarc{DB}\right)$$ $$m\angle 3 = m\angle 4 = \frac{1}{2}\left(m\overarc{BC} + m\overarc{AD}\right)$$
Vertex on the circle 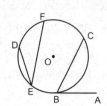	angle measure $= \frac{1}{2}$ of the intercepted arcs $$m\angle ABC = \frac{1}{2}m\overarc{BC}$$ $$m\angle DEF = \frac{1}{2}m\overarc{DF}$$
Vertex outside the circle	angle measure $= \frac{1}{2}$ the *difference* of the intercepted arcs $$m\angle P = \frac{1}{2}\left(m\overarc{ACB} - m\overarc{AB}\right)$$ $$m\angle P = \frac{1}{2}\left(m\overarc{CD} - m\overarc{AB}\right)$$ $$m\angle P = \frac{1}{2}\left(m\overarc{AC} - m\overarc{AB}\right)$$

SEGMENT RELATIONSHIPS IN INTERSECTING CHORDS, TANGENTS, AND SECANTS

The relationships between segment lengths formed by intersecting chords, tangents, and secants are shown in the following table.

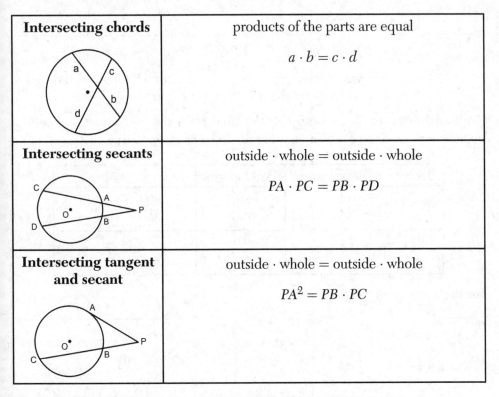

Intersecting chords	products of the parts are equal $a \cdot b = c \cdot d$
Intersecting secants	outside · whole = outside · whole $PA \cdot PC = PB \cdot PD$
Intersecting tangent and secant	outside · whole = outside · whole $PA^2 = PB \cdot PC$

RADIAN MEASURE, ARC LENGTH, AND SECTORS

The **radian** is a unit of angle measure. 2π radians is equal to one complete revolution around a circle, or $360°$. Convert between units using:

$$\text{radians} = \frac{\pi}{180°} \cdot \text{degrees}$$

$$\text{degrees} = \frac{180°}{\pi} \cdot \text{radians}$$

- Besides having an angle measure, an arc has a length. The arc and its central angle also enclose a region called a sector.

Arc with length S
intercepted by angle θ

Sector APB intercepted
by an angle θ

Arc length and sector area can be calculated from the measure of the central angle and the radius of the circle using the following formulas:

Unit of Angle Measure	Arc Length	Sector Area
Degrees	$\dfrac{\pi}{180^\circ} R \cdot \theta$	$\dfrac{1}{360^\circ} \pi R^2 \cdot \theta$
Radians	$R \cdot \theta$	$\dfrac{1}{2} R^2 \cdot \theta$

θ is the measure of the central angle.

Practice Exercises

1 Which of the following is a precise definition of a circle?

  (1) the set of points a fixed distance away from a given point

  (2) the set of points equidistant from the two endpoints of a given diameter

  (3) the set of points equidistant from a given center point and a given line

  (4) a closed figure with no angles

2 $\triangle STY$ is inscribed in circle P. Which of the following is *not necessarily* true?

  (1) Point P is equidistant from S, T, and Y.

  (2) The perpendicular bisector of $\overline{ST}$ passes through P.

  (3) The measure of angle Y equals one-half the measure of $\overset{\frown}{ST}$.

  (4) A tangent to the circle at S forms a right angle with $\overline{ST}$.

3 $\triangle FLY$ is inscribed in circle O. If $FL = LY$ and $m\overset{\frown}{FY} = 80°$, what is the measure of $\angle F$?

  (1) 10°   (3) 70°

  (2) 40°   (4) 80°

4 In $\odot P$, $\angle APB$ is a central angle and $\angle ADB$ is an inscribed angle. If $m\angle APB = 110°$, find $m\angle ADB$.

  (1) 55°   (3) 70°

  (2) 60°   (4) 80°

5 The clock in the tower of the Ridgefield town hall is in the shape of a circle 15 feet in diameter. What is the length of the arc around the clock whose central angle is defined by the hour and minute hands at 4:00 P.M.?

(1) 15.7 ft   (3) 31.4 ft
(2) 18.6 ft   (4) 60 ft

6 In circle Z, arc $\overset{\frown}{KP}$ is intercepted by central angle $\angle KZP$. If $m\angle KZP = 1.2$ radians and $ZK = 7$ cm, what is the length of $\overset{\frown}{KP}$?

(1) 5.8 cm   (3) 10.1 cm
(2) 8.4 cm   (4) 26.4 cm

7 Circle O has a radius of 14 inches. If $\angle HOT$ is a central angle in circle O that measures $132°$, what is the area of sector HOT?

(1) 4.7 in^2 (3) 226 in^2
(2) 10 in^2 (4) 294 in^2

8 Given ⊙P with radius $\overline{PB}$, tangent $\overline{AB}$, and m∠$P = 53°$, what is the measure of ∠A?

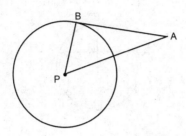

(1) 26.5°   (3) 53°

(2) 37°   (4) 60°

9 In ⊙P, chords $\overline{AB}$ and $\overline{CD}$ are parallel. If m$\overarc{AB} = 118°$ and m$\overarc{CD} = 78°$, find m$\overarc{AC}$.

10 In ⊙P, chords $\overline{AB}$ and $\overline{CD}$ are congruent, m$\overarc{BC} = 60°$, and m$\overarc{CD} = 80°$. Find m∠ABD.

11 $\overline{BDP}$ is a radius of $\odot P$ and is perpendicular to $\overline{AC}$. If $AC = 24$ and $PC = 13$, find PD and DB.

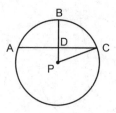

12 $\triangle RST$ is circumscribed about circle P. If $SX = 6$, $XR = 7$, and $TZ = 8$, what is the perimeter of $\triangle RST$?

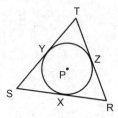

13 Secant $\overline{BEA}$ intersects circle O at E and A. Secant $\overline{BDR}$ intersects circle O at points D and R. If m$\widehat{AR} = 78°$ and m$\widehat{ED} = 35°$, what is the measure of $\angle B$?

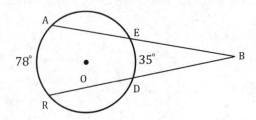

14 Secant $\overline{FGH}$ intersects circle O at G and H. Secant $\overline{FPQ}$ intersects circle O
   at P and Q. If $FG = 6$, $GH = 10$, and $FP = 8$, find the length of PQ.

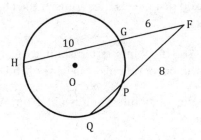

15 Tangents $\overline{TR}$ and $\overline{TQ}$ are drawn from point T to circle A. $\overline{RAS}$ is a diameter
   of circle A and m$\angle T = 46°$, what is the measure of $\overset{\frown}{SQ}$?

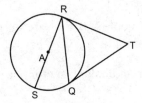

16 In $\odot O$, chords $\overline{HLI}$ and $\overline{KLJ}$ intersect at L. Prove: $HL \cdot LI = KL \cdot LJ$

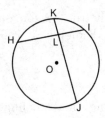

Solutions

1 Examine each choice:

(1) Correct—The fixed distance is the radius of the circle, and the given point is the center of the circle.

(2) Incorrect—The set of points equidistant from the two endpoints of a diameter would be the perpendicular bisector of the diameter.

(3) Incorrect—The set of points equidistant from a center point and line would be a parabola.

(4) Incorrect—Circles are closed figures with no angles, but not all closed figures with no angles are circles. A counterexample would be an ellipse.

The correct choice is (**1**).

2 Choice (1) is true—The center of the circumscribed circle is equidistant from the vertices of the triangle.

Choice (2) is true—A perpendicular bisector of a chord always passes through the center of the circle.

Choice (3) is true—The measure of the inscribed angle equals half the measure of the intercepted arc.

Choice (4) is not necessarily true—A tangent will make a right angle with a chord only if the chord is a diameter. $\overline{ST}$ is not necessarily a diameter.

The correct choice is (**4**).

3 Sketching the isosceles triangle and labeling the known arc, we see $\angle L$ is the inscribed angle that intercepts the $80°$ arc.

An inscribed angle measures half the intercepted arc:

$$m\angle L = \frac{1}{2}m\widehat{FY}$$
$$= \frac{1}{2}\left(80°\right)$$
$$= 40°$$

Since $FL = LY$, $m\angle F = m\angle Y$ from the isosceles triangle theorem, and we can use the angle sum theorem to find the measures.

$$m\angle F + m\angle L + m\angle Y = 180°$$
$$2x + 40° = 180°$$
$$2x = 140°$$
$$x = 70°$$
$$m\angle F = 70°$$

The correct choice is **(3)**.

4 The strategy is to use the central angle to find $m\widehat{AB}$, and then use the arc to find the inscribed angle.

$m\widehat{AB} = m\angle APB = 110°$ The central angle and arc have equal measure.

$m\angle ADB = \frac{1}{2}m\widehat{AB} = 55°$ The inscribed angle measure is half the intercepted arc.

The correct choice is **(1)**.

5 The numbers on the face of the clock divide the circle into 12 congruent central angles. Each central angle measures $\dfrac{360°}{12}$, or 30°. Therefore, the angle formed by the two hands on the clock measures $4 \cdot 30°$, or 120°. The radius $= \dfrac{1}{2}(15)$, or 7.5 ft. Using the formula for arc length,

$$S = 2\pi R\frac{\theta}{360°}$$
$$= 2\pi(7.5\,\text{ft})\frac{120°}{360°}$$
$$= 15.7\,\text{ft}$$

The correct choice is **(1)**.

6 Arc length, S, is given by the equation $S = R\theta$, where θ is the measure of the central angle in radians.

$$S = R\theta$$
$$= 7\,\text{cm} \cdot 1.2$$
$$= 8.4\,\text{cm}$$

The correct choice is **(2)**.

7 The area of a sector is given by $A = \dfrac{\theta}{360°}\pi R^2$, where θ is the central angle measured in degrees.

$$A = \frac{132°}{360°}\pi \cdot (14\,\text{in})^2$$
$$= 225.775\,\text{in}^2$$
$$= 226\,\text{in}^2$$

The correct choice is **(3)**.

8
$$m\angle B = 90°$$ radius perpendicular to a tangent at the point of tangency

$$m\angle P + m\angle B + m\angle A = 180°$$ triangle angle sum theorem

$$53° + 90° + m\angle A = 180°$$

$$143° + m\angle A = 180°$$

$$m\angle A = 37°$$

The correct choice is (2).

9 Arcs $\overset{\frown}{AC}$ and $\overset{\frown}{DB}$ are the congruent arcs between a pair of parallel chords.

$$m\overset{\frown}{AB} + m\overset{\frown}{BD} + m\overset{\frown}{CD} + m\overset{\frown}{AC} = 360°$$ circle–arc sum theorem

$$118° + x + 78° + x = 360°$$ parallel chord theorem

$$196° + 2x = 360°$$

$$2x = 164°$$

$$x = 82°$$

$$m\overset{\frown}{AC} = 82°$$

10 $\angle ABD$ is an inscribed angle, so its measure will be half of $m\overset{\frown}{AD}$.

$$m\overset{\frown}{AB} + m\overset{\frown}{BC} + m\overset{\frown}{CD} + m\overset{\frown}{AD} = 360°$$ circle–arc sum theorem

$$m\overset{\frown}{AB} = m\overset{\frown}{CD} = 80°$$ congruent chord theorem

$$80° + 60° + 80° + m\overset{\frown}{AD} = 360°$$

$$220° + m\overset{\frown}{AD} = 360°$$

$$m\overset{\frown}{AD} = 140°$$

$$m\angle ABD = \frac{1}{2} m\overset{\frown}{AD}$$

$$= 70°$$

11 Radius $\overline{BDP}$ is perpendicular to $\overline{AC}$, so it must also bisect $\overline{AC}$. The strategy is to calculate DC and then apply the Pythagorean theorem in $\triangle PDC$.

$$DC = \frac{1}{2}AC$$

$$DC = \frac{1}{2}(24)$$

$$DC = 12$$

$$(PD)^2 + (DC)^2 = PC^2 \qquad \text{Pythagorean theorem}$$

$$(PD)^2 + 12^2 = 13^2$$

$$(PD)^2 + 144 = 169$$

$$(PD)^2 = 25$$

$$PD = 5$$

$$PB = PC = 13 \quad \text{all radii of a circle are congruent}$$

$$PD + DB = PB \qquad\quad \text{partition}$$

$$5 + DB = 13$$

$$DB = 8$$

12 $SY = SX$, $TY = TZ$, and $ZR = XR$ because each pair are tangents from the same point.

The perimeter is, therefore, $2 \cdot 6 + 2 \cdot 7 + 2 \cdot 8 = 42$.

13 The angle formed by two secants is equal to half the difference of the intercepted arcs.

$$m\angle B = \frac{1}{2}\left(m\widehat{AB} - m\widehat{ED}\right)$$

$$= \frac{1}{2}\left(78° - 35°\right)$$

$$= 21.5°$$

14 When two tangents from the same point intersect a circle, the segments formed follow the relationship:

$$\text{outside} \cdot \text{whole} = \text{outside} \cdot \text{whole}$$
$$FG \cdot FH = FP \cdot FQ$$
$$6 \cdot (6 + 10) = 8 \cdot (8 + PQ)$$
$$6 \cdot 16 = 8(8 + PQ)$$
$$96 = 64 + 8PQ$$
$$32 = 8PQ$$
$$PQ = 4$$

15 This problem requires applying multiple circle concepts. The strategy is to first use the fact $\triangle TRQ$ is isosceles to find $m\angle TRQ$ and $m\angle SRQ$ and then to apply the relationship that the measure of $\overset{\frown}{SQ}$ is twice the inscribed angle that intercepts it.

$$\overline{TR} \cong \overline{TQ} \qquad \text{congruent tangent theorem}$$
$$m\angle TRQ = \tfrac{1}{2}(180° - m\angle T) \quad \text{isosceles triangle theorem in } \triangle TRQ$$
$$m\angle TRQ = \tfrac{1}{2}(180° - 46°)$$
$$= 67°$$

$$m\angle SRT = 90° \qquad \text{tangent } RT \perp \text{diameter } \overline{RAS}$$
$$m\angle SRQ + m\angle TRQ = 90° \qquad \angle SRQ \text{ and } \angle TRQ \text{ are complementary}$$
$$m\angle SRQ + 67° = 90°$$
$$m\angle SRQ = 23°$$
$$m\overset{\frown}{SQ} = 2m\angle SRQ \qquad \text{inscribed angle theorem}$$
$$m\overset{\frown}{SQ} = 2(23°)$$
$$m\overset{\frown}{SQ} = 46°$$

16 The strategy is to sketch $\overline{HJ}$ and $\overline{KI}$, show $\triangle HLJ \sim \triangle KLI$, and then write a proportion between corresponding sides.

Statements	Reasons
1. Chords $\overline{HLI}$ and $\overline{KLJ}$ intersect at L	1. Given
2. Construct $\overline{IK}$ and $\overline{HJ}$	2. Two points define a segment
3. $\angle H \cong \angle K$ $\angle J \cong \angle I$	3. Inscribed angles that intercept the same arc are congruent
4. $\triangle HLJ \sim \triangle KLI$	4. AA
5. $\dfrac{HL}{KL} = \dfrac{LJ}{LI}$	5. Corresponding parts of similar triangles are proportional
6. $HL \cdot LI = KL \cdot LJ$	6. Cross products of a proportion are equal

3.12 SOLIDS

DEFINITIONS

- A *solid* is any 3-D figure that is fully enclosed.
- A *face* of a solid is any of the surfaces that bound the solid.
- A *polyhedron* is any solid whose faces are polygons.
- An *edge* is the intersection of two faces in a polyhedron.

PRISMS

A *prism* is a polyhedron with two congruent, parallel polygons for bases. The bases of a prism can have any shape. The accompanying figure shows 6 different prisms.

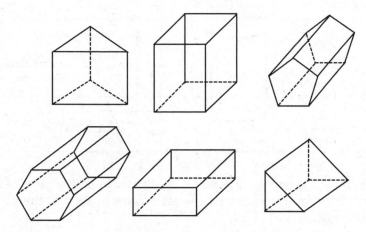

When working with prisms, keep in mind the following facts:

- The height of a prism, h, is the distance between the two bases shown in the accompanying figure.
- The lateral faces are all the faces other than the two parallel bases.
- A right prism has lateral edges that are perpendicular to the bases and lateral faces that are rectangles.
- The lateral edges of an oblique prism are not perpendicular to the bases, and the lateral faces are parallelograms.
- The volume of any prism is found with the formula $V = Bh$, where B is the area of the base.

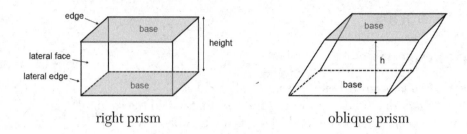

right prism   oblique prism

CYLINDERS

A *circular cylinder* is a solid figure with two parallel and congruent circular bases and a curved lateral area. The height is the perpendicular distance between the bases. As with the prisms, cylinders can be right or oblique as shown in the accompanying figure.

right circular cylinder   oblique circular cylinder

The volume of a cylinder is given by $V = Bh$, where B is the area of the base. In a circular cylinder, the circular base has an area of πR^2, so the volume formula can be rewritten as $V = \pi R^2 h$.

CONES AND PYRAMIDS

A *circular cone* is a solid with one circular base that comes to a point at an apex. A *pyramid* is a polyhedron having one polygonal base and triangles for lateral faces. The base of a pyramid can be any polygon, and the lateral faces are all triangles. The height of cones and pyramids is the perpendicular distance from the apex to the base. The slant height is the distance along the lateral surface perpendicular to the perimeter of the base.

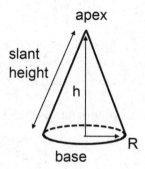

The volume of pyramids and cones are found from the same formula: $V = \frac{1}{3}Bh$ where B is the area of the base.

SPHERES

A *sphere* is the set of points a fixed distance from a center point, and its volume is given by the formula $V = \frac{4}{3}\pi R^3$.

SURFACE AREA AND LATERAL AREA

The *surface area* is the area of all faces of a solid. Surface area can be found by calculating the area of all the faces of a solid individually and then summing them. You should be able to calculate the surface area of cubes, prisms, and pyramids since the faces of these solids are all polygons. Surface area of cones, cylinders, and spheres involve curved surfaces and are outside the scope of this course.

The lateral area of a solid excludes the bases. For a prism, do not include the two parallel bases. For a pyramid, exclude the one base.

Lateral face—any face of a solid other than its bases
Lateral area—the area of all the lateral faces of a solid
Surface area—the total area of a solid, lateral area + area of bases

CROSS-SECTIONS

A cross-section of a solid is the two-dimensional figure created when a plane intercepts a solid. Cross-sections are often taken parallel or perpendicular to the base of a figure. The shape of the cross-section depends on the angle at which the plane intersects the solid.

Cross-sections of a pyramid, sphere, and cylinder are shown in the accompanying figure.

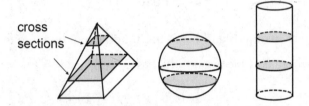

A cross-section of a pyramid taken perpendicular to the base will be shaped triangular, as shown in the figure, or trapezoidal. The cross-section of a cylinder perpendicular to the base is a rectangle.

SOLIDS OF REVOLUTION

Some solids can be generated by rotating a planar figure around a line. These are called *solids of revolution*. Some examples of solids of revolution and the figures that generate them are given in the following table.

Figure Rotated	Solid
Right triangle, rotate 360° about leg $\overline{AB}$	Cone
Rectangle $ABCD$, rotate 360° around side $\overline{CD}$	Cylinder
Circle P, rotate 180° about diameter $\overline{AB}$	Sphere

CAVALIERI'S PRINCIPLE

If two solids are contained between two parallel planes, and every parallel plane between these two planes intercepts regions of equal area, then the solids have equal volume. Also, any two parallel planes intercept two solids of equal volume.

Both the triangular prism and the rectangular prism shown in the accompanying figure are bounded by planes R and T. "Bounded" means that the solids have at least one point contained in each of planes R and T, and that the solids do not pass through to the other sides of the planes. As plane S sweeps upward from R toward T, corresponding cross-sections are formed in the triangular prism and the rectangular prism. If every pair of corresponding cross-sections are equal in area (though not necessarily congruent), then the two solids have the same volume.

The second part of Cavalieri's principle is illustrated by parallel planes R and S. If the conditions of Cavalieri's principle are true, then the two volumes between planes R and S are equal.

Example

Two stacks of quarters are shown in the accompanying figure. If each stack contains 7 identical quarters, what must be true about the volume of the two stacks? Explain your answer in terms of Cavalieri's principle.

Solution:

Each of the quarters has a uniform and congruent cross-section, and any plane parallel to the bases of the two stacks must intercept cross-sections of equal area. Cavalieri's principle states that the volumes of the stacks must be equal.

JUSTIFICATION OF THE AREA AND VOLUME FORMULAS

Circumference of a Circle

The formula for the area of a circle can be derived using a limit argument by considering a polygon inscribed in a circle and letting the number of sides increase toward infinity.

Consider a polygon inscribed in circle O, as shown in the figure. The central angle, $\angle AOB$, has a measure equal to $\dfrac{360°}{n}$, where n is the number of sides. Altitude $\overline{OP}$ will form $\triangle AOP$, where $m\angle AOP$ equals half the central angle, or $\left(\dfrac{360°}{2n}\right)$, and AO has a length equal to the radius of the circle, R. Length AP can be found by applying the sin ratio.

$$\sin(\theta) = \frac{\text{opposite}}{\text{hypotenuse}}$$

$$\sin\left(\frac{360°}{2n}\right) = \frac{AP}{R}$$

$$AP = R\sin\left(\frac{360°}{2n}\right)$$

The perimeter of the polygon is equal to $2n \cdot AP$, since there are n sides and AP is equal to half a side length.

$$\text{Perimeter} = 2n \cdot AP$$

$$= 2nR\sin\left(\frac{360°}{2n}\right)$$

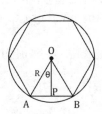

In the limit as the number of sides of the polygon approaches infinity, n approaches infinity, and the perimeter approaches the circumference of the circle. The accompanying table shows the value $n\sin\left(\dfrac{360°}{2n}\right)$ for increasing values of n. As n approaches infinity, $n\sin\left(\dfrac{360°}{2n}\right)$ approaches π.

number of sides n	$n \sin\left(\dfrac{360°}{2n}\right)$
6	3
20	3.1286
100	3.1410
1000	3.1415
Infinite	π

The previous equation for perimeter can be rearranged to

$$\text{Perimeter} = 2n \sin\left(\frac{360°}{2n}\right)R$$

- Substituting circumference for perimeter and π for $n \sin\left(\dfrac{360°}{2n}\right)$ gives the familiar result, circumference $= 2\pi R$

Area of a Circle

The area of a circle can also be derived from a limit argument and inscribed polygons. The formula for the area of a polygon is $A = \frac{1}{2}$ perimeter $\times$ apothem. As the number of sides increases toward infinity, the perimeter approaches the circumference of the circle, and the apothem approaches the radius of the circle.

$$A = \frac{1}{2}\text{perimeter} \times \text{apothem} \rightarrow \frac{1}{2}(2\pi R)R$$

$$A_{\text{circle}} = \frac{1}{2}(2\pi R)R = \pi R^2$$

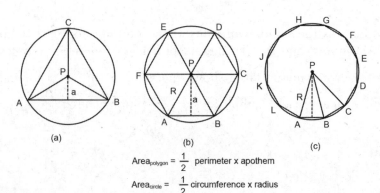

(a)   (b)   (c)

$$\text{Area}_{\text{polygon}} = \frac{1}{2} \ \text{perimeter x apothem}$$

$$\text{Area}_{\text{circle}} = \frac{1}{2} \ \text{circumference x radius}$$

Volume of a Cylinder

The formula for volume of a cylinder can be derived from the formula for the volume of a prism. As the number of sides of the base of the prism increases toward infinity, the base takes the shape of a circle with area πR^2 resulting in the cylinder formula of $\pi R^2 h$.

Volume of a Pyramid

The volume of a square-based pyramid can be derived by *dissection* of a cube with side length s, shown in the accompanying figure (a similar method can be used for pyramids of other bases). Let point P be the center of the cube. The four bottom vertices and point P form a square-based pyramid. Five more congruent pyramids can be formed using point P and four vertices at the top, front, back, left, and right. None of the pyramids overlap, so the volume of the cube must be equal to the volume of the 6 pyramids:

$$6\text{ Volume}_{\text{pyramid}} = \text{Volume}_{\text{cube}} = s^3$$

$$\text{Volume}_{\text{pyramid}} = \frac{1}{6}s^3$$

The height of each pyramid is equal to $\frac{1}{2}s$, or $s = 2h$. Substitute to get

$$\text{Volume}_{\text{pyramid}} = \frac{1}{6}s^3$$
$$= \frac{1}{6}s^2 \cdot s$$
$$= \frac{1}{6}\left(s^2\right)2h$$
$$= \frac{1}{3}\left(s^2\right)h$$

The area of the base of each pyramid is the same as the area of the base of the cube, or s^2, so s^2 can be replaced with B to give:

$$\text{Volume}_{\text{pyramid}} = \frac{1}{3}Bh$$

Volume of a Cone

The same approach that was used for the volume of a cylinder can be applied to derive the formula for the volume of a cone. Starting with a pyramid, whose volume is $\frac{1}{3}Bh$, let the number of sides in the polygonal base approach infinity. The shape of the base will approach a circle, but the height will be unchanged, so the formula remains the same:

$$V_{\text{cone}} = \frac{1}{3}Bh$$

MODELING WITH SOLIDS

Many physical objects can be modeled using the basic solids discussed in this chapter. Some examples are:

Cylinders—trees, cans, barrels, people
Spheres—balls, balloons, planets
Prisms—bricks, boxes, aquariums, swimming pools, rooms, books

Volumes or areas can also be used to calculate some other quantity. Some common relationships to look for are

- Mass = density × volume
- Total cost = cost per unit volume × volume

    or

Total cost = cost per unit area × area
- Energy contained in a material = heat content × volume
- Total population = population density × area

For example, if a can has a volume of 100 cm^3 and is filled with a liquid whose density is 1.2 grams/cm^3, then the mass of the liquid inside the can is

$$\text{Mass} = \text{density} \cdot \text{volume}$$
$$\text{Mass} = 1.2 \text{ grams/cm}^3 \cdot 100 \text{ cm}^3$$
$$= 120 \text{ grams}$$

In geometry design problems, we are asked to find the dimensions of some figure or solid that lets the object meet some given condition. Conditions that may be specified include

- maximum or minimum size
- a specified size
- maximum or minimum cost

Approach these problems in the same way as any other modeling problem. The difference comes in using the model to find a maximum or minimum. For example, you may have an equation that gives volume in terms of a length. A maximum or minimum can be found by

- *Trial and error/table*—Using a best first guess for the variable, evaluate the quantity. Change the value of the variable and evaluate the quantity again—it will increase or decrease. Continue until you see the quantity change from decreasing to increasing (a minimum) or increasing to decreasing (a maximum). The table feature of the graphing calculator is excellent for calculating many trials quickly. The table increment can be set to the desired precision.
- *Graphing*—The equation that described the quantity can be graphed, and the maximum/minimum feature of the calculator can then be applied to the graph.
- *Axis of symmetry*—If the equation that describes the quantity is *quadratic* in the form $y = ax^2 + bx + c$, then the maximum or minimum will always occur at the axis of symmetry $x = -b/2a$.

Practice Exercises

1 A basketball has a diameter of 9.6 inches. What is its volume?

(1) 72.3 in^3 (3) 463 in^3

(2) 347 in^3 (4) 3706 in^3

2 A right circular cylinder has a volume of 200 cm^3. If the height of the cylinder is 6 cm, what is the radius of the cylinder? Round your answer to the nearest hundredth.

(1) 2.9 cm   (3) 3.3 cm

(2) 3.0 cm   (4) 3.8 cm

3 Jenna is preparing to plant a new lawn and has a pile of topsoil delivered to her house. A dump truck delivers the topsoil and dumps it in a pile that is cone shaped. The radius of the pile is 9 ft, and the height is 3 ft. If the topsoil has a density of 95 lb/ft^3, what is the weight of the topsoil?

(1) 18,764 lb   (3) 28,546 lb

(2) 24,175 lb   (4) 32,676 lb

4 A right circular cylinder has a radius of 4 and a height of 10. What is the area of a cross-section taken parallel to, and halfway between, the bases? Write your answer in terms of π.

(1) 4π (3) 12π

(2) 8π (4) 16π

5 A cylinder and a cone have the same volume. If each solid has the same height, what is the ratio of the radius of the cylinder to the radius of the cone?

(1) 2:1   (3) $\sqrt{2}$:2

(2) 3:1   (4) $\sqrt{3}$:3

6 $\triangle JKL$ is a right triangle with a right angle at K, $JK = 6$, and $KL = 10$. Which of the following solids is generated with $\triangle JKL$ rotated 360° about KL?

(1) a cylinder with a height of 10 and a diameter of 12

(2) a cylinder with a height of 6 and a diameter of 20

(3) a right circular cone with a height of 10 and a diameter of 12

(4) a right circular cone with a height of 6 and a diameter of 20

7 Solids A and B, shown in the figure below, each have uniform cross-sections perpendicular to the shaded bases.

If the heights of the two solids are equal, which of the following statements represents an application of Cavalieri's principle?

(1) The surface areas of solid A and solid B may be different even though their volumes are equal.

(2) Solid A can be generated by rotating a rectangle, whereas solid B is not a solid of revolution.

(3) If the area of base A equals the area of base B, then the volumes of the solids are equal.

(4) If the volume and height of solid A and B are equal, then their surface areas must be equal.

8 The base of a cone is described by the curve $x^2 + y^2 = 16$. If the altitude of the cone intersects the center of its base, and the volume of the cone is 72π, what is the height of the cone?

9 A cylindrical piece of metal has a radius of 15 in and a height of 2 in. It goes through a hot-press machine that reduces the height of the cylinder to $\frac{1}{4}$ in. What is the new radius of the cylinder, assuming no material is lost?

10 Jack is making a scaled drawing of the floor plan of his home. The scale factor is 1 in:4 ft. The drawing of his living room is a rectangle measuring 5 inches by 3 inches. He is planning to purchase new carpet for the living room that costs $4 per square foot. How much will the carpet cost?

(1) $720   (3) $960

(2) $840   (4) $1,040

11 The city of Deersfield sits along the bank of the Fox Run River. A map of the city is shown below. The downtown region, indicated by region A, is $\frac{1}{2}$ mile wide and 1 mile long. The population density of Deersfield, except for the downtown region, is 800 people per square mile. The population density of the downtown region is 4,000 people per square mile. What is the population of Deersfield? Round to the nearest whole number.

12 An engineer for a construction company needs to calculate the total volume of a home that is under construction so the proper sized heating and air conditioning equipment can be installed. The house is modeled as a rectangular prism with a triangular prism for the roof as shown in the accompanying figure. The roof is symmetric with $EF = ED$. The dimensions are $AB = 30$ ft, $BC = 50$ ft, $BD = 25$ ft, and the measure of $\angle EDF = 40°$. What is the total volume of the house? Round to the nearest cubic foot.

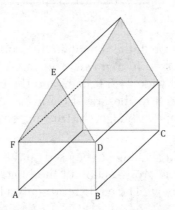

13 The Red Ribbon Orchard sorts their apple harvest by size. The three size categories are shown in the accompanying table.

	Grade B	**Grade A**	**Grade AA**
Average diameter	3 inches	4 inches	5 inches

After one day's harvest, there were 3,000 pounds of grade B apples, 4,200 pounds of grade A apples, and 3,200 pounds of grade AA apples. If the average density of an apple is 0.016 lb/in^3, what is the total number of apples that were harvested?

14 A construction company is working on plans for a project that call for digging a straight 0.25 mile tunnel through a mountain. The cross-section of the tunnel is shown in the accompanying figure.

As the workers dig through the mountain, the rubble is brought to a gravel company to be processed into gravel. The cargo beds of the dump trucks are 18 ft long by 10 ft wide by 6 ft high, and the trucking company charges $450 per load delivered to the gravel company. Approximately how much money should the engineer budget for trucking costs for the project?

15 An engineer is designing a hinge to be used on the landing gear of a new airplane. She sketches the cross-section of the hinge on a computer, which is shown in the figure below. The circle represents a hole for a hinge pin.

The engineer next uses a function on the computer that creates a solid by translating the cross-section in a direction perpendicular to the sketch. The engineer enters a translation distance of 10 units. In the final step, the engineer enters a scale factor of 1 unit = 2 centimeters. The resulting computer image of the solid is then sent to a factory to be manufactured out of a high-strength metal. If the density of the metal is 8.44 grams/cm^3 and the metal costs $32 per kilogram, what is the expected cost for the metal to make one hinge? Round to the nearest dollar.

Solutions

1 The basketball can be modeled as a sphere whose volume is $\frac{4}{3}\pi R^3$. The radius equals half the diameter, or 4.8 in.

$$V = \frac{4}{3}\pi R^3$$

$$= \frac{4}{3}\pi (4.8 \text{ in})^3$$

$$= 463 \text{ in}^3$$

The correct choice is **(3)**.

2 Start with the formula for volume, substitute the known values, and then solve for the radius.

$$V = B \cdot h$$

$$V = \pi R^2 h$$

$$200 = \pi (R^2)(6)$$

$$R^2 = 10.61032$$

$$R = \sqrt{10.61032}$$

$$= 3.3 \text{ cm}$$

The correct choice is **(3)**.

3 To find the weight of the topsoil, we need to first find the volume of the pile. The volume of a cone is $\frac{1}{3}Bh$, where B is the area of the base. This requires finding the area of the circular base.

$$B = \pi R^2$$

$$= \pi (9^2)$$

$$= 254.4690 \text{ ft}^2$$

We are now ready to find the volume.

$$V = \frac{1}{3}B \cdot h$$
$$= \frac{1}{3}(254.4690)(3)$$
$$= 254.469 \text{ ft}^3$$

Now calculate the weight using the volume and density.

$$\text{weight} = V \cdot \text{density}$$
$$= 254.469 \text{ ft}^3 \cdot 95 \text{ lb/ft}^3$$
$$= 24,174.55 \text{ lb}$$
$$= 24,175 \text{ lb}$$

The correct choice is **(2)**.

4 The cross-section will be congruent to the base, so it is a circle with radius of 4. The area is πR^2, or 16π.

The correct choice is **(4)**.

5 The strategy is to set the two volume formulas equal to each other and rearrange, solving for the ratio of the radii.

$$V_{\text{cone}} = \frac{1}{3}\pi R_{\text{cone}}^2$$
$$V_{\text{cylinder}} = \pi R_{\text{cylinder}}^2 h$$
$$V_{\text{cylinder}} = V_{\text{cone}}$$
$$\pi R_{\text{cylinder}}^2 h = \frac{1}{3}\pi R_{\text{cone}}^2 h$$

Now solve for the ratio $\dfrac{R_{\text{cylinder}}}{R_{\text{cone}}}$

$$R^2_{\text{cylinder}} = \frac{1}{3}R^2_{\text{cone}} \quad \text{divide by } \pi h$$

$$\frac{R^2_{\text{cylinder}}}{R^2_{\text{cone}}} = \frac{1}{3} \quad \text{divide by } R^2_{\text{cone}}$$

$$\frac{R_{\text{cylinder}}}{R_{\text{cone}}} = \frac{1}{\sqrt{3}} \quad \text{take square root of each side}$$

The ratio does not match any of the choices, so try rationalizing the denominator. To do so, multiply the numerator and denominator $\sqrt{3}$.

$$\frac{R_{\text{cylinder}}}{R_{\text{cone}}} = \frac{1}{\sqrt{3}} \cdot \frac{\sqrt{3}}{\sqrt{3}}$$

$$\frac{R_{\text{cylinder}}}{R_{\text{cone}}} = \frac{\sqrt{3}}{3}$$

The correct choice is (**4**).

6 A right triangle rotated 360° about one of its legs will generate a cone.

The leg aligned with the axis of rotation becomes the height, so the height equals 10. The leg perpendicular to the axis of rotation becomes the radius, so the radius equals 6 and the diameter equals 12.

The correct choice is (**3**).

7 Cavalieri's principle states that two parallel planes will intercept the same volume in two solids if the cross-sectional areas are uniform and equal. Choice 3 represents this principle.

The correct choice is (**3**).

8 $x^2 + y^2 = 16$ describes a circle centered at the origin with a radius of 4. Its area is

$$A = \pi R^2 = 16\pi$$

Apply the volume formula of the cone to find the height.

$$V = \frac{1}{3}Bh$$

$$72\pi = \frac{1}{3}16\pi h$$

$$h = 72 \cdot 3 \cdot \frac{1}{16}$$

$$= 13.5$$

9 Find the volume of the cylinder before pressing and set it equal to the expression for the volume after pressing. Use this equation to solve for the radius after pressing.

$$V = \pi R^2 h$$

$$V_{before} = \pi \cdot 15^2 \cdot 2$$

$$= 450\pi \text{ in}^3$$

$$V_{after} = \pi \cdot R^2 \cdot \frac{1}{4} \qquad \text{we don't know the new radius}$$

$$V_{before} = V_{after}$$

$$\pi \cdot R^2 \cdot \frac{1}{4} = 450\pi \qquad \text{volumes must be equal}$$

$$\frac{R^2}{4} = 450$$

$$R^2 = 1{,}800$$

$$R = \sqrt{1{,}800}$$

$$= 42.426 \text{ in}$$

10 The actual length and width of the living room are found by applying the scale factor to the drawing dimensions.

$$5 \text{ inches} \cdot \frac{4 \text{ ft}}{1 \text{ in}} = 20 \text{ ft}$$

$$3 \text{ inches} \cdot \frac{4 \text{ ft}}{1 \text{ in}} = 12 \text{ ft}$$

The area of the rectangular living room is

$$A = \text{length} \cdot \text{width}$$
$$= 20 \text{ ft} \cdot 12 \text{ ft}$$
$$= 240 \text{ ft}^2$$

The cost of the carpet is

$$\text{cost} = \text{area} \cdot \text{cost per ft}^2$$
$$\text{cost} = 240 \text{ ft}^2 \cdot \frac{\$4}{\text{ft}^2}$$
$$= \$960$$

The correct choice is **(3)**.

11 We can model the town as a semicircle and the downtown region as a rectangle. To calculate the population, we first find the area of each region. Let the area of the downtown region be represented by A_A and the area of the remainder of the town be represented by A_B.

$$A_A = \text{length} \cdot \text{width} \qquad \text{area of circle}$$
$$= \frac{1}{2} \cdot 1$$
$$= 0.5 \text{ mi}^2$$

The area of the remainder of the town is a semicircle minus the area of the downtown region.

$$A_B = \frac{1}{2}\pi R^2 - 0.5 \qquad \text{area of semicircle circle} - \text{area of rectangle}$$
$$= \frac{1}{2}\pi (2)^2 - 0.5$$
$$= 5.78318 \text{ mi}^2$$

The population of each region is the product of the area and the population density

$$\text{Population} = A_A \cdot 800 \text{ people/mi}^2 + A_B \cdot 4{,}000 \text{ people/mi}^2$$
$$= 5.78318 \text{ mi}^2 \cdot 800 \text{ people/mi}^2 + 0.5 \text{ mi}^2 \cdot 4{,}000 \text{ people/mi}^2$$
$$= 6{,}627 \text{ people.}$$

The population of Deerfield is 6,627 people.

12 Calculate the volume of each section.

Rectangular prism:

$$\text{length} = BC \quad \text{width} = AB \quad \text{height} = BD$$
$$\text{length} = 50 \text{ ft} \quad \text{width} = 30 \text{ ft} \quad \text{height} = 25 \text{ ft}$$

$$\text{Volume}_{\text{rectangular prism}} = \text{length} \cdot \text{width} \cdot \text{height}$$
$$= (50 \text{ ft})(30 \text{ ft})(25 \text{ ft})$$
$$= 37{,}500 \text{ ft}^3$$

Triangular prism:

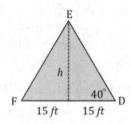

The first step is to find the area of the triangular base, $\triangle FED$. FD has a length equal to AB, or 30 ft. The height of the isosceles triangle divides the base into two 15 ft lengths. The tangent ratio can be used to find the height of $\triangle FED$.

$$\tan = \frac{\text{opposite}}{\text{adjacent}}$$

$$\tan(40°) = \frac{h}{15}$$

$$h = 15 \tan(40°)$$

$$= 12.586 \text{ ft}$$

Now multiply the formula for the area of a triangle.

$$A_{\triangle FED} = 2 \cdot \frac{1}{2} \text{base} \cdot \text{height}$$

$$= 15 \text{ ft} \cdot 12.586 \text{ ft}$$

$$= 188.79 \text{ ft}^2$$

The volume of a triangular prism is equal to Bh, where B is the area of the triangular base and h is the length of the prism, which is equal to BC.

$$V_{\text{triangular prism}} = Bh$$

$$= A_{\triangle FED} \cdot BC$$

$$= 188.79 \text{ ft}^2 \cdot 50 \text{ ft}$$

$$= 9,439.5 \text{ ft}^3$$

Sum the two volumes to get the total volume of the house.

$$\text{Volume} = 37,500 \text{ ft}^3 + 9,439 \text{ ft}^3$$

$$= 46,939.5 \text{ ft}^3$$

$$= 46,940 \text{ ft}^3$$

13 The number of apples of each type equals the total volume of that type divided by the volume of one apple. Model the apples as spheres to calculate the volume of one apple. Find the total volume by dividing the weight by the density. The following table summarizes the calculations.

	Grade B	**Grade A**	**Grade AA**
volume of 1 apple = $V = \frac{4}{3}\pi R^3$	$= \frac{4}{3}\pi \cdot (1.5)^3$ $= 14.1372 \text{ in}^3$	$= \frac{4}{3}\pi \cdot (2)^3$ $= 33.5103 \text{ in}^3$	$= \frac{4}{3}\pi \cdot (2.5)^3$ $= 65.4498 \text{ in}^3$
Total volume of each type = $\frac{\text{weight}}{\text{density}}$	$= \frac{3,000 \text{ lb}}{0.016 \text{ lb/in}^3}$ $= 187,500 \text{ in}^3$	$= \frac{4,200 \text{ lb}}{0.016 \text{ lb/in}^3}$ $= 262,500 \text{ in}^3$	$= \frac{3,000 \text{ lb}}{0.016 \text{ lb/in}^3}$ $= 200,000 \text{ in}^3$
number of each type = $\frac{V_{\text{total}}}{V_{\text{one apple}}}$	$= \frac{187,500}{14.1372}$ $= 13,263$	$= \frac{262,500}{33.5013}$ $= 7,833$	$= \frac{200,000}{65.4498}$ $= 3,056$

The total number of apples is $13{,}263 + 7{,}833 + 3{,}056 = 24{,}152$ apples.

14 First, find the volume of the tunnel, and divide by the volume of the dump truck bed to find the number of dump-truck loads needed. The cross-section of the tunnel is a semicircle, which makes the tunnel in the shape of a half cylinder. The cylinder radius is 10 ft and the length is 0.25 miles.

First, convert 0.25 miles to feet using the conversion factor from the formula sheet.

$$0.25 \text{ mi} \cdot \frac{5,280 \text{ ft}}{\text{mi}} = 1,320 \text{ ft}$$

$$V_{\text{tunnel}} = \frac{1}{2}\pi R^2 h$$

$$= \frac{1}{2}\pi \cdot 10^2 \cdot 1,320$$

$$= 207,345.11 \text{ ft}^3$$

Model the dump-truck bed as a rectangular prism.

$$V_{truck} = \text{length} \cdot \text{width} \cdot \text{height}$$
$$= 18 \cdot 10 \cdot 6 = 1{,}080 \text{ ft}^3$$

The number of dump-truck loads is the ratio of the two volumes.

Number of loads $= V_{tunnel}/V_{truck}$

$$= \frac{207{,}345.11}{1{,}080}$$

$$= 191.98 \text{ dump-truck loads}$$

192 dump-truck loads are needed.

The estimated cost $= \dfrac{\$450}{\text{load}} \cdot 192 \text{ loads} = \$86{,}400$

15 The solid produced will be a prism with the cross-section shown in the figure. The first step is to calculate the area of the cross-section. Divide the cross-section into the 5 regions shown below. The 5th region represents the hole and must be subtracted from the other areas.

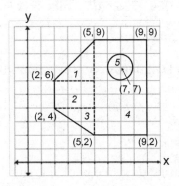

$A_1 = \frac{1}{2}b \cdot h$	$A_2 = l \cdot w$	$A_3 = \frac{1}{2}b \cdot h$	$A_4 = l \cdot w$	$A_5 = \pi R^2$
$= \frac{1}{2} \cdot 3 \cdot 3$	$= 3 \cdot 2$	$= \frac{1}{2} \cdot 3 \cdot 2$	$= 4 \cdot 7$	$= \pi(1)^2$
$= 4.5$	$= 6$	$= 3$	$= 28$	$= 3.14159$

$$A_{\text{drawing}} = A_1 + A_2 + A_3 + A_4 - A_5$$

$$= 4.5 + 6 + 3 + 28 - 3.14159 = 38.35841 \text{ units}^2$$

To find the cross-sectional area of the actual part in centimeters, use the scale factor and the fact that area is proportional to the scale factor *squared*.

$$A_{\text{cross-section}} = 38.35841 \text{ units}^2 \cdot \left(\frac{2 \text{ cm}}{1 \text{ unit}}\right)^2$$

$$= 153.43364 \text{ cm}^2$$

The solid is formed by translating the cross-section 10 units, or 20 cm. Use these dimensions to find the volume of the prism.

$$V = B \times h \qquad \text{where } B \text{ is the cross-sectional area}$$

$$= 153.43364 \text{ cm}^2 \cdot 20 \text{ cm}$$

$$= 3{,}068.6728 \text{ cm}^3$$

Now calculate the mass of the metal

$$= 8.44 \frac{\text{grams}}{\text{cm}^3} \cdot 3{,}068.6728 \text{ cm}^3 \qquad \text{mass} = \text{density} \times \text{volume}$$

$$= 25{,}899.5984 \text{ grams}$$

The price is given per kilogram, so convert grams to kilograms.

$$\text{Mass} = 25{,}899.5984 \text{ grams} \cdot \frac{1 \text{ kg}}{1{,}000 \text{ grams}}$$

$$= 25.899 \text{ kg}$$

Finally, find the cost using the price per kilogram.

$$\text{Cost} = 25.899 \text{ kg} \cdot \$32/\text{kg} \quad \text{total cost} = \text{mass (price per kg)}$$

$$= \$829$$

Glossary of Geometry Terms

Acute angle An angle whose measure is greater than 0° and less than 90°.

Acute triangle A triangle whose angles are all acute angles.

Adjacent angles Two angles that share a common vertex and one side, but do not share interior points.

Altitude (of a triangle) A segment from a vertex of a triangle perpendicular to the opposite side.

Angle A figure formed by two rays with a common endpoint. Symbol is $\angle$.

Angle bisector A line, segment, or ray that divides an angle into two congruent angles.

Angle measure The amount of opening of an angle, measured in degrees or radians.

Angle of depression The angle formed by the horizontal and the line of sight when looking downward to an object.

Angle of elevation The angle formed by the horizontal and the line of sight when looking upward to an object.

Angle of rotation The angle measure by which a figure or point spins around a center point.

Arc A portion of a circle, with two endpoints on the circle. A **major arc** measures more than 180°, a **minor arc** measures less than 180°, and a **semicircle** measures 180°. Symbol $\frown$.

Arc length The distance between the endpoints of an arc. Arc length equals $R\theta$ where θ is the arc measure in radians and R is the radius.

Arc measure The angle measure of an arc equal to the measure of the central angle that intercepts the arc.

Base (of a circular cone) The circular face of a cone; it is opposite the apex.

Base (of an isosceles triangle) The noncongruent side of an isosceles triangle.

Base (of a pyramid) The polygonal face of a pyramid that is opposite the apex.

Bases (of a prism) A pair of faces of a prism that are parallel, congruent polygons.

Bases (of a circular cylinder) The pair of congruent parallel circular faces of a cylinder.

Base angles The angles formed at each end of the base of an isosceles triangle.

Bisector, or segment bisector A line, segment, or ray that passes through the midpoint of a segment.

Biconditional A compound statement in the form "If and only if *hypothesis* then *conclusion*." The hypothesis and conclusion are statements that can be true or false. The biconditional is true when the hypothesis and conclusion have the same truth value.

Cavalieri's principle If two solids are contained between two parallel planes, and every parallel plane between these two planes intercepts regions of equal area, then the solids have equal volume. Also, any two parallel planes intercept two solids of equal volume.

Center of dilation The fixed reference point used to determine the expansion or contraction of lengths in a dilation. The center of dilation is the only invariant point in a dilation.

Center of a regular polygon The center of the inscribed or circumscribed circle of a regular polygon.

Center–radius equation of a circle A circle can be represented on the coordinate plane by $(x - h)^2 + (y - k)^2 = r^2$, where the point (h, k) is the center of the circle and r is the radius.

Central angle An angle in a circle formed by two distinct radii.

Centroid of a triangle The point of concurrency of the three medians of a triangle.

Chord A segment whose endpoints lie on a circle.

Circle The set of points that are a fixed distance from a fixed center point. Symbol $\odot$.

Circumcenter The point that is the center of the circle circumscribed about a polygon, equidistant from the vertices of a polygon, and the point of concurrency of the perpendicular bisectors of a triangle.

Circumference The distance around a circle. Circumference $= 2\pi \cdot$ radius.

Coincide (coincident) Figures that lay entirely one on the other.

Collinear Points that lie on the same line.

Compass A tool for drawing accurate circles and arcs.

Complementary angles Angles whose measures sum to 90°.

Completing the square A method used to rewrite a quadratic expression of the form $ax^2 + bx + c$ as a squared binomial of the form $(dx + e)^2$.

Composition of transformations A sequence of transformations in a specified order.

Concave polygon A polygon with at least one diagonal outside the polygon.

Concentric circles Circles with the same center.

Concurrent When three or more lines all intersect at a single point.

Cone (circular) A solid figure with a single circular base, and a curved lateral face that tapers to a point called the apex.

Congruent Figures with the same size and shape. Symbol $\cong$.

Construction A figure drawn with only a compass and straightedge.

Convex polygon A polygon whose diagonals all lie within the polygon.

Coplanar Figures that lie in the same plane.

Corresponding parts A pair of parts (usually points, sides, or angles) of two figures that are paired together through a specified relationship, such as a congruence or similarity statement or a transformation function.

Cosine of an angle In a right triangle, the ratio of the length of the side adjacent to an acute angle to the length of the hypotenuse.

CPCTC Corresponding parts of congruent triangles are congruent.

Cross-section The intersection of a plane with a solid.

Cube A prism whose faces are all squares.

Cubic unit The amount of space occupied by a cube 1 unit of length in each dimension. (1 ft^3 is the amount of space occupied by a 1 ft $\times$ 1 ft $\times$ 1 ft cube.)

Cylinder The solid figure formed when a rectangle is rotated 360° about one of its sides.

Cylinder (circular) A solid with two parallel, congruent circular bases and a curved lateral face.

Degree A unit of angle measure equal to $\frac{1}{360}$th of a complete rotation.

Diagonal A segment in a polygon whose endpoints are nonconsecutive vertices of the polygon.

Diameter A chord that passes through the center of a circle.

Dilation A similarity transformation about a center point O, which maps pre-image P to image P' such that O, P, and P' are collinear and $OP' = k \cdot OP$. Notation is D_k.

Direct transformation A transformation that preserves orientation.

Distance from a point to a line The length of the segment from the point perpendicular to the line.

Edge of a solid The intersection of two polygonal faces of a solid.

Endpoint The two points that bound a segment.

Equiangular A figure whose angles all have the same measure.

Equidistant The same distance from two or more points.

Equilateral A figure whose sides all have the same length.

Exterior angle of a polygon The angle formed by a side of a polygon and the extension of an adjacent side.

Face of a solid Any of the surfaces that bound a solid.

Figure Any set of points. The points may form a segment, line, ray, plane, polygon, curve, solid, etc.

Glide reflection The composition of a line reflection and a translation along a vector parallel to the line of reflection.

Great circle The largest circle that can be drawn on a sphere. The circle formed by the intersection of a sphere and a plane passing through the center of the sphere.

Hexagon A 6-sided polygon.

Hypotenuse The longest side of a right triangle; it is always opposite the right angle.

Identity transformation A transformation in which the pre-image and image are coincident.

Image The figure that results from applying a transformation to an initial figure called the pre-image.

Incenter of a triangle The point that is the center of the inscribed circle of a triangle, equidistant from the 3 sides of a triangle, and the point of concurrency of the three angle bisectors of a triangle.

Inscribed angle An angle in a circle formed by two chords with a common endpoint.

Inscribed circle of a triangle A circle that is tangent to all three sides of the triangle.

Intersecting Figures that share at least one common point.

Isometry See *Rigid motion*.

Isosceles trapezoid A trapezoid with congruent legs.

Isosceles triangle A triangle with at least two congruent sides.

Kite A polygon with two distinct pairs of adjacent congruent sides. The opposite sides are not congruent.

Lateral edge The intersection between two lateral faces of a polyhedron.

Lateral face Any face of a polyhedron that is not a base.

Line One of the undefined terms in geometry. An infinitely long set of points that has no width or thickness. Symbol ↔.

Linear pair of angles Two adjacent supplementary angles.

Line symmetry A line over which one half of a figure can be reflected and mapped onto the other half of the figure.

Line reflection A rigid motion in which every point P is transformed to a point P' such that the line is the perpendicular bisector of $\overline{PP'}$.

Major arc An arc whose degree measure is greater than 180°.

Map (mapping) A pairing of every point in a pre-image with a point in an image. Reflections, rotations, translations, and dilations are one-to-one mappings because every point in the pre-image maps to exactly one point in the image, and every point in the image maps to exactly one point in the pre-image.

Mean proportional (geometric mean) The square root of the product of two numbers, a and b. If $\dfrac{a}{m} = \dfrac{m}{b}$, then m is the geometric mean.

Median of a triangle The segment from a vertex of a triangle to the midpoint of the opposite side.

Midpoint A point that divides a segment into two congruent segments.

Midsegment (median) of a trapezoid A segment joining the midpoints of the two legs of a trapezoid.

Midsegment of a triangle A segment joining the midpoints of two sides of a triangle.

Minor arc An arc whose degree measure is less than 180°.

Noncollinear Points that do not lie on the same line.

Noncoplanar Points or lines that do not lie on the same plane.

Obtuse angle An angle whose measure is greater than 90° and less than 180°.

Octagon A polygon with 8 sides.

Opposite rays Two rays with a common endpoint that together form a straight line.

Opposite transformation A transformation that changes the orientation of a figure.

Orientation The order in which the points of a figure are encountered when moving around a figure. The orientation can be clockwise or counterclockwise.

Orthocenter The point of concurrence of the three altitudes of a triangle.

Parallel lines Coplanar lines that do not intersect. Symbol //.

Parallelogram A quadrilateral with two pairs of opposite parallel sides.

Parallel planes Planes that do not intersect.

Pentagon A polygon with 5 sides.

Perimeter The sum of the lengths of the sides of a polygon.

Perpendicular Intersecting at right angles. Symbol ⊥.

Perpendicular bisector A line, segment, or ray perpendicular to another segment at its midpoint.

Pi The ratio of the circumference of a circle to its diameter. Pi is an irrational number whose value is approximately 3.14159. Symbol π.

Plane One of the undefined terms in geometry. A set of points with no thickness that extends infinitely in all directions. It is usually visualized as a flat surface.

Point An undefined term in geometry. A location in space with no length, width, or thickness. Symbol •.

Point of tangency The point where a tangent intersects a curve.

Point reflection A transformation in which a specified center point is the midpoint of each of the segments connecting any point in the pre-image with its corresponding point in the image. It is equivalent to a rotation of 180°.

Point-slope equation of a line A form of the equation of a line, $y - y_1 = m(x - x_1)$, where m is the slope and (x_1, y_1) are the coordinates of a point on the line.

Polygon A closed planar figure whose sides are segments that intersect only at their endpoints (do not overlap).

Polyhedron A solid figure in which each face is a polygon. Plural *polyhedra*. Prisms and pyramids are examples of polyhedra.

Postulate A statement that is accepted to be true without proof.

Pre-image The original figure that is acted on by a transformation.

Preserves Remains unchanged.

Prism A polyhedron with two congruent, parallel, polygons for bases, and whose lateral faces are parallelograms.

Proportion An equation that states two ratios are equal. For example, $\frac{a}{b} = \frac{c}{d}$. The cross-products of a proportion are equal, $a \cdot d = b \cdot c$.

Pyramid A polyhedron having one polygonal base and triangles for lateral faces.

Pythagorean theorem In a right triangle, the sum of the squares of the two legs equals the square of the hypotenuse, or $a^2 + b^2 = c^2$, where a and b are the lengths of the two legs and c is the length of the hypotenuse.

Quadratic equation An equation in the form $ax^2 + bx + c = 0$, where $a \neq 0$.

Quadratic formula A formula for finding the two solutions to a quadratic equation of the form $ax^2 + bx + c = 0$, $x = \dfrac{-b \pm \sqrt{b^2 - 4ac}}{2a}$.

Quadrilateral A polygon with four sides.

Radian An angle measure in which one full rotation is 2π radians. Also, 1 radian is the measure of an arc such that the arc's length is equal to the radius of that circle.

Radius A segment from the center of a circle to a point on the circle.

Ray A portion of a line starting at an endpoint and including all points on one side of the endpoint. Symbol →.

Rectangle A parallelogram with right angles.

Reflection See *Line reflection* and *Point reflection*.

Reflexive property of equality Any quantity is equal to itself. Also, for figures, any figure is congruent to itself.

Regular polygon A polygon in which all sides are congruent and all angles are congruent.

Rhombus A parallelogram with 4 congruent sides.

Right angle An angle that measures 90°.

Right circular cone A cone with a circular base and whose altitude passes through the center of the base.

Right circular cylinder A cylinder with circular bases and whose altitude passes through the center of the bases.

Right pyramid A pyramid whose faces are isosceles triangles.

Right triangle A triangle that contains a right angle.

Rigid motion (isometry) A transformation that preserves distance. The image and pre-image are congruent under a rigid motion. Translations, reflections, and rotations are isometries.

Rotation A rigid motion in which every point, P, in the pre-image spins by a fixed angle around a center point, C, to point P'. The distance to the center point is preserved.

Rotational symmetry A figure has rotational symmetry if it maps onto itself after a rotation of less than 360°.

Scale drawing A drawing of a figure or object that represents a dilation of the actual figure or object. A drawing of an object in which every length is enlarged or reduced by the same scale factor.

Scale factor The ratio by which a figure is enlarged or reduced by a dilation.

Scalene triangle A triangle in which no sides have the same length.

Secant of a circle A line that intersects a circle in exactly two points.

Sector of a circle A region bounded by a central angle of a circle and the arc it intersects.

Segment A portion of a line bounded by two endpoints.

Semicircle An arc that measures 180°.

Similarity transformation A transformation in which the pre-image and image are similar. A transformation that includes a dilation.

Similar polygons Two polygons with the same shape but not necessarily the same size.

Sine of an angle In a right triangle, the ratio of the length of the side opposite an acute angle to the length of the hypotenuse.

Skew lines Two lines that are not coplanar.

Slant height The distance along a lateral face of a solid from the apex perpendicular to the opposite edge.

Slope A numerical measure of the steepness of a line. In the coordinate plane, the slope of a line equals the change in the y-coordinates divided by the change in the x-coordinates between any two points. The slope of a vertical line is undefined.

Slope-intercept equation of a line A form of the equation of a line, $y = mx + b$, where m is the slope and b is the y-intercept.

Solid figure A 3-dimensional figure fully enclosed by surfaces.

Sphere A solid comprised of the set of points in space that are a fixed distance from a center point.

Square A parallelogram with 4 right angles and 4 congruent sides.

Straightedge A ruler with no length marking, used for constructing straight lines.

Substitution property of equality The property that a quantity can be replaced by an equal quantity in an equation.

Subtraction property of equality If the same or equal quantities are subtracted from the same or equal quantities, then the differences are equal.

Supplementary angles Two angles whose measures sum to 180°.

Surface area The sum of the areas of all the faces or curved surfaces of a solid figure.

Tangent of an angle In a right triangle, the ratio of the length of the side opposite an acute angle to the length of the side adjacent to the angle.

Tangent to a circle A line, coplanar with a circle, that intersects the circle at only one point.

Theorem A general statement that can be proven.

Transformation A one-to-one function that maps a set of points, called the pre-image, to a new set of points, called the image.

Transitive property of congruence The property that states: if figure $A \cong$ figure B and figure $B \cong$ figure C, then figure $A \cong$ figure C.

Transitive property of equality The property that states: if $a = b$ and $b = c$, then $a = c$.

Translation A rigid motion that slides every point in the pre-image in the same direction and by the same distance.

Transversal A line that intersects two or more other lines at different points.

Trapezoid A quadrilateral that has at least one pair of parallel sides.

Undefined terms Terms which cannot be formally defined using previously defined terms. In geometry these traditionally include point, line, and plane.

Vector A quantity that has both magnitude and direction; represented geometrically by a directed line segment. Symbol $\rightarrow$.

Vertex The point of intersection of two consecutive sides of a polygon or the two rays of an angle.

Vertex angle The angles formed by the two congruent sides in an isosceles triangle.

Vertical angles The pairs of opposite angles formed by two intersecting lines.

Volume The amount of space occupied by a solid figure measured in cubic units (in^3, cm^3, etc.). The number of nonoverlapping unit cubes that can fit in the interior of a solid.

Zero product property The property that states: if $a \cdot b = 0$, then either $a = 0$, or $b = 0$, or a and $b = 0$.

Regents Examinations, Answers, and Self-Analysis Charts

Examination
August 2017
Geometry (Common Core)

GEOMETRY REFERENCE SHEET

1 inch = 2.54 centimeters

1 meter = 39.37 inches

1 mile = 5280 feet

1 mile = 1760 yards

1 mile = 1.609 kilometers

1 kilometer = 0.62 mile

1 pound = 16 ounces

1 pound = 0.454 kilogram

1 kilogram = 2.2 pounds

1 ton = 2000 pounds

1 cup = 8 fluid ounces

1 pint = 2 cups

1 quart = 2 pints

1 gallon = 4 quarts

1 gallon = 3.785 liters

1 liter = 0.264 gallon

1 liter = 1000 cubic centimeters

Triangle	$A = \frac{1}{2}bh$
Parallelogram	$A = bh$
Circle	$A = \pi r^2$
Circle	$C = \pi d$ or $C = 2\pi r$

General Prisms	$V = Bh$
Cylinder	$V = \pi r^2 h$
Sphere	$V = \frac{4}{3}\pi r^3$
Cone	$V = \frac{1}{3}\pi r^2 h$
Pyramid	$V = \frac{1}{3}Bh$
Pythagorean Theorem	$a^2 + b^2 = c^2$
Quadratic Formula	$x = \dfrac{-b \pm \sqrt{b^2 - 4ac}}{2a}$
Arithmetic Sequence	$a_n = a_1 + (n-1)d$
Geometric Sequence	$a_n = a_1 r^{n-1}$
Geometric Series	$S_n = \dfrac{a_1 - a_1 r^n}{1 - r}$ where $r \neq 1$
Radians	$1 \text{ radian} = \dfrac{180}{\pi} \text{ degrees}$
Degrees	$1 \text{ degrees} = \dfrac{\pi}{180} \text{ radians}$
Exponential Growth/Decay	$A = A_0 e^{k(t-t_0)} + B_0$

180 Minutes–36 Questions

PART I

Answer all **24** questions in this part. Each correct answer will receive **2 credits.** No partial credit will be allowed. For each statement or question, write in the space provided the numeral preceding the word or expression that best completes the statement or answers the question. **[48 credits]**

1 A two-dimensional cross section is taken of a three-dimensional object. If this cross section is a triangle, what can *not* be the three-dimensional object?

   (1) cone   (3) pyramid

   (2) cylinder   (4) rectangular prism   1 _____

2 The image of $\triangle DEF$ is $\triangle D'E'F'$. Under which transformation will the triangles *not* be congruent?

   (1) a reflection through the origin

   (2) a reflection over the line $y = x$

   (3) a dilation with a scale factor of 1 centered at $(2, 3)$

   (4) a dilation with a scale factor of $\frac{3}{2}$ centered at the origin   2 _____

3 The vertices of square *RSTV* have coordinates $R(-1, 5)$, $S(-3, 1)$, $T(-7, 3)$, and $V(-5, 7)$. What is the perimeter of *RSTV*?

   (1) $\sqrt{20}$ (3) $4\sqrt{20}$

   (2) $\sqrt{40}$ (4) $4\sqrt{40}$ 3 _____

4 In the diagram below of circle O, chord $\overline{CD}$ is parallel to diameter $\overline{AOB}$ and $m\widehat{CD} = 130°$.

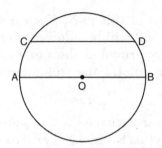

What is $m\widehat{AC}$?

(1) 25°   (3) 65°

(2) 50°   (4) 115°   4 _____

5 In the diagram below, $\overline{AD}$ intersects $\overline{BE}$ at C, and $\overline{AB} // \overline{DE}$.

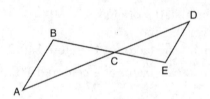

If $CD = 6.6$ cm, $DE = 3.4$ cm, $CE = 4.2$ cm, and $BC = 5.25$ cm, what is the length of $\overline{AC}$, to the *nearest hundredth of a centimeter*?

(1) 2.70   (3) 5.28

(2) 3.34   (4) 8.25   5 _____

6 As shown in the graph below, the quadrilateral is a rectangle.

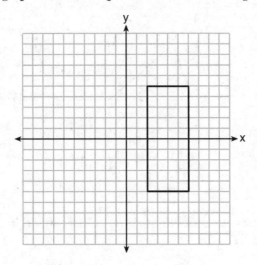

Which transformation would *not* map the rectangle onto itself?

(1) a reflection over the *x*-axis
(2) a reflection over the line $x = 4$
(3) a rotation of 180° about the origin
(4) a rotation of 180° about the point (4, 0)   6 _____

7 In the diagram below, triangle *ACD* has points *B* and *E* on sides
$\overline{AC}$ and $\overline{AD}$, respectively, such that $\overline{BE} \, / /\overline{CD}$, $AB = 1$, $BC = 3.5$,
and $AD = 18$.

What is the length of $\overline{AE}$, to the *nearest tenth*?

(1) 14.0   (3) 3.3
(2) 5.1   (4) 4.0   7 _____

8 In the diagram below of parallelogram *ROCK*, m∠*C* is 70° and
m∠*ROS* is 65°.

What is m∠*KSO*?

(1) 45°   (3) 115°

(2) 110°   (4) 135°   8 _____

9 In the diagram below, ∠*GRS* ≅ ∠*ART*, *GR* = 36, *SR* = 45,
AR = 15, and *RT* = 18.

Which triangle similarity statement is correct?

(1) △*GRS* ~ △*ART* by AA.

(2) △*GRS* ~ △*ART* by SAS.

(3) △*GRS* ~ △*ART* by SSS.

(4) △*GRS* is not similar to △*ART*.   9 _____

10 The line represented by the equation $4y = 3x + 7$ is transformed
by a dilation centered at the origin. Which linear equation could
represent its image?

(1) $3x - 4y = 9$ (3) $4x - 3y = 9$

(2) $3x + 4y = 9$ (4) $4x + 3y = 9$ 10 _____

11 Given △ABC with m∠B = 62° and side $\overline{AC}$ extended to D, as shown below.

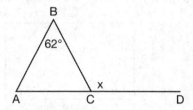

What value of x makes $\overline{AB} \cong \overline{CB}$?

(1) 59°   (3) 118°

(2) 62°   (4) 121°   11 _____

12 In the diagram shown below, $\overline{PA}$ is tangent to circle T at A, and secant $\overline{PBC}$ is drawn where point B is on circle T.

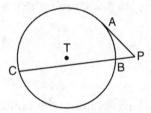

If PB = 3 and BC = 15, what is the length of $\overline{PA}$?

(1) $3\sqrt{5}$ (3) 3

(2) $3\sqrt{6}$ (4) 9   12 _____

13 A rectangle whose length and width are 10 and 6, respectively, is shown below. The rectangle is continuously rotated around a straight line to form an object whose volume is 150π.

6

10

Which line could the rectangle be rotated around?

(1) a long side
(2) a short side
(3) the vertical line of symmetry
(4) the horizontal line of symmetry   13 _____

14 If *ABCD* is a parallelogram, which statement would prove that *ABCD* is a rhombus?

(1) $\angle ABC \cong \angle CDA$ (3) $\overline{AC} \perp \overline{BD}$

(2) $\overline{AC} \cong \overline{BD}$ (4) $\overline{AB} \perp \overline{CD}$ 14 _____

15 To build a handicapped-access ramp, the building code states that for every 1 inch of vertical rise in height, the ramp must extend out 12 inches horizontally, as shown in the diagram below.

x°

12

1

What is the angle of inclination, *x*, of this ramp, to the *nearest hundredth of a degree*?

(1) 4.76   (3) 85.22

(2) 4.78   (4) 85.24   15 _____

16 In the diagram below of $\triangle ABC$, D, E, and F are the midpoints of $\overline{AB}$, $\overline{BC}$, and $\overline{CA}$, respectively.

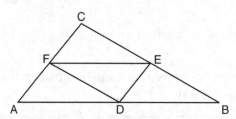

What is the ratio of the area of $\triangle CFE$ to the area of $\triangle CAB$?

(1) $1:1$ (3) $1:3$

(2) $1:2$ (4) $1:4$ 16 _____

17 The coordinates of the endpoints of $\overline{AB}$ are $A(-8, -2)$ and $B(16, 6)$. Point P is on $\overline{AB}$. What are the coordinates of point P, such that $AP : PB$ is $3 : 5$?

(1) $(1, 1)$ (3) $(9.6, 3.6)$

(2) $(7, 3)$ (4) $(6.4, 2.8)$ 17 _____

18 Kirstie is testing values that would make triangle KLM a right triangle when $\overline{LN}$ is an altitude, and $KM = 16$, as shown below.

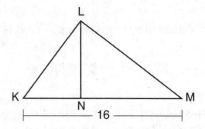

Which lengths would make triangle KLM a right triangle?

(1) $LM = 13$ and $KN = 6$ (3) $KL = 11$ and $KN = 7$

(2) $LM = 12$ and $NM = 9$ (4) $LN = 8$ and $NM = 10$ 18 _____

19 In right triangle ABC, m$\angle A = 32°$, m$\angle B = 90°$, and $AC = 6.2$ cm. What is the length of $\overline{BC}$, to the *nearest tenth of a centimeter*?

   (1) 3.3   (3) 5.3

   (2) 3.9   (4) 11.7   19 _____

20 The 2010 U.S. Census populations and population densities are shown in the table below.

State	Population Density $\left(\frac{\text{people}}{\text{mi}^2}\right)$	Population in 2010
Florida	350.6	18,801,310
Illinois	231.1	12,830,632
New York	411.2	19,378,102
Pennsylvania	283.9	12,702,379

Based on the table above, which list has the states' areas, in square miles, in order from largest to smallest?

   (1) Illinois, Florida, New York, Pennsylvania
   (2) New York, Florida, Illinois, Pennsylvania
   (3) New York, Florida, Pennsylvania, Illinois
   (4) Pennsylvania, New York, Florida, Illinois   20 _____

21 In a right triangle, $\sin(40 - x)° = \cos(3x)°$. What is the value of x?

   (1) 10   (3) 20

   (2) 15   (4) 25   21 _____

22 A regular decagon is rotated n degrees about its center, carrying the decagon onto itself. The value of n could be

   (1) 10°   (3) 225°

   (2) 150°   (4) 252°   22 _____

23 In a circle with a diameter of 32, the area of a sector is $\dfrac{512\pi}{3}$. The measure of the angle of the sector, in radians, is

(1) $\dfrac{\pi}{3}$

(3) $\dfrac{16\pi}{3}$

(2) $\dfrac{4\pi}{3}$

(4) $\dfrac{64\pi}{3}$

23 _____

24 What is an equation of the perpendicular bisector of the line segment shown in the diagram below?

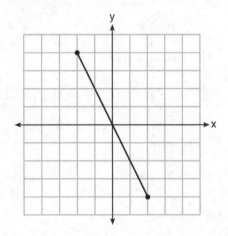

(1) $y + 2x = 0$

(3) $2y + x = 0$

(2) $y - 2x = 0$

(4) $2y - x = 0$

24 _____

PART II

Answer all 7 questions in this part. Each correct answer will receive 2 credits. Clearly indicate the necessary steps, including appropriate formula substitutions, diagrams, graphs, charts, etc. For all questions in this part, a correct numerical answer with no work shown will receive only 1 credit.   [14 credits]

25 Sue believes that the two cylinders shown in the diagram below have equal volumes.

Is Sue correct? Explain why.

26 In the diagram of rhombus *PQRS* below, the diagonals $\overline{PR}$ and $\overline{QS}$ intersect
at point *T*, *PR* = 16, and *QS* = 30. Determine and state the perimeter of
PQRS.

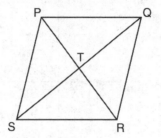

27 Quadrilateral *MATH* and its image *M″A″T″H″* are graphed on the set of axes below.

Describe a sequence of transformations that maps quadrilateral *MATH* onto quadrilateral *M″A″T″H″*.

28 Using a compass and straightedge, construct a regular hexagon inscribed in circle O.

[Leave all construction marks.]

29 The coordinates of the endpoints of $\overline{AB}$ are $A(2, 3)$ and $B(5, -1)$.
    Determine the length of $\overline{A'B'}$, the image of $\overline{AB}$, after a dilation of $\frac{1}{2}$ centered at the origin.

    [The use of the set of axes below is optional.]

30 In the diagram below of $\triangle ABC$ and $\triangle XYZ$, a sequence of rigid motions maps $\angle A$ onto $\angle X$, $\angle C$ onto $\angle Z$, and $\overline{AC}$ onto $\overline{XZ}$.

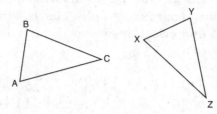

Determine and state whether $\overline{BC} \cong \overline{YZ}$. Explain why.

31 Determine and state the coordinates of the center and the length of the radius of a circle whose equation is $x^2 + y^2 - 6x = 56 - 8y$.

PART III

Answer all 3 questions in this part. Each correct answer will receive 4 credits. Clearly indicate the necessary steps, including appropriate formula substitutions, diagrams, graphs, charts, etc. For all questions in this part, a correct numerical answer with no work shown will receive only 1 credit.   [12 credits]

32 Triangle PQR has vertices $P(-3, -1)$, $Q(-1, 7)$, and $R(3, 3)$, and points A and B are midpoints of $\overline{PQ}$ and $\overline{RQ}$, respectively. Use coordinate geometry to prove that $\overline{AB}$ is parallel to $\overline{PR}$ and is half the length of $\overline{PR}$.

[The use of the set of axes below is optional.]

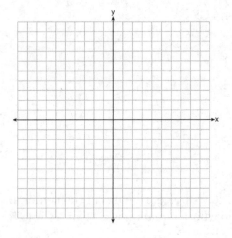

33 In the diagram below of circle O, tangent $\overleftrightarrow{EC}$ is drawn to diameter $\overline{AC}$. Chord $\overline{BC}$ is parallel to secant $\overrightarrow{ADE}$, and chord $\overline{AB}$ is drawn.

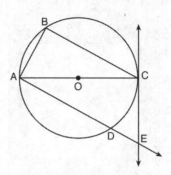

$$\text{Prove: } \frac{BC}{CA} = \frac{AB}{EC}$$

34 Keira has a square poster that she is framing and placing on her wall. The poster has a diagonal 58 cm long and fits exactly inside the frame. The width of the frame around the picture is 4 cm.

Determine and state the total area of the poster and frame to the *nearest tenth of a square centimeter.*

PART IV

Answer the 2 questions in this part. Each correct answer will receive 6 credits. Clearly indicate the necessary steps, including appropriate formula substitutions, diagrams, graphs, charts, etc. For all questions in this part, a correct numerical answer with no work shown will receive only 1 credit.   [12 credits]

35 Isosceles trapezoid $ABCD$ has bases $\overline{DC}$ and $\overline{AB}$ with nonparallel legs $\overline{AD}$ and $\overline{BC}$. Segments AE, BE, CE, and DE are drawn in trapezoid $ABCD$ such that $\angle CDE \cong \angle DCE$, $\overline{AE} \perp \overline{DE}$, and $\overline{BE} \perp \overline{CE}$.

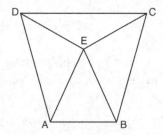

Prove $\triangle ADE \cong \triangle BCE$ and prove $\triangle AEB$ is an isosceles triangle.

36 A rectangular in-ground pool is modeled by the prism below. The inside of
the pool is 16 feet wide and 35 feet long. The pool has a shallow end and a
deep end, with a sloped floor connecting the two ends. Without water, the
shallow end is 9 feet long and 4.5 feet deep, and the deep end of the pool is
12.5 feet long.

If the sloped floor has an angle of depression of 16.5 degrees, what is the
depth of the pool at the deep end, to the *nearest tenth of a foot*?

Find the volume of the inside of the pool to the *nearest cubic foot*.

A garden hose is used to fill the pool. Water comes out of the hose at a rate
of 10.5 gallons per minute. How much time, to the *nearest hour*, will it take
to fill the pool 6 inches from the top? [1 ft³ = 7.48 gallons]

HOW TO CONVERT YOUR RAW SCORE TO YOUR GEOMETRY REGENTS EXAMINATION SCORE

The conversion chart below can be used to determine your final score on the August 2017 Regents Exam in Geometry. To find your final exam score, locate in the "Raw Score" column the total number of points you scored out of a possible 86. Then locate in the adjacent column to the right the scale score that corresponds to your raw score. The scale score is your final score.

Raw Score	Scale Score	Performance Level	Raw Score	Scale Score	Performance Level	Raw Score	Scale Score	Performance Level
86	100	5	57	79	3	28	59	2
85	99	5	56	79	3	27	58	2
84	97	5	55	78	3	26	57	2
83	96	5	54	78	3	25	55	2
82	95	5	53	77	3	24	54	1
81	94	5	52	77	3	23	53	1
80	93	5	51	76	3	22	51	1
79	92	5	50	76	3	21	50	1
78	91	5	49	75	3	20	48	1
77	90	5	48	75	3	19	47	1
76	90	5	47	74	3	18	45	1
75	89	5	46	74	3	17	43	1
74	88	5	45	73	3	16	42	1
73	87	5	44	73	3	15	40	1
72	87	5	43	72	3	14	38	1
71	86	5	42	72	3	13	36	1
70	86	5	41	71	3	12	34	1
69	86	5	40	70	3	11	32	1
68	85	5	39	70	3	10	30	1
67	84	4	38	69	3	9	28	1
66	83	4	37	68	3	8	25	1
65	83	4	36	67	3	7	23	1
64	82	4	35	66	3	6	20	1
63	82	4	34	66	3	5	18	1
62	82	4	33	65	3	4	15	1
61	81	4	32	64	2	3	12	1
60	81	4	31	63	2	2	8	1
59	80	4	30	61	2	1	5	1
58	80	3	29	60	2	0	0	1

Answers
August 2017
Geometry (Common Core)

Answer Key

PART I

1. 2	**5.** 4	**9.** 4	**13.** 3	**17.** 1	**21.** 4
2. 4	**6.** 3	**10.** 1	**14.** 3	**18.** 2	**22.** 4
3. 3	**7.** 4	**11.** 4	**15.** 1	**19.** 1	**23.** 2
4. 1	**8.** 4	**12.** 2	**16.** 4	**20.** 1	**24.** 4

PART II

25. The heights and cross-sectional areas are equal, so the volumes must be equal.

26. 68

27. A rotation of 180° about the origin followed by a translation of 1 unit up and 1 unit left

28. See detailed solution for the construction.

29. $\frac{5}{2}$

30. Rigid motions preserve angle measure and length, so the triangles are congruent and $\overline{BC} \cong \overline{YZ}$ by CPCTC.

31. Center $(3, -4)$ radius $= 3$

PART III

32. Slopes of $\overline{AB}$ and $\overline{PR}$ are both $\frac{2}{3}$. $AB = \sqrt{13}$ and $PR = 2\sqrt{13}$.

33. Prove $\triangle ABC$ is similar to $\triangle ECA$, then form a proportion from corresponding sides.

34. 2402.2 cm^2

PART IV

35. See the detailed solution for the proof.

36. Depth $= 8.5$ ft, volume $= 3{,}752$ ft^3, time $= 41$ hours

In **PARTS II–IV,** you are required to show how you arrived at your answers. For sample methods of solutions, see the *Answers Explained* section.

Answers and Explanations

PART I

1. The only choice that does not have a triangular cross-section is the cylinder. The triangular cross-sections of the other three choices are shown in the accompanying figure.

   The correct choice is (**2**).

2. Reflections are rigid motions that always result in a congruent image. Dilations maintain the original shape but may change the size if the scale factor is different from 1. A dilation with a scale factor of $\frac{3}{2}$ will result in an image larger than the original.

   The correct choice is (**4**).

3. Use the distance formula and the given coordinates $(-1, 5)$ and $(-3, 1)$ to find the length of side $\overline{RS}$.

$$\begin{aligned} d &= \sqrt{\left(x_2 - x_1\right)^2 + \left(y_2 - y_1\right)^2} \\ &= \sqrt{\left(-3(-1)\right)^2 + (1 - 5)^2} \\ &= \sqrt{(-2)^2 + (-4)^2} \\ &= \sqrt{20} \end{aligned}$$

Since we know the figure is a square with four sides of equal length, we can calculate the perimeter by multiplying the side length by 4. The perimeter is $4\sqrt{20}$.

The correct choice is (**3**).

4. The arcs between parallel chords are congruent, therefore $m\overset{\frown}{AC} = m\overset{\frown}{BD}$ and each can be represented by x. The diameter $\overline{AOB}$ intercepts a semi-circle whose arcs sum to $180°$.

$$\begin{aligned} m\overset{\frown}{AC} + m\overset{\frown}{CD} + m\overset{\frown}{BD} &= 180 \\ x + 130° + x &= 180° \\ 2x + 130° &= 180° \\ 2x &= 50° \\ x &= 25° \end{aligned}$$

The correct choice is (**1**).

5. $\angle A \cong \angle D$ because they are alternate interior angles formed by the parallel segments $\overline{AB}$ and $\overline{DE}$. Also, the vertical angles $\angle ACB$ and $\angle DCE$ are congruent. Therefore, $\triangle ACB$ is similar to $\triangle DCE$ by the AA theorem and corresponding sides lengths are proportional.

$$\frac{CD}{AC} = \frac{CE}{BC}$$

$$\frac{6.6}{AC} = \frac{4.2}{5.25}$$

$$4.2\,AC = 6.6(5.25)$$

$$AC = 8.25$$

The correct choice is **(4)**.

6. A reflection over the x-axis would map the top and bottom halves of the rectangle onto each other. The line $x = 4$ is a vertical line through the center of the rectangle, and it would map the left and right halves onto each other. The point $(4, 0)$ is in the center of the rectangle. A rotation of $180°$ about that center point would also result in the rectangle mapping onto itself. The only choice that does not map the rectangle onto itself is a $180°$ rotation about the origin.

The correct choice is **(3)**.

7. Sketch the two triangles separately as shown to help identify congruent angles and corresponding sides. The parallel segments $\overline{BE}$ and $\overline{CD}$ form congruent alternate interior angles $\angle ABE$ and $\angle ACD$. $\angle A$ is a shared angle and is congruent to itself. Therefore, the two triangles are similar by the AA postulate. Write a proportion using corresponding sides to solve for the length AE.

$$\frac{AC}{AB} = \frac{AD}{AE}$$

$$\frac{4.5}{1} = \frac{18}{AE}$$

$$4.5\,AE = (1)(18)$$

$$AE = 4$$

The correct choice is (**4**).

8. Opposite angles of a parallelogram are congruent, so $m\angle R = m\angle C = 70°$. $\angle KSO$ is an exterior angle of $\triangle RSO$. From the exterior angle theorem we know the measure of the exterior angle is equal to the sum of the two further interior angles.

$$m\angle KSO = m\angle R + m\angle ROS$$
$$= 70° + 65°$$
$$= 135°$$

The correct choice is (**4**).

9. Check if the two pairs of corresponding sides are proportional. Form ratios using the two shorter sides and the two longer sides. If it is a true proportion the cross products will be equal.

$$\frac{AR}{GR} = \frac{TR}{SR}$$

$$\frac{15}{36} = \frac{18}{45}$$

$$15 \cdot 45 = 18 \cdot 36$$

$$675 \neq 638$$

The cross products are not equal, so the two pairs of sides are not proportional. The triangles are not similar.

The correct choice is **(4)**.

10. Dilating a line through the origin will multiply the y-intercept by the scale factor, but leave the slope unchanged. We need to find which choice has the same slope as the given line. Rewrite the given equation in $y = mx + b$ form to identify the slope.

$$4y = 3x + 7$$

$$y = \frac{3}{4}x + \frac{7}{4}$$

The y-intercept is $\frac{7}{4}$ and the slope is $\frac{3}{4}$. Now check the slope of each choice by writing in $y = mx + b$ form. For choice (1) we have

$$3x - 4y = 9$$

$$-4y = -3x + 9$$

$$y = \frac{3}{4}x - \frac{9}{4}$$

Equation (1) also has a slope of $\frac{3}{4}$, so this must be the dilation.

The correct choice is **(1)**.

11. The isosceles triangle theorem states that if $\overline{AB} \cong \overline{BC}$ then the opposite angles must also be congruent. Therefore, $\angle BAC \cong \angle BCA$ and we can represent both angles with the variable y (not x since that variable is already used for another angle!). Apply the angle sum theorem in $\triangle ABC$.

$$m\angle A + m\angle B + m\angle ACB = 180°$$
$$y + 62° + y = 180°$$
$$2y + 62° = 180°$$
$$2y = 180°$$
$$y = 59°$$

$m\angle A$ and $m\angle ACB$ both measure 59°. The angle labeled x is an exterior angle, so its measure is equal to the sum of the two furthest interior angles.

$$x = m\angle A + m\angle B$$
$$x = 59° + 62°$$
$$x = 121°$$

The correct choice is (**4**).

12. The lengths of the secant and tangent are related by the formula

    (secant)^2 = (outside part of tangent) · (entire tangent)

$$PA^2 = PB(PB + PC)$$
$$PA^2 = 3(3 + 15)$$
$$PA^2 = 3(18)$$
$$PA^2 = 54$$
$$PA = \sqrt{54}$$

Since this is not one of the choices we simplify the radical. Look for a factor of 54 that is a perfect square. The desired factors are 9 and 6, where 9 is the perfect square.

$$\sqrt{54} = \sqrt{9 \cdot 6}$$
$$= 3\sqrt{6}$$

The correct choice is **(2)**.

As an alternative to simplifying the radical, you can type the radical into your calculator to find the decimal approximation 7.34846. Check each of the choices to find the one with the matching decimal approximation.

13. Rotating a rectangle around a side will form a cylinder. The length of the side it is rotated around is the height and the other side gives the radius as shown in figures (a) and (b) below. The situation is similar if the rectangle is rotated about a line of symmetry, except that one of the sides gives the diameter instead of the radius as shown in figure (c) below.

(a) rotation about the short side   (b) rotation about the long side   (c) rotation about the line of symmetry

Calculate the volume for each choice using the volume formula $V = \pi R^2 h$, which is provided on your reference sheet.

Choice 1:
$h = 10, R = 6$

$V = \pi R^2 h$

$\quad = \pi \cdot 6^2 \cdot 10$

$\quad = 360\pi$

Choice 3:
$h = 6, R = 5$

$V = \pi R^2 h$

$\quad = \pi \cdot 5^2 \cdot 6$

$\quad = 150\pi$

Choice 2:
$h = 6, R = 10$

$V = \pi R^2 h$

$\quad = \pi \cdot 10^2 \cdot 6$

$\quad = 600\pi$

Choice 4:
$h = 10, R = 3$

$V = \pi R^2 h$

$\quad = \pi \cdot 3^2 \cdot 10$

$\quad = 90\pi$

14. A parallelogram whose diagonals are perpendicular is a rhombus. $\overline{AC} \perp \overline{BD}$ represents the two perpendicular diagonals.

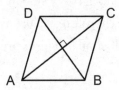

The correct choice is (**3**).

The other possible rhombus properties that could be used are consecutive sides congruent or diagonals bisect the vertex angles. However, none of the choices represents these properties.

15. Find the unknown angle by setting up a trig ratio using an inverse trig function. Relative to the angle x, the opposite side has length 1 and the adjacent side has length 12. These suggest using the tangent.

$$\tan(x) = \frac{\text{opposite}}{\text{adjacent}}$$
$$= \frac{1}{12}$$
$$x = \tan^{-1}\left(\frac{1}{12}\right)$$
$$= 4.76$$

The correct choice is (**1**).

16. The three sides of $\triangle DEF$ are all midsegments because they are formed by joining midpoints of the sides of $\triangle ABC$. A midsegment has a length equal to $\frac{1}{2}$ of the opposite side, making every side in $\triangle DEF$ half the length of a corresponding side in $\triangle ABC$. Therefore, $\triangle DEF$ is the image after a dilation with a scale factor of $\frac{1}{2}$. Areas of similar figures are proportional to the scale factor squared.

$$\text{Ratio of areas} = \left(\frac{1}{2}\right)^2$$

$$= \frac{1}{4}$$

The correct choice is (**4**).

17. Find the x- and y-coordinates of P with the formula

$$\text{ratio} = \frac{x - x_1}{x_2 - x} \qquad \text{ratio} = \frac{y - y_1}{y_2 - y}$$

Using the coordinates $A(-8, -2)$ and $B(16, 6)$, we have $x_1 = -8$, $y_1 = -2$, $x_2 = 16$, and $y_2 = 6$. The ratio is $\frac{3}{5}$.

Solve for the x-coordinate:

$$\text{ratio} = \frac{x - x_1}{x_2 - x}$$

$$\frac{3}{5} = \frac{x - (-8)}{16 - x}$$

$$5(x + 8) = 3(16 - x) \qquad \text{cross-multiply}$$

$$5x + 40 = 48 - 3x$$

$$8x = 8$$

$$x = 1$$

Solve for the y-coordinate:

$$\text{ratio} = \frac{y - y_1}{y_2 - y}$$

$$\frac{3}{5} = \frac{y - (-2)}{6 - y}$$

$$5(y + 2) = 3(6 - y) \qquad \text{cross-multiply}$$

$$5y + 10 = 18 - 3y$$

$$8y = 8$$

$$y = 1$$

The coordinates of P are $(1, 1)$.

The correct choice is (**1**).

18. One possible approach for this problem is to calculate the measures of $\angle K$ and $\angle M$ using the given lengths and assuming $\angle KLM$ is a right angle. If the sum of $\angle K$ and $\angle M$ is 90°, then our assumption was correct and $\angle KLM$ must be a right angle. Using choice (2), we find

$$\cos M = \left(\frac{9}{12}\right) \text{ in } \triangle KLM \quad \sin K = \frac{12}{16} \text{ in } \triangle KLM$$

$$M = \cos^{-1}\frac{9}{12} \qquad\qquad K = \sin^{-1}\frac{12}{16}$$

$$= 41.4° \qquad\qquad\qquad = 48.6°$$

Since $41.4° + 48.6° = 90°$, we know the third angle is 90° as well.

The correct choice is (**2**).

19. Apply a trig ratio to find BC. Relative to $\angle A$, $\overline{AC}$ is the hypotenuse and $\overline{BC}$ is the opposite. We use the sine ratio.

$$\sin(32) = \frac{\text{opp}}{\text{hyp}}$$

$$\sin(32) = \frac{BC}{6.2}$$

$$BC = 6.2\sin(32)$$

$$= 3.3$$

The correct choice is **(1)**.

20. The population density, area, and population are related by the formula

$$\text{population density} = \frac{\text{population}}{\text{area}}$$

Since we need to find the area of four states, it will be easiest to solve for area in terms of the other variables.

$$\text{population density} = \frac{\text{population}}{\text{area}}$$

$$\text{area} \cdot \text{population density} = \text{population}$$

$$\text{area} = \frac{\text{population}}{\text{population density}}$$

Florida

$$\text{area} = \frac{18{,}801{,}310}{350.6}$$

$$= 53{,}626 \text{ miles}^2$$

New York

$$\text{area} = \frac{19{,}378{,}102}{411.2}$$

$$= 47{,}126 \text{ miles}^2$$

Illinois

$$\text{area} = \frac{12{,}830{,}632}{231.1}$$

$$= 55{,}520$$

Pennsylvania

$$\text{area} = \frac{12{,}702{,}379}{283.9}$$

$$= 44{,}742$$

The order of states from largest area to smallest is Illinois, Florida, New York, and Pennsylvania.

The correct choice is **(1)**.

21. The co-function relationship states that if a sine and cosine are equal, then the angles must sum to 90°. Set the sum of the two angles to 90° and then solve for x.

$$40 - x + 3x = 90$$
$$40 + 2x = 90$$
$$2x = 50$$
$$x = 25$$

The correct choice is **(4)**.

22. A rotation of any multiple of $\dfrac{360}{n}$, where n is the number of sides, will map a regular polygon onto itself. A decagon has ten sides, so the angle must be a multiple of $\dfrac{360}{10}$ or 36°. Divide each choice by 36°. The one with an integer quotient is a multiple of 36°.

$$\frac{10}{36} = 0.27 \qquad \frac{225}{36} = 6.25$$

$$\frac{150}{36} = 4.17 \qquad \frac{252}{36} = 7$$

A rotation of 252° will map a decagon onto itself.

The correct choice is **(4)**.

23. The radius of the circle is half the diameter, or 16. Apply the formula for the area of a sector to find the angle.

$$A_{\text{sector}} = \frac{1}{2}R^2 , \text{where } \theta \text{ is the central angle measured in radians}$$

$$\frac{512\pi}{3} = \frac{1}{2} \cdot 16^2$$

$$\frac{1024\pi}{3} = 16^2$$

$$= \frac{1024}{3 . 16^2}$$

$$= \frac{4\pi}{3}$$

The correct choice is **(2)**.

24. The perpendicular bisector passes through the midpoint of the given segment, and has a slope equal to the negative reciprocal of the segment's slope. Find the midpoint using the endpoints $(-2, 4)$ and $(2, -4)$ and the midpoint formula.

$$\text{midpoint} = \left(\frac{x_1 + x_2}{2}, \frac{y_1 + y_2}{2} \right)$$

$$= \left(\frac{-2 + 2}{2}, \frac{4 + (-4)}{2} \right)$$

$$= (0, 0)$$

Next find the slope of the segment.

$$\text{slope} = \frac{y_2 - y_1}{x_2 - x_1}$$

$$= \frac{-4 - 4}{2 - (-2)}$$

$$= \frac{-8}{4}$$

$$= -2$$

The negative reciprocal of the slope is $\frac{1}{2}$. Using the point–slope form of a line with a slope of $\frac{1}{2}$ and the point $(0, 0)$ we have

$$y - y_1 = m(x - x_1)$$

$$y - 0 = \frac{1}{2}(x - 0)$$

$$y = \frac{1}{2}x$$

$$y - \frac{1}{2}x = 0$$

$$2y - x = 0 \qquad \text{multiply both sides by 2 to}$$
$$\text{eliminate the fraction}$$

The correct choice is (**4**).

PART II

25. The volumes are the same. Each cylinder has a radius of 5 meters, so the cross-sectional areas are equal. Also they each have a height of 11.5 meters. Cavalieri's principle states that if two solids have equal cross-sectional areas and heights, then their volumes are equal. *Note that calculating the volume of each cylinder would earn only 1 of the 2 possible credits.*

26. The diagonals of a rhombus, like all parallelograms, bisect each other. Calculate the lengths of $\overline{TS}$ and $\overline{TR}$ by dividing each diagonal by 2. $TS = 15$ and $TR = 8$. The diagonals of rhombuses are also perpendicular, so $\triangle STR$ is a right triangle. SR can be found using the Pythagorean theorem.

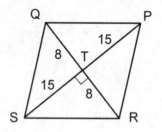

$$a^2 + b^2 = c^2$$
$$15^2 + 8^2 = SR^2$$
$$289 = SR^2$$
$$SR = 17$$

All four sides of a rhombus are congruent, so the perimeter is equal to four times the side length.

$$\text{perimeter} = 4(17)$$
$$= 68$$

27. A rotation of 180° about the origin followed by a translation of 1 unit up and 1 unit left will map *MATH* to *M″A″T″H″*.

28. Mark point *A* anywhere on the circle. Using the same radius as the circle and the point of the compass at *A*, make an arc intersecting the circle at *B*. With the same compass opening, place the point of the compass at *B* and make an arc at *C*. Continue making arcs intersecting at *D*, *E*, and *F*. *ABCDEF* is a regular hexagon.

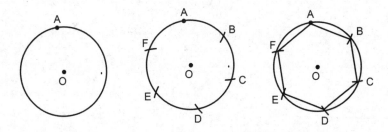

29. The length of $\overline{AB}$ can be calculated using the distance formula and coordinates $A(2, 3)$ and $B(5, -1)$.

$$d = \sqrt{(x_2 - x_1)^2 + (y_2 - y_1)^2}$$
$$= \sqrt{(5 - 2)^2 + (-1 - 3)^2}$$
$$= \sqrt{25}$$
$$= 5$$

A dilation of $\frac{1}{2}$ will multiply the length of the segment by $\frac{1}{2}$, so the length of the image is $\frac{5}{2}$.

30. Rigid motions preserve angle measure and length, so $\angle A \cong \angle X$, $\angle C \cong \angle Z$, and $\overline{AC} \cong \overline{XZ}$. The triangles are therefore congruent by the ASA postulate. $\overline{BC} \cong \overline{YZ}$ because corresponding parts of congruent triangles are congruent.

31. Use completing the square to rewrite the equation in the form $(x - h)^2 + (y - k)^2 = r^2$. First group the x-terms and the y-terms on the left, and move the constant term to the right.

$$x^2 + y^2 - 6x = 56 - 8y$$
$$x^2 - 6x + y^2 + 8y = 56$$

Next find the constant terms needed to complete the square:

constant for the x – terms	constant for the y – terms
$\left(\dfrac{1}{2} \text{ coefficient of } x\right)^2$	$\left(\dfrac{1}{2} \text{ coefficient of } y\right)^2$
$= \left(\dfrac{1}{2} \cdot (-6)\right)^2$	$= \left(\dfrac{1}{2} \cdot 8\right)^2$
$= (-3)^2$	$= (4)^2$
$= 9$	$= 16$

Add the required constants to each side of the equation and factor the left-hand side.

$x^2 - 6x + 9 + y^2 + 8y + 16 = 56 + 9 + 16$ add 9 and 16 to each side

$x^2 - 6x + 9 + y^2 + 8y + 16 = 81$ simplify

$\qquad (x - 3)^2 + (y + 4)^2 = 9$ factor

The values of h and k are 3 and -4. Be careful with the signs because h and k are the values subtracted from x and y. The center is located at $(3, -4)$. Since $r^2 = 9$, the radius is equal to 3.

center: $(\mathbf{3}, -\mathbf{4})$ radius: $(\mathbf{3})$.

PART III

32. Find the coordinates of the midpoints A and B using the midpoint formula.

midpoint of $P(-3,-1)Q(-1,7)$

$$\text{midpoint} = \left(\frac{x_1 + x_2}{2}, \frac{y_1 + y_2}{2}\right)$$

$$= \left(\frac{-3 + (-1)}{2}, \frac{-1 + 7}{2}\right)$$

$$= (-2, 3)$$

midpoint of $Q(-1,7)R(3,3)$

$$\text{midpoint} = \left(\frac{x_1 + x_2}{2}, \frac{y_1 + y_2}{2}\right)$$

$$= \left(\frac{-1 + 3}{2}, \frac{7 + 3}{2}\right)$$

$$= (1, 5)$$

Next find the length and slope of segments $\overline{AB}$ and $\overline{PR}$.

$$\text{slope of } \overline{AB} = \frac{y_2 - y_1}{x_2 - x_1}$$

$$= \frac{5 - 3}{1 - (-2)}$$

$$= \frac{2}{3}$$

$$\text{slope of } \overline{PR} = \frac{y_2 - y_1}{x_2 - x_1}$$

$$= \frac{3 - (-1)}{3 - (-3)}$$

$$= \frac{4}{6}$$

$$= \frac{2}{3}$$

$$AB = \sqrt{\left(x_2 - x_1\right)^2 + \left(y_2 - y_1\right)^2}$$
$$= \sqrt{\left(1 - (-2)\right)^2 + (5 - 3)^2}$$
$$= \sqrt{9 + 4}$$
$$= \sqrt{13}$$

$$PR = \sqrt{\left(x_2 - x_1\right)^2 + \left(y_2 - y_1\right)^2}$$
$$= \sqrt{\left(3 - (-3)\right)^2 + \left(3 - (-1)\right)^2}$$
$$= \sqrt{36 + 16}$$
$$= \sqrt{52}$$
$$= \sqrt{4 \cdot 13}$$
$$= 2\sqrt{13}$$

$\overline{AB}$ is parallel to $\overline{PR}$ because they have the same slope. $\overline{AB}$ is half the length of $\overline{PR}$ because $\sqrt{13}$ is one-half of $2\sqrt{13}$.

33. To prove two of the pairs of sides are proportional we can prove $\triangle ABC$ is similar to $\triangle ECA$. $\angle B$ is a right angle because it is an inscribed angle that intercepts a diameter. $\angle ECA$ is a right angle because it is formed by a tangent and radius. Therefore, $\angle B \cong \angle ECA$. We also have $\angle BCA \cong \angle ECA$ because they are the alternate interior angles formed by the given parallel lines. Since two pairs of angles are congruent, $\triangle ABC$ is similar to $\triangle ECA$ by the AA postulate. $\dfrac{BC}{CA} = \dfrac{AB}{EC}$ because the corresponding parts of similar triangles are proportional.

34. Let s represent the side length of the poster. The 58 centimeter diagonal forms a right isosceles triangle with side length s. Use the Pythagorean theorem to solve for s.

$$a^2 + b^2 = c^2$$
$$s^2 + s^2 = 58^2$$
$$2s^2 = 3,364$$
$$s^2 = 1,682$$
$$s^2 = \sqrt{1,682}$$
$$s = 41.012$$

The length of the frame L can be found by adding 4 centimeters to each side of the poster.

$$L = 41.012 + 4 + 4$$
$$= \mathbf{49.012}$$

The area of the square frame is found using the area formula for a square.

$$\text{Area} = L^2$$
$$= 49.012^2$$
$$= 2402.2195$$
$$= \mathbf{2402.2 \ cm^2}$$

PART IV

35. The strategy for this proof is to show △DEC is isosceles. The congruent legs $\overline{DE}$ and $\overline{EC}$ can be used to prove △ADE and △BCE are congruent. CPCTC can then be used to show △AEB is isosceles.

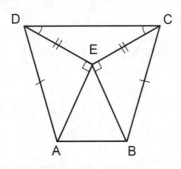

Statement	Reason
1. ∠CDE ≅ ∠DEC	1. Given
2. $\overline{DE} ≅ \overline{EC}$	2. Sides of a triangle opposite congruent angles are congruent
3. $\overline{AE}$ and $\overline{DE}$, $\overline{BE}$ and $\overline{CE}$	3. Given
4. AED and BEC are right angles	4. Perpendicular lines form right angles
5. △AED and △BEC are right triangles	5. A triangle with a right angle is a right triangle
6. Isosceles trapezoid ABCD	6. Trigonometric Expressions and Equations
7. $\overline{AD} ≅ \overline{BC}$	7. Legs of an isosceles trapezoid are congruent
8. △AED ≅ △BEC	8. HL
9. $\overline{AE} ≅ \overline{BE}$	9. CPCTC
10. △AEB is isosceles	10. A triangle with two congruent sides is isosceles

36. The pool is composed of three rectangular prisms and a triangular prism as shown in the figure below. The dimensions x and y are needed in the triangular prism in order to calculate its base area. The dimension y is also needed to determine the height of prism D.

$$x = 35 - 9 - 12.5$$
$$= 13.5$$

Apply the tangent ratio to find the dimension x.

$$\tan(16.5) = \frac{y}{13.5}$$
$$y = 13.5\tan(16.5)$$
$$= 3.9988$$

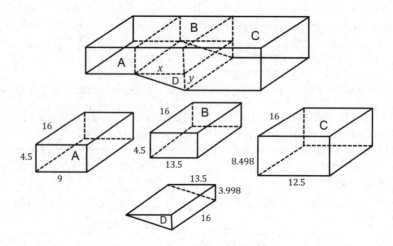

Now calculate the volume of each prism.

<table>
<tr><td align="center">Prism A</td><td align="center">Prism C</td></tr>
</table>

Prism A

$$V = Bh$$
$$= (9 \cdot 16)4.5$$
$$= 648 \text{ ft}^3$$

Prism C

$$V = Bh$$
$$= \left(\frac{1}{2}13.5 \cdot 3.9988\right)16$$
$$= 431.8704 \text{ ft}^3$$

Prism B

$$V = Bh$$
$$= (13.5 \cdot 16)4.5$$
$$= 972 \text{ ft}^3$$

Prism D

$$V = Bh$$
$$= (12.5 \cdot 16)(4.5 + 3.9988)$$
$$= 1699.76 \text{ ft}^3$$

The total volume of the pool is

$$648 + 972 + 431.8704 + 1699.76 = 3751.6304.$$

Rounded to the nearest cubic foot, the volume is 3752 ft^3.

To find how long it will take to fill the pool 6 inches from the top we need to subtract the volume of a thin 6-inch-high rectangular prism from the total volume. This volume is taken from the very top of the pool, so its dimensions are 16 feet $\times$ 35 feet $\times$ 6 inches. Convert the 6 inches to feet before calculating the volume.

$$6 \text{ inches} \cdot \frac{1 \text{ foot}}{12 \text{ inches}} = 0.5 \text{ foot}$$

$$\text{The volume of the top layer} = B\,h$$
$$= (16.35)\,0.5$$
$$= 280 \text{ ft}^3$$

$$\text{The volume to be filled} = 3752 \text{ ft}^3 - 280 \text{ ft}^3$$
$$= 3472 \text{ ft}^3$$

Next, convert the volume from cubic feet to gallons using the conversion ratio provided in the problem.

$$\text{Number of gallons} = 3472 \text{ ft}^3 \cdot \frac{7.48 \text{ gallons}}{\text{ft}^3}$$
$$= 25970.56 \text{ gallons}$$

Now calculate the time required to fill the pool using the fill rate $\frac{10.5 \text{ gallons}}{\text{minute}}$. Since the gallons per minute ratio has gallons in the numerator, we need to *divide* by this ratio.

$$\text{time} = 25970.56 \text{ gallons} \div \frac{10.5 \text{ gallons}}{\text{minute}}$$
$$= 25970.56 \cdot \frac{1 \text{ minute}}{10.5 \text{ gallons}}$$
$$= 2473.38 \text{ minutes}$$

Finally, convert the time in minutes to time in hours using the ratio $\frac{1 \text{ hour}}{60 \text{ minutes}}$

$$\text{time} = 2473.38 \text{ minutes} \cdot \frac{1 \text{ hour}}{60 \text{ minutes}}$$
$$= 41.22 \text{ hours}$$

Rounded to the nearest hour, it will take 41 hours to fill the pool.

Examination
June 2018
Geometry

GEOMETRY REFERENCE SHEET

1 inch = 2.54 centimeters	1 ton = 2000 pounds
1 meter = 39.37 inches	1 cup = 8 fluid ounces
1 mile = 5280 feet	1 pint = 2 cups
1 mile = 1760 yards	1 quart = 2 pints
1 mile = 1.609 kilometers	1 gallon = 4 quarts
1 kilometer = 0.62 mile	1 gallon = 3.785 liters
1 pound = 16 ounces	1 liter = 0.264 gallon
1 pound = 0.454 kilogram	1 liter = 1000 cubic centimeters
1 kilogram = 2.2 pounds	

Triangle	$A = \frac{1}{2}bh$
Parallelogram	$A = bh$
Circle	$A = \pi r^2$
Circle	$C = \pi d$ or $C = 2\pi r$

General Prisms	$V = Bh$
Cylinder	$V = \pi r^2 h$
Sphere	$V = \frac{4}{3}\pi r^3$
Cone	$V = \frac{1}{3}\pi r^2 h$
Pyramid	$V = \frac{1}{3}Bh$
Pythagorean Theorem	$a^2 + b^2 = c^2$
Quadratic Formula	$x = \dfrac{-b \pm \sqrt{b^2 - 4ac}}{2a}$
Arithmetic Sequence	$a_n = a_1 + (n-1)d$
Geometric Sequence	$a_n = a_1 r^{n-1}$
Geometric Series	$S_n = \dfrac{a_1 - a_1 r^n}{1 - r}$ where $r \neq 1$
Radians	$1 \text{ radian} = \dfrac{180}{\pi} \text{ degrees}$
Degrees	$1 \text{ degree} = \dfrac{\pi}{180} \text{ radians}$
Exponential Growth/Decay	$A = A_0 e^{k(t-t_0)} + B_0$

180 Minutes–35 Questions

PART I

Answer all 24 questions in this part. Each correct answer will receive 2 credits. No partial credit will be allowed. For each statement or question, write in the space provided the numeral preceding the word or expression that best completes the statement or answers the question.   [48 credits]

1 After a counterclockwise rotation about point X, scalene triangle ABC maps onto $\triangle RST$, as shown in the diagram below.

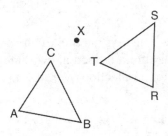

Which statement must be true?

(1) $\angle A \cong \angle R$

(2) $\angle A \cong \angle S$

(3) $\overline{CB} \cong \overline{TR}$

(4) $\overline{CA} \cong \overline{TS}$ 1 _____

2 In the diagram below, $\overline{AB} \parallel \overline{DEF}$, $\overline{AE}$ and $\overline{BD}$ intersect at C, $m\angle B = 43°$, and $m\angle CEF = 152°$.

Which statement is true?

(1) $m\angle D = 28°$ (3) $m\angle ACD = 71°$

(2) $m\angle A = 43°$ (4) $m\angle BCE = 109°$ 2 _____

3 In the diagram below, line m is parallel to line n. Figure 2 is the image of Figure 1 after a reflection over line m. Figure 3 is the image of Figure 2 after a reflection over line n.

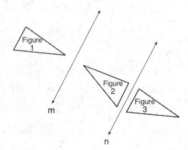

Which single transformation would carry Figure 1 onto Figure 3?

(1) a dilation   (3) a reflection

(2) a rotation   (4) a translation   3 _____

4 In the diagram below, $\overline{AF}$ and $\overline{DB}$ intersect at C, and $\overline{AD}$ and $\overline{FBE}$ are drawn such that m∠D = 65°, m∠CBE = 115°, DC = 7.2, AC = 9.6, and FC = 21.6.

What is the length of $\overline{CB}$?

(1) 3.2   (3) 16.2

(2) 4.8   (4) 19.2   4 _____

5 Given square RSTV, where RS = 9 cm. If square RSTV is dilated by a scale factor of 3 about a given center, what is the perimeter, in centimeters, of the image of RSTV after the dilation?

(1) 12   (3) 36

(2) 27   (4) 108   5 _____

6 In right triangle ABC, hypotenuse $\overline{AB}$ has a length of 26 cm, and side $\overline{BC}$ has a length of 17.6 cm. What is the measure of angle B, to the *nearest degree*?

(1) 48°   (3) 43°

(2) 47°   (4) 34°   6 _____

7 The greenhouse pictured below can be modeled as a rectangular
   prism with a half-cylinder on top. The rectangular prism is 20 feet
   wide, 12 feet high, and 45 feet long. The half-cylinder has a diameter of 20 feet.

   To the *nearest cubic foot*, what is the volume of the greenhouse?

   (1) 17,869   (3) 39,074

   (2) 24,937   (4) 67,349   7 _____

8 In a right triangle, the acute angles have the relationship
   $\sin(2x + 4) = \cos(46)$.

   What is the value of x?

   (1) 20   (3) 24

   (2) 21   (4) 25   8 _____

9 In the diagram below, $\overline{AB} \parallel \overline{DFC}$, $\overline{EDA} \parallel \overline{CBG}$, and $\overline{EFB}$ and $\overline{AG}$ are drawn.

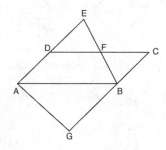

Which statement is always true?

(1) $\triangle DEF \cong \triangle CBF$

(3) $\triangle BAG \sim \triangle AEB$

(2) $\triangle BAG \cong \triangle BAE$

(4) $\triangle DEF \sim \triangle AEB$ 9 _____

10 The base of a pyramid is a rectangle with a width of 4.6 cm and a length of 9 cm. What is the height, in centimeters, of the pyramid if its volume is 82.8 cm^3?

(1) 6

(3) 9

(2) 2

(4) 18   10 _____

11 In the diagram below of right triangle AED, $\overline{BC} \parallel \overline{DE}$.

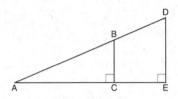

Which statement is always true?

(1) $\dfrac{AC}{BC} = \dfrac{DE}{AE}$ (3) $\dfrac{AC}{CE} = \dfrac{BC}{DE}$

(2) $\dfrac{AB}{AD} = \dfrac{BC}{DE}$ (4) $\dfrac{DE}{BC} = \dfrac{DB}{AB}$ 11____

12 What is an equation of the line that passes through the point $(6, 8)$ and is perpendicular to a line with equation $y = \dfrac{3}{2}x + 5$?

(1) $y - 8 = \dfrac{3}{2}(x - 6)$ (3) $y + 8 = \dfrac{3}{2}(x + 6)$

(2) $y - 8 = -\dfrac{2}{3}(x - 6)$ (4) $y + 8 = -\dfrac{2}{3}(x + 6)$ 12____

13 The diagram below shows parallelogram $ABCD$ with diagonals $\overline{AC}$ and $\overline{BD}$ intersecting at E.

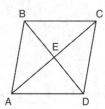

What additional information is sufficient to prove that parallelogram $ABCD$ is also a rhombus?

(1) $\overline{BD}$ bisects $\overline{AC}$.

(2) $\overline{AB}$ is parallel to $\overline{CD}$.

(3) $\overline{AC}$ is congruent to $\overline{BD}$.

(4) $\overline{AC}$ is perpendicular to $\overline{BD}$.   13 _____

14 Directed line segment DE has endpoints $D(-4, -2)$ and $E(1, 8)$. Point F divides $\overline{DE}$ such that $DF : FE$ is $2 : 3$. What are the coordinates of F?

(1) $(-3, 0)$ (3) $(-1, 4)$

(2) $(-2, 2)$ (4) $(2, 4)$ 14 _____

15 Triangle *DAN* is graphed on the set of axes below. The vertices of
△*DAN* have coordinates $D(-6, -1)$, $A(6, 3)$, and $N(-3, 10)$.

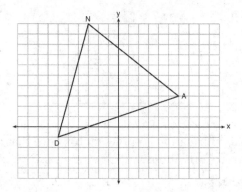

What is the area of △*DAN*?

(1) 60

(2) 120

(3) $20\sqrt{13}$

(4) $40\sqrt{13}$

15 _____

16 Triangle ABC, with vertices at $A(0, 0)$, $B(3, 5)$, and $C(0, 5)$, is
graphed on the set of axes shown below.

Which figure is formed when $\triangle ABC$ is rotated continuously
about $\overline{BC}$?

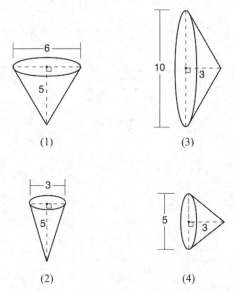

(1)   (3)

(2)   (4)

16 _____

17 In the diagram below of circle O, chords $\overline{AB}$ and $\overline{CD}$ intersect at E.

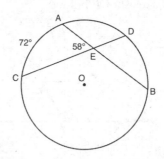

If m $\overarc{AC}$ = 72° and m∠AEC = 58°, how many degrees are in m $\overarc{DB}$?

(1) 108°   (3) 44°

(2) 65°   (4) 14°   17 _____

18 In triangle SRK below, medians $\overline{SC}$, $\overline{KE}$, and $\overline{RL}$ intersect at M.

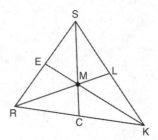

Which statement must always be true?

(1) $3(MC) = SC$ (3) $RM = 2MC$

(2) $MC = \frac{1}{3}(SM)$ (4) $SM = KM$ 18 _____

19 The regular polygon below is rotated about its center.

   Which angle of rotation will carry the figure onto itself?

   (1) 60°   (3) 216°

   (2) 108°   (4) 540°   19 _____

20 What is an equation of circle O shown in the graph below?

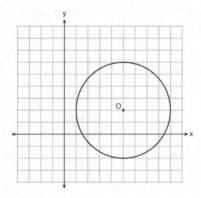

   (1) $x^2 + 10x + y^2 + 4y = -13$
   (2) $x^2 - 10x + y^2 - 4y = -13$
   (3) $x^2 + 10x + y^2 + 4y = -25$
   (4) $x^2 - 10x + y^2 - 4y = -25$ 20 _____

21 In the diagram below of $\triangle PQR$, $\overline{ST}$ is drawn parallel to $\overline{PR}$, $PS = 2$, $SQ = 5$, and $TR = 5$.

What is the length of $\overline{QR}$?

(1) 7

(3) $12\frac{1}{2}$

(2) 2

(4) $17\frac{1}{2}$ 21 _____

22 The diagram below shows circle O with radii $\overline{OA}$ and $\overline{OB}$. The measure of angle AOB is 120°, and the length of a radius is 6 inches.

Which expression represents the length of arc AB, in inches?

(1) $\frac{120}{360}(6\pi)$

(3) $\frac{1}{3}(36\pi)$

(2) $120(6)$

(4) $\frac{1}{3}(12\pi)$ 22 _____

23 Line segment CD is the altitude drawn to hypotenuse $\overline{EF}$ in right triangle ECF. If $EC = 10$ and $EF = 24$, then, to the *nearest tenth*, the length of ED is

(1) 4.2   (3) 15.5

(2) 5.4   (4) 21.8   23 _____

24 Line MN is dilated by a scale factor of 2 centered at the point $(0, 6)$. If $\overleftrightarrow{MN}$ is represented by $y = -3x + 6$, which equation can represent $\overleftrightarrow{M'N'}$, the image of $\overleftrightarrow{MN}$?

(1) $y = -3x + 12$ (3) $y = -6x + 12$

(2) $y = -3x + 6$ (4) $y = -6x + 6$ 24 _____

PART II

Answer all 7 questions in this part. Each correct answer will receive 2 credits. Clearly indicate the necessary steps, including appropriate formula substitutions, diagrams, graphs, charts, etc. For all questions in this part, a correct numerical answer with no work shown will receive only 1 credit. [14 credits]

25 Triangle $A'B'C'$ is the image of triangle ABC after a translation of 2 units to the right and 3 units up. Is triangle ABC congruent to triangle $A'B'C'$? Explain why.

26 Triangle *ABC* and point *D*(1, 2) are graphed on the set of axes below.

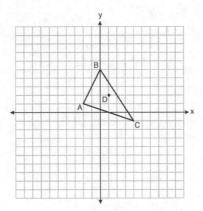

Graph and label △*A'B'C'*, the image of △*ABC*, after a dilation of scale factor 2 centered at point *D*.

27 Quadrilaterals *BIKE* and *GOLF* are graphed on the set of axes below.

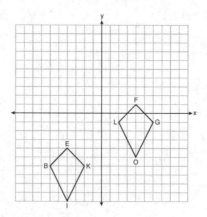

Describe a sequence of transformations that maps quadrilateral *BIKE* onto quadrilateral *GOLF*.

28 In the diagram below, secants $\overline{RST}$ and $\overline{RQP}$, drawn from point R, intersect circle O at S, T, Q, and P.

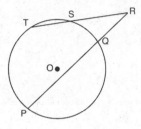

If $RS = 6$, $ST = 4$, and $RP = 15$, what is the length of $\overline{RQ}$?

29 Using a compass and straightedge, construct the median to side $\overline{AC}$ in $\triangle ABC$ below. [Leave all construction marks.]

30 Skye says that the two triangles below are congruent. Margaret says that the two triangles are similar.

Are Skye and Margaret both correct? Explain why.

31 Randy's basketball is in the shape of a sphere with a maximum circumference of 29.5 inches. Determine and state the volume of the basketball, to the *nearest cubic inch*.

PART III

Answer all 3 questions in this part. Each correct answer will receive 4 credits. Clearly indicate the necessary steps, including appropriate formula substitutions, diagrams, graphs, charts, etc. For all questions in this part, a correct numerical answer with no work shown will receive only 1 credit.   [12 credits]

32 Triangle ABC has vertices with coordinates $A(-1, -1)$, $B(4, 0)$, and $C(0, 4)$. Prove that $\triangle ABC$ is an isosceles triangle but *not* an equilateral triangle. [The use of the set of axes below is optional.]

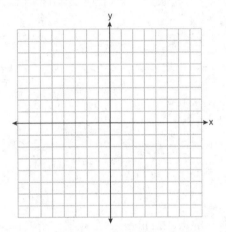

33 The map of a campground is shown below. Campsite C, first aid station F, and supply station S lie along a straight path. The path from the supply station to the tower, T, is perpendicular to the path from the supply station to the campsite. The length of path $\overline{FS}$ is 400 feet. The angle formed by path $\overline{TF}$ and path $\overline{FS}$ is 72°. The angle formed by path $\overline{TC}$ and path $\overline{CS}$ is 55°.

Determine and state, to the *nearest foot*, the distance from the campsite to the tower.

34 Shae has recently begun kickboxing and purchased training equipment as
    modeled in the diagram below. The total weight of the bag, pole, and
    unfilled base is 270 pounds. The cylindrical base is 18 inches tall with a
    diameter of 20 inches. The dry sand used to fill the base weighs 95.46 lbs
    per cubic foot.

To the *nearest pound*, determine and state the total weight of the training
equipment if the base is filled to 85% of its capacity.

PART IV

Answer the question in this part. A correct answer will receive 6 credits. Clearly indicate the necessary steps, including appropriate formula substitutions, diagrams, graphs, charts, etc. A correct numerical answer with no work shown will receive only 1 credit. [6 credits]

35 Given: Parallelogram $ABCD$, $\overline{BF} \perp \overline{AFD}$, and $\overline{DE} \perp \overline{BEC}$

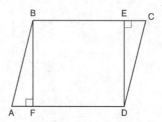

Prove: $BEDF$ is a rectangle

Statements	Reasons
1. Parallelogram $ABCD$, $\overline{BF} \perp \overline{AFD}$, $\overline{DE} \perp \overline{BEC}$	1. Given
2. $\overline{BEC} /\!/ \overline{AFD}$	2. Opposite sides of a parallelogram are parallel
3. $\overline{BE} /\!/ \overline{FD}$	3. Segments contained in parallel lines are parallel
4. $\overline{DE} \perp \overline{AFD}$	4. A segment perpendicular to one of two parallel segments is also perpendicular to the other
5. $\overline{BF} /\!/ \overline{DE}$	5. Two segments perpendicular to the same segment are parallel
6. $BEDF$ is a parallelogram	6. A quadrilateral with opposite sides parallel is a parallelogram
7. $\angle BFD$ is a right angle	7. Perpendicular segments intersect at right angles
8. $BEDF$ is a rectangle	8. A parallelogram with a right angle is a rectangle

Topic	Question Numbers	Number of Points	Your Points	Your Percentage
1. Basic Angle and Segment Relationships	2	2		
2. Angle and Segment Relationships in Triangles and Polygons	18	2		
3. Constructions	29	2		
4. Transformations	1, 3, 19, 25, 26, 27	2 + 2 + 2 + 2 + 2 + 2 = 12		
5. Triangle Congruence	30	2		
6. Lines, Segments and Circles on the Coordinate Plane	12, 14, 15, 20, 24	2 + 2 + 2 + 2 + 2 = 10		
7. Similarity	4, 5, 9, 11, 21, 23	2 + 2 + 2 + 2 + 2 + 2 = 12		
8. Trigonometry	6, 8, 33	2 + 2 + 4 = 8		
9. Parallelograms	13, 35	2 + 6 = 8		
10. Coordinate Geometry Proofs	32	4		
11. Volume and Solids	7, 10, 16, 31	2 + 2 + 2 + 2 = 8		
12. Modeling	34	4		
13. Circles	17, 22, 28	2 + 2 + 2 = 6		

HOW TO CONVERT YOUR RAW SCORE TO YOUR GEOMETRY REGENTS EXAMINATION SCORE

The conversion chart below can be used to determine your final score on the June 2018 Regents Exam in Geometry. To find your final exam score, locate in the "Raw Score" column the total number of points you scored out of a possible 86. Then locate in the adjacent column to the right the scale score that corresponds to your raw score. The scale score is your final score.

Raw Score	Scale Score	Performance Level	Raw Score	Scale Score	Performance Level	Raw Score	Scale Score	Performance Level
80	100	5	53	80	4	26	61	2
79	99	5	52	80	4	25	60	2
78	98	5	51	79	3	24	59	2
77	97	5	50	79	3	23	58	2
76	96	5	49	78	3	22	56	2
75	95	5	48	78	3	21	55	2
74	94	5	47	77	3	20	53	1
73	94	5	46	77	3	19	51	1
72	93	5	45	76	3	18	50	1
71	92	5	44	76	3	17	48	1
70	91	5	43	75	3	16	46	1
69	90	5	42	74	3	15	44	1
68	90	5	41	74	3	14	42	1
67	89	5	40	73	3	13	40	1
66	88	5	39	73	3	12	38	1
65	88	5	38	72	3	11	35	1
64	87	5	37	71	3	10	33	1
63	86	5	36	71	3	9	30	1
62	86	5	35	70	3	8	28	1
61	85	5	34	69	3	7	25	1
60	84	4	33	68	3	6	22	1
59	84	4	32	67	3	5	19	1
58	83	4	31	67	3	4	15	1
57	82	4	30	66	3	3	12	1
56	82	4	29	65	3	2	8	1
55	81	4	28	64	2	1	4	1
54	81	4	27	63	2	0	0	1

Answers
June 2018
Geometry

Answer Key

PART I

1. 1	**5.** 4	**9.** 4	**13.** 4	**17.** 3	**21.** 4
2. 3	**6.** 2	**10.** 1	**14.** 2	**18.** 1	**22.** 4
3. 4	**7.** 1	**11.** 2	**15.** 1	**19.** 3	**23.** 1
4. 3	**8.** 1	**12.** 2	**16.** 3	**20.** 2	**24.** 2

PART II

25. $\overline{AB} \cong \overline{A'B'}$, $\overline{BC} \cong \overline{B'C'}$, and so $\overline{CD} \cong \overline{C'D'}$ the triangles are congruent by SSS.

26. $A'(-5, 0)$ $B'(-1, 8)$ $C'(7, -4)$

27. A reflection over the y-axis followed by a translation 5 units up

28. $RQ = 4$

29. Construct the midpoint of side $\overline{AC}$. Connect this midpoint to point B to form the median.

PART III

30. Use the Pythagorean theorem to show the triangles both have side lengths of 5, 12, and 13. The triangles are congruent by SSS and similar with a scale factor of 1. Skye and Meghan are both correct.

31. Volume = 434 in^3

32. Use the distance formula to show $AB = \sqrt{26}$, $BC = \sqrt{32}$, and $CA = \sqrt{26}$. Only two sides are congruent so the triangle is isosceles but not equilateral.

33. The distance from the campsite to the tower is 1503 ft.

34. The total weight is 536 lbs.

PART IV

35. Prove *BEDF* is a parallelogram with a right angle. See the complete solutions for the proof.

In **PARTS II–IV**, you are required to show how you arrived at your answers. For sample methods of solutions, see the *Answers Explained* section.

Answers and Explanations

PART I

1. A rotation is a rigid motion that results in an image congruent to the preimage. All corresponding sides and angles are congruent. By using the given congruence statement $\triangle ABC \cong \triangle RST$, we see that $\angle A$ and $\angle R$ are corresponding congruent angles.

   The correct choice is (**1**).

2. Use the given information to start finding any angle measures by using basic angle relationships. $\angle B$ and $\angle D$ are congruent alternate interior angles formed by the parallel segments, so $m\angle D = 43°$. Since $\angle DEC$ and $\angle FEC$ are a linear pair, they sum to 180°, making $m\angle DEC = 28°$. By applying the angle sum theorem in $\triangle CDE$, we can calculate $m\angle DCE$:

$$m\angle D + m\angle DEC + m\angle DCE = 180°$$
$$43° + 28° + m\angle DCE = 180°$$
$$71° + m\angle DCE = 180°$$
$$m\angle DCE = 109°$$

   $\angle DCE$ and $\angle ACD$ are another linear pair. So they sum to 180°. Calculate $m\angle ACD$:

$$m\angle DCE + m\angle ACD = 180°$$
$$109° + m\angle ACD = 180°$$
$$m\angle ACD = 71°$$

   The correct choice is (**3**).

3. Two reflections over parallel lines are always equivalent to a single translation.

   The correct choice is (**4**).

4. To solve this problem, first show that the two triangles are similar. Then apply the theorem that corresponding sides of similar triangles are proportional. We know $\angle DCA$ and $\angle FCB$ are congruent vertical angles. Next find $m\angle CBF$ using the linear pair relationship:

$$m\angle EBC + m\angle CBF = 180°$$
$$115° + m\angle CBF = 180°$$
$$m\angle CBF = 65°$$

We now know that the two triangles have two pairs of congruent angles. So the triangles are similar by AA. Write a similarity statement by arranging corresponding angles in the same order. A valid similarity statement is $\triangle ADC \sim \triangle FBC$. Now set up a proportion using corresponding sides and cross-multiply:

$$\frac{CB}{CD} = \frac{CF}{AC}$$
$$\frac{CB}{7.2} = \frac{21.6}{9.6}$$
$$(9.6)(CB) = (21.6)(7.2)$$
$$CB = 16.2$$

The correct choice is (**3**).

5. Side lengths are multiplied by the scale factor in a dilation. The dilated square will have side lengths of $9 \times 3 = 27$. The perimeter is equal to the sum of the side lengths. Since a square has four sides, the perimeter is $27 \times 4 = 108$.

The correct choice is (**4**).

6. Sketch the triangle based on the information provided in the question:

Solve for m∠B using an inverse trig ratio. Relative to ∠B, $\overline{BC}$ is the adjacent side and $\overline{AB}$ is the hypotenuse. Adjacent and hypotenuse indicate the cosine ratio, so use the inverse cosine function:

$$\cos B = \frac{\text{adjacent}}{\text{hypotenuse}}$$

$$\cos B = \frac{17.6}{26}$$

$$B = \cos^{-1}\left(\frac{17.6}{26}\right)$$

$$B = 47°$$

The correct choice is (**2**).

7. The volume of the greenhouse can be found by adding the volume of the prism and the volume of the half-cylinder. The diameter of the half-cylinder is 20 feet, which makes its radius 10 feet. The height of the cylinder is 45 feet. Note that the formulas for the volume of a prism and the volume of a cylinder can be found on the reference sheet. Be sure to divide the cylinder formula by 2 to account for the greenhouse having only half a cylinder.

Rectangular prism:

$$\begin{aligned} \text{Volume} &= \text{length} \cdot \text{width} \cdot \text{height} \\ &= 45 \text{ ft} \cdot 20 \text{ ft} \cdot 12 \text{ ft} \\ &= 10{,}800 \text{ ft}^3 \end{aligned}$$

Half-cylinder:

$$\begin{aligned} \text{Volume} &= \frac{1}{2}\pi r^2 h \\ &= \frac{1}{2}\pi\left(10^2\right)(45) \\ &= 7068.58 \text{ ft}^3 \end{aligned}$$

$$\begin{aligned} \text{Total Volume} &= 10{,}800 \text{ ft}^3 + 7068.58 \text{ ft}^3 \\ &= 17{,}868.58 \text{ ft}^3 \end{aligned}$$

To the *nearest cubic foot*, the volume is $17{,}869 \text{ ft}^3$.

The correct choice is (**1**).

8. The trigonometric co-function relationship states that if the sine of an angle is equal to the cosine of another angle, the two angles sum to 90°. In this case, the two angle measures are $(2x + 4)$ and 46, so we set their sum equal to 90:

$$\begin{aligned} 2x + 4 + 46 &= 90 \\ 2x + 50 &= 90 \\ 2x &= 40 \\ x &= 20 \end{aligned}$$

The correct choice is (**1**).

9. When a line is drawn in a triangle parallel to one of the sides, it intercepts a triangle that is similar to the original triangle. In the figure, we see $\triangle ABE$ has $\overline{DFC}$ drawn parallel to side $\overline{AB}$. $\overline{DFC}$ intercepts $\triangle DEF$, making $\triangle DEF$ similar to original triangle $\triangle AEB$.

The correct choice is (**4**).

10. Apply the formula for the volume of a pyramid found on the reference sheet. The area of the rectangular base, represented by B in the formula, is found using area $=$ length $\cdot$ width:

$$\text{Volume} = \frac{1}{3}Bh$$

$$\text{Volume} = \frac{1}{3}(\text{length})(\text{width})h$$

$$82.8 = \frac{1}{3}(9)(4.6)h$$

$$82.8 = (13.8)h$$

$$h = 6$$

The correct choice is (**1**).

11. A line parallel to a side of a triangle will always intercept a smaller triangle that is similar to the original. The two similar triangles are $\triangle ABC$ and $\triangle ADE$. Pairs of corresponding sides are proportional. So the similarity statement $\triangle ABC \sim \triangle ADE$ can be used to write the proportion $\dfrac{AB}{AD} = \dfrac{BC}{DE}$. $\overline{AB}$ and $\overline{AD}$ are formed from the first two vertices of each triangle, while $\overline{BC}$ and $\overline{DE}$ are formed from the second and third vertices of each triangle.

Note that if none of the choices matched using the similarity statement, you could also make valid proportions using the side splitter theorem.

For example, $\dfrac{AB}{BD} = \dfrac{AC}{CE}$ would be a valid proportion.

The correct choice is (**2**).

12. Perpendicular lines have negative reciprocal slopes. The slope of the given line is $\frac{3}{2}$, so the slope of the perpendicular line is $-\frac{2}{3}$. All the choices are in point-slope form, so use that to set up the equation. From the given point $(6, 8)$, use $x_1 = 6$ and $y_1 = 8$:

$$y - y_1 = m\left(x - x_1\right)$$
$$y - 8 = -\frac{2}{3}(x - 6)$$

The correct choice is (**2**).

13. A parallelogram can be proven to be a rhombus if it has one of the special rhombus properties:

- Diagonals are perpendicular.

- Consecutive sides are congruent.

- Diagonals bisect the vertex angles.

Choice (4) states that diagonals $\overline{AC}$ and $\overline{BD}$ are perpendicular to each other. None of the other choices would let us conclude a special property from the list above.

The correct choice is (**4**).

14. Find the x- and y-coordinates of F with the following formulas:

$$\text{ratio} = \frac{x - x_1}{x_2 - x} \text{ and ratio} = \frac{y - y_1}{y_2 - y}$$

Using the coordinates $D(-4, -2)$ and $E(1, 8)$, we have $x_1 = -4$, $y_1 = -2$, $x_2 = 1$, and $y_2 = 8$. The ratio for $DF : FE$ is $2 : 3$ or $\frac{2}{3}$.

Solve for the x-coordinate of point F:

$$\text{ratio} = \frac{x - x_1}{x_2 - x}$$
$$\frac{2}{3} = \frac{x - (-4)}{1 - x}$$
$$3(x + 4) = 2(1 - x) \quad \text{Cross-multiply}$$
$$3x + 12 = 2 - 2x$$
$$5x + 12 = 2$$
$$5x = -10$$
$$x = -2$$

Solve for the y-coordinate of point F:

$$\text{ratio} = \frac{y - y_1}{y_2 - y}$$
$$\frac{2}{3} = \frac{y - (-2)}{8 - y}$$
$$3(y + 2) = 2(8 - y) \quad \text{Cross-multiply}$$
$$3y + 6 = 16 - 2y$$
$$5y + 6 = 16$$
$$5y = 10$$
$$y = 2$$

The coordinates of point F are $(-2, 2)$.

The correct choice is (**2**).

15. The easiest method to find the area of $\triangle DAN$ is to find the area of a bounding rectangle and then subtract those areas that are not part of $\triangle DAN$. Start by sketching a rectangle that bounds $\triangle DAN$ at each vertex. Then divide the bounding rectangle into four regions—three right triangles and $\triangle DAN$.

Triangles I, II, and III are each right triangles whose areas can be found using $A = \frac{1}{2}bh$. The bases and heights of triangles I, II, and III are all horizontal and vertical segments whose lengths can be found by counting units along the x- and y-axes. The area of $\triangle DAN$ is equal to the area of the bounding rectangle minus the areas of triangles I, II, and III.

Bounding Rectangle:

$$l = 12, \quad w = 11$$
$$A = lw$$
$$= (12)(11)$$
$$= 132$$

Triangle I:

$$b = 3, \quad h = 11$$
$$A = \frac{1}{2}bh$$
$$= \frac{1}{2}(3)(11)$$
$$= 16.5$$

Triangle II:

$$b = 9, \quad h = 7$$
$$A = \frac{1}{2}bh$$
$$= \frac{1}{2}(9)(7)$$
$$= 31.5$$

Triangle III:

$$b = 12, \quad h = 4$$
$$A = \frac{1}{2}bh$$
$$= \frac{1}{2}(12)(4)$$
$$= 24$$

Area of $\triangle DAN$ = Area of rectangle $-$ (Area of $\triangle$I $+$ Area of $\triangle$II $+$ Area of $\triangle$III)
$$= 132 - (16.5 + 31.5 + 24)$$
$$= 60$$

The correct choice is (**1**).

16. A right triangle rotated along one of its legs will always sweep out into a cone. The leg it is rotated around, in this case $\overline{BC}$, is the height of the cone. The other leg, $\overline{AC}$, is the radius. Count units vertically to find $\overline{AC} = 5$, and count units horizontally to find $\overline{BC} = 3$. The height of the cone is 3, and the radius is 5. The diameter is twice the radius, so the diameter is 10. Choice (3) shows a cone with a height of 3 and a diameter of 10.

The correct choice is (**3**).

17. The angle formed by two intersecting chords in a circle is always equal to half the sum of the intercepted arcs. The two intercepted arcs are $\overset{\frown}{AC}$ and $\overset{\frown}{BD}$, and the angle formed by the chords is $\angle AEC$:

$$m\angle AEC = \frac{1}{2}\left(m\overset{\frown}{AC} + m\overset{\frown}{BD}\right)$$

$$58° = \frac{1}{2}\left(72° + m\overset{\frown}{BD}\right)$$

$$116° = 72° + m\overset{\frown}{BD}$$

$$m\overset{\frown}{BD} = 44°$$

The correct choice is (**3**).

18. Medians of a triangle intersect at the centroid, which is point M. The centroid divides each median in a 1:2 ratio. The larger part of the median is 2 times the length of the shorter part, and the entire median is 3 times the length of the shorter part. $\overline{MC}$ is the shorter part of median $\overline{SC}$, therefore $SC = 3(MC)$.

The correct choice is (**1**).

19. The minimum number of degrees required to rotate a regular polygon onto itself is given by $\dfrac{360°}{n}$, where n is the number of sides. The polygon shown has 5 sides, so the minimum rotation angle is $\dfrac{360°}{5}$, or 72°. Any multiple of 72° will also rotate the polygon onto itself: 72°, 144°, 216°, 288°, and 360°. The only choice that is a multiple of 72° is 216°.

The correct choice is (**3**).

20. Circle O has its center located at $(5, 2)$. Its radius can be found by counting units horizontally or vertically from point O to the circle. The radius is 4 units. Use the center-radius form of the equation of a circle:

$$(x - h)^2 + (y - k)^2 = r^2$$

In this equation, h and k are the coordinates of the center of the circle and r is the radius. Substitute values to write the equation in center-radius form:

$$(x - 5)^2 + (y - 2)^2 = 4^2$$

None of the choices are in this form. So rewrite the equation by expanding each of the squared terms:

$$(x - 5)^2 = x^2 - 10x + 25$$
$$(y - 2)^2 = y^2 - 4y + 4$$

Now rewrite the expanded equation:

$$x^2 - 10x + 25 + y^2 - 4y + 4 = 16$$

Finally, rewrite the equation with all constant terms on the right:

$$x^2 - 10x + y^2 - 4y + 29 = 16$$
$$x^2 - 10x + y^2 - 4y = -13$$

The correct choice is (**2**).

21. The side splitter theorem states that a line parallel to a side of a triangle divides the other two sides proportionally. The four parts of the sides are QS, SP, QT, and TR. Start by writing a proportion involving these four parts. Then solve for QT:

$$\frac{QS}{SP} = \frac{QT}{TR}$$

$$\frac{5}{2} = \frac{QT}{5}$$

$$(5)(5) = 2QT$$

$$QT = 12\frac{1}{2}$$

Now sum QT and TR to get QR:

$$QR = QT + QR$$

$$= 12\frac{1}{2} + 5$$

$$= 17\frac{1}{2}$$

The correct choice is (**4**).

22. Apply the formula for arc length. The central angle θ is 120°, and the radius r is 6:

$$\text{arc length} = \left(\frac{\theta}{360}\right) 2\pi r$$

$$= \left(\frac{120}{360}\right)(2\pi)(6)$$

$$= \frac{1}{3}(12\pi)$$

The correct choice is (**4**).

23. An altitude to the hypotenuse of a right triangle will divide the triangle into two similar triangles, which are also similar to the original. We can use that theorem to set up a proportion from the corresponding sides of two of the triangles. Sketch the figure and label the known lengths.

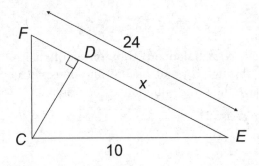

Look at $\triangle CDE$ in the figure. We know the length of hypotenuse $\overline{CE}$ and are looking for the length of longer leg $\overline{ED}$. In $\triangle FCE$, we know the length of both the hypotenuse $\overline{FE}$ and longer leg $\overline{CE}$. Use $\triangle CDE$ and $\triangle FCE$ to set up a proportion and solve for $\overline{ED}$:

$$\frac{\text{hypotenuse}}{\text{long leg}} = \frac{\text{hypotenuse}}{\text{long leg}}$$
$$\frac{CE}{ED} = \frac{FE}{CE}$$
$$\frac{10}{ED} = \frac{24}{10}$$
$$24ED = 100$$
$$ED = 4.1666$$

Rounded to the *nearest tenth*, ED is 4.2.

The correct choice is (**1**).

24. The point $(0, 6)$ lies on line MN. We can confirm this by substituting $x = 0$ and $y = 6$ into the equation of the line:

$$y = -3x + 6$$
$$6 = -3(0) + 6$$
$$6 = 6$$

Since the equation is balanced, the center of dilation lies on the line. Dilating a line through a point on the line does not change the line. The dilation must have the same equation as the original.

The correct choice is (**2**).

PART II

25. A translation is a rigid motion that preserves lengths, so all pairs of corresponding sides are congruent:

$$\overline{AB} \cong \overline{A'B'} \qquad \overline{BC} \cong \overline{B'C'} \qquad \overline{CD} \cong \overline{C'D'}$$

Since all three pairs of sides are congruent, **the two triangles are congruent by the SSS postulate**.

26. A dilation increases the size of the original figure by the scale factor. It also increases the distance of each point from the center. We can use this second feature of a dilation to find the coordinates of A', B', and C'. To find A', count the number of horizontal and vertical units from D to A. Point A is 3 units left and 1 unit down from point D. Since we need a scale factor of 2, double those numbers to get 6 units left and 2 units down. Counting 6 left and 2 down from D puts A' at $(-5, 0)$. Repeat this process to find B' and C'.

D to A: 3 left, 1 down
D to A': 6 left, 2 down

D to B: 1 left, 3 up
D to B': 2 left, 6 up

D to C: 3 right, 3 down
D to C': 6 right, 6 down

The new coordinates are $A'(-5, 0)$, $B'(-1, 8)$, $C'(7, -4)$.

27. A good strategy for any problem that asks you to find a sequence of transformations is first to identify if a reflection is needed. The orientation is the direction, clockwise or counterclockwise, that the vertices are encountered. If the orientation changes, then a reflection is required since only a reflection can change the orientation. *BIKE* has a counterclockwise orientation, while *GOLF* has a clockwise orientation. Therefore, we need a reflection. Each figure is symmetric about a vertical line, so we can reflect about the y-axis. Following the reflection, we see that we just need to translate $B'I'K'E'$ up 5 units to map it onto *GOLF*.

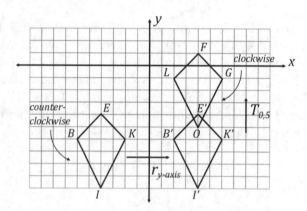

The complete sequence is a **reflection over the y-axis followed by a translation 5 units up, or $T_{0,5} \circ r_{y\text{-axis}}$**.

28. When two secants intersect, the relationship between the parts of the secants can be shown by the following formula:

$$\text{Outside} \cdot \text{Whole} = \text{Outside} \cdot \text{Whole}$$

The outside of each secant is the part that lies outside the circle, in this case $\overline{RS}$ and $\overline{RQ}$. The whole part represents the entire length of the secant, or $\overline{RT}$ and $\overline{RP}$. To find the length of $\overline{RQ}$, apply the formula, using the sum of $RS + ST$ for RT:

$$\text{Outside} \cdot \text{Whole} = \text{Outside} \cdot \text{Whole}$$
$$RS \cdot RT = RQ \cdot RT$$
$$6 \cdot 10 = RQ \cdot 15$$
$$60 = 15RQ$$
$$RQ = 4$$

The length of $\overline{RQ}$ is 4.

29. A median of a triangle is a segment from a vertex to the opposite midpoint. Begin by constructing the midpoint of $\overline{AC}$. With the point of the compass at A, open the compass more than half the distance of $\overline{AC}$ and make an arc. Without changing the compass setting, move the compass point to C. Make another arc that intersects the first arc as shown in figure (a). Connect the two points of intersection using a straightedge. Figure (b) shows this new segment intersecting $\overline{AC}$ at midpoint D. Finally, use the straightedge to draw a segment from point B to midpoint D. This is shown in figure (c). Segment $\overline{BD}$ is the median to side $\overline{AC}$.

(a)

(b)

(c)
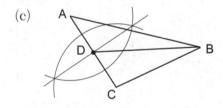

PART III

30. Triangles can be proven to be similar using the AA, SAS, or SSS postulates. In this example, apply the Pythagorean theorem to each triangle to find the missing side. Then check if each pair of corresponding sides are in the same ratio. If they are, then the triangles are similar by SSS.

Left triangle:

$$a^2 + b^2 = c^2$$
$$5^2 + 12^2 = c^2$$
$$25 + 144 = c^2$$
$$169 = c^2$$
$$c = 13$$

Right triangle:

$$a^2 + b^2 = c^2$$
$$5^2 + b^2 = 13^2$$
$$25 + b^2 = 169$$
$$b^2 = 144$$
$$b = 12$$

The three sides of the left triangle are 5, 12, and 13. The three sides of the right triangle are also 5, 12, and 13. All three pairs of sides are in the same ratio:

$$\frac{5}{5} = \frac{12}{12} = \frac{13}{13} = 1$$

Since all three pairs of sides are in the same ratio, **the triangles are similar by SSS. The triangles are also congruent by the SSS postulate because all three pairs of sides are congruent. Therefore, both Skye and Margaret are correct.**

31. The radius of the basketball is needed to find the volume. Use the given circumference to calculate the radius. The circumference formula can be found on the reference sheet:

$$C = 2\pi r$$
$$29.5 = 2\pi r$$
$$r = \frac{29.5}{2\pi}$$
$$= 4.695070821 \text{ in.}$$

Now apply the formula for the volume of a sphere, which is also found on the reference sheet:

$$V = \frac{4}{3}\pi r^3$$
$$V = \frac{4}{3}\pi (4.695070821)^3$$
$$= 433.5259036 \text{ in.}^3$$

The volume of the basketball to the *nearest cubic inch* is **434 in.**^{3}

32. Use the distance formula to find the length of each side of the triangle. If two sides have the same length and the third length is different, we can conclude the triangle is isosceles but not equilateral:

$$d = \sqrt{(x_2 - x_1)^2 + (y_2 - y_1)^2}$$

$$AB = \sqrt{(4 - (-1))^2 + (0 - (-1))^2}$$

$$= \sqrt{5^2 + 1^2}$$

$$= \sqrt{26}$$

$$BC = \sqrt{(0 - 4)^2 + (4 - 0)^2}$$

$$= \sqrt{(-4)^2 + 4^2}$$

$$= \sqrt{32}$$

$$CA = \sqrt{(-1 - 0)^2 + (-1 - 4)^2}$$

$$= \sqrt{(-1)^2 + (-5)^2}$$

$$= \sqrt{26}$$

$\triangle ABC$ **is isosceles because** $AB = CA$. $\triangle ABC$ **is not equilateral because** BC **is not equal to the other two side lengths.**

33. Start by sketching the figure as two separate right triangles.

The distance from the campsite to the tower is side length CT in $\triangle CST$. We can apply a trig ratio to find the length of ST in $\triangle FST$. Then the length of CT can be found using a trig ratio in $\triangle CST$.

Find ST using $\triangle FST$. Relative to $\angle F$, $\overline{ST}$ is the opposite and $\overline{FS}$ is the adjacent. Therefore, use the tangent ratio:

$$\text{tangent} = \frac{\text{opposite}}{\text{adjacent}}$$

$$\tan F = \frac{ST}{FS}$$

$$\tan 72° = \frac{ST}{400}$$

$$ST = (400)(\tan 72°) \quad \text{Cross-multiply}$$

$$ST = 1231.073415$$

Find CT using $\triangle CST$. Relative to $\angle C$, $\overline{ST}$ is the opposite and $\overline{CT}$ is the hypotenuse. Therefore, use the sine ratio:

$$\text{sine} = \frac{\text{opposite}}{\text{hypotenuse}}$$

$$\sin C = \frac{ST}{CT}$$

$$\sin 55° = \frac{1231.073415}{CT}$$

$$(CT)(\sin 55°) = 1231.073415 \quad \text{Cross-multiply}$$

$$CT = \frac{1231.073415}{\sin 55°}$$

$$CT = 1502.863142$$

Rounded to the *nearest foot*, the distance from the campsite to the tower is **1503 feet**.

34. The total weight of the equipment is the given weight of the bag, pole, and unfilled base plus the weight of the sand in the base. Find the weight of the sand by first calculating the volume of the base and by then applying the following relationship:

$$\text{weight} = \text{density} \cdot \text{volume}$$

The given density is in lbs per cubic foot, while the dimensions of the base are in inches. Convert the base's dimensions to feet in order to have consistent units. Remember that 1 foot is equal to 12 inches:

$$\text{height} = 18 \text{ in.} \cdot \frac{1 \text{ ft}}{12 \text{ in.}}$$

$$= 1.5 \text{ ft}$$

$$\text{diameter} = 20 \text{ in.} \cdot \frac{1 \text{ ft}}{12 \text{ in.}}$$

$$= 1.666666667 \text{ ft}$$

Find the volume of the base using the volume formula of a cylinder found on the reference sheet. Remember to divide the diameter by 2 to get the radius:

$$V = \pi r^2 h$$
$$= \pi (0.833333333 \text{ ft})^2 (1.5 \text{ ft})$$
$$= 3.272492347 \text{ ft}^3$$

Calculate the mass of the sand, multiplying by 0.85 to account for the base not being filled completely:

$$\text{weight} = \text{density} \cdot \text{volume} \cdot 0.85$$
$$= 95.46 \; \frac{\text{lbs}}{\text{ft}^3} \cdot 3.272492347 \text{ ft}^3 \cdot 0.85$$
$$= 265.5333015 \text{ lbs}$$

Sum the weight of the sand and the equipment:

$$\text{total weight} = 265.5333015 \text{ lbs} + 270 \text{ lbs}$$
$$= 535.5333015$$

Rounded to the *nearest pound*, the total weight of the training equipment is **536 lbs**.

PART IV

35. A parallelogram can be proven to be a rectangle if it has one of the special properties of rectangles. These properties are right angles (just one right angle need be proven) and congruent diagonals. Since we are given parallel and perpendicular segments, it makes sense to prove *BEDF* is a parallelogram by showing it has opposite sides parallel and then show that *BEDF* has a right angle. The key to showing that opposite sides are parallel is to realize that $\overline{BF}$ and $\overline{ED}$ are both perpendicular to $\overline{FD}$, so they must be parallel to each other.

Examination
August 2018
Geometry

HIGH SCHOOL MATH REFERENCE SHEET

Conversions

1 inch = 2.54 centimeters

1 meter = 39.37 inches

1 mile = 5280 feet

1 mile = 1760 yards

1 mile = 1.609 kilometers

1 kilometer = 0.62 mile

1 pound = 16 ounces

1 pound = 0.454 kilogram

1 kilogram = 2.2 pounds

1 ton = 2000 pounds

1 cup = 8 fluid ounces

1 pint = 2 cups

1 quart = 2 pints

1 gallon = 4 quarts

1 gallon = 3.785 liters

1 liter = 0.264 gallon

1 liter = 1000 cubic centimeters

Formulas

Triangle	$A = \frac{1}{2}bh$
Parallelogram	$A = bh$
Circle	$A = \pi r^2$
Circle	$C = \pi d$ or $C = 2\pi r$

Formulas (continued)

General Prisms	$V = Bh$
Cylinder	$V = \pi r^2 h$
Sphere	$V = \frac{4}{3}\pi r^3$
Cone	$V = \frac{1}{3}\pi r^2 h$
Pyramid	$V = \frac{1}{3}Bh$
Pythagorean Theorem	$a^2 + b^2 = c^2$
Quadratic Formula	$x = \dfrac{-b \pm \sqrt{b^2 - 4ac}}{2a}$
Arithmetic Sequence	$a_n = a_1 + (n-1)d$
Geometric Sequence	$a_n = a_1 r^{n-1}$
Geometric Series	$S_n = \dfrac{a_1 - a_1 r^n}{1 - r}$ where $r \neq 1$
Radians	$1 \text{ radian} = \dfrac{180}{\pi} \text{ degrees}$
Degrees	$1 \text{ degree} = \dfrac{\pi}{180} \text{ radians}$
Exponential Growth/Decay	$A = A_0 e^{k(t - t_0) + B_0}$

180 Minutes–35 Questions

PART I

Answer all 24 questions in this part. Each correct answer will receive 2 credits. No partial credit will be allowed. For each statement or question, write in the space provided, the numeral preceding the word or expression that best completes the statement or answers the question. [48 credits]

1 In the diagram below, $\overline{AEFB} \parallel \overline{CGD}$, $\overline{GE}$, and $\overline{GF}$ are drawn.

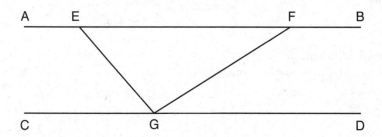

If m∠EFG = 32° and m∠AEG = 137°, what is m∠EGF?

(1) 11°   (3) 75°

(2) 43°   (4) 105°   1 _____.

2 If △ABC is mapped onto △DEF after a line reflection and △DEF is mapped onto △XYZ after a translation, the relationship between △ABC and △XYZ is that they are always

(1) congruent and similar

(2) congruent but not similar

(3) similar but not congruent

(4) neither similar nor congruent   2 _____

3 An isosceles right triangle whose legs measure 6 is continuously rotated about one of its legs to form a three-dimensional object. The three-dimensional object is a

(1) cylinder with a diameter of 6

(2) cylinder with a diameter of 12

(3) cone with a diameter of 6

(4) cone with a diameter of 12   3 _____

4 In regular hexagon *ABCDEF* shown below, $\overline{AD}$, $\overline{BE}$, and $\overline{CF}$ all intersect at *G*.

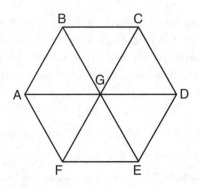

When △*ABG* is reflected over $\overline{BG}$ and then rotated 180° about point *G*, △*ABG* is mapped onto

(1) △*FEG*   (3) △*CBG*

(2) △*AFG*   (4) △*DEG*   4 _____

5 A right cylinder is cut perpendicular to its base.
   The shape of the cross section is a

(1) circle   (3) rectangle

(2) cylinder   (4) triangular prism   5 _____

6 Yolanda is making a springboard to use for gymnastics. She has 8-inch-tall springs and wants to form a 16.5° angle with the base, as modeled in the diagram below.

To the *nearest tenth of an inch*, what will be the length of the springboard, *x*?

(1) 2.3   (3) 27.0

(2) 8.3   (4) 28.2   6 _____

7 In the diagram below of right triangle ABC, altitude $\overline{BD}$ is drawn to hypotenuse $\overline{AC}$.

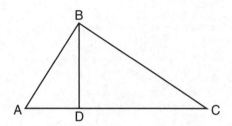

If $BD = 4$, $AD = x - 6$, and $CD = x$, what is the length of $\overline{CD}$?

(1) 5   (3) 8
(2) 2   (4) 11   7____

8 Rhombus $STAR$ has vertices $S(-1, 2)$, $T(2, 3)$, $A(3, 0)$, and $R(0, -1)$. What is the perimeter of rhombus $STAR$?

(1) $\sqrt{34}$ (3) $\sqrt{10}$
(2) $4\sqrt{34}$ (4) $4\sqrt{10}$ 8____

9 In the diagram below of $\triangle HAR$ and $\triangle NTY$, angles H and N are right angles, and $\triangle HAR \sim \triangle NTY$.

If $AR = 13$ and $HR = 12$, what is the measure of angle Y, to the *nearest degree*?

(1) 23°   (3) 65°
(2) 25°   (4) 67°   9____

10 In the diagram below, $\overline{AKS}$, $\overline{NKC}$, $\overline{AN}$, and $\overline{SC}$ are drawn such that $\overline{AN} \cong \overline{SC}$.

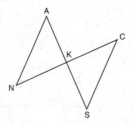

Which additional statement is sufficient to prove $\triangle KAN \cong \triangle KSC$ by AAS?

(1) $\overline{AS}$ and $\overline{NC}$ bisect each other.   (3) $\overline{AS} \perp \overline{CN}$

(2) K is the midpoint of $\overline{NC}$.   (4) $\overline{AN} \parallel \overline{SC}$ 10 _____

11 Which equation represents a line that is perpendicular to the line represented by $y = \frac{2}{3}x + 1$?

(1) $3x + 2y = 12$ (3) $y = \frac{3}{2}x + 2$

(2) $3x - 2y = 12$ (4) $y = -\frac{2}{3}x + 4$ 11 _____

12 In the diagram of $\triangle ABC$ below, points D and E are on sides $\overline{AB}$ and $\overline{CB}$ respectively, such that $\overline{DE} \parallel \overline{AC}$.

If EB is 3 more than DB, $AB = 14$, and $CB = 21$, what is the length of $\overline{AD}$?

(1) 6   (3) 9

(2) 8   (4) 12   12 _____

13 Quadrilateral *MATH* has both pairs of opposite sides congruent and parallel. Which statement about quadrilateral *MATH* is always true?

(1) $\overline{MT} \cong \overline{AH}$

(3) $\angle MHT \cong \angle ATH$

(2) $\overline{MT} \perp \overline{AH}$

(4) $\angle MAT \cong \angle MHT$ 13 _____

14 In the figure shown below, quadrilateral *TAEO* is circumscribed around circle *D*. The midpoint of $\overline{TA}$ is *R*, and $\overline{HO} \cong \overline{PE}$.

If *AP* = 10 and *EO* = 12, what is the perimeter of quadrilateral *TAEO*?

(1) 56

(3) 72

(2) 64

(4) 76   14 _____

15 The coordinates of the endpoints of directed line segment *ABC* are *A*(−8, 7) and *C*(7, −13). If *AB:BC* = 3:2, the coordinates of *B* are

(1) (1, −5)

(3) (−3, 0)

(2) (−2, −1)

(4) (3, −6)   15 _____

16 In triangle ABC, points D and E are on sides $\overline{AB}$ and $\overline{BC}$, respectively, such that $\overline{DE} \parallel \overline{AC}$, and $AD{:}DB = 3{:}5$.

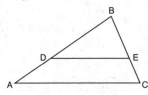

If $DB = 6.3$ and $AC = 9.4$, what is the length of $\overline{DE}$, to the *nearest tenth?*

(1) 3.8

(2) 5.6

(3) 5.9

(4) 15.7

16 _____

17 In the diagram below, rectangle $ABCD$ has vertices whose coordinates are $A(7, 1)$, $B(9, 3)$, $C(3, 9)$, and $D(1, 7)$.

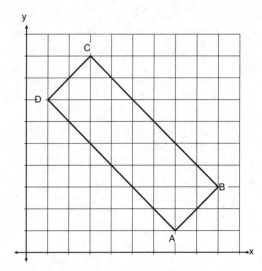

Which transformation will *not* carry the rectangle onto itself?

(1) a reflection over the line $y = x$

(2) a reflection over the line $y = -x + 10$

(3) a rotation of 180° about the point $(6, 6)$

(4) a rotation of 180° about the point $(5, 5)$

17 _____

18 A circle with a diameter of 10 cm and a central angle of 30° is
    drawn below.

What is the area, to the *nearest tenth of a square centimeter*, of the
sector formed by the 30° angle?

(1) 5.2   (3) 13.1
(2) 6.5   (4) 26.2   18 _____

19 A child's tent can be modeled as a pyramid with a square base
    whose sides measure 60 inches and whose height measures 84
    inches. What is the volume of the tent, to the *nearest cubic foot*?

(1) 35   (3) 82
(2) 58   (4) 175   19 _____

20 In the accompanying diagram of right triangle ABC, altitude $\overline{BD}$ is
    drawn to hypotenuse $\overline{AC}$.

Which statement must always be true?

(1) $\dfrac{AD}{AB} = \dfrac{BC}{AC}$ (3) $\dfrac{BD}{BC} = \dfrac{AB}{AD}$

(2) $\dfrac{AD}{AB} = \dfrac{AB}{AC}$ (4) $\dfrac{AB}{BC} = \dfrac{BD}{AC}$ 20 _____

21 An equation of circle O is $x^2 + y^2 + 4x - 8y = -16$. The statement that best describes circle O is the

(1) center is $(2, -4)$ and is tangent to the x-axis
(2) center is $(2, -4)$ and is tangent to the y-axis
(3) center is $(-2, 4)$ and is tangent to the x-axis
(4) center is $(-2, 4)$ and is tangent to the y-axis   21 _____

22 In $\triangle ABC$, $\overline{BD}$ is the perpendicular bisector of $\overline{ADC}$. Based upon this information, which statements below can be proven?

  I. $\overline{BD}$ is a median.
  II. $\overline{BD}$ bisects $\angle ABC$.
  III. $\triangle ABC$ is isosceles.

(1) I and II, only   (3) II and III, only
(2) I and III, only   (4) I, II, and III   22 _____

23 Triangle RJM has an area of 6 and a perimeter of 12. If the triangle is dilated by a scale factor of 3 centered at the origin, what are the area and perimeter of its image, triangle $R'J'M'$?

(1) area of 9 and perimeter of 15
(2) area of 18 and perimeter of 36
(3) area of 54 and perimeter of 36
(4) area of 54 and perimeter of 108   23 _____

24 If $\sin (2x + 7)° = \cos (4x - 7)°$, what is the value of x?

(1) 7   (3) 21
(2) 15   (4) 30   24 _____

PART II

Answer all 7 questions in this part. Each correct answer will receive 2 credits. Clearly indicate the necessary steps, including appropriate formula substitutions, diagrams, graphs, charts, etc. For all questions in this part, a correct numerical answer with no work shown will receive only 1 credit.   [14 credits]

25 In the circle below, $\overline{AB}$ is a chord. Using a compass and straightedge, construct a diameter of the circle. [Leave all construction marks.]

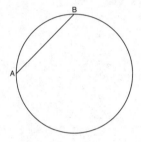

26 In parallelogram $ABCD$ shown below, the bisectors of $\angle ABC$ and $\angle DCB$ meet at E, a point on $\overline{AD}$.

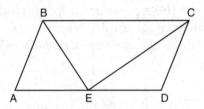

If $m\angle A = 68°$, determine and state $m\angle BEC$.

27 In circle *A* below, chord $\overline{BC}$ and diameter $\overline{DAE}$ intersect at *F*.

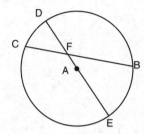

If m$\overset{\frown}{CD}$ = 46° and m$\overset{\frown}{DB}$ = 102°, what is m∠*CFE*?

28 Trapezoids *ABCD* and *A″ B″ C″ D″* are graphed on the set of axes below.

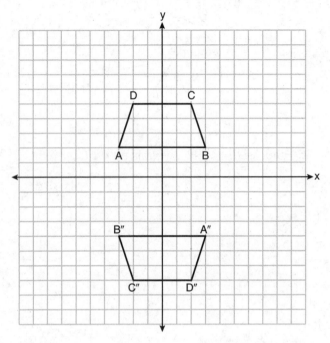

Describe a sequence of transformations that maps trapezoid *ABCD* onto trapezoid *A″ B″ C″ D″*.

29 In the model below, a support wire for a telephone pole is attached to the
    pole and anchored to a stake in the ground 15 feet from the base of the tele-
    phone pole. Jamal places a 6-foot wooden pole under the support wire par-
    allel to the telephone pole, such that one end of the pole is on the ground
    and the top of the pole is touching the support wire. He measures the dis-
    tance between the bottom of the pole and the stake in the ground.

Jamal says he can approximate how high the support wire attaches to the
telephone pole by using similar triangles. Explain why the triangles are
similar.

30 Aliyah says that when the line $4x + 3y = 24$ is dilated by a scale factor of 2 centered at the point $(3, 4)$, the equation of the dilated line is $y = -\frac{4}{3}x + 16$. Is Aliyah correct? Explain why.

[The use of the set of axes below is optional.]

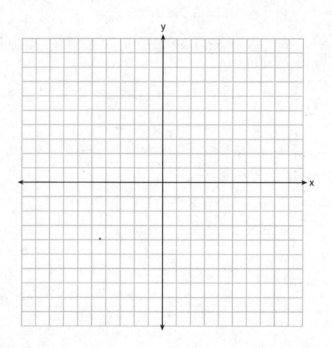

31 Ian needs to replace two concrete sections in his sidewalk, as modeled below.
    Each section is 36 inches by 36 inches and 4 inches deep. He can mix his
    own concrete for $3.25 per cubic foot.

How much money will it cost Ian to replace the two concrete sections?

PART III

Answer all 3 questions in this part. Each correct answer will receive 4 credits. Clearly indicate the necessary steps, including appropriate formula substitutions, diagrams, graphs, charts, etc. For all questions in this part, a correct numerical answer with no work shown will receive only 1 credit.   [12 credits]

32   Given: $\triangle ABC$, $\overline{AEC}$, $\overline{BDE}$ with $\angle ABE \cong \angle CBE$, and $\angle ADE \cong \angle CDE$

     Prove: $\overline{BDE}$ is the perpendicular bisector of $\overline{AC}$

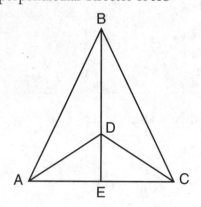

Fill in the missing statement and reasons below.

Statements	Reasons
(1) $\triangle ABC$, $\overline{AEC}$, $\overline{BDE}$ with $\angle ABE \cong \angle CBE$ and $\angle ADE \cong \angle CDE$	(1) Given
(2) $\overline{BD} \cong \overline{BD}$	(2) _____
(3) $\angle BDA$ and $\angle ADE$ are supplementary. $\angle BDC$ and $\angle CDE$ are supplementary.	(3) Linear pairs of angles are supplementary.
(4) _____	(4) Supplements of congruent angles are congruent.
(5) $\triangle ABD \cong \triangle CBD$	(5) ASA
(6) $\overline{AD} \cong \overline{CD}$, $\overline{AB} \cong \overline{CB}$	(6) _____
(7) $\overline{BDE}$ is the perpendicular bisector of $\overline{AC}$.	(7) _____ _____ _____ _____

33   A homeowner is building three steps leading to a deck, as modeled by the diagram below. All three step rises, $\overline{HA}$, $\overline{FG}$, and $\overline{DE}$, are congruent, and all three step runs, $\overline{HG}$, $\overline{FE}$, and $\overline{DC}$, are congruent. Each step rise is perpendicular to the step run it joins. The measure of $\angle CAB = 36°$ and m$\angle CBA = 90°$.

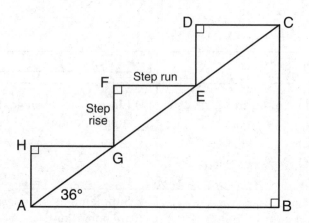

If each step run is parallel to $\overline{AB}$ and has a length of 10 inches, determine and state the length of each step rise, to the *nearest tenth of an inch*.

Determine and state the length of $\overline{AC}$, to the *nearest inch*.

34 A bakery sells hollow chocolate spheres. The larger diameter of each sphere is 4 cm. The thickness of the chocolate of each sphere is 0.5 cm. Determine and state, to the *nearest tenth of a cubic centimeter*, the amount of chocolate in each hollow sphere.

The bakery packages 8 of them into a box. If the density of the chocolate is 1.308 g/cm^3, determine and state, to the *nearest gram*, the total mass of the chocolate in the box.

PART IV

Answer the question in this part. Each correct answer will receive 6 credits. Clearly indicate the necessary steps, including appropriate formula substitutions, diagrams, graphs, charts, etc. For all questions in this part, a correct numerical answer with no work shown will receive only 1 credit. [6 credits]

35 The vertices of quadrilateral *MATH* have coordinates $M(-4, 2)$, $A(-1, -3)$, $T(9, 3)$, and $H(6, 8)$.

Prove that quadrilateral *MATH* is a parallelogram.

[The use of the set of axes on the next page is optional.]

Question 35 is continued on the next page.

Question 35 continued

Prove that quadrilateral *MATH* is a rectangle.

[The use of the set of axes below is optional.]

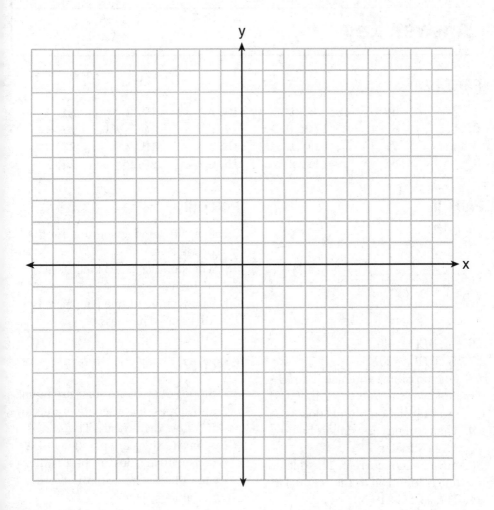

Answers
August 2018
Geometry

Answer Key

PART I

1. 4	5. 3	9. 1	13. 4	17. 3	21. 4
2. 1	6. 4	10. 4	14. 2	18. 2	22. 4
3. 4	7. 3	11. 1	15. 1	19. 2	23. 3
4. 1	8. 4	12. 2	16. 3	20. 2	24. 2

PART II

25.

26. m∠BEC = 90°
27. m∠CFE = 118°
28. One possible sequence is a reflection over the line $y = 2$ followed by a translation 6 units down.
29. The triangles are similar by the AA postulate.
30. Aliyah is not correct. The dilated line is identical to the original line.
31. The cost is $19.50

PART III

32. See the detailed solution for the proof.
33. The step rise is 7.3 inches. $AC = 37$ inches.
34. The volume of one chocolate is 19.4 cm^3. The mass of one box is 203 g.

PART IV

35. Slopes of opposite sides are equal, and slopes of consecutive sides are negative reciprocals. Therefore, opposite sides are parallel and consecutive sides are perpendicular. Other correct solutions are possible.

In **PARTS II–IV**, you are required to show how you arrived at your answers. For sample methods of solutions, see the *Answers Explained* section.

Answers Explained

PART I

1. $\angle AEG$ and $\angle GEF$ are a linear pair, which sum to 180°.

$$m\angle AEG + m\angle GEF = 180°$$
$$137° + m\angle GEF = 180°$$
$$m\angle GEF = 43°$$

Next, apply the angle sum theorem in $\triangle EFG$.

$$m\angle GEF + m\angle EFG + m\angle EGF = 180°$$
$$43° + 32° + m\angle EGF = 180°$$
$$75° + m\angle EGF = 180°$$
$$m\angle EGF = 105°$$

The correct choice is (**4**).

2. Line reflections and translations are both rigid motions that result in congruent images. Therefore, $\triangle ABC \cong \triangle DEF$ and $\triangle DEF \cong \triangle XYZ$, which implies that $\triangle ABC \cong \triangle XYZ$. Congruent images are also always similar with a scale factor of 1. $\triangle ABC$ and $\triangle XYZ$ are both congruent and similar.

The correct choice is (**1**).

3. A right triangle rotated about one of its legs will always form a cone, with the radius of the cone equal to one leg length and the height of the cone equal to the other leg length. In this case, the height of the cone is 6 and the radius is 6, making the diameter 12.

The correct choice is (**4**).

4. The reflection over $\overline{BG}$ maps $\triangle ABG$ onto $\triangle CBG$. The $180°$ rotation then maps it onto $\triangle FEG$.

   The correct choice is **(1)**.

5. The cross section of a right cylinder cut perpendicular to its base is always a rectangle. It may also be a square if the diameter is equal to the height.

   The correct choice is **(3)**.

6. We have a right triangle with a given angle and side and are looking for a second side. Relative to the $16.5°$ angle, the unknown length x is the hypotenuse and the 8-inch length is the opposite. The sine ratio can be used to solve for x.

$$\sin\left(16.5°\right) = \frac{opposite}{hypotenuse}$$
$$\sin\left(16.5°\right) = \frac{8}{x}$$
$$x = \frac{8}{\sin\left(16.5°\right)}$$
$$x = 28.16749$$

   To the nearest tenth of an inch, $x = 28.2$.

   The correct choice is **(4)**.

7. When an altitude is drawn to the hypotenuse of a right triangle, the altitude is the mean proportional to the parts of the hypotenuse. The relationship in this case is:

$$\frac{AD}{BD} = \frac{BD}{CD}$$

$$\frac{x-6}{4} = \frac{4}{x}$$

Next, cross multiply and solve the quadratic equation by factoring.

$$x(x-6) = 4 \times 4$$

$$x^2 - 6x = 16$$

$$x^2 - 6x - 16 = 0$$

$$(x-8)(x+2) = 0$$

$$x - 8 = 0 \qquad x + 2 = 0$$

$$x = 8 \qquad x = -2$$

There are two solutions to the quadratic equation, but the length of CD cannot be negative, so the only possible answer is $x = 8$.

The correct choice is **(3)**.

8. The perimeter of a polygon is the sum of the lengths of all its sides. Since a rhombus has four congruent sides, we need to find the length of only one side and then multiply that by 4. The length of a side can be found using the distance formula. Using the coordinates for side $\overline{ST}$, we have:

$$ST = \sqrt{(x_2 - x_1)^2 + (y_2 - y_1)^2}$$

$$= \sqrt{(2 - (-1))^2 + (3 - 2)^2}$$

$$= \sqrt{(3)^2 + (1)^2}$$

$$= \sqrt{10}$$

The perimeter is $4\sqrt{10}$.

The correct choice is **(4)**.

9. From the similarity statement $\triangle HAR \sim \triangle NTY$, we know that corresponding angles $\angle R$ and $\angle Y$ are congruent. We can find $\angle R$ using an inverse trigonometric ratio. Relative to $\angle R$, $\overline{AR}$ is the hypotenuse and $\overline{HR}$ is the adjacent, so we need an inverse cosine function.

$$\cos(R) = \frac{adjacent}{hypotenuse}$$

$$\cos(R) = \frac{12}{13}$$

$$m\angle R = \cos^{-1}\frac{12}{13}$$

$$m\angle R = 22.619°$$

Rounded to the nearest degree, $m\angle R = 23°$ and $m\angle Y = 23°$.

The correct choice is **(1)**.

10. The given congruent sides $\overline{AN}$ and $\overline{SC}$, along with congruent vertical angles $\angle AKN$ and $\angle CKS$, provide one pair of sides and one pair of angles. To use AAS, we need one more pair of angles. Choices (1) and (2) lead us to congruent sides, not angles, so those can be eliminated. Choice (3) would let us conclude that $\angle AKN$ and $\angle CKS$ are congruent right angles, but that does not provide the additional pair of congruent angles needed since we already know those angles are congruent. Choice (4) tells us that $\overline{AN}$ and $\overline{SC}$ are parallel. The parallel sides let us conclude two pairs of congruent alternate interior angles, $\angle N \cong \angle C$ and $\angle A \cong \angle S$, which gives us sufficient information to use the AAS postulate.

The correct choice is **(4)**.

11. Perpendicular lines have negative reciprocal slopes. The slope of the given line is $\frac{2}{3}$, so the slope of a perpendicular line must be $-\frac{3}{2}$. Rearrange each equation into $y = mx + b$ form to find the slope if necessary.

Choice (1):

$$3x + 2y = 12$$
$$2y = -3x + 12$$
$$y = -\frac{3}{2}x + 6$$

This has the correct slope, so the correct choice is (**1**).

12. First, the statement "*EB* is 3 more than *DB*" needs to be converted into an algebraic expression. If we let $DB = x$, then EB is $x + 3$. A line parallel to a side of a triangle forms two similar triangles, so $\triangle ABC \sim \triangle DBE$. We can use the fact that pairs of corresponding sides are proportional to solve for x. Sketching the triangles separately helps identify the parts we want to use in our proportion.

$$\frac{DB}{AB} = \frac{EB}{CB}$$
$$\frac{x}{14} = \frac{x + 3}{21}$$
$$21x = 14(x + 3)$$
$$21x = 14x + 42$$
$$7x = 42$$
$$x = 6$$

Next find AD.

$$AD = AB - DB$$
$$= 14 - 6$$
$$= 8$$

The correct choice is (**2**).

13. A quadrilateral with opposite sides congruent and parallel is a parallelogram. Evaluate each choice to determine which one is consistent with the properties of a parallelogram.

   - Choice (1): $\overline{MT} \cong \overline{AH}$ states that the diagonals are congruent. This is true of rectangles but not all parallelograms.
   - Choice (2): $\overline{MT} \cong \overline{AH}$ states that the diagonals are perpendicular. This is true of rhombuses but not all parallelograms.
   - Choice (3): $\angle MHT \cong \angle ATH$ states that a pair of consecutive angles are congruent. This is true of rectangles but not all parallelograms.
   - Choice (4): $\angle MAT \cong \angle MHT$ states that a pair of opposite angles are congruent. This is true of all parallelograms.

The correct choice is (**4**).

14. Tangents to a circle from a common point are congruent, so $TH = TR$ and $RA = AP$. R is the midpoint of $\overline{TA}$, which makes $TR = RA$. Combining these congruence statements yields:

$$TH = TR = RA = AP = 10$$

In the bottom half of the figure, we can combine the given statement $\overline{HO} \cong \overline{PE}$ with the congruent tangents $HO = OZ$ and $ZE = PE$ to find:

$$HO = OZ = ZE = PE = 6$$

The perimeter is the sum of all the side lengths.

$$\text{Perimeter} = 10 + 10 + 10 + 10 + 6 + 6 + 6 + 6$$

$$= 64$$

The correct choice is (**2**).

15. We can calculate the coordinates of the point (x, y) that divides a segment in a given ratio using the formula:

$$\frac{x - x_1}{x_2 - x} = \text{ratio and } \frac{y - y_1}{y_2 - y} = \text{ratio}$$

The coordinates of $A(-8, 7)$ are (x_1, x_2), the coordinates of $C(7, -13)$ are (y_1, y_2), and the ratio is $\frac{3}{2}$.

$$\frac{x - (-8)}{7 - x} = \frac{3}{2}$$
$$2(x + 8) = 3(7 - x)$$
$$2x + 16 = 21 - 3x$$
$$5x = 5$$
$$x = 1$$
$$\frac{y - 7}{-13 - y} = \frac{3}{2}$$
$$2(y - 7) = 3(-13 - y)$$
$$2y - 14 = -39 - 3y$$
$$5y = -25$$
$$y = -5$$

The coordinates of B are $(1, -5)$.

The correct choice is **(1)**.

16. Parallel segment $\overline{DE}$ divides $\triangle ABC$ into similar triangles $\triangle ABC$ and $\triangle DBE$. Corresponding segments of similar triangles are proportional.

$$\frac{AB}{DB} = \frac{AC}{DE}$$

The given ratio $AD{:}DB = 3{:}5$ can be used to find the ratio $AB{:}DB$. If AD is 3 parts and DB is 5 parts, then AD is the sum of these, or 8 parts. The ratio $AB{:}DB$ is therefore 8:5. Now use the proportion to solve for DE.

$$\frac{8}{5} = \frac{9.4}{DE}$$
$$8(DE) = 5(9.4)$$
$$8(DE) = 47$$
$$DE = 5.875$$
$$DE = 5.9, \text{ rounded to the nearest tenth}$$

The correct choice is **(3)**.

17. A reflection over a line of symmetry would map a rectangle onto itself. Using the figure, sketch the two lines of symmetry. Their equations are $y = -x + 10$ and $y = x$. Choices (1) and (2) represent reflections over these lines. A 180° rotation about the center of the rectangle will also map it onto itself. The center point $P(5, 5)$ is the point where the two lines of symmetry intersect. This rotation is represented by choice (4). Choice (3) will not map the rectangle onto itself.

The correct choice is **(3)**.

18. The radius of the circle is half the diameter, or 5 cm. Substitute the radius and 30° central angle into the formula for the area of a sector.

$$\text{Area of Sector} = \frac{\theta}{360}\pi r^2$$
$$= \frac{30}{360}\pi(5)^2$$
$$= 6.544$$

Rounded to the nearest tenth, the area of the sector is 6.5 cm^2.

The correct choice is **(2)**.

19. First, find the area of the square base using the formula for the area of a square.

$$B = (\text{side})^2$$
$$= (60 \text{ in})^2$$
$$= 3{,}600 \text{ in}^2$$

Next, apply the formula for the volume of a pyramid, using the area of the square base for B and 84 inches for the height.

$$V = \frac{1}{3}Bh$$

$$= \frac{1}{3}(3600 \text{ in}^2)(84 \text{ in})$$

$$= 100,800 \text{ in}^3$$

Now convert cubic inches to cubic feet using the ratio $\frac{1 \text{ ft}}{12 \text{ in}}$. Remember to cube the conversion ratio since we are working with cubic feet and cubic inches.

$$V = \left(100,800 \text{ in}^3\right)\left(\frac{1 \text{ ft}}{12 \text{ ft}}\right)^3$$

$$= 58.333 \text{ ft}^3$$

Rounded to the nearest cubic foot, the volume is 58 ft^3.

The correct choice is (**2**).

20. An altitude to the hypotenuse of a right triangle forms three similar triangles with proportional sides. Sketch each of the three triangles (small $\triangle$, medium $\triangle$, large $\triangle$), and label their sides to aid in evaluating which choice represents a correct proportion. Each side is either the hypotenuse (across from the right angle), the short leg, or the long leg.

- Choice (1): $\frac{AD}{AB} = \frac{BC}{AC}$ is equivalent to $\frac{\text{short leg}(\text{small } \triangle)}{\text{short leg}(\text{large } \triangle)}$

$$= \frac{\text{hyp}(\text{medium } \triangle)}{\text{hyp}(\text{large } \triangle)}.$$ These are not corresponding parts.

- Choice (2): $\frac{AD}{AB} = \frac{AB}{AC}$ is equivalent to $\frac{\text{short leg}(\text{small } \triangle)}{\text{short leg}(\text{large } \triangle)}$

$$= \frac{\text{hyp}(\text{small } \triangle)}{\text{hyp}(\text{large } \triangle)}.$$ This is a valid proportion.

- Choice (3): $\dfrac{BD}{BC} = \dfrac{AB}{AD}$ is equivalent to $\dfrac{\text{long leg(small } \triangle)}{\text{long leg(large } \triangle)}$

$= \dfrac{\text{hyp(small } \triangle)}{\text{short leg(small } \triangle)}$. These are not corresponding parts.

- Choice (4): $\dfrac{AB}{BC} = \dfrac{BD}{AC}$ is equivalent to $\dfrac{\text{hyp(small } \triangle)}{\text{hyp(medium } \triangle)}$

$= \dfrac{\text{long leg(small } \triangle)}{\text{hyp(large } \triangle)}$. These are not corresponding parts.

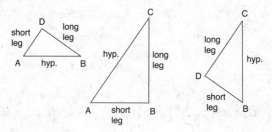

The correct choice is (2).

21. The center-radius equation of a circle is $(x - h)^2 + (y - k)^2 = r^2$, where the center is (h, k) and the radius is r. Rewrite the equation of the circle in center-radius form using the completing-the-square method.

$x^2 + y^2 + 4x - 8y = -16$

$x^2 + 4x + y^2 - 8y = -16$ Rearrange to group x and y terms

$x^2 + 4x + 4 + y^2 - 8y = -16 + 4$ Add (1/2 coefficient of x)^2 to each side

$x^2 + 4x + 4 + y^2 - 8y + 16 =$

$\qquad\qquad\qquad -16 + 4 + 16$ Add (1/2 coefficient of y)^2 to each side

$(x + 2)^2 + (y - 4)^2 = 4$ Factor

The center of the circle is $(-2, 4)$ and the radius is 2. We can eliminate choices (1) and (2) because they do not have the correct center. The center of the circle has an x-coordinate of -2, so it is 2 units left of the y-axis. The radius is 2, so the circle just reaches the y-axis and is tangent to it. You can also graph the circle to see that it is tangent to the y-axis.

The correct choice is (4).

22. Perpendicular bisector $\overline{BD}$ is shown in $\triangle ABC$. D is a midpoint, which makes $\overline{BD}$ a median, so statement I is true. $\angle ADB$ and $\angle CDB$ are congruent right angles and $\overline{BD}$ is congruent to itself, so $\triangle ADB \cong \triangle CDB$ by the SAS postulate. This leads us to $\angle ABD \cong \angle CBD$ by CPCTC, making $\overline{BD}$ an angle bisector. Statement II is also true. We can also conclude that $\overline{AB} \cong \overline{BC}$ by CPCTC, which makes $\triangle ABC$ isosceles. Statement III is also true.

The correct choice is (**4**).

23. Dilations always create similar figures. Perimeters of similar figures are proportional to the scale factor, and areas are proportional to the scale factor squared. Given a scale factor of 3, the perimeter of the dilated image must be 3 times greater and the area is 9 times greater.

Perimeter of $R' J' M' = 12 \cdot 3 = 36$

Area of $R' J' M' = 6 \cdot 3^2 = 54$

The correct choice is (**3**).

24. The trigonometric cofunction relationship states that if $\sin(A) = \cos(B)$, then $A + B = 90°$. Given $\sin(2x + 7)° = \cos(4x - 7)°$, we can conclude:

$$2x + 7 + 4x - 7 = 90$$
$$6x = 90$$
$$x = 15$$

The correct choice is (**2**).

PART II

25. Given any chord in a circle, the perpendicular bisector of the chord passes through the center of the circle. Construct a diameter by constructing the perpendicular bisector of $\overline{AB}$:

- With compass point at A, open the compass more than half the length of $\overline{AB}$.

- Make arc on both sides of $\overline{AB}$.

- Move the compass point to B.

- With the same compass opening, make a second pair of arcs that intersect the first pair.

- Connect the two arc intersections.

- $\overline{MN}$ is a diameter of the circle.

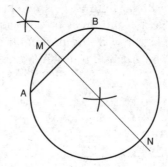

26. Starting with given $\angle A = 68°$, we know that opposite angles of a parallelogram are congruent, so $\angle BCD = 68°$. Bisector $\overline{CE}$ divides $\angle BCD$ in half, so m$\angle BCE = 34°$. Next, we know that consecutive angles of a parallelogram are supplementary, so $\angle ABC$ and $\angle A$ sum to 180°.

$$m\angle ABC + m\angle A = 180°$$
$$m\angle ABC = 180° - m\angle A$$
$$= 180° - 68°$$
$$= 112°$$

Bisector $\overline{BE}$ divides $\angle ABC$ in half, so m$\angle EBC = 56°$. Now we can apply the angle sum theorem in $\triangle BEC$.

$$m\angle EBC + m\angle BCE + m\angle BEC = 180°$$
$$56° + 34° + m\angle BEC = 180°$$
$$90 + m\angle BEC = 180°$$
$$m\angle BEC = 90°$$

27. To find m$\angle CFE$, we need arcs $\overset{\frown}{CE}$ and $\overset{\frown}{DB}$. Calculate $\overset{\frown}{CE}$ using semicircle $\overset{\frown}{DCE}$, which we know measures 180°:

$$m\overset{\frown}{CD} + m\overset{\frown}{CE} = 180°$$
$$46° + m\overset{\frown}{CE} = 180°$$
$$m\overset{\frown}{CE} = 134°$$

∠*CFE* is an angle formed by intersecting chords and is equal in measure to half the sum of the intercepted arcs.

$$m\angle CFE = \frac{1}{2}\left(\overset{\frown}{CE} + \overset{\frown}{DB}\right)$$

$$= \frac{1}{2}(134 + 102)$$

$$= 118°$$

28. There are a many different transformation sequences that would map *ABCD* to *A″ B″ C″ D″*. One possible sequence is a reflection over the line $y = 2$ followed by a translation 6 units down.

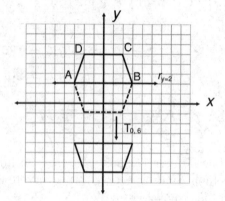

29. The stake and telephone pole are parallel. The line along the ground is a transversal that forms congruent corresponding angles. The wire is another transversal forming another pair of congruent corresponding angles. The two triangles are similar by the AA postulate.

30. Begin by rearranging the line in point-slope form so we can graph it.

$$4x + 3y = 24$$
$$3y = -4x + 24$$
$$y = -\frac{4}{3}x + 8$$

Now graph the line and the center of dilation $(3, 4)$. From the graph, we see that the line passes through the center of dilation. This indicates that the dilated line is identical to the original line, so Aliyah is not correct.

31. Each section of concrete is a rectangular prism 36 inches long, 36 inches wide, and 4 inches high. Since the cost of the concrete is given in dollars per cubic foot, convert the dimensions to feet.

$$36 \text{ in} \times \frac{1 \text{ ft}}{12 \text{ in}} = 3 \text{ ft} \qquad 4 \text{ in} \times \frac{1 \text{ ft}}{12 \text{ in}} = \frac{1}{3} \text{ ft}$$

Next, calculate the volume using the volume formula for a rectangular prism.

$$V = \text{length} \times \text{width} \times \text{height}$$
$$= (3 \text{ ft})(3 \text{ ft})\left(\frac{1}{3} \text{ ft}\right)$$
$$= 3 \text{ ft}^3$$

There are two sections of concrete, so the total volume is 6 ft³. Finally, calculate the cost.

$$\text{cost} = \text{volume} \times \text{cost/ft}^3$$
$$= \left(6 \text{ ft}^3\right)\left(\$3.25 \, / \, \text{ft}^3\right)$$
$$= \$19.50$$

PART III

32. The strategy for this proof is to use the theorem that all points on the perpendicular bisector of a segment are equidistant from the endpoints of the segment. Start by proving $\triangle ABD \cong \triangle CBD$. Then $\overline{AD} \cong \overline{CD}$ by CPCTC. We now have $\overline{AD} \cong \overline{CD}$ and $\overline{AB} \cong \overline{CB}$, which indicate that points B and D are equidistant from endpoints A and C. This is sufficient to prove $\overline{BDE}$ is the perpendicular bisector of $\overline{AC}$.

Statements	Reasons
(1) $\triangle ABC$, $\overline{AEC}$, $\overline{BDE}$ with $\angle ABE \cong \angle CBE$ and $\angle ADE \cong \angle CDE$	(1) Given
(2) $\overline{BD} \cong \overline{BD}$	(2) Reflexive Property
(3) $\angle BDA$ and $\angle ADE$ are supplementary. $\angle BDC$ and $\angle CDE$ are supplementary.	(3) Linear pairs of angles are supplementary.
(4) $\angle BDA \cong \angle BDC$	(4) Supplements of congruent angles are congruent.
(5) $\triangle ABD \cong \triangle CBD$	(5) ASA
(6) $\overline{AD} \cong \overline{CD}$, $\overline{AB} \cong \overline{CB}$	(6) CPCTC
(7) $\overline{BDE}$ is the perpendicular bisector of $\overline{AC}$.	(7) The perpendicular bisector of a segment is the set of points equidistant from the endpoints of the segment.

33. Focusing just on $\triangle AHG$ and segment $\overline{AB}$, we have transversal $\overline{AG}$ intersecting parallel segments $\overline{HG}$ and $\overline{AB}$. $\angle AHG$ and $\angle BAG$ are congruent alternate interior angles, so m$\angle G = 36°$. We can use a trigonometric ratio to find the rise HA. Relative to $\angle G$, $\overline{HA}$ is the opposite and $\overline{HG}$ is the adjacent, so we use the tangent ratio.

$$\tan G = \frac{opposite}{adjacent}$$

$$\tan G = \frac{HA}{HG}$$

$$\tan 36 = \frac{HA}{10}$$

$$HA = 7.265$$

Rounded to the nearest tenth of an inch, the run is 7.3 inches.

To find AC, we can calculate AG and multiply by 3. AG is a hypotenuse relative to $\angle G$, so we can use a cosine ratio:

$$\cos G = \frac{adjacent}{hypotenuse}$$

$$\cos G = \frac{HG}{AG}$$

$$\cos 36 = \frac{10}{AG}$$

$$AG = 12.360$$

Now calculate AC:

$$AC = 3AG$$
$$= 3(12.36)$$
$$= 37.082$$

Rounded to the nearest inch, $AC = 37$ inches.

34. Each chocolate sphere is actually a spherical shell with an inner and outer radius. The outer radius is half the diameter, or 2 cm. The inner radius is found by subtracting the wall thickness from the outer radius.

$$r_{in} = r_{out} - \text{thickness}$$
$$= 2\,\text{cm} - 0.5\,\text{cm}$$
$$= 1.5\,\text{cm}$$

Now use the formula for the volume of a sphere to find each volume.

$$V_{out} = \frac{4}{3}\pi r^3$$
$$= \frac{4}{3}\pi (2)^3$$
$$= 33.51032\,\text{cm}^3$$
$$V_{in} = \frac{4}{3}\pi r^3$$
$$= \frac{4}{3}\pi (1.5)^3$$
$$= 14.13716\,\text{cm}^3$$

The volume of chocolate is equal to the difference between these two volumes.

$$V_{chocolate} = V_{out} - V_{in}$$
$$= 33.51032\,\text{cm}^3 - 14.13716\,\text{cm}^3$$
$$= 19.37316$$

Rounded to the nearest tenth of a cubic centimeter, the volume is 19.4 cm³.

Use the relationship *mass = volume × density* to find the mass of 1 chocolate, and multiply by 8 because there are 8 chocolates per box.

$$\text{mass} = 8 \times \text{volume} \times \text{density}$$
$$= 8(19.4\,\text{cm}^3)(1.308\text{g/cm}^3)$$
$$= 203.0016\,\text{g}$$

To the nearest gram, the mass of chocolate in one box is 203 g.

PART IV

35. There are many different combinations of properties we can use to prove that a figure is both a parallelogram and rectangle. One option is to find the slopes of the four sides, which can be used to show that the opposite sides and parallel and consecutive sides are perpendicular.

a) Calculate each slope using coordinates $M(-4, 2)$, $A(-1, -3)$, $T(9, 3)$, and $H(6, 8)$.

$$\text{slope} = \frac{y_2 - y_1}{x_2 - x_1}$$

$$\text{slope of } \overline{MA} : \frac{-3 - 2}{-1 - (-4)} = -\frac{5}{3}$$

$$\text{slope of } \overline{AT} : \frac{3 - (-3)}{9 - (-1)} = \frac{6}{10} = \frac{3}{5}$$

$$\text{slope of } \overline{TH} : \frac{8 - 3}{6 - 9} = -\frac{5}{3}$$

$$\text{slope of } \overline{HM} : \frac{8 - 2}{6 - (-4)} = \frac{6}{10} = \frac{3}{5}$$

Opposite sides of *MATH* have equal slopes, which makes them parallel. A quadrilateral with opposite sides parallel is a parallelogram, therefore *MATH* is a parallelogram.

b) The slopes of consecutive sides are negative reciprocals, which makes them perpendicular. A parallelogram with perpendicular consecutive sides is a rectangle.

Topic	Question Numbers	Number of Points	Your Points	Your Percentage
1. Basic Angle and Segment Relationships	1	2		
2. Angle and Segment Relationships in Triangles and Polygons	22	2		
3. Constructions	25	2		
4. Transformations	2, 4, 17, 28	$2 + 2 + 2 + 2 = 8$		
5. Triangle Congruence	10, 32	$2 + 4 = 6$		
6. Lines, Segments, and Circles on the Coordinate Plane	8, 11, 15, 21, 30	$2 + 2 + 2 + 2 + 2 = 10$		
7. Similarity	7, 12, 16, 20, 23, 29	$2 + 2 + 2 + 2 + 2 + 2 = 12$		
8. Trigonometry	6, 9, 24, 33	$2 + 2 + 2 + 4 = 10$		
9. Parallelograms	13, 26	$2 + 2 = 4$		
10. Coordinate Geometry Proofs	35	6		
11. Volume, Solids, and Modeling	3, 5, 19, 31, 34	$2 + 2 + 2 + 2 + 4 = 12$		
12. Circles	14, 18, 27	$2 + 2 + 2 = 6$		

Regents Examination in Geometry—August 2018
Chart for Converting Total Test Raw Scores to Final
Examination Scores (Scaled Scores)

Raw Score	Scaled Score	Performance Level	Raw Score	Scaled Score	Performance Level	Raw Score	Scaled Score	Performance Level
80	100	5	53	80	4	26	61	2
79	99	5	52	80	4	25	60	2
78	98	5	51	79	3	24	58	2
77	97	5	50	79	3	23	57	2
76	96	5	49	78	3	22	56	2
75	95	5	48	78	3	21	55	2
74	95	5	47	77	3	20	53	1
73	94	5	46	77	3	19	51	1
72	93	5	45	76	3	18	50	1
71	92	5	44	76	3	17	48	1
70	91	5	43	75	3	16	46	1
69	91	5	42	74	3	15	44	1
68	90	5	41	74	3	14	42	1
67	89	5	40	73	3	13	40	1
66	88	5	39	73	3	12	38	1
65	88	5	38	72	3	11	36	1
64	87	5	37	71	3	10	33	1
63	86	5	36	70	3	9	31	1
62	86	5	35	70	3	8	28	1
61	85	5	34	69	3	7	26	1
60	84	4	33	68	3	6	23	1
59	84	4	32	67	3	5	19	1
58	83	4	31	66	3	4	16	1
57	83	4	30	65	3	3	3	1
56	82	4	29	64	2	2	9	1
55	81	4	28	63	2	1	5	1
54	81	4	27	62	2	0	0	1

Examination
June 2019
Geometry

HIGH SCHOOL MATH REFERENCE SHEET

Conversions

1 inch = 2.54 centimeters	1 cup = 8 fluid ounces
1 meter = 39.37 inches	1 pint = 2 cups
1 mile = 5280 feet	1 quart = 2 pints
1 mile = 1760 yards	1 gallon = 4 quarts
1 mile = 1.609 kilometers	1 gallon = 3.785 liters
	1 liter = 0.264 gallon
1 kilometer = 0.62 mile	1 liter = 1000 cubic centimeters
1 pound = 16 ounces	
1 pound = 0.454 kilogram	
1 kilogram = 2.2 pounds	
1 ton = 2000 pounds	

Formulas

Triangle	$A = \frac{1}{2}bh$
Parallelogram	$A = bh$
Circle	$A = \pi r^2$
Circle	$C = \pi d$ or $C = 2\pi r$

Formulas (continued)

General Prisms	$V = Bh$
Cylinder	$V = \pi r^2 h$
Sphere	$V = \frac{4}{3}\pi r^3$
Cone	$V = \frac{1}{3}\pi r^2 h$
Pyramid	$V = \frac{1}{3}Bh$
Pythagorean Theorem	$a^2 + b^2 = c^2$
Quadratic Formula	$x = \dfrac{-b \pm \sqrt{b^2 - 4ac}}{2a}$
Arithmetic Sequence	$a_n = a_1 + (n-1)d$
Geometric Sequence	$a_n = a_1 r^{n-1}$
Geometric Series	$S_n = \dfrac{a_1 - a_1 r^n}{1 - r}$ where $r \neq 1$
Radians	1 radian $= \dfrac{180}{\pi}$ degrees
Degrees	1 degree $= \dfrac{\pi}{180}$ radians
Exponential Growth/Decay	$A = A_0 e^{k(t-t_0)} + B_0$

180 Minutes–35 Questions

PART I

Answer all 24 questions in this part. Each correct answer will receive 2 credits. No partial credit will be allowed. pick up after "allowed." with the following: For each statement or question, write in the space provided the numeral preceding the word or expression that best completes the statement or answers the question.   [48 credits]

1 On the set of axes below, triangle *ABC* is graphed. Triangles *A'B'C'* and *A"B"C"*, the images of triangle *ABC*, are graphed after a sequence of rigid motions.

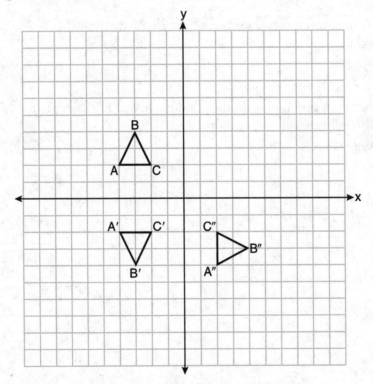

Identify which sequence of rigid motions maps △*ABC* onto △*A'B'C'* and then maps △*A'B'C'* onto △*A"B"C"*.

(1) a rotation followed by another rotation
(2) a translation followed by a reflection
(3) a reflection followed by a translation
(4) a reflection followed by a rotation

1 _____

2 The table below shows the population and land area, in square miles, of four counties in New York State at the turn of the century.

County	2000 Census Population	2000 Land Area (mi²)
Broome	200,536	706.82
Dutchess	280,150	801.59
Niagara	219,846	522.95
Saratoga	200,635	811.84

Which county had the greatest population density?

(1) Broome   (3) Niagara
(2) Dutchess   (4) Saratoga   2 _____

3 If a rectangle is continuously rotated around one of its sides, what is the three-dimensional figure formed?

(1) rectangular prism   (3) sphere
(2) cylinder   (4) cone   3 _____

4 Which transformation carries the parallelogram below onto itself?

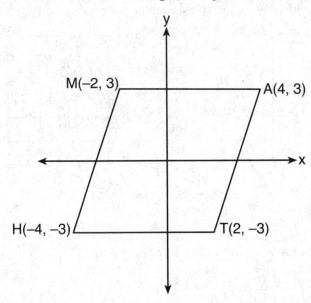

(1) a reflection over $y = x$
(2) a reflection over $y = -x$
(3) a rotation of 90° counterclockwise about the origin
(4) a rotation of 180° counterclockwise about the origin   4 ____

5 After a dilation centered at the origin, the image of $\overline{CD}$ is $\overline{C'D'}$. If the coordinates of the endpoints of these segments are $C(6, -4)$, $D(2, -8)$, $C'(9, -6)$, and $D'(3, -12)$, the scale factor of the dilation is

(1) $\dfrac{3}{2}$ (3) 3

(2) $\dfrac{2}{3}$ (4) $\dfrac{1}{3}$ 5 ____

6 A tent is in the shape of a right pyramid with a square floor. The square floor has side lengths of 8 feet. If the height of the tent at its center is 6 feet, what is the volume of the tent, in cubic feet?

(1) 48   (3) 192
(2) 128   (4) 384   6 ____

7 The line $-3x + 4y = 8$ is transformed by a dilation centered at the origin. Which linear equation could represent its image?

(1) $y = \dfrac{4}{3}x + 8$ (3) $y = -\dfrac{3}{4}x - 8$

(2) $y = \dfrac{3}{4}x + 8$ (4) $y = -\dfrac{4}{3}x - 8$ 7 _____

8 In the diagram below, $\overline{AC}$ and $\overline{BD}$ intersect at E.

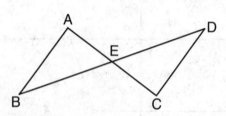

Which information is always sufficient to prove $\triangle ABE \cong \triangle CDE$?

(1) $\overline{AB} \parallel \overline{CD}$

(2) $\overline{AB} \cong \overline{CD}$ and $\overline{BE} \cong \overline{DE}$

(3) E is the midpoint of $\overline{AC}$.

(4) $\overline{BD}$ and $\overline{AC}$ bisect each other.   8 _____

9 The expression $\sin 57°$ is equal to

(1) $\tan 33°$ (3) $\tan 57°$

(2) $\cos 33°$ (4) $\cos 57°$ 9 _____

10 What is the volume of a hemisphere that has a diameter of 12.6 cm, to the *nearest tenth of a cubic centimeter*?

(1) 523.7   (3) 4189.6

(2) 1047.4   (4) 8379.2   10 _____

11 In the diagram below of △*ABC*, *D* is a point on $\overline{BA}$, *E* is a point on $\overline{BC}$, and $\overline{DE}$ is drawn.

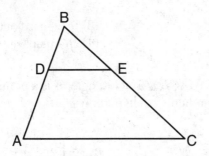

If *BD* = 5, *DA* = 12, and *BE* = 7, what is the length of $\overline{BC}$ so that $\overline{AC}$ ∥ $\overline{DE}$?

(1) 23.8   (3) 15.6
(2) 16.8   (4) 8.6   11 _____

12 A quadrilateral must be a parallelogram if
    (1) one pair of sides is parallel and one pair of angles is congruent
    (2) one pair of sides is congruent and one pair of angles is congruent
    (3) one pair of sides is both parallel and congruent
    (4) the diagonals are congruent   12 _____

13 In the diagram below of circle *O*, chords $\overline{JT}$ and $\overline{ER}$ intersect at *M*.

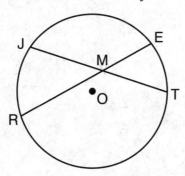

If *EM* = 8 and *RM* = 15, the lengths of $\overline{JM}$ and $\overline{TM}$ could be

(1) 12 and 9.5   (3) 16 and 7.5
(2) 14 and 8.5   (4) 18 and 6.5   13 _____

14 Triangles *JOE* and *SAM* are drawn such that $\angle E \cong \angle M$ and $\overline{EJ} \cong \overline{MS}$. Which mapping would *not* always lead to $\triangle JOE \cong \triangle SAM$?

(1) $\angle J$ maps onto $\angle S$

(2) $\angle O$ maps onto $\angle A$

(3) $\overline{EO}$ maps onto $\overline{MA}$

(4) $\overline{JO}$ maps onto $\overline{SA}$

14 _____

15 In $\triangle ABC$ shown below, $\angle ACB$ is a right angle, E is a point on $\overline{AC}$, and $\overline{ED}$ is drawn perpendicular to hypotenuse $\overline{AB}$.

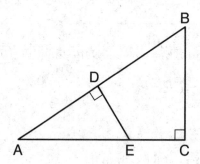

If $AB = 9$, $BC = 6$, and $DE = 4$, what is the length of $\overline{AE}$?

(1) 5

(2) 6

(3) 7

(4) 8

15 _____

16 Which equation represents a line parallel to the line whose equation is $-2x + 3y = -4$ and passes through the point $(1, 3)$?

(1) $y - 3 = -\dfrac{3}{2}(x - 1)$

(2) $y - 3 = \dfrac{2}{3}(x - 1)$

(3) $y + 3 = -\dfrac{3}{2}(x + 1)$

(4) $y + 3 = \dfrac{2}{3}(x + 1)$

16 _____

17 In rhombus *TIGE*, diagonals $\overline{TG}$ and $\overline{IE}$ intersect at *R*. The perimeter of *TIGE* is 68, and *TG* = 16.

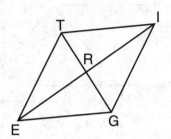

What is the length of diagonal $\overline{IE}$?

(1) 15   (3) 34

(2) 30   (4) 52   17 _____

18 In circle *O* two secants, $\overline{ABP}$ and $\overline{CDP}$, are drawn to external point *P*. If $m\widehat{AC} = 72°$, and $m\widehat{BD} = 34°$, what is the measure of $\angle P$?

(1) 19°   (3) 53°

(2) 38°   (4) 106°   18 _____

19 What are the coordinates of point *C* on the directed segment from $A(-8, 4)$ to $B(10, -2)$ that partitions the segment such that *AC*:*CB* is 2:1?

(1) $(1, 1)$ (3) $(2, -2)$

(2) $(-2, 2)$ (4) $(4, 0)$ 19 _____

20 The equation of a circle is $x^2 + 8x + y^2 - 12y = 144$. What are the coordinates of the center and the length of the radius of the circle?

(1) center $(4, -6)$ and radius 12

(2) center $(-4, 6)$ and radius 12

(3) center $(4, -6)$ and radius 14

(4) center $(-4, 6)$ and radius 14   20 _____

21 In parallelogram $PQRS$, $\overline{QP}$ is extended to point T and $\overline{ST}$ is drawn.

If $\overline{ST} \cong \overline{SP}$ and m$\angle R = 130°$, what is m$\angle PST$?

(1) 130°   (3) 65°
(2) 80°   (4) 50°   21 ____

22 A 12-foot ladder leans against a building and reaches a window
10 feet above ground. What is the measure of the angle, to the
nearest degree, that the ladder forms with the ground?

(1) 34   (3) 50
(2) 40   (4) 56   22 ____

23 In the diagram of equilateral triangle ABC shown below, E and F
are the midpoints of $\overline{AC}$ and $\overline{BC}$, respectively.

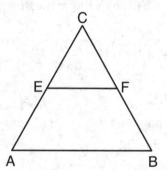

If $EF = 2x + 8$ and $AB = 7x - 2$, what is the perimeter of
trapezoid $ABFE$?

(1) 36   (3) 100
(2) 60   (4) 120   23 ____

24 Which information is *not* sufficient to prove that a parallelogram is a square?

  (1) The diagonals are both congruent and perpendicular.

  (2) The diagonals are congruent and one pair of adjacent sides are congruent.

  (3) The diagonals are perpendicular and one pair of adjacent sides are congruent.

  (4) The diagonals are perpendicular and one pair of adjacent sides are perpendicular.   24 ____

PART II

Answer all 7 questions in this part. Each correct answer will receive 2 credits. Clearly indicate the necessary steps, including appropriate formula substitutions, diagrams, graphs, charts, etc. For all questions in this part, a correct numerical answer with no work shown will receive only 1 credit.   [14 credits]

25 Triangle $A'B'C'$ is the image of triangle ABC after a dilation with a scale factor of $\frac{1}{2}$ and centered at point A. Is triangle ABC congruent to triangle $A'B'C'$? Explain your answer.

26 Determine and state the area of triangle PQR, whose vertices have coordinates $P(-2, -5)$, $Q(3, 5)$, and $R(6, 1)$.

[The use of the set of axes below is optional.]

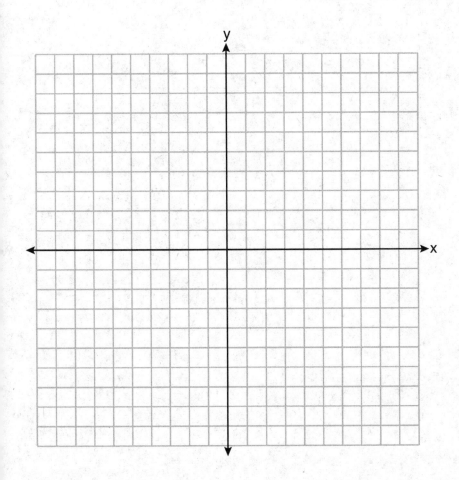

27 A support wire reaches from the top of a pole to a clamp on the ground. The pole is perpendicular to the level ground and the clamp is 10 feet from the base of the pole. The support wire makes a 68° angle with the ground. Find the length of the support wire to the *nearest foot*.

28 In the diagram below, circle O has a radius of 10.

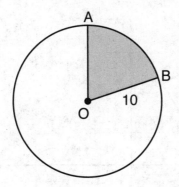

If $m\widehat{AB} = 72°$, find the area of shaded sector AOB, in terms of π.

29 On the set of axes below, △ABC ≅ △STU.

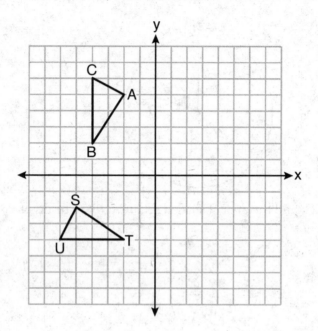

Describe a sequence of rigid motions that maps △ABC onto △STU.

30 In right triangle *PRT*, m∠*P* = 90°, altitude $\overline{PQ}$ is drawn to hypotenuse $\overline{RT}$, *RT* = 17, and *PR* = 15.

Determine and state, to the *nearest tenth*, the length of $\overline{RQ}$.

31 Given circle O with radius $\overline{OA}$, use a compass and straightedge to construct an equilateral triangle inscribed in circle O. [Leave all construction marks.]

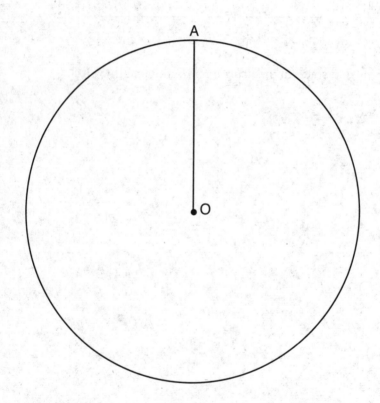

PART III

Answer all 3 questions in this part. Each correct answer will receive 4 credits. Clearly indicate the necessary steps, including appropriate formula substitutions, diagrams, graphs, charts, etc. Utilize the information provided for each question to determine your answer. Note that diagrams are not necessarily drawn to scale. For all questions in this part, a correct numerical answer with no work shown will receive only 1 credit.   [12 credits]

32 Riley plotted $A(-1, 6)$, $B(3, 8)$, $C(6, -1)$, and $D(1, 0)$ to form a quadrilateral. Prove that Riley's quadrilateral $ABCD$ is a trapezoid.
[The use of the set of axes on the next page is optional.]

Question 32 is continued on the next page.

Question 32 continued

Riley defines an isosceles trapezoid as a trapezoid with congruent diagonals. Use Riley's definition to prove that *ABCD* is *not* an isosceles trapezoid.

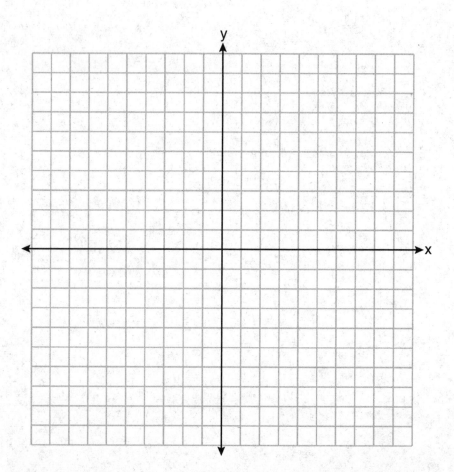

33 A child-sized swimming pool can be modeled by a cylinder. The pool has a diameter of $6\frac{1}{2}$ feet and a height of 12 inches. The pool is filled with water to $\frac{2}{3}$ of its height. Determine and state the volume of the water in the pool, to the *nearest cubic foot*.

One cubic foot equals 7.48 gallons of water. Determine and state, to the *nearest gallon*, the number of gallons of water in the pool.

34 Nick wanted to determine the length of one blade of the windmill
pictured below. He stood at a point on the ground 440 feet from
the windmill's base. Using surveyor's tools, Nick measured the angle
between the ground and the highest point reached by the top blade
and found it was 38.8°. He also measured the angle between the
ground and the lowest point of the top blade, and found it was 30°.

Determine and state a blade's length, x, to the *nearest foot*.

PART IV

Answer the question in this part. Each correct answer will receive 6 credits. Clearly indicate the necessary steps, including appropriate formula substitutions, diagrams, graphs, charts, etc. For all questions in this part, a correct numerical answer with no work shown will receive only 1 credit.   [6 credits]

35   Given: Quadrilateral $MATH$, $\overline{HM} \cong \overline{AT}$, $\overline{HT} \cong \overline{AM}$, $\overline{HE} \perp \overline{MEA}$ and $\overline{HA} \perp \overline{AT}$

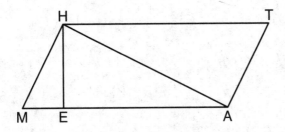

Prove: $TA \bullet HA = HE \bullet TH$

Work space for question 35 is continued on the next page.

Question 35 continued

Answers Explained

PART I

1. The sequence of transformations needs to map the following the points:

    $$A(-4, 2) \rightarrow A'(-4, -2) \rightarrow A''(2, -4).$$

    A reflection over the x-axis changes the sign of the y-coordinate, and takes A to A'. A counterclockwise rotation of $90°$ centered at the origin changes coordinates (x, y) to $(-y, x)$, and takes A' to A''. This same sequence also maps B to B'' and C to C''.

    The correct choice is (**4**).

2. Find the population density of each county by dividing the population by the area.

    Broome: population density $= \dfrac{200,536}{706.82} = 283.716$

    Dutchess: population density $= \dfrac{280,150}{801.59} = 349.493$

    Niagara: population density $= \dfrac{219,846}{522.95} = 428.396$

    Saratoga: population density $= \dfrac{200,635}{811.84} = 247.136$

    Niagara county has the greatest population density.

    The correct choice is (**3**).

3. The solid formed by rotating a rectangle about one of its sides is a cylinder.

    The correct choice is (**2**).

4. One approach is to choose a vertex and apply each transformation. We are looking to map the chosen vertex to one of the other vertices. If the vertex does not map to one of the original vertices, that transformation can be eliminated. Using vertex $M(-2, 3)$ we apply each transformation.

Choice 1 – A reflection over the line $y = x$ swaps the coordinates with the rule $(x, y) \rightarrow (y, x)$. $M(-2, 3) \rightarrow M'(3, -2)$ which is not a vertex, so this choice is eliminated.

Choice 2 – A reflection over the line $y = -x$ swaps the coordinates and changes the signs with the rule $(x, y) \rightarrow (-y, -x)$. $M(-2, 3) \rightarrow M'(-3, 2)$ which is not a vertex, so this choice is eliminated.

Choice 3 – A counterclockwise rotation of $90°$ about the origin transforms the coordinates with the rule $(x, y) \rightarrow (-y, x)$. $M(-2, 3) \rightarrow M'(-3, -2)$ which is not a vertex, so this choice is eliminated.

Choice 4 – A counterclockwise rotation of $180°$ about the origin transforms the coordinates with the rule $(x, y) \rightarrow (-x, -y)$. $M(-2, 3) \rightarrow M'(2, -3)$ which is vertex T. You should confirm that the same rule maps A to H, H to A, and T to M.

The correct choice is (**4**).

5. A dilation always multiples each coordinate by the scale factor. Work backwards by dividing the new coordinate by the original to calculate the scale factor. We can use any point and either coordinate. Using the x-coordinates of points C and C' we find:

$$\text{scale factor} = \frac{x\text{-coordinate of } C'}{x\text{-coordinate of } C}$$

$$= \frac{9}{6}$$

$$= \frac{3}{2}$$

You can confirm this result using the other coordinates.

The correct choice is (**1**).

6. The volume of a pyramid is given by $V = \frac{1}{3}Bh$ where B is the area of the base and h is the height. First find the area of the square base.

$$B = \text{side}^2$$
$$= 8^2$$
$$= 64$$

Now apply the volume formula with a height of 6.

$$V = \frac{1}{3}Bh$$
$$= \frac{1}{3}(64)(6)$$
$$= 128$$

The volume of the tent is 128 ft^2.

The correct choice is (**2**).

7. Dilating a line through the origin multiplies the y-intercept by the scale factor, but leaves the slope unchanged. The first step is to identify the slope of the original line by rewriting it in $y = mx + b$ form.

$$-3x + 4y = 8$$
$$4y = 3x + 8$$
$$y = \frac{3}{4}x + 2$$

The slope of the line is $\frac{3}{4}$ and the y-intercept is 2. A dilated line must also have a slope of $\frac{3}{4}$. The only choice with the correct slope is $y = \frac{3}{4}x + 8$.

The correct choice is (**2**).

8. The only congruent parts we are given are the vertical angles, $\angle AEB \cong \angle DEC$. Examine each of the choices and determine if it provides enough information to apply one of the triangle congruence postulates.

Choice 1 – Parallel sides $\overline{AB}$ and $\overline{CD}$ result in congruent alternate interior angles, so we can conclude $\angle A \cong \angle C$ and $\angle B \cong \angle D$. All three angles are congruent, but AAA is not a congruence postulate.

Choice 2 – $\overline{AB} \cong \overline{CD}$ and $\overline{BE} \cong \overline{DE}$ give 2 pairs of congruent sides, but the congruent angles are not the included angles. This choice results in ASS which is not a congruence postulate.

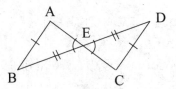

Choice 3 – Midpoint E lets us conclude $\overline{AE} \cong \overline{AC}$. This one additional pair of sides results in SA, which is not a congruence postulate.

Choice 4 – $\overline{BD}$ and $\overline{AC}$ bisecting each other implies E is the midpoint of $\overline{AC}$ and $\overline{BD}$.

Therefore, $\overline{BE} \cong \overline{ED}$ and $\overline{AE} \cong \overline{EC}$ for SAS, which is one of the triangle congruence postulates.

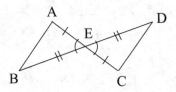

The correct choice is (**4**).

9. The co-function relationship states that if two angles A and B sum to $90°$, then $\sin(A) = \cos(B)$. Since $57° + 33° = 90°$, $\sin(57°)$ must equal $\cos(33°)$. Alternatively, you can use your calculator to evaluate each of the expressions. Both $\sin(57°)$ and $\cos(33°)$ are equal to 0.838671.

The correct choice is (**2**).

10. The radius of the hemisphere is found by dividing the 12.6 cm diameter by 2, so the radius is 6.3 cm. A hemisphere is half of a sphere, so multiply the volume formula of a sphere by $\frac{1}{2}$.

$$V = \frac{1}{2}\left(\frac{4}{3}\pi R^3\right)$$

$$= \frac{1}{2}\left(\frac{4}{3}\pi\left(6.3^3\right)\right)$$

$$= 523.7$$

The correct choice is (**1**).

11. A segment parallel to a side of a triangle forms a triangle similar to the original. $\triangle BDE \sim \triangle BAC$, and we can use the side splitter theorem to solve for EC.

$$\frac{BD}{DA} = \frac{BE}{EC}$$

$$\frac{5}{12} = \frac{7}{EC}$$

$$5EC = 7\cdot 12$$

$$EC = \frac{84}{5}$$

$$= 16.8$$

Now calculate BC.

$$BC = BE + EC$$

$$= 7 + 16.8$$

$$= 23.8$$

The correct choice is (**1**).

12. The only choice that always specifies a parallelogram is choice (3). A counter-example can be identified for each of the other choices as shown in the figures.

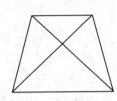

1 pair of parallel sides   1 pair of congruent sides

1 pair of congruent angles   1 pair of congruent angles   congruent diagonals

The correct choice is (**3**).

13. When two chords intersect in a circle, the products of the parts of the chords are equal.

$$JM \cdot TM = EM \cdot RM$$
$$JM \cdot TM = 8 \cdot 15$$
$$JM \cdot TM = 120$$

Now check which pair has a product of 120. The only choice that works is $16 \cdot 7.5 = 120$.

The correct choice is (**3**).

14. Sketch the two triangles with the given congruent parts marked. Check each choice to determine which additional pair of parts would not result in a valid triangle congruence postulate.

$\angle J \cong \angle S$ results in triangles congruent by ASA.

$\angle O \cong \angle A$ results in triangles congruent by AAS.

$\overline{EO} \cong \overline{MA}$ results in triangles congruent by SAS.

$\overline{JO} \cong \overline{SA}$ results in ASS, which is not a triangles congruence postulate.

The correct choice is (**4**).

15. Sketching $\triangle ABC$ and $\triangle AED$ separately, we see that right angles C and D are congruent, and the shared angle A is also congruent. Therefore, the triangles are similar by the AA postulate. The similarity statement is $\triangle ABC \sim \triangle AED$. Corresponding sides of similar triangles are proportional, so we can set up a proportion and solve for AE.

$$\frac{AE}{AB} = \frac{ED}{BC}$$

$$\frac{AE}{9} = \frac{4}{6}$$

$$6AE = 9 \cdot 4$$

$$AE = 6$$

The correct choice is (**2**).

16. First rewrite the original line in $y = mx + b$ form to identify the slope.

$$-2x + 3y = -4$$

$$3y = 2x - 4$$

$$y = \frac{2}{3}x - \frac{4}{3}$$

The slope of the original line is $\frac{2}{3}$. Parallel lines have equal slopes, so the new line must also have a slope of $\frac{2}{3}$. Write the equation of the parallel line using the point-slope form $y - y_1 = m(x - x_1)$.

Substitute $\frac{2}{3}$ for m and the given coordinates $(1, 3)$ for x_1 and y_1 to get

$$y - 3 = \frac{2}{3}(x - 1).$$

The correct choice is **(2)**.

17. All sides of a rhombus are congruent, so divide the perimeter by 4 to find each side length.

$$\begin{aligned} \text{side length} &= \frac{1}{4} \text{ perimeter} \\ &= \frac{68}{4} \\ &= 17 \end{aligned}$$

The diagonals of a rhombus bisect each other, so divide TG by 2 to find TR and RG

$$TR = RG = \frac{16}{2}$$
$$TR = RG = 8$$

The diagonals of a rhombus are perpendicular, making $\triangle GRE$ a right triangle. Apply the Pythagorean theorem in $\triangle GRE$ to find RE.

$$\begin{aligned} a^2 + b^2 &= c^2 \\ RE^2 + RG^2 &= EG^2 \\ RE^2 + 8^2 &= 17^2 \\ RE^2 + 64 &= 289 \\ RE^2 &= 225 \\ RE &= \sqrt{225} \\ &= 15 \end{aligned}$$

Diagonal $\overline{IE}$ is twice the length of $\overline{RE}$, so $IE = 30$.

The correct choice is (**2**).

18. The angle formed by two secants from an external point is equal to half the difference of the intercepted arcs.

$$m\angle P = \frac{1}{2}\left(m\widehat{AC} - m\widehat{BD}\right)$$

$$= \frac{1}{2}(72 - 34)$$

$$= 19$$

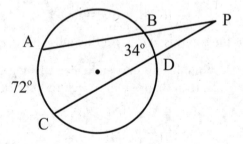

The correct choice is (**1**).

19. There are several different methods for solving a directed line segment problem. A graphical approach is to sketch the starting and ending x-coordinates on a number line. Then using the given ratio, counts jumps from the left and right. The two ends will meet at the desired x-coordinate. In this case the ratio is 2:1, so count two jumps from the left and 1 jump from the right.

After 6 jumps the two ends meet at an x-coordinate of 4. The only choice with an x-coordinate of 4 is (4, 0). If needed, you can repeat this procedure for the y-coordinate.

x-coordinate is 4

The correct choice is **(4)**.

20. Use the completing the square procedure to rewrite the equation in the form $(x - h)^2 + (y - k)^2 = r^2$.

First find the constant terms needed to complete the square:

constant for the x-terms $= (\frac{1}{2} \cdot$ coefficient of $x)^2$

$$= \left(\frac{1}{2} \cdot 8\right)^2$$

$$= (4)^2$$

$$= 16$$

The constant for the y-terms $= (\frac{1}{2} \cdot$ coefficient of $y)^2$

$$= \left(\frac{1}{2} \cdot (-12)\right)^2$$

$$= (-6)^2$$

$$= 36$$

Add the required constant to each side of the equation and factor.

$$x^2 + 8x + 16 + y^2 - 12y + 36 = 144 + 16 + 36$$

$$(x + 4)^2 + (y - 6)^2 = 196$$

The values of h and k are -4 and 6. Be careful with the signs because h and k are the values subtracted from x and y. The center is located at $(-4, 6)$. Since $r^2 = 196$, the radius is equal to $\sqrt{196}$, or 14.

The correct choice is **(4)**.

21. Starting with m∠R = 130°, we know that m∠QPS = 130° because opposite angles of a parallelogram are congruent. ∠QPS and ∠SPT are a linear pair which sum to 180°, making m∠SPT = 50°. △SPT is isosceles with ST = SP. The angles opposite the congruent sides of a triangle are congruent, therefore m∠PTS is also 50°. Finally, the angle sum theorem states that the sum of the angles in a triangle equals 180°.

$$m\angle PST + m\angle SPT + m\angle STP = 180°$$
$$m\angle PST + 50° + 50° = 180°$$
$$m\angle PST = 80°$$

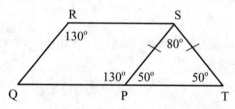

The correct choice is **(2)**.

22. The ladder, building and ground form a right triangle as shown. Trigonometry can be used to find the unknown angle using the two known sides. Relative to the unknown angle, the 12 foot ladder is the hypotenuse and the 10 foot height is the opposite. The opposite and hypotenuse indicate that a sine ratio is needed. After setting up the ratio, apply the inverse sine function to find the angle.

$$\sin(x) = \frac{\text{opposite}}{\text{hypotenuse}}$$
$$\sin(x) = \frac{10}{12}$$
$$x = \sin^{-1}\left(\frac{10}{12}\right)$$
$$= 56°$$

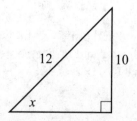

The correct choice is **(4)**.

23. Midpoint E and F form midsegment $\overline{EF}$, which is half the length of the opposite side $\overline{AB}$.

$$EF = \frac{1}{2}AB$$
$$2EF = AB$$
$$2(2x + 8) = 7x - 2$$
$$4x + 16 = 7x - 2$$
$$16 = 3x - 2$$
$$18 = 3x$$
$$x = 6$$

Substitute to find EF and AB.

$$EF = 2(6) + 8$$
$$= 20$$
$$AB = 7(6) - 2$$
$$= 40$$

Because $\triangle ABC$ is equilateral $AB = AC = 40$. AE is half of AC, so AE is 20. FB is also 20. Now add the sides length to find the perimeter of $ABFE$.

$$\text{perimeter} = AE + EF + FB + AB$$
$$= 20 + 20 + 20 + 40$$
$$= 100$$

The correct choice is **(3)**.

24. To show a parallelogram is a square, you must show it has a special rectangle property and a special rhombus property. These properties are:

  rhombus – adjacent sides congruent, diagonals bisect each other, and diagonals bisect the vertex angles
  rectangle – congruent diagonals, and adjacent sides perpendicular.

Perpendicular diagonals is a rectangle property, and congruent adjacent sides is a rhombus property.

The correct choice is **(3)**.

PART II

25. A dilation changes side lengths by multiplying them by the scale factor. Since all the side lengths of $\triangle A'BC'$ are half the side lengths of $\triangle ABC$, the triangles cannot be congruent. Congruent figures must have congruent corresponding sides.

26. The area of triangle is found using $A = \frac{1}{2}bh$, where b is the base of the triangle and h is the height. $\triangle PQR$ appears to be a right triangle. We can confirm this by looking at the slopes of QR and PR. If they are negative reciprocals the sides are perpendicular, and the triangle is a right triangle.

$$\text{slope} = \frac{rise}{run}$$

$$\text{slope}_{QR} = -\frac{4}{3}$$

$$\text{slope}_{PR} = \frac{6}{8} = \frac{3}{4}$$

The two slopes are negative reciprocals, making $\triangle PQR$ a right triangle. $\overline{PR}$ is the base of the triangle and $\overline{QR}$ is the height. We can find the lengths using the distance formula.

$$d = \sqrt{(x_2 - x_1)^2 + (y_2 - y_1)^2}$$

$$QR = \sqrt{(6-3)^2 + (1-5)^2}$$

$$= \sqrt{(3)^2 + (-4)^2}$$

$$= \sqrt{9 + 16}$$

$$= 5$$

$$PR = \sqrt{(6-(-2))^2 + (1-(-5))^2}$$

$$= \sqrt{(8)^2 + (6)^2}$$

$$= \sqrt{64 + 36}$$

$$= 10$$

Now apply the area formula.

$$A = \frac{1}{2}bh$$

$$= \frac{1}{2}10 \cdot 5$$

$$= 25$$

The area of $\triangle PQR$ is 25.

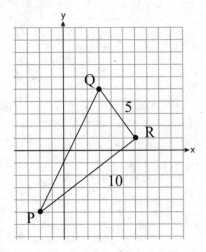

27. The wire, pole and ground form a right triangle as shown. Relative to the 68° angle, the length of 10 along the ground is the adjacent and the unknown wire length is the hypotenuse. Therefore, we should use a cosine ratio to find the wire length.

$$\cos(68°) = \frac{10}{x}$$

$$x\cos(68°) = 10$$

$$x = \frac{10}{\cos(68)}$$

$$= 26.694$$

Rounded to the nearest foot the wire length is 27 ft.

28. The area of a sector is found using $A_{sector} = \pi R^2 \dfrac{\theta}{360}$ where θ is the measure of the central angle in degrees and R is the radius. The central angle is equal in measure to the intercepted arc, which in this case is 72°. Substitute the $\theta = 72$ and $R = 10$ to find the area.

$$A_{sector} = \pi R^2 \frac{\theta}{360}$$

$$= \pi 10^2 \frac{72}{360}$$

$$= 100\pi$$

The area of the shaded sector is 20π.

29. Start with one pair of coordinates and look for one or more transformations that will map one point to another. Be sure to confirm your transformations work for the other pairs of points. Taking points $A(-2, 5)$ and $T(-5, -2)$, we see that a rotation of 90° clockwise about the origin would map A to T since the rotation transforms (x, y) to $(-y, x)$. The same transformation also maps C to U and B to T. A rotation of 90° counterclockwise about the origin maps $\triangle ABC$ to $\triangle STU$.

Note that there are numerous other correct answers, such as a rotation of 90° counterclockwise about point A followed by a translation 6 down and left 3.

30. Altitude $\overline{PQ}$ drawn to a hypotenuse $\overline{RT}$ forms two triangles similar to the original. From the given side lengths, we are working with $\triangle PRT$ and $\triangle QRP$. Sketch the triangles, matching up corresponding angles. Angles $\angle TPR$ and $\angle PQR$ are congruent right angles. The shared angle at R is also congruent in each. The similarity statement is $\triangle PRT \sim \triangle QRP$. Corresponding sides of similar triangles are proportional, so we can set up a proportion and solve for RQ.

$$\frac{RQ}{PR} = \frac{PR}{RT}$$

$$\frac{RQ}{15} = \frac{15}{17}$$

$$17RQ = 15 \cdot 15$$

$$RQ = \frac{225}{17}$$

$$= 13.23529$$

$RQ = 13.2$ rounded to the nearest tenth.

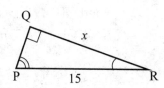

31. Use your compass to measure the radius $\overline{OA}$ by putting the point at O and the pencil at A. Keep the same compass opening and put the point at A and make an arc at B. Move the compass point to B and make an arc at C. Continue making arcs at D, E, and F. If you are accurate, a final arc should bring you back to point A. Connect points A, C and E. $\triangle ACE$ is an equilateral triangle.

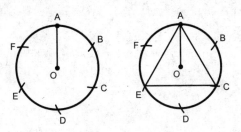

PART III

32. A trapezoid has at least one pair of parallel sides. From the graph, it appears that $\overline{AD}$ and $\overline{BC}$ may be parallel. We can check if they are parallel by calculating the slope of each.

$$\text{slope} = \frac{y_2 - y_1}{x_2 - x_1}$$

$$\text{slope of } \overline{AD} = \frac{0 - 6}{1 - (-1)} = -3$$

$$\text{slope of } \overline{BC} = \frac{-1 - 8}{6 - 3} = -3$$

The slopes of $\overline{AD}$ and $\overline{BC}$ are equal, therefore they are parallel. ABCD is a trapezoid because it has a pair of parallel sides.

We can check if the trapezoid is isosceles by calculating the lengths of the diagonals $\overline{AC}$ and $\overline{BD}$.

$$\text{length} = \sqrt{(x_2 - x_1)^2 + (y_2 - y_1)^2}$$

$$\text{length of } \overline{BD} = \sqrt{(1 - 3)^2 + (0 - 8)^2}$$

$$= \sqrt{(-2)^2 + (-8)^2}$$

$$= \sqrt{4 + 64}$$

$$= \sqrt{68}$$

$$\text{length of } \overline{AC} = \sqrt{(6 - (-1))^2 + (-1 - 6)^2}$$

$$= \sqrt{(7)^2 + (-7)^2}$$

$$= \sqrt{49 + 49}$$

$$= \sqrt{98}$$

The lengths of $\overline{BD}$ and $\overline{AC}$ are not equal. The diagonals are not congruent so $ABCD$ is not isosceles.

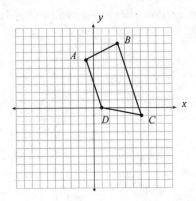

33. The volume of a cylinder is found using $V = \pi R^2 h$ where R is the radius and h is the height. The radius is half the diameter, or 3.25 ft. Since we want our answer in cubic feet, we need to convert the height to feet. 12 inches is equivalent to 1 ft, so the height equals 1 ft.

$$V = \pi R^2 h$$

$$= \pi \, (3.25)^2 \, (1)$$

$$= 33.183072 \text{ ft}^3$$

The pool is filled only $\frac{2}{3}$ full, so multiple the volume by $\frac{2}{3}$.

$$V_{water} = \frac{2}{3}(33.183072)$$

$$= 22.122048 \text{ ft}^3$$

The volume of water is 22 ft^3 rounded the nearest cubic foot.

Find the number of gallons by multiplying the volume by the given conversion factor.

$$22 \text{ ft}^3 \cdot 7.48 \, \frac{\text{gallons}}{\text{ft}^3} = 164.56 \text{ gallons.}$$

To the nearest gallon, the amount of water is 165 gallons.

34. The figure can be broken up into the two right triangles shown below. The elevation of the bottom of the blade is y, and the elevation of the top of the blade z. Use two trig ratios to calculate y and z. In each triangle, the 440 ft distance is the adjacent and the unknown side is the opposite, so we use tangent ratios.

calculate y:

$$\tan(30°) = \frac{y}{440}$$
$$y = 440\tan(30°)$$
$$= 254.034118$$

calculate z:

$$\tan(38.8°) = \frac{z}{440}$$
$$z = 440\tan(38.8°)$$
$$= 353.769082$$

The blade length x is found by subtracting y from z.

$$x = z - y$$
$$= 353.769082 - 254.034118$$
$$= 99.734964$$

Rounded to the nearest foot the blade length is 100 ft.

PART IV

35. The strategy for this proof is to show that △*HEA* is similar to △*TAH*. Corresponding sides in these triangles are proportional, and the proportion leads to $TA \cdot HA = HE \cdot TH$ when cross multiplied. In order to prove the triangles similar, we start by proving *MHTA* is a parallelogram so we can make use of its parallel sides.

Statements	Reasons
1. $\overline{HM} \cong \overline{AT}, \overline{HT} \cong \overline{AM}$	1. Given
2. *MHTA* is a parallelogram	2. A quadrilateral with opposite sides congruent is a parallelogram
3. $\overline{HT} \parallel \overline{MA}$	3. Opposite sides of a parallelogram are parallel
4. ∠*THA* ≅ ∠*EAH*	4. Alternate interior angles formed by parallel lines are congruent
5. $\overline{HE} \perp \overline{MEA}$, $\overline{HA} \perp \overline{AT}$	5. Given
6. ∠*HEA* and ∠*TAH* are right angles	6. Perpendicular lines form right angles
7. ∠*HEA* ≅ ∠*TAH*	7. Right angles are congruent
8. △*HEA* ~ △*TAH*	8. AA
9. $\dfrac{TA}{HE} = \dfrac{TH}{HA}$	9. Corresponding sides in similar triangles are proportional
10. $TA \cdot HA = HE \cdot TH$	10. Cross products of a proportion are equal

Topic	Question Numbers	Number of Points	Your Points	Your Percentage
1. Basic Angle and Segment Relationships				
2. Angle and Segment Relationships in Triangles and Polygons	21, 23	4		
3. Constructions	31	2		
4. Transformations	1, 4, 5, 29	8		
5. Triangle Congruence	8, 14, 25	6		
6. Lines, Segments, and Circles on the Coordinate Plane	7, 16, 19, 20, 26	10		
7. Similarity	11, 15, 30, 35	12		
8. Trigonometry	9, 22, 27, 34	10		
9. Parallelograms	12, 17, 24	6		
10. Coordinate Geometry Proofs	32	4		
11. Volume and Solids	3, 6, 10, 33	10		
12. Modeling	2	2		
13. Circles	13, 18, 28	6		

Examination
August 2019
Geometry

HIGH SCHOOL MATH REFERENCE SHEET

1 inch = 2.54 centimeters

1 meter = 39.37 inches

1 mile = 5280 feet

1 mile = 1760 yards

1 mile = 1.609 kilometers

1 kilometer = 0.62 mile

1 pound = 16 ounces

1 pound = 0.454 kilogram

1 kilogram = 2.2 pounds

1 ton = 2000 pounds

1 cup = 8 fluid ounces

1 pint = 2 cups

1 quart = 2 pints

1 gallon = 4 quarts

1 gallon = 3.785 liters

1 liter = 0.264 gallon

1 liter = 1000 cubic centimeters

Triangle	$A = \frac{1}{2}bh$
Parallelogram	$A = bh$
Circle	$A = \pi r^2$
Circle	$C = \pi d \text{ or } C = 2\pi r$

General Prisms	$V = Bh$
Cylinder	$V = \pi r^2 h$
Sphere	$V = \dfrac{4}{3}\pi r^3$
Cone	$V = \dfrac{1}{3}\pi r^2 h$
Pyramid	$V = \dfrac{1}{3}Bh$
Pythagorean Theorem	$a^2 + b^2 = c^2$
Quadratic Formula	$x = \dfrac{-b \pm \sqrt{b^2 - 4ac}}{2a}$
Arithmetic Sequence	$a_n = a_1 + (n-1)d$
Geometric Sequence	$a_n = a_1 r^{n-1}$
Geometric Series	$S_n = \dfrac{a_1 - a_1 r^n}{1 - r}$ where $r \neq 1$
Radians	$1 \text{ radian} = \dfrac{180}{\pi} \text{ degrees}$
Degrees	$1 \text{ degree} = \dfrac{\pi}{180} \text{ radians}$
Exponential Growth/Decay	$A = A_0 e^{k(t - t_0)} + B_0$

180 Minutes–35 Questions

PART I

Answer all 24 questions in this part. Each correct answer will receive 2 credits. No partial credit will be allowed. For each statement or question, write in the space provided the numeral preceding the word or expression that best completes the statement or answers the question.   [48 credits]

1 On the set of axes below, triangle ABC is graphed. Triangles $A'B'C'$ and $A''B''C''$, the images of triangle ABC, are graphed after a sequence of rigid motions.

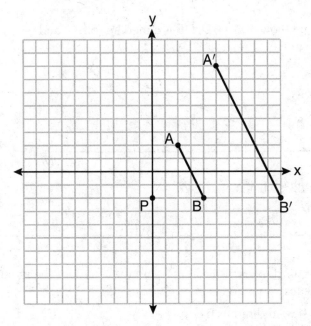

Which statement is always true?

(1) $\overline{PA} \cong \overline{AA'}$ (3) $AB = A'B'$

(2) $\overline{AB} \parallel \overline{A'B'}$ (4) $\frac{5}{2}(A'B') = AB$ 1 ___

2 The coordinates of the vertices of parallelogram $CDEH$ are $C(-5,5)$, $D(2,5)$, $E(-1,-1)$, and $H(-8,-1)$. What are the coordinates of P, the point of intersection of diagonals $\overline{CE}$ and $\overline{DH}$

(1) $(-2,3)$ (3) $(-3,2)$

(2) $(-2,2)$ (4) $(-3,-2)$ 2 ___

3 The coordinates of the endpoints of $\overline{QS}$ are $Q(-9,8)$ and $S(9,-4)$. Point R is on $\overline{QS}$ such that $QR{:}RS$ is in the ratio of 1:2. What are the coordinates of point R?

(1) (0,2)   (3) (−3,4)
(2) (3,0)   (4) (−6,6)   3 _____

4 If the altitudes of a triangle meet at one of the triangle's vertices, then the triangle is

(1) a right triangle   (3) an obtuse triangle
(2) an acute triangle   (4) an equilateral triangle   4 _____

5 In the diagram below of $\triangle ACD$. $\overline{DB}$ is a median to $\overline{AC}$, and $AB \simeq DB$.

If $m\angle DAB = 32°$, what is $m\angle BDC$?

(1) 32°   (3) 58°
(2) 52°   (4) 64°   5 _____

6 What are the coordinates of the center and the length of the radius of the circle whose equation is $x^2 + y^2 = 8x - 6y + 39$?

(1) center (−4,3) and radius 64
(2) center (4,−3) and radius 64
(3) center (−4,3) and radius 8
(4) center (4,−3) and radius 8   6 _____

7 In the diagram below of parallelogram $ABCD$, $\overline{AFGB}, \overline{CF}$ bisects $\angle DCB$, $\overline{DG}$ bisects $\angle ADC$, and $\overline{CF}$ and $\overline{DG}$ intersect at E.

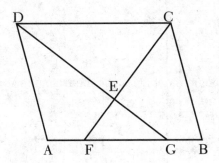

If $m\angle B = 75°$, then the measure of $\angle EFA$ is

(1) 142.5°

(3) 52.5°

(2) 127.5°

(4) 37.5°   7 _____

8 What is an equation of a line that is perpendicular to the line whose equation is $2y + 3x = 1$?

(1) $y = \frac{2}{3}x + \frac{5}{2}$

(3) $y = -\frac{2}{3}x + 1$

(2) $y = \frac{3}{2}x + 2$

(4) $y = -\frac{3}{2}x + \frac{1}{2}$ 8 _____

9 Triangles *ABC* and *RST* are graphed on the set of axes below.

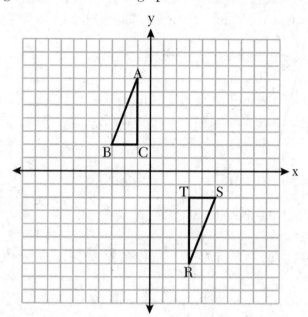

Which sequence of rigid motions will prove △*ABC* ≅ △*RST*?

(1) a line reflection over $y = x$

(2) a rotation of 180° centered at (1,0)

(3) a line reflection over the *x*-axis followed by a translation of 6 units right

(4) a line reflection over the *x*-axis followed by a line reflection over $y = x$ 9 _____

10 If the line represented by $y = -\frac{1}{4}x - 2$ is dilated by a scale factor of 4 centered at the origin, which statement about the image is true?

(1) The slope is $-\frac{1}{4}$ and the *y*-intercept is −8.

(2) The slope is $-\frac{1}{4}$ and the *y*-intercept is −2.

(3) The slope is −1 and the *y*-intercept is −8.

(4) The slope is −1 and the *y*-intercept is −2.   10 _____

11 Square *MATH* has a side length of 7 inches. Which three-dimensional object will be formed by continuously rotating square *MATH* around side $\overline{AT}$?

   (1) a right cone with a base diameter of 7 inches

   (2) a right cylinder with a diameter of 7 inches

   (3) a right cone with a base radius of 7 inches

   (4) a right cylinder with a radius of 7 inches   11 _____

12 Circle *O* with a radius of 9 is drawn below. The measure of central angle *AOC* is 120°.

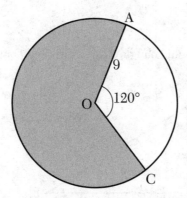

   What is the area of the shaded sector of circle *O*?

   (1) 6π (3) 27π

   (2) 12π (4) 54π 12 _____

13 In quadrilateral *QRST*, diagonals $\overline{QS}$ and $\overline{RT}$ intersect at *M*. Which statement would always prove quadrilateral *QRST* is a parallelogram?

   (1) $\angle TQR$ and $\angle QRS$ are supplementary.

   (2) $\overline{QM} \cong \overline{SM}$ and $\overline{QT} \cong \overline{RS}$

   (3) $\overline{QR} \cong \overline{TS}$ and $\overline{QT} \cong \overline{RS}$

   (4) $\overline{QR} \cong \overline{TS}$ and $\overline{QT} \parallel \overline{RS}$ 13 _____

14 A standard-size golf ball has a diameter of 1.680 inches. The material used to make the golf ball weighs 0.6523 ounce per cubic inch. What is the weight, to the *nearest hundredth of an ounce*, of one golf ball?

(1) 1.10   (3) 2.48

(2) 1.62   (4) 3.81   14 _____

15 Chelsea is sitting 8 feet from the foot of a tree. From where she is sitting, the angle of elevation of her line of sight to the top of the tree is 36°. If her line of sight starts 1.5 feet above ground, how tall is the tree, to the *nearest foot*?

(1) 8   (3) 6

(2) 7   (4) 4   15 _____

16 In the diagram below of right triangle ABC, altitude $\overline{CD}$ intersects hypotenuse $\overline{AB}$ at D.

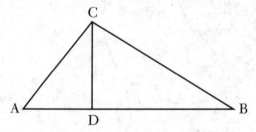

Which equation is always true?

(1) $\dfrac{AD}{AC} = \dfrac{CD}{BC}$ (3) $\dfrac{AC}{CD} = \dfrac{BC}{CD}$

(2) $\dfrac{AD}{CD} = \dfrac{BD}{CD}$ (4) $\dfrac{AD}{AC} = \dfrac{AC}{BD}$ 16 _____

17 A countertop for a kitchen is modeled with the dimensions shown
    below. An 18-inch by 21-inch rectangle will be removed for the
    installation of the sink.

What is the area of the top of the installed countertop, to the
nearest square foot?

(1) 26   (3) 22
(2) 23   (4) 19   17 ____

18 In the diagram below, $\overline{BC}$ connects points B and C on the congru-
    ent sides of isosceles triangle ADE, such that $\triangle ABC$ is isosceles with
    vertex angle A.

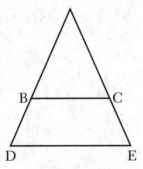

If $AB = 10$, $BD = 5$, and $DE = 12$, what is the length of $\overline{BC}$?

(1) 6   (3) 8
(2) 7   (4) 9   18 ____

19 In △ABC below, angle C is a right angle.

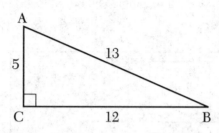

Which statement must be true?

(1) $\sin A = \cos B$ (3) $\sin B = \tan A$
(2) $\sin A = \tan B$ (4) $\sin B = \cos B$ 19 _____

20 In right triangle RST, altitude $\overline{TV}$ is drawn to hypotenuse $\overline{RS}$. If
$RV = 12$ and $RT = 18$, what is the length of $\overline{SV}$?

(1) $6\sqrt{5}$ (3) $6\sqrt{6}$
(2) 15   (4) 27   20 _____

21 What is the volume, in cubic centimeters, of a right square pyramid
with base edges that are 64 cm long and a slant height of 40 cm?

(1) 8192.0   (3) 32,768.0
(2) 13,653.$\overline{3}$ (4) 54,613.$\overline{3}$ 21 _____

22 In the diagram below, chords $\overline{PQ}$ and $\overline{RS}$ of circle O intersect at T.

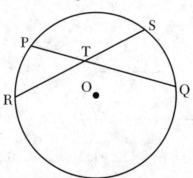

Which relationship must always be true?

(1) $RT = TQ$ (3) $RT + TS = PT + TQ$
(2) $RT = TS$ (4) $RT \times TS = PT \times TQ$ 22 _____

23 A rhombus is graphed on the set of axes below.

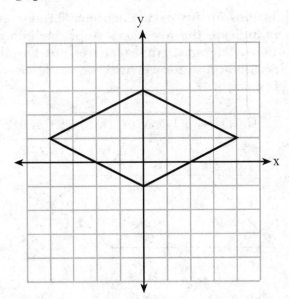

Which transformation would carry the rhombus onto itself?

(1) 180° rotation counterclockwise about the origin

(2) reflection over the line $y = \frac{1}{2}x + 1$

(3) reflection over the line $y = 0$

(4) reflection over the line $x = 0$ 23 _____

24 A 15-foot ladder leans against a wall and makes an angle of 65° with the ground. What is the horizontal distance from the wall to the base of the ladder, to the *nearest tenth of a foot*?

(1) 6.3 (3) 12.9

(2) 7.0 (4) 13.6 24 _____

PART II

Answer all 7 questions in this part. Each correct answer will receive 2 credits. Clearly indicate the necessary steps, including appropriate formula substitutions, diagrams, graphs, charts, etc. For all questions in this part, a correct numerical answer with no work shown will receive only 1 credit. [14 credits]

25 In parallelogram $ABCD$ shown below, m$\angle DAC = 98°$ and m$\angle ACD = 36°$.

What is the measure of angle B? Explain why.

26 An airplane took off at a constant angle of elevation. After the
plane traveled for 25 miles, it reached an altitude of 5 miles, as
modeled below.

To the *nearest tenth of a degree*, what was the angle of elevation?

27 On the set of axes below, $\triangle ABC \cong \triangle DEF$.

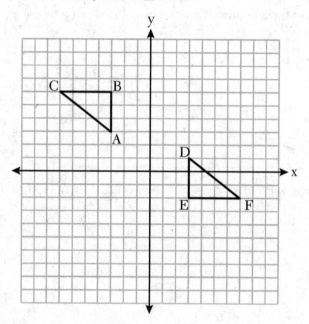

Describe a sequence of rigid motions that maps $\triangle ABC$ onto $\triangle DEF$.

28 The vertices of △ABC have coordinates $A(-2,-1)$, $B(10,-1)$, and $C(4,4)$. Determine and state the area of △ABC. [The use of the set of axes below is optional.]

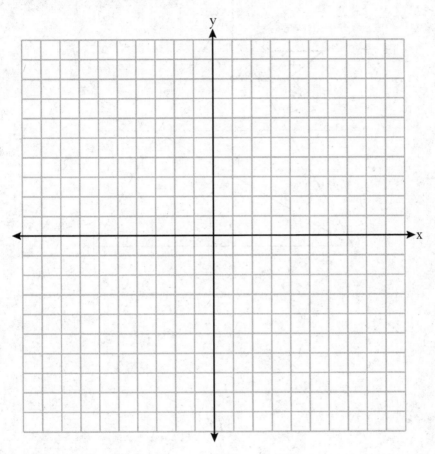

29 Using the construction below, state the degree measure of ∠*CAD*.
Explain why.

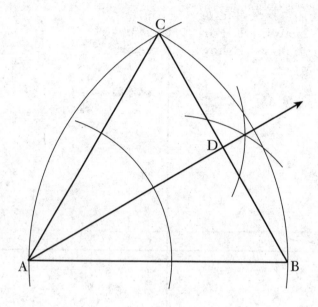

30 In the diagram below of circle K, secant $\overline{PLKE}$ and tangent $\overline{PZ}$ are drawn from external point P.

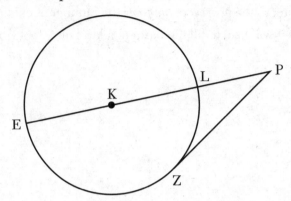

If $m\widehat{LZ} = 56°$, determine and state the degree measure of angle P.

31 A large water basin is in the shape of a right cylinder. The inside of the basin has a diameter of $8\frac{1}{4}$ feet and a height of 3 feet. Determine and state, to the *nearest cubic foot*, the number of cubic feet of water that it will take to fill the basin to a level of $\frac{1}{2}$ foot from the top.

PART III

Answer all 3 questions in this part. Each correct answer will receive 4 credits. Clearly indicate the necessary steps, including appropriate formula substitutions, diagrams, graphs, charts, etc. For all questions in this part, a correct numerical answer with no work shown will receive only 1 credit.   [12 credits]

32 Triangle *ABC* is shown below. Using a compass and straightedge, construct the dilation of △*ABC* centered at *B* with a scale factor of 2. [Leave all construction marks.]

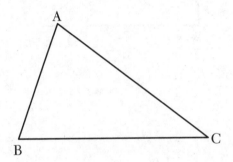

Is the image of △*ABC* similar to the original triangle? Explain why.

33 In the diagram below, $\triangle ABE \cong \triangle CBD$.

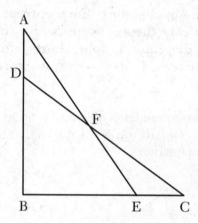

Prove: $\triangle AFD \cong \triangle CFE$

34 A cargo trailer, pictured below, can be modeled by a rectangular prism and a triangular prism. Inside the trailer, the rectangular prism measures 6 feet wide and 10 feet long. The walls that form the triangular prism each measure 4 feet wide inside the trailer. The diagram below is of the floor, showing the inside measurements of the trailer.

Cargo Trailer

Cargo Trailer Floor

If the inside height of the trailer is 6.5 feet, what is the total volume of the inside of the trailer, to the *nearest cubic foot*?

PART IV

Answer the question in this part. A correct answer will receive 6 credits. Clearly indicate the necessary steps, including appropriate formula substitutions, diagrams, graphs, charts, etc. A correct numerical answer with no work shown will receive only 1 credit. [6 credits]

35 The coordinates of the vertices of $\triangle ABC$ are $A(1,2)$, $B(-5,3)$, and $C(-6,-3)$.

Prove that $\triangle ABC$ is isosceles.
[The use of the set of axes on the next page is optional.]

State the coordinates of point D such that quadrilateral $ABCD$ is a square.

Question 35 is continued on the next page.

Question 35 continued

Prove that your quadrilateral *ABCD* is a square.
[The use of the set of axes below is optional.]

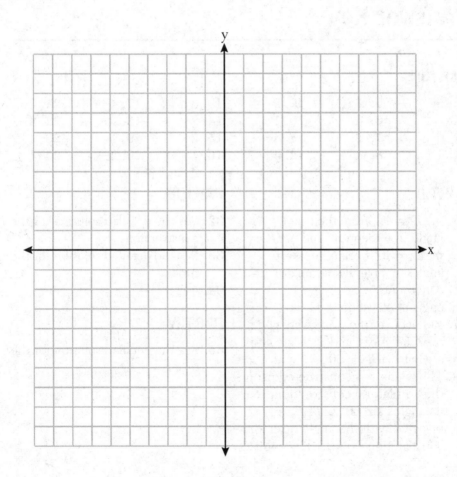

464

Answers
August 2019
Geometry

Answer Key

PART I

1. 2	5. 3	9. 2	13. 3	17. 4	21. 3
2. 3	6. 4	10. 1	14. 2	18. 3	22. 4
3. 3	7. 2	11. 4	15. 2	19. 1	23. 4
4. 1	8. 1	12. 4	16. 1	20. 2	24. 1

PART II

25. m∠D = 46° from the triangle angle sum theorem, and m∠B = m∠D = 46° because opposite angles of a parallelogram are congruent.
26. angle of elevation = 11.5°
27. Reflect over the x-axis, reflect over the y-axis, and translate 4 up.
28. area of △ABC = 30
29. The triangle is constructed using the equilateral construction, so its angles are 60°. ∠BAC is bisected with the angle bisector construction, so m∠CAD = 30°.
30. m∠P = 34°
31. 134 ft³

PART III

32. See the detailed solution for the construction.
33. See the detailed solution for the proof.
34. 442 ft³

PART IV

35. See the detailed solution for the proof.

In **PARTS II–IV**, you are required to show how you arrived at your answers. For sample methods of solutions, see the *Answers Explained* section.

Answers Explained

PART I

1. The dilation of a line segment is always parallel to the original segment, so $\overline{AB} \parallel \overline{A'B'}$. Examining the other choices, we see that choice (1) is incorrect because $\overline{PA} \cong \overline{AA'}$ implies a 2:1 ratio instead of the 5:2 ratio. Choices (3) and (4) are incorrect because. $A'B' = \dfrac{5}{2}AB$.

   The correct choice is **(2)**.

2. The diagonals of a parallelogram bisect each other, so the point of intersection must be the midpoint of the diagonals. Using the coordinates $(-5, 5)$ and $(-1, -1)$ for diagonal $\overline{CE}$, apply the midpoint formula.

$$\text{midpoint} = \left(\frac{x_1 + x_2}{2}, \frac{y_1 + y_2}{2} \right)$$

$$= \left(\frac{-5 + (-1)}{2}, \frac{5 + (-1)}{2} \right)$$

$$= (-3, 2)$$

   The correct choice is **(3)**. Note that you would get the same result using diagonal $\overline{DH}$.

3. The coordinates of the point that divide a segment in a specified ratio can be calculated using the following formula:

$$\text{ratio} = \frac{x - x_1}{x_2 - x} \qquad \text{ratio} = \frac{y - y_1}{y_2 - y}$$

   where the coordinates of $Q(-9, 8)$ are (x_1, y_1) the coordinates of $S(9, -4)$ are (x_2, y_2).

$$\frac{1}{2} = \frac{x - (-9)}{9 - x} \qquad\qquad \frac{1}{2} = \frac{y - 8}{-4 - y}$$

$$2x + 18 = 9 - x \qquad\qquad 2y - 16 = -4 - y$$

$$3x = -9 \qquad\qquad\qquad 3y = 12$$

$$x = -3 \qquad\qquad\qquad\quad y = 4$$

   The coordinates of R are $(-3, 4)$.

   The correct choice is **(3)**.

4. Altitudes of a triangle always form a right angle with the opposite side. If the altitudes meet at a vertex, then the triangle must be a right triangle.

altitudes

The correct choice is **(1)**.

5. Since $\overline{AB} \cong \overline{BD}$, we know $\triangle ABD$ is isosceles and $\angle DAB \cong \angle ADB$. Apply the angle sum theorem in $\triangle ABD$ to find m$\angle ABD$.

$$m\angle ABD + m\angle DAB + m\angle ADB = 180°$$

$$m\angle ABD + 32° + 32° = 180°$$

$$m\angle ABD = 116°.$$

Next, we see that $\angle ABD$ and $\angle CBD$ are a linear pair and sum to 180°.

$$m\angle ABD + m\angle CBD = 180°$$

$$116° + m\angle CBD = 180°$$

$$m\angle CBD = 64°.$$

A median divides a side of a triangle into two congruent segments, so $\overline{AB} \cong \overline{BC}$. Combining this with $\overline{AB} \cong \overline{BD}$, we can conclude that $\overline{BC} \cong \overline{BD}$. We now know that $\triangle BDC$ is isosceles with $\angle BDC \cong \angle BCD$. Apply the triangle sum theorem to find m$\angle BDC$.

$$m\angle BDC + m\angle BCD + m\angle CBD = 180°$$

$$2\, m\angle BDC + 64° = 180°$$

$$2\, m\angle BDC = 116°$$

$$m\angle BDC = 58°$$

The correct choice is **(3)**.

6. Use the completing the square procedure to convert the equation to center-radius form, $(x - h)^2 + (y - k^2) = R^2$. The center has coordinates (h, k) and the radius is R. Start by rearranging the equation so all the x and y terms are grouped together on the left.

$$x^2 + y^2 = 8x - 6y + 39$$
$$x^2 - 8x + y^2 + 6y = 39$$

Now find the constant terms we need to complete the square:

constant for the x-terms $= (\frac{1}{2} \cdot \text{coefficient of } x)^2$

$$= \left(\frac{1}{2} \cdot (-8)\right)^2$$
$$= 16$$

constant for the y-terms $= (\frac{1}{2} \cdot \text{coefficient of } y)^2$

$$= \left(\frac{1}{2} \cdot 6\right)^2$$
$$= 9$$

Add the required constant to each side of the equation and factor.

$$x^2 - 8x + 16 + y^2 + 6y + 9 = 39 + 16 + 9$$
$$(x - 4)^2 + (y + 3)^2 = 64$$

The values of h and k are 4 and 3. Be careful with the signs because h and k are the values subtracted from x and y. The center is located at $(4, -3)$. Since $R^2 = 64$, the radius is equal to $\sqrt{64}$, or 8.

The correct choice is (**4**).

7. Consecutive angles of a parallelogram are supplementary, so we can find $m\angle BCD$.

$$m\angle BCD + m\angle B = 180°$$
$$m\angle BCD + 75° = 180°$$
$$m\angle BCD = 105°$$

Next, use angle bisector $\overline{CF}$ to find m$\angle DCF$.

$$m\angle DCF = \frac{1}{2}\, m\angle BCD$$

$$= \frac{1}{2}(105°)$$

$$= 52.5°$$

$\angle DCF$ and $\angle CFB$ are congruent alternate interior angles, so m$\angle CFG$ is also equal to 52.5°. Finally, $\angle EFA$ and $\angle CFG$ are a linear pair that sum to 180°.

$$m\angle EFA + m\angle CFG = 180°$$

$$m\angle EFA + 52.5° = 180°$$

$$m\angle EFA = 127.5°$$

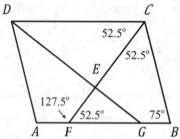

The correct choice is (**2**).

8. First, identify the slope by rewriting the equation in $y = mx + b$ form.

$$2y + 3x = 1$$

$$2y = -3x + 1$$

$$y = -\frac{3}{2}x + \frac{1}{2}$$

The slope of the given line is $-\frac{3}{2}$. Slopes of perpendicular lines are negative reciprocals, which we can find by changing the sign and flipping the fraction. The negative reciprocal of $-\frac{3}{2}$ is $\frac{2}{3}$. The only choice with slope of $\frac{2}{3}$ is $y = \frac{2}{3}x + \frac{5}{2}$.

The correct choice is (**1**).

9.

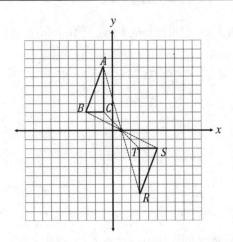

Two figures are congruent if a sequence of rigid motions maps one figure onto another. To check if a 180° rotation was used, sketch segments from each vertex of △ABC to the corresponding vertex of △RST as shown in the figure. All three segments are concurrent at (1, 0), so this point is the center of a 180° rotation that maps △ABC to △RST.

The correct choice is (2).

10. When a line is dilated about the origin, the slope remains the same, but the y-intercept is multiplied by the scale factor. The slope remains $-\frac{1}{4}$. The y-intercept transforms to $4 \cdot (-2)$, or -8.

The correct choice is (1).

11. A square rotated about any of its sides always sweeps out a cylinder. The side length of the square is the height, as well as the radius of the cylinder. The solid is a cylinder with a radius of 7 inches.

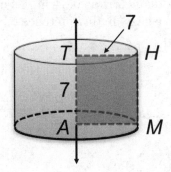

The correct choice is (4).

12. The area of a sector is calculated using the formula $A_{\text{sector}} = \pi R^2 \dfrac{\theta}{360}$, where θ is the measure of the central angle and R is the radius. First, calculate the central angle that intercepts the shaded arc.

$$\theta = 360° - 120°$$

$$= 240°$$

$$= A_{\text{sector}} = \pi R^2 \frac{\theta}{360}$$

$$= \pi \cdot 9^2 \cdot \frac{240}{360}$$

$$= 54\pi$$

The correct choice is **(4)**.

13. One possible property that would prove a quadrilateral is a parallelogram is congruent opposite sides. For quadrilateral $QRST$, congruent opposite sides would be $\overline{QR} \cong \overline{TS}$ and $\overline{QT} \cong \overline{RS}$.

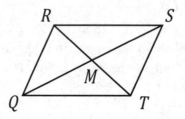

The correct choice is **(3)**.

14. The golf ball can be modeled as a sphere with a diameter of 1.680 in. Its radius is one half the diameter, or 0.84 in. Calculate the volume of the sphere using the formula given in the reference table.

$$V = \frac{4}{3}\pi R^3$$

$$= \frac{4}{3}\pi (0.84)^3$$

$$= 2.482713 \text{ in}^3$$

The weight of an object is equal to the product of its volume and density.

$$\text{weight} = \text{volume} \cdot \text{density}$$

$$= 2.482713 \text{ in}^3 \left(0.6523 \frac{\text{ounces}}{\text{in}^3} \right)$$

$$= 1.62 \text{ ounces}$$

The correct choice is (**2**).

15. The tree and Chelsea's line of sight form a right triangle as shown in the figure. Relative to the 36° angle of elevation, the 8-foot distance is the adjacent and the observed tree height is the opposite. A tangent ratio can be used to find the unknown height.

$$\tan(36°) = \frac{\text{opposite}}{\text{adjacent}}$$

$$\tan(36°) = \frac{x}{8}$$

$$x = 8 \tan(36°)$$

$$= 5.8123 \text{ ft}$$

Add the 1.5-ft distance from the ground to Chelsea's line of sight to get the total height.

$$\text{total height} = 5.8123 + 1.5$$

$$= 7.3123 \text{ ft}$$

To the nearest foot the height of the tree is 7 feet.

The correct choice is (**2**).

16. When an altitude is drawn to the hypotenuse of a right triangle, three similar triangles are formed. Sketch the three triangles separately and label the congruent angles. We know that corresponding sides of similar triangles are congruent, so we can write the following proportion:

$$\frac{AD}{AC} = \frac{CD}{BC}$$

$\overline{AD}$ and $\overline{CD}$ are corresponding sides of the small and medium triangles. $\overline{AC}$ and $\overline{BC}$ are also corresponding sides of the same two triangles. None of the other choices match up corresponding sides of the same two triangles.

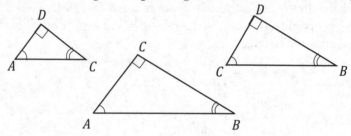

The correct choice is (**1**).

17. The countertop can be broken up into three rectangles—a solid 2 ft by 8 ft rectangle, a solid 2 ft by 3 ft rectangle, and a 21 in × 18 in cutout. Calculate the area of two solid rectangles.

$$\text{Area}_{\text{solid}} = \text{length}_1 \cdot \text{width}_1 + \text{length}_2 \cdot \text{width}_2$$

$$= 2 \cdot 8 + 2 \cdot 3$$

$$= 22 \text{ ft}^2$$

Before finding its area, the sink cutout dimensions need to be converted to feet by multiplying by the conversion factor $\frac{1 \text{ ft}}{12 \text{ in}}$.

Sink width $= 21 \text{ in} \cdot \dfrac{1 \text{ ft}}{12 \text{ in}}$ $\qquad$ Sink length $= 18 \text{ in} \cdot \dfrac{1 \text{ ft}}{12 \text{ in}}$

$= 1.75 \text{ ft}$ $\qquad\qquad\qquad\qquad$ $= 1.5 \text{ ft}$

$\text{Area}_{\text{sink}} = (1.75)(1.5)$

$= 2.625 \text{ ft}^2$

Now subtract the sink cutout for the sink from the solid area.

$$\text{Area}_{\text{countertop}} = 22 - 2.625$$

$$= 19.375 \text{ ft}^2$$

Rounded to the nearest square foot the area is 19 ft^2.

The correct choice is **(4)**.

18. Using the definition of isosceles triangles, we know $\overline{AB} \cong \overline{AC}$ and $\overline{AD} \cong \overline{AE}$. Dividing these congruence statements, we get $\frac{AB}{AD} \cong \frac{AC}{AE}$. The shared angle $\angle A$ is congruent in each triangle as well, so $\triangle BAC$ is similar to $\triangle DAE$ by the SAS postulate. Apply the theorem that corresponding sides in similar triangles are proportional.

$$\frac{BA}{DA} = \frac{BC}{DE}$$

$$\frac{10}{10+5} = \frac{BC}{12}$$

$$\frac{10}{15} = \frac{BC}{12}$$

$$15BC = 120$$

$$BC = 8$$

The correct choice is **(3)**.

19. Given two complementary angles A and B, the co-function relationship states that $\sin(A) = \cos(B)$. In right triangle ABC, angle C is a right triangle, so $\angle A$ and $\angle B$ are complementary. Therefore, $\sin(A) = \cos(B)$.

    The correct choice is (**1**).

20. There are several ways to approach this problem. One method is to apply trigonometry to find $\angle R$, and then RS.

    Using $\triangle TVR$, relative to $\angle R$ side $\overline{RT}$ is the hypotenuse and $\overline{RV}$ is the adjacent. We can apply a cosine ratio to find $\angle R$.

    $$\cos(R) = \frac{\text{adjacent}}{\text{hypotenuse}}$$

    $$\cos(R) = \frac{12}{18}$$

    $$R = \cos^{-1}\left(\frac{12}{18}\right)$$

    $$R = 48.189685°$$

    Now use $\triangle TRS$ to find RS. Relative to $\angle R$ side $\overline{RS}$ is the hypotenuse and side $\overline{TR}$ is the adjacent. Use a cosine ratio to find $\overline{RS}$.

    $$\cos(R) = \frac{\text{adjacent}}{\text{hypotenuse}}$$

    $$\cos(48.189685) = \frac{18}{RS}$$

    $$RS = \frac{18}{\cos(48.189685)}$$

    $$= 27$$

    From the figure,

    $$SV = RS - 12$$

    $$= 27 - 12$$

    $$= 15$$

The correct choice is (**2**).

21. The height, h, of the pyramid forms a right triangle with the base and the 40 cm slant height. Use the Pythagorean theorem to find the height h. Be sure to use half of the side length for the bottom of the right triangle.

$$a^2 + b^2 = c^2$$
$$h^2 + 32^2 = 40^2$$
$$h^2 + 1024 = 1600$$
$$h^2 = 576$$
$$h = 24$$

Now find the volume of the pyramid using the volume formula from the reference table. The area of the base, B, is the area of the square base.

$$V = \frac{1}{3}Bh$$
$$= \frac{1}{3}\left(64^2\right)24$$
$$= 32{,}768$$

The volume of the pyramid is 32,768 cm³.

The correct choice is **(3)**.

22. When two chords intersect within a circle, the products of the parts of the chords are equal. In this case $RT \cdot TS = PT \cdot TQ$.

The correct choice is **(4)**.

23. The rhombus is symmetric about the y-axis, so a reflection over that axis will map the rhombus onto itself. The equation of the y-axis is $x = 0$.

The correct choice is **(4)**.

24. The ladder, wall, and ground form a right triangle. We can use trigonometry to calculate the distance from the ladder to the wall. Relative to the $65°$ angle, the ladder is the hypotenuse and the distance to the wall is the adjacent. This indicates a cosine ratio.

$$\cos(65°) = \frac{\text{adjacent}}{\text{hypotenuse}}$$

$$\cos(65°) = \frac{x}{15}$$

$$x = 15\cos(65°)$$

$$= 6.33927$$

Rounded to the nearest foot, the distance to the wall is 6.3 ft.

The correct choice is **(1)**.

PART II

25. Two of the three angles are given in $\triangle ADC$. Apply the angle sum theorem to find $\angle D$.

$$m\angle D + 36° + 98° = 180°$$

$$m\angle D + 134° = 180°$$

$$m\angle D = 46°$$

Opposite angles of a parallelogram are congruent, so $m\angle B = m\angle D$. Therefore, $m\angle B = 46°$.

26. The angle of elevation is the angle formed between the ground and the 25-mile flight path. Relative to the angle of elevation, the 25-mile flight path is the hypotenuse and the 5-mile altitude is the opposite. This indicates using a sine ratio. Set up the correct ratio and then apply the inverse sine function to find the angle.

$$\sin(\theta) = \frac{5}{25}$$

$$\theta = \sin^{-1}\left(\frac{5}{25}\right)$$

$$= 11.53695°$$

Rounded to the nearest tenth of a degree, the angle of elevation is $11.5°$.

27. There are many possible sequences of transformations that will map $\triangle ABC$ to $\triangle DEF$. One possible sequence is two reflections followed by a translation. First, reflect $\triangle ABC$ over the x-axis to triangle I. Then, reflect triangle I over the y-axis to triangle II. Finally, translate triangle II up 4 to $\triangle DEF$.

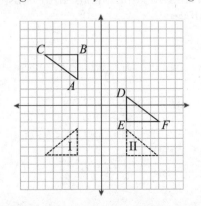

28. After graphing the triangle, sketch altitude $\overline{CD}$ to side $\overline{AB}$. We can calculate their lengths by counting grid units since they are vertical and horizontal segments. We find that $AB = 12$ and $CD = 5$. Now apply the formula for the area of a triangle.

$$A = \frac{1}{2}bh$$

$$= \frac{1}{2}(12)(5)$$

$$= 30$$

The area of $\triangle ABC$ is 30.

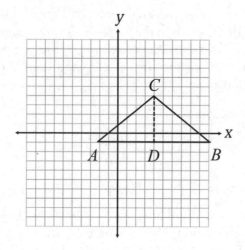

29. $\triangle ABC$ is constructed using the equilateral triangle construction. Since all angles of an equilateral triangle measure 60°, we know that $m\angle CAB = 60°$. The angle bisector construction was used to construct $\overrightarrow{AD}$. Bisector $\overrightarrow{AD}$ divides $\angle CAB$ in half, so $m\angle CAD = 30°$.

30. First find $m\widehat{EZ}$ using diameter $\overline{LE}$, which intercepts a 180° semicircle.

$$m\widehat{EZ} + m\widehat{LZ} = 180°$$

$$m\widehat{EZ} + 56° = 180°$$

$$m\widehat{EZ} = 124°$$

When a tangent and secant intersect, the angle formed is equal to one-half the difference of the intercepted arcs.

$$m\angle P = \frac{1}{2}\left(m\widehat{EZ} - m\widehat{LZ}\right)$$

$$= \frac{1}{2}(124° - 56°)$$

$$= 34°$$

31. Calculate the height of the water in the cylinder by subtracting $\frac{1}{2}$ ft from the height of the basin.

$$h_{\text{water}} = 3 - \frac{1}{2}$$

$$= 2.5 \text{ ft}$$

The radius of the cylinder is half the diameter, or 4.125 ft.

Next, calculate the volume of the water using the volume formula for a cylinder.

$$V = \pi R^2 h$$

$$= \pi(4.125^2)(2.5)$$

$$= 133.64041 \text{ ft}^3$$

To the nearest cubic foot, the volume of water is 134 ft³.

PART III

32. (a) With the compass point at B, open the compass so the pencil is at A and make an arc to show that you measured that distance.

(b) Move the point to A and, with the same opening, make a second arc.

(c) Extend $\overline{BA}$ to the second arc. Label the intersection point E.

(d) Extend $\overline{BC}$ to F by repeating steps (a)–(c).

$\triangle EBF$ is a dilation of $\triangle ABC$ by a scale factor of 2 centered at B. The image is similar to the original because dilations are a similarity transformation. They increase length without changing any of the angle measures.

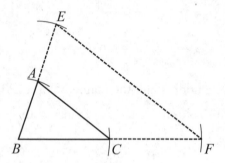

33. Prove the two triangles congruent by AAS. We will use a pair of congruent angles from the given congruent triangles, a pair of vertical angles, and sides $\overline{AD}$ and $\overline{CE}$. The sides are shown to be congruent by subtracting two pairs of congruent sides from the given triangles.

Statement	Reason
1. $\triangle ABE \cong \triangle CBD$	1. Given
2. $\angle A \cong \angle C$	2. CPCTC
3. $\angle AFD \cong \angle CFE$	3. Vertical angles are congruent
4. $\overline{AB} \cong \overline{CB}$	4. CPCTC
5. $\overline{BD} \cong \overline{BE}$	5. CPCTC
6. $\overline{AB} - \overline{BD} \cong \overline{CB} - \overline{BE}$	6. Subtraction Postulate
7. $\overline{AD} \cong \overline{CE}$	7. Partition Postulate
8. $\triangle AFD \cong \triangle CFE$	8. AAS

34. The volume of a prism is calculated using $V = Bh$, where B is the area of the base and h is the height of the prism. First, calculate the volume of the rectangular prism. The base is a 10 ft by 6 ft rectangle and the height is 6.5 ft.

$$B = 10 \cdot 6$$
$$= 60 \text{ ft}^2$$

$$V = Bh$$
$$= (60)(6.5)$$
$$= 390 \text{ ft}^3$$

The base of the triangular prism is an isosceles triangle with 4-ft legs and a base of 6 ft. To find its area we need to calculate the altitude, a, of the triangle. The altitude forms two right triangles as shown in the figure, and we can use the Pythagorean theorem to calculate a.

$$a^2 + b^2 = c^2$$
$$a^2 + 3^2 = 4^2$$
$$a^2 = 7$$
$$a = \sqrt{7}$$

The area of the triangle base can now be calculated.

$$B = \frac{1}{2}(6)\left(\sqrt{7}\right)$$
$$= 3\sqrt{7}$$

Now calculate the volume of the triangular prism.

$$V = Bh$$
$$= \left(3\sqrt{7}\right)(6.5)$$
$$= 51.592 \text{ ft}^2$$

Finally combine the two volumes.

$$V_{total} = 390 \text{ ft}^3 + 51.592 \text{ ft}^3$$
$$441.592 \text{ ft}^3$$

Rounded to the nearest cubic foot the volume is 442 ft^3.

PART IV

35. An isosceles triangle has two congruent sides, so the proof requires us to show the lengths of two sides are equal. $\overline{AB}$ and $\overline{BC}$ appear to be the congruent sides, so start with those.

Length of AB:

$$d = \sqrt{(x_1 - x_2)^2 + (y_1 - y_2)^2}$$
$$= \sqrt{(1 - (-5))^2 + (2 - 3)^2}$$
$$= \sqrt{37}$$

Length of BC:

$$d = \sqrt{(x_1 - x_2)^2 + (y_1 - y_2)^2}$$
$$= \sqrt{(-5 - (-6))^2 + (3 - (-3))^2}$$
$$= \sqrt{37}$$

$\overline{AB}$ and $\overline{BC}$ have the same length, so they are congruent and $\triangle ABC$ is isosceles.

To find point D that completes the square, count the horizontal and vertical translations from point B to point A. Point A is 6 units to the right and 1 unit down from point B. Count those same distances from point C to locate point D. Starting from C $(-6, -3)$, 6 units right and 1 unit down puts point D at $(0, -4)$.

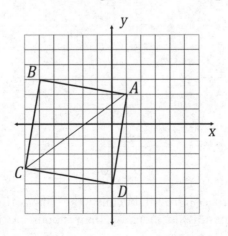

We need to demonstrate one parallelogram property, one rhombus property, and one rectangle property. One possible approach is to show:

- opposite sides are congruent (parallelogram)
- consecutive sides are congruent (rhombus)
- consecutive sides form a right angle (rectangle)

Since we already have the lengths of $\overline{AB}$ and $\overline{BC}$, we just need $\overline{CD}$ and $\overline{AD}$.

Length of CD:

$$d = \sqrt{(x_1 - x_2)^2 + (y_1 - y_2)^2}$$
$$= \sqrt{(-6 - 0)^2 + (-3 - (-4))^2}$$
$$= \sqrt{37}$$

Length of AD:

$$d = \sqrt{(x_1 - x_2)^2 + (y_1 - y_2)^2}$$
$$= \sqrt{(1 - 0)^2 + (2 - (-4))^2}$$
$$= \sqrt{37}$$

Now calculate the slopes of $\overline{AB}$ and $\overline{BC}$.

Slope of AB:

$$\text{slope} = \frac{y_2 - y_1}{x_2 - x_1}$$
$$= \frac{3 - 2}{-5 - 1}$$
$$= -\frac{1}{6}$$

Slope of BC:

$$\text{slope} = \frac{y_2 - y_1}{x_2 - x_1}$$
$$= \frac{-3 - 3}{-6 - (-5)}$$
$$= 6$$

All four sides are congruent because they have the same lengths. $\overline{AB}$ and $\overline{BC}$ are perpendicular because their slopes are negative reciprocals. Therefore, $ABCD$ is a square.

The Geometry Learning Standards

The New York State Geometry Regents is aligned to the Geometry Learning Standards. All topics listed below may be found on the Regents exam.

CONGRUENCE G-CO

Experiment with Transformations in the Plane

1. Know precise definitions of angle, circle, perpendicular line, parallel line, and line segment, based on the undefined notions of point, line, distance along a line, and distance around a circular arc.

2. Represent transformations in the plane using, for example, transparencies and geometry software; describe transformations as functions that take points in the plane as inputs and give other points as outputs. Compare transformations that preserve distance and angle to those that do not (e.g., translation versus horizontal stretch).

3. Given a rectangle, parallelogram, trapezoid, or regular polygon, describe the rotations and reflections that carry it onto itself.

4. Develop definitions of rotations, reflections, and translations in terms of angles, circles, perpendicular lines, parallel lines, and line segments.

5. Given a geometric figure and a rotation, reflection, or translation, draw the transformed figure using, for example, graph paper, tracing paper, or geometry software. Specify a sequence of transformations that will carry a given figure onto another.

Understand Congruence in Terms of Rigid Motions

6. Use geometric descriptions of rigid motions to transform figures and to pre-dict the effect of a given rigid motion on a given figure; given two figures, use the definition of congruence in terms of rigid motions to decide if they are congruent.

7. Use the definition of congruence in terms of rigid motions to show that two triangles are congruent if and only if corresponding pairs of sides and corre-sponding pairs of angles are congruent.

8. Explain how the criteria for triangle congruence (ASA, SAS, and SSS) fol-low from the definition of congruence in terms of rigid motions.

Prove Geometric Theorems

9. Prove theorems about lines and angles. *Theorems include: vertical angles are congruent; when a transversal crosses parallel lines, alternate interior angles are congruent and corresponding angles are congruent; points on a perpendicular bisector of a line segment are exactly those equidistant from the segment's endpoints.*

10. Prove theorems about triangles. *Theorems include: measures of interior angles of a triangle sum to 180°; base angles of isosceles triangles are con-gruent; the segment joining midpoints of two sides of a triangle is parallel to the third side and half the length; the medians of a triangle meet at a point.*

11. Prove theorems about parallelograms. *Theorems include: opposite sides are congruent; opposite angles are congruent; the diagonals of a parallelogram bisect each other; and conversely, rectangles are parallelograms with con-gruent diagonals.*

Make Geometric Constructions

12. Make formal geometric constructions with a variety of tools and methods (compass and straightedge, string, reflective devices, paper folding, dynamic geometric software, etc.). *Copying a segment; copying an angle; bisecting a segment; bisecting an angle; constructing perpendicular lines, including the perpendicular bisector of a line segment; and constructing a line parallel to a given line through a point not on the line.*

13. Construct an equilateral triangle, a square, and a regular hexagon inscribed in a circle.

SIMILARITY, RIGHT TRIANGLES, AND TRIGONOMETRY G-SRT

Understand Similarity in Terms of Similarity Transformations

1. Verify experimentally the properties of dilations given by a center and a scale factor:

   a. A dilation takes a line not passing through the center of the dilation to a parallel line and leaves a line passing through the center unchanged.

   b. The dilation of a line segment is longer or shorter in the ratio given by the scale factor.

2. Given two figures, use the definition of similarity in terms of similarity transformations to decide if they are similar; explain using similarity transformations the meaning of similarity for triangles as the equality of all corresponding pairs of angles and the proportionality of all corresponding pairs of sides.

3. Use the properties of similarity transformations to establish the AA criterion for two triangles to be similar.

Prove Theorems Involving Similarity

4. Prove theorems about triangles. *Theorems include: a line parallel to one side of a triangle divides the other two proportionally and conversely; the Pythagorean theorem proved using triangle similarity.*

5. Use congruence and similarity criteria for triangles to solve problems and to prove relationships in geometric figures.

Define Trigonometric Ratios and Solve Problems Involving Right Triangles

6. Understand that by similarity, side ratios in right triangles are properties of the angles in the triangle, leading to definitions of trigonometric ratios for acute angles.

7. Explain and use the relationship between the sine and cosine of complementary angles.

8. Use trigonometric ratios and the Pythagorean theorem to solve right triangles in applied problems.

CIRCLES G-C

Understand and Apply Theorems About Circles

1. Prove that all circles are similar.

2. Identify and describe relationships among inscribed angles, radii, and chords. *Include the relationship between central, inscribed, and circumscribed angles; inscribed angles on a diameter as right angles; the radius of a circle being perpendicular to the tangent where the radius intersects the circle.*

3. Construct the inscribed and circumscribed circles of a triangle, and prove properties of angles for a quadrilateral inscribed in a circle.

Find Arc Lengths and Areas of Sectors of Circles

4. Derive using similarity the fact that the length of the arc intercepted by an angle is proportional to the radius, and define the radian measure of the angle as the constant of proportionality; derive the formula for the area of a sector.

EXPRESSING GEOMETRIC PROPERTIES WITH EQUATIONS G-GPE

Translate Between the Geometric Description and the Equation for a Conic Section

1. Derive the equation of a circle of given center and radius using the Pythagorean theorem; complete the square to find the center and radius of a circle given by an equation.

2. Derive the equation of a parabola given a focus and directrix.

Use Coordinates to Prove Simple Geometric Theorems Algebraically

3. Use coordinates to prove simple geometric theorems algebraically. *For example, prove or disprove that a figure defined by four given points in the coordinate plane is a rectangle; prove or disprove that the point $(1, \sqrt{3})$ lies on the circle centered at the origin and containing the point $(0, 2)$.*

4. Prove the slope criteria for parallel and perpendicular lines and use them to solve geometric problems (e.g., find the equation of a line parallel or perpendicular to a given line that passes through a given point).

5. Find the point on a directed line segment between two given points that partitions the segment in a given ratio.

6. Use coordinates to compute perimeters of polygons and areas of triangles and rectangles (e.g., using the distance formula).

GEOMETRIC MEASUREMENT AND DIMENSION G-GMD

Explain Volume Formulas and Use Them to Solve Problems

1. Give an informal argument for the formulas for the circumference of a circle, area of a circle, volume of a cylinder, pyramid, and cone. *Use dissection arguments, Cavalieri's principle, and informal limit arguments.*

2. Use volume formulas for cylinders, pyramids, cones, and spheres to solve problems.

Visualize Relationships Between Two-dimensional and Three-dimensional Objects

3. Identify the shapes of 2-dimensional cross-sections of 3-dimensional objects, and identify 3-dimensional objects generated by rotations of 2-dimensional objects.

MODELING WITH GEOMETRY G-MG

Apply Geometric Concepts in Modeling Situations

1. Use geometric shapes, their measures, and their properties to describe objects (e.g., modeling a tree trunk or a human torso as a cylinder).

2. Apply concepts of density based on area and volume in modeling situations (e.g., persons per square mile, BTUs per cubic foot).

3. Apply geometric methods to solve design problems (e.g., designing an object or structure to satisfy physical constraints or minimize cost; working with typographic grid systems based on ratios).

NOTES

NOTES

NOTES

NOTES

NOTES

NOTES

NOTES

NOTES